AF752846

Wildlife Conservation on Farmland

Wildlife Conservation on Farmland

Volume 2

Conflict in the Countryside

EDITED BY

David W. Macdonald

and

Ruth E. Feber

OXFORD
UNIVERSITY PRESS

Wildlife Conservation on Farmland. Conflict in the Countryside. Edited by David W. Macdonald and Ruth E. Feber.
© Oxford University Press 2015. Published 2015 by Oxford University Press.

OXFORD
UNIVERSITY PRESS

Great Clarendon Street, Oxford, OX2 6DP,
United Kingdom

Oxford University Press is a department of the University of Oxford.
It furthers the University's objective of excellence in research, scholarship,
and education by publishing worldwide. Oxford is a registered trade mark of
Oxford University Press in the UK and in certain other countries

© Oxford University Press 2015

The moral rights of the authors have been asserted

First Edition published in 2015

Impression: 1

All rights reserved. No part of this publication may be reproduced, stored in
a retrieval system, or transmitted, in any form or by any means, without the
prior permission in writing of Oxford University Press, or as expressly permitted
by law, by licence or under terms agreed with the appropriate reprographics
rights organization. Enquiries concerning reproduction outside the scope of the
above should be sent to the Rights Department, Oxford University Press, at the
address above

You must not circulate this work in any other form
and you must impose this same condition on any acquirer

Published in the United States of America by Oxford University Press
198 Madison Avenue, New York, NY 10016, United States of America

British Library Cataloguing in Publication Data
Data available

Library of Congress Control Number: 2015935287

ISBN 978–0–19–874550–1

Printed in Great Britain by
Clays Ltd, St Ives plc

Links to third party websites are provided by Oxford in good faith and
for information only. Oxford disclaims any responsibility for the materials
contained in any third party website referenced in this work.

Preface

Wildlife Conservation on Farmland. Volume 2. Conflict in the Countryside

This book, and her sister volume, *Wildlife Conservation on Farmland: Managing for Nature on Lowland Farms*, offer perspectives on the same landscape, but viewed from different vantage points. Both books consider farmland, as the bulk, the matrix, and the engine of the countryside, and both address the detailed science that informs a big societal choice, namely what does society want from farmers and farmland and, considering the diversity of conflicting pressures on the farmscape, how do we make the best of it? Indeed, what do we, as citizens and as a nation, think is 'the best of it'? But the perspectives of these books, while from different hilltops, are through the same eyes. That is, while Volume 1, *Wildlife Conservation on Farmland: Managing for Nature on Lowland Farms*, focuses on understanding Nature (and all that affects her) with the purpose of fortifying the evidence base for managing outcomes, and while this volume, *Wildlife Conservation on Farmland: Conflict in the Countryside*, focuses on understanding the very same Nature with the purpose of untangling and mitigating the tensions between people and wildlife in the countryside, both focus through a particular set of eyes, those of the broadly based team of Oxford University's Wildlife Conservation Research Unit—the WildCRU—and the colleagues and collaborators that comprise its extended family.

What journey, personal and professional, led us to pose the questions these books seek to answer? For one of us, the ten-year-old DWM, the roots of this enterprise may have lain in that first week's work on my Uncle Iain's sheep farm, Thornihill Farm, at Colvend in Dumfriesshire—a week spent learning to hand-milk Mary the house cow, and to rescue upturned ewes from the eye-pecking attentions of carrion crows (and that first ever wage packet containing, as I recall, a ten shilling note). Those roots later may have been nourished in the hours spent drinking welcome cups of tea while debating the love of, and conflict with, nature, in the kitchens of farms across which I followed the movements of some of the first foxes in Europe to be radio-tracked. But professionally, the roots of our work certainly lie in the conversations with many farmers in the early 1980s, when DWM asked each of them to list, and explain, the issues with wildlife that most impacted their farming and which, if solved, would facilitate a better compromise between farming and wildlife—conversations that put the onus on farmers and conservationists to work together rather than in opposition, and led directly in 1986 to the Wytham field margins experiment, near Oxford. This was then the biggest farm-scale, controlled and replicated experiment in farmland conservation, and, subsequently, the framework of REF's doctoral thesis on butterflies of field margins, and the beginning of her captivation with farmland wildlife.

That work was a first step in much of the journey related here. Over the past 30 years, as we describe in Volume 1, our work has broadened in scale as thinking about agro-ecology has developed: the field margins experiment manipulated patches of habitat around fields; later, for example, on the farms around the Royal Agricultural College (now University), we magnified the scale to whole-field comparisons; in our work on farming systems in the 1990s and 2000s we contrasted both fields and whole farms (under organic or conventional management); and, most recently, we have focused on the landscape scale—first on the Chichester coastal plain and latterly, with Wytham once again our hub, in the Upper Thames region, with projects on organisms as varied as bats, toads, hedgehogs, and dragonflies, and habitats from ditches and ponds to rivers, set in farmland, hedgerow, and woodland.

High among the list of WildCRU's aspirations is to conduct long-term studies, especially of the autecology of species, the understanding of which can create the foundation for evidence-based policy and practical, problem-solving action. Thus, here, in Volume 2, we report in depth on our 25-year study of badger biology, setting the scene for our studies of the control of bovine tuberculosis (Chapters 4 and 5), and around 20 years of study of foxes (Chapter 2), shedding light on rabies control and, more parochially, the hunting debate (Chapter 3). Similarly, our long-term studies of American mink underpin our understanding of mustelid guilds and the conservation of water voles (Chapters 6 and 7). In every case we have in mind the WildCRU's conservation quartet[1], whereby each project seeks to embrace, to various extents, research to crack the problem, education to spread the word, stake-holder engagement to help change behaviour, and implementation to bring about action.

That this book is explicitly a WildCRU perspective is an important point, because, if readers do not grasp it, they may be puzzled by the selection and topics of the chapters that follow. Let us be clear what these books are not. Neither of these books is comprehensive, but both are compendious. Neither the books, nor the chapters within them, individually or in aggregate, pretend to be a complete, or systematically impartial, review of a whole literature (although they certainly aspire to be fully informed of the salient realities that create the framework for understanding each topic). Rather, each chapter summarizes, synthesizes, and builds upon the research that the WildCRU team (and its antecedent the Oxford Foxlot) has undertaken on each topic between 1972 and 2015, with the hope that our findings shed light that illuminates the wider subject, advances understanding, and fortifies the evidence base for policy and action. However, while we have set out to write with authority about what we have done, and to link that with a wider literature as is necessary or fruitful, we have not set out to review that wider literature with an even hand. Of field experiments, we report on ours at Wytham, rather than on all similar experiments everywhere; of mustelids we report on badgers, rather than on pine martens; of raptors on buzzards rather than on red kites, not because we think or imply the one is more important than the other, but because it is what we have chosen (we believe, with good reason) to work on, and thus to report in these syntheses.

Mention of syntheses raises another point which may affect the reader's understanding, and thus enjoyment, of these two companion volumes. Not only is it plain to all that 'you can't do everything', but anybody with even a glancing acquaintance with the process of science will know that a hurly-burly of scholarship, expediency, urgency, practicality, funding, and luck dictates that sliver of 'everything' which one actually does at a given time, and indeed the way in which these slivers assemble over time. Thus, over the years, as the WildCRU team worked on water voles or bats or toads, or on field margins or ponds or hedgerows, some onlookers may have presumed that we were flitting, much like the gadflies we have also studied, between worthwhile but largely unrelated topics. Nothing could be further from the truth, and one purpose in writing these syntheses has been to demonstrate that. From the start, and growing organically and logically as the opportunity allowed, our purpose has been to link projects, themes, and principles. Consider how, for example, the conservation of water voles can be effective only if it integrates a number of linked themes—local habitat management (riverbanks), landscape scale (reducing isolation of populations), control of invasive species (mink), and health and welfare of reintroduced water voles. So it is that you will find in these volumes that the factors which affect the movements of hedgehogs and of toads, in the contexts of linear habitats, fragmentation, and dispersal, are insightful to, and illustrative of, similar ecological principles and have similar policy relevance. You will read of how the same principle, we call it perturbation, affects the transmission of infectious disease among both badgers and foxes, and makes relevant the study of the societies of both. Indeed, although only touched on in these books (which concentrate on the UK), these same linkages create the weave and waft of the entire, international, WildCRU research agenda, applying the same principles to the corridors through which moths move in the Oxfordshire countryside[2] to the movements of lions in Zimbabwe[3] or tigers in the Western Ghats[4], just as the principles governing the group-formation of badgers in Wytham seem to us to be aligned with those facilitating social groups of giant otters in Peru[5]. On this topic of common threads, readers will find cross-references herein to other shared principles, such as the importance of resource

[1] Macdonald, D.W. (2013). From ethology to biodiversity: case studies of wildlife conservation. *Nova Acta Leopoldina*, **111**:380, 111–156.

[2] Merckx, T., Marini, L., Feber, R.E., and Macdonald, D.W. (2012). Hedgerow trees and extended field margins enhance macro-moth diversity: implications for management. *Journal of Applied Ecology*, **49**, 1396–1404.

[3] Elliot, N., Cushman, S., Macdonald, D.W., and Loveridge, A. (2014). The devil is in the dispersers: the metrics of landscape connectivity change with demography. *Journal of Applied Ecology*, **51**, 1169–1178.

[4] Gopalaswamy, A.M., Mohan, D., Karanth, K.U., Kumar, N.S., and Macdonald, D.W. (2015). An examination of index-calibration experiments: counting tigers at macroecological scales. *Methods in Ecology and Evolution. doi: 10.1111/2041-210X.12351.*

[5] Groenendijk, J., Hajek, F., Schenck, C., Staib, E., Johnson, P.J., and Macdonald, D.W. (2015). Effects of territory size on the reproductive success and social system of the giant otter, south-eastern Peru. *Journal of Zoology*. doi: 10.1111/jzo.12231.

dispersion[6], or of economics and substitutability[7, 8], or payments for ecosystem services[9], or culture[10, 11], and indeed wider inter-disciplinarity[12], all wrapped in the fabric of inconvenient truths and the elephants in the room that reveal conservation to be inherently about biology, but ultimately a branch of politics[13].

This mention of resonance, indeed synergy, between WildCRU's work on the farmland of lowland Britain, and our wider work, prompts the question of to whom these books might be useful. Much of what we have written is in the form of case studies and it goes without saying that we expect aficionados of water voles or damselflies to find interest in our chapters on these organisms, and it is a modest hope that enthusiasts of riparian or pond conservation would also find merit in these chapters. However, we hope for greater reach. Not only do we intend these cases to illustrate lessons that can be applied to different species, but also to different landscapes, and even different countries. Globally, we have asked how culture[11] and the media[14] affect attitudes to lions and leopards in India, and how societal factors can influence education with regard to farmers tolerating, for example, jaguars in Brazil[10]. Further, globally it is clear that conservation priorities vary from country to country, as too do national capacities to deliver[15], and the citizens' preferences for different species[16]. All of these issues and principles are transferrable to British farmland, where the world's conservation dilemmas play out in microcosm. Here, our examples are drawn largely from the lowlands, and surely there are aspects of upland ecology even within Britain on which we do not touch at all. However, not only do we have some upland case histories—the affairs of grey and red squirrels in Perthshire and the rewilding activities of wild boar in Sutherland come to mind, alongside the beavers of Knapdale (Volume 1)—but our lack of focus on peatbogs or heather moors does not disqualify most of our material from nationwide relevance. Indeed, it seems to us that our case studies apply directly and in detail and in principle to landscapes across much of continental Europe and indeed throughout the temperate world. Further, we hope our chapters will be read from Patagonia to Parak by conservationists and policy makers seeking insight into the issues of ecosystem management and policy formulation that are common themes globally[15, 17].

The conservation of biodiversity on farmland raises dilemmas as complex as those besieging conservation anywhere in the world—issues of principle and practice always with ethical dimensions. For example, what approaches should be taken to control alien species that have exploited agricultural habitats, especially if they are having impacts on native species? How, and to what extent, should wildlife be managed on farmland, to reduce human–wildlife conflict? Crucial to answering questions of wildlife management are studies of fundamental ecology of the species concerned. In this volume we explore how such studies contribute to wildlife management solutions. The introduction explores challenges and moral dilemmas associated with wildlife conservation interventions and the types of conflicts that occur on farmland. Foxes provide a vivid example of such conflicts: knowledge of fox ecology

[6] Macdonald, D.W. and Johnson, D.D.P. (2015). Patchwork planet: the resource dispersion hypothesis, society and the ecology of life. *Journal of Zoology*, **295**, 75–107.

[7] De Barros, A.E., Macdonald, E.A., Matsumoto, M.H., Paula, R.C., Nijhawan, S., Malhi, Y., and Macdonald, D.W. (2013). Identification of areas in Brazil that optimize conservation of forest carbon, jaguars, and biodiversity. *Conservation Biology*, **28**, 580–593.

[8] Dutton, A.J., Hepburn, C., and Macdonald, D.W. (2011). A stated preference investigation into the Chinese demand for farmed vs. wild bear bile. *PLOS ONE* 6: e21243. doi:10.1371/journal.pone.0021243.

[9] Dickman, A.J., Macdonald, E.A., and Macdonald, D.W. (2011). A review of financial instruments to pay for predator conservation and encourage human-carnivore coexistence. *Proceedings of the National Academy of Sciences of the United States of America*, **108**, 13937–13944.

[10] Marchini, S. and Macdonald, D.W. (2012). Predicting ranchers' intention to kill jaguars: case studies in Amazonia and Pantanal. *Biological Conservation*, **147**, 213–221.

[11] Meena, V., Macdonald, D.W., and Montgomery, R. (2014). Managing success: Asiatic lion conservation, interface problems and peoples' perceptions in the Gir Protected Area. *Biological Conservation*, **174**, 120–126.

[12] Macdonald, D.W., Loveridge, A.J., and Rabinowitz, A. (2010). Felid futures: crossing disciplines, borders and generations. In: *The Biology and Conservation of Wild Felids*, pp. 599–649. Oxford University Press, Oxford.

[13] Macdonald, D.W. and Willis, K.J. (2013). Elephants in the room: tough choices for a maturing discipline. In D.W. Macdonald and K. Willis, eds., Key Topics in Conservation Biology 2, pp. 469–494. Wiley-Blackwell, John Wiley & Sons Ltd., Oxford.

[14] Bhatia, S.N. and Macdonald, D.W. (2013). Understanding the role of representation of human-leopard conflict in Mumbai through media content analysis. *Conservation Biology*, **27**, 588–594.

[15] Dickman, A.J., Hinks, A.E., Macdonald, E.A., Burnham, D., and Macdonald, D.W. (2015). Priorities for global felid conservation. *Conservation Biology*, **29**, 854–864.

[16] Macdonald, E.A., Burnham, D., Hinks, A.E., Dickman, A.J., Malhi, Y., and Macdonald, D.W. (2015). Conservation inequality and the charismatic cat: *Felis felicis*. *Global Ecology & Conservation*, **3**, 851–866.

[17] Macdonald, D.W., Collins, N.M., and Wrangham, R. (2006). Principles, practice and priorities: the quest for alignment. In D. Macdonald and K. Service, eds., *Key Topics in Conservation Biology*, pp. 273–292. Blackwell Publishing, Oxford.

and how it can be used to address issues of foxes as pests is the subject of Chapter 2, which prepares the ground for Chapter 3's focus on the controversy surrounding hunting with dogs and the bio-economics of fox control. Chapter 4 considers how studies of badger ecology can contribute to our understanding of impacts of climate change, information which underpins Chapter 5, which focuses on a key challenge for farmers, policy makers, and conservationists: how an understanding of badger behavioural ecology can inform the controversial debate of the role of badgers in transmitting bovine tuberculosis.

The next three chapters cover the increasingly difficult challenges posed by the spread of alien species, their impacts on native species, and how to control them. The examples used are the American mink and the American crayfish, both widespread in many areas of lowland English farmland, and particularly prevalent in the Upper Thames. The ecology of mink and the complex relationships between mink and native mustelids are explored in Chapters 6 and 7, and Chapter 8 focuses on whether the huge increases in American crayfish in the Upper Thames can be controlled.

The pheasant is another non-native species, but a commercially important one. Releasing pheasants for shooting is perhaps a less perceptible type of farming, but one that may have impacts on wildlife—in Chapter 9 we explore the potential consequences for small seed-eating farmland birds of the annual release of huge numbers of pheasants on farmland and the widespread provision of their supplementary feed. In Britain, wild raptors have often been at the heart of controversy and conflict and, in Chapter 10, we delve into the knotty problem of raptor management, focusing on the chequered history of the buzzard in Britain. Chapter 11 describes fascinating studies of the behaviour of wild brown rats and the implications of the results for pest control strategies. How attitudes and actions of farmers, householders, and amenity managers differ with respect to moles and mole control is the subject of Chapter 12, and Chapter 13 demonstrates the importance of understanding animal behaviour to better manage human–wildlife conflict, using novel examples of non-lethal control of wildlife.

From conflict to engagement, the last two chapters in this volume centre on two WildCRU projects which have focused on the human dimension. In Chapter 14, we discuss, using examples of landscape-scale projects, how it is possible to achieve stakeholder partnerships and effectively deliver habitat restoration and management across a catchment-scale farmed landscape. Lastly, Chapter 15 describes WildCRU projects that have been designed, in partnership with Earthwatch, specifically to maximize opportunities for benefiting, and benefiting from, 'Citizen Scientists': volunteers assisting with studies of a diverse range of animal species.

As we draw this project to an end, we are awed by the number of people who have helped to bring it to fruition. This has been a long and arduous process and so we thank first and fervently our partners, Dawn and Guy, who have supported us throughout, keeping us afloat as we navigated steadfastly to our purpose during the choppiest months. It's a lofty aspiration, but we hope that the findings in these books will help lead to a more richly biodiverse, beautiful, but nonetheless functional, countryside for our children.

Each of the chapters is written by a small subset of the people who undertook the research, and we thank so gratefully the authorial teams, and the community of researchers who contributed to the underlying research. Equally, the books are a product of the friendships and tolerance of large numbers of farmers who have endured, indeed, encouraged, our antics on their land, and been those most valuable of allies: the critical friends. Our editorial task was hugely aided by Dr Eva Raebel, and we thank also the many members and friends of WildCRU who have helped in myriad ways, from providing photographs to critically reviewing drafts. We thank also our editors at Oxford University Press: Ian Sherman and, successively, Helen Eaton then Lucy Nash, and, during the production process, Cherline Daniel of Integra. Our work on these books has been an immense task, and was only possible due to the support of the Holly Hill Trust and The Rivers Trust. We are grateful to the many sponsors of our projects, and they are mostly acknowledged in individual chapters, but several donors have kept the enterprise afloat more generally: we have in mind especially the Tubney Charitable Trust (and we remember fondly the particular enthusiasm of the late Terry Collins), Defra, the Esmée Fairbairn Foundation, the People's Trust for Endangered Species, Earthwatch, and the Environment Agency. Latterly, WildCRU has been based at The Recanati-Kaplan Centre at Tubney House—reverberating through these names is our gratitude to Tom and Daphne Kaplan for their friendship and for securing the future of WildCRU's headquarters.

It has been a privilege to bring together the work of so many people in these two books. We hope you will enjoy them.

David W. Macdonald and Ruth E. Feber

Wildlife Conservation Research Unit, Recanati-Kaplan Centre, Department of Zoology, University of Oxford and Lady Margaret Hall, University of Oxford.

Contents

List of Contributors

Eduardo M. Arraut Wildlife Conservation Research Unit, The Recanati-Kaplan Centre, Department of Zoology, University of Oxford, UK

Joanna Bagniewska School of Biological Sciences, University of Reading, UK

Sandra E. Baker Wildlife Conservation Research Unit, The Recanati-Kaplan Centre, Department of Zoology, University of Oxford, UK

Manuel Berdoy Oxford University Biomedical Services, Oxford, UK

Christina D. Buesching Wildlife Conservation Research Unit, The Recanati-Kaplan Centre, Department of Zoology, University of Oxford, UK

Patrick Doncaster School of Biological Sciences, University of Southampton, UK

Ruth E. Feber Wildlife Conservation Research Unit, The Recanati-Kaplan Centre, Department of Zoology, University of Oxford, UK

Lauren A. Harrington Wildlife Conservation Research Unit, The Recanati-Kaplan Centre, Department of Zoology, University of Oxford, UK

Heribert Hofer Leibniz Institute for Zoo and Wildlife Research, Berlin, Germany

Paul J. Johnson Wildlife Conservation Research Unit, The Recanati-Kaplan Centre, Department of Zoology, University of Oxford, UK

Robert E. Kenward Centre for Ecology and Hydrology, Wallingford, UK

Alan Larkman Larkman Consulting Ltd, Bampton, UK

David W. Macdonald Wildlife Conservation Research Unit, The Recanati-Kaplan Centre, Department of Zoology, University of Oxford, UK

Fiona Mathews College of Life and Environmental Sciences, University of Exeter, UK

Tom P. Moorhouse Wildlife Conservation Research Unit, The Recanati-Kaplan Centre, Department of Zoology, University of Oxford, UK

Malcolm Newdick Riverbank IT Management, UK

Chris Newman Wildlife Conservation Research Unit, The Recanati-Kaplan Centre, Department of Zoology, University of Oxford, UK

Ian Newton Centre for Ecology and Hydrology, Wallingford, UK

Eva M. Raebel Wildlife Conservation Research Unit, The Recanati-Kaplan Centre, Department of Zoology, University of Oxford, UK

Philip Riordan Department of Zoology, University of Oxford, UK

Terhi Riutta Environmental Change Institute, School of Geography and the Environment, University of Oxford, UK

Eleanor M. Slade Department of Zoology, University of Oxford, UK

Michael D.F. Thom School of Biological Sciences, University of Plymouth, UK

Joanne P. Webster Department of Infectious Disease Epidemiology, St Mary's Hospital, Imperial College Faculty of Medicine, London, UK

Rosie Woodroffe Zoological Society of London, London, UK

Nobuyuki Yamaguchi Department of Biological and Environmental Sciences, University of Qatar, Qatar

CHAPTER 1

War or peace: farming and wildlife

David W. Macdonald, Eva M. Raebel, Chris Newman, and Ruth E. Feber

We shall never achieve harmony with the land, any more than we shall achieve absolute justice or liberty for people. In these higher aspirations the important thing is not to achieve but to strive.

Aldo Leopold

1.1 Introduction

One of us, DWM, remembers a personal journey that sets the scene for this book: 'My own introduction to the love–hate relationship of farming and wildlife was made through foxes. My prying into their private lives led me deep, and wide-eyed, into the underworld of agriculture. There, I learnt from men deeply knowledgeable about nature, who became my closest friends more often despite, rather than because of, their actions towards it. For example, Westmoreland shepherd, Edwin Dargue who, taking pity on my derided efforts to radio-tag foxes in sheep country, persuaded a hostile community to offer me a deal: they'd enter their terriers into the fox earth, and I'd start digging—if I got to the fox before the dogs killed it, they'd spare it. After one such life-and-death excavation, choked on dirt and hands bleeding, men who'd measured themselves by the foxes they'd killed introduced me to their companions in a tiny fell-side pub—"This is David, he chews nails and spits rust"—then they spoke for hours, almost reverentially, of foxes against which they'd waged war—always concluding with the consolation "but I'd never kill the last one". So, I was introduced to the war and peace that is farming and wildlife, and learnt, to quote Leo Tolstoy, that "One of the first conditions of happiness is that the link between Man and Nature shall not be broken".' How is that link holding up?

In *Wildlife Conservation on Farmland: Managing for Nature on Lowland Farms* (Macdonald and Feber 2015) we delivered some answers, describing elements of the agro-ecosystem, from field margins to woodlands, revealing aspects of their functioning, and using evidence from our team's research (the Wildlife Conservation Research Unit—WildCRU), to inform management actions that foster biodiversity on farmland, from butterflies to bats, and wood mice to water voles. Now, in *Wildlife Conservation on Farmland: Conflict in the Countryside*, we draw on further case studies from the WildCRU's work to illustrate that, while farmers can be powerful allies in delivering nature, there are also tensions, even wars, between farmers and species that are, or are perceived to be, problematic. In Macdonald and Feber (2015: Chapter 1), we used the device of a hot-air balloon journey, rising and falling to see the farmscape at different spatial scales. To use that metaphor briefly here, we note that almost all the issues we tackle in this book would, if the balloon floated up to take a global view, be examples of what is known in the conservation literature as human–wildlife conflict (e.g. Sillero-Zubiri et al. 2007; Macdonald et al. 2013). Oddly, lessons, even vocabulary, drawn from the wealth of international evidence, are not always transferred to instances in our own backyard (and nor, at the risk of double standards, are attitudes and ideas about wildlife in our backyards always applied consistently to the backyards of people in far-flung lands)—in this chapter we will draw attention to principles and practices that link the issues faced by British farmers to their counterparts in less developed countries.

What are the issues on which we present our team's case studies? *Cave Cave Deus Videt* ('Beware, Beware, God Sees') inscribed Hieronymus Bosch beneath his 1485 depiction of an all-seeing divine eye, vigilant over the seven deadly sins (wrath, envy, avarice, gluttony, idleness, lust, and pride). Actually, this is a pretty good rendition of contemporary environmental follies and, replacing the religious catechism with an eye to sustainability, we may reflect nervously on which of Bosch's 'four last things'—death, judgment, hell, or

Wildlife Conservation on Farmland. Conflict in the Countryside. Edited by David W. Macdonald and Ruth E. Feber.
© Oxford University Press 2015. Published 2015 by Oxford University Press.

glory—are the more likely fates for humanity. We offer, for the purposes of this book, a more agricultural list of seven deadly risk factors: *habitat loss*, *chemical inputs*, *non-native invasive species*, *persecution*, and *infectious disease*, set on a stage inhabited by an over-exploitative *human population*, against a backdrop of *climate change*. Of course, the pack of problems could be shuffled in different ways, and Diamond's famous evil quartet (habitat destruction, over-hunting/exploitation, introduced species, and trophic cascades leading to secondary extinctions: Diamond 1984) is well matched by the WildCRU's conservation quartet of research to understand the problem, education to explain it, community involvement to ensure participation, and implementation of solutions (Fig. 1.1; Macdonald 2013).

So, to whet the appetite for the 14 chapters that follow, we will begin by considering, first, the deadly protagonist, humans—the principal actors on the stage upon which these dramas are set—and how they have come to change the landscape to what it is today. The habitat loss wrought as a result of these changes is our second deadly risk factor, affecting the population processes of all the species discussed in *Wildlife Conservation on Farmland*, Volumes 1 and 2. Paradoxically, we humans also hold the key to solving these conflicts between people and wildlife on farmland; so we illustrate in this book how informed management and compromise, striking a balance between productivity and sustainability, can conserve wildlife, particularly

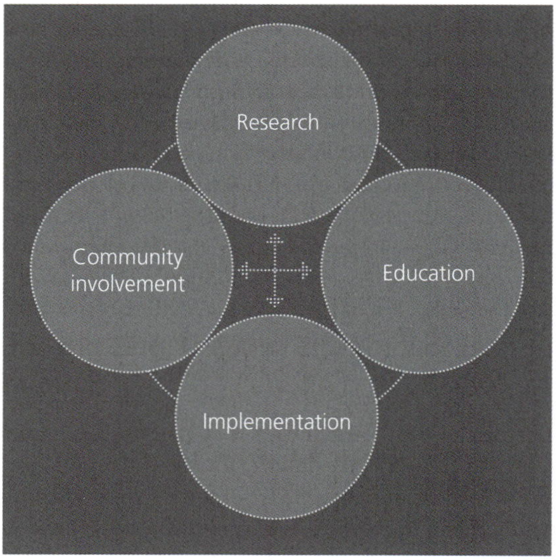

Figure 1.1 WildCRU's conservation quartet. © Wildlife Conservation Research Unit.

so if this joined-up thinking is undertaken at the landscape scale (Chapter 14, this volume) and is able to harness the power of Citizen Science (Chapter 15, this volume). Successive acts in this theatre of conflict resolution will include the conflict born of livestock losses and leading to persecution, of which red foxes *Vulpes vulpes*, offer a vivid example (Chapters 2 and 3, this volume), where foxes, through rabies, lead us to infectious disease, the topic of chapters on brown rats *Rattus norvegicus* (Chapter 11, this volume), and badgers *Meles meles* (Chapters 4 and 5, this volume). Rats themselves, although perhaps now ecological citizens in the UK (*sensu* Macdonald and Burnham 2010), were once invaders, which leads us to American mink *Neovison vison* (Chapters 6 and 7, this volume) and American signal crayfish *Pacifastacus leniusculus* (Chapter 8, this volume), both non-native species exerting destructive cascades upon natives of the riparian ecosystem. The incorporation of mink into the riverside predator guild touches on another deadly risk factor, chemical inputs, insofar as the door had been opened for arriving mink due to the impact of agrochemicals on otters *Lutra lutra*. A startlingly numerous invasive, albeit by invitation, is the common pheasant *Phaisanus colchicus*, of which the colossal annual release of 35 million individuals must surely impact the farmland ecosystem. We therefore offer a speculative exploration of how these game birds, or more accurately the food provided for them, may trigger changes in farmland guild structure that might derail some objectives of agri-environment schemes (Chapter 9, this volume). Invasive species remind us that human interventions often involve killing a pest (or a perceived one, sometimes a ghost of pests past!), illustrating that an important arrow in the conservation scientist's quiver is concern for animal welfare (Box 1.1). We highlight this in the case of moles *Talpa europaea* (Chapter 12, this volume), and consider more widely the quest for non-lethal control (Chapter 13, this volume). Finally, all of these acts play out against a backdrop to the stage that is unsettlingly labile, a changing climate, the impacts of which are clearly exemplified by badgers (Chapter 4, this volume). Their current abundance in the British countryside is substantially due to farming practice, and their ability to adapt to changing climatic conditions illustrates processes, and constraints, relevant to many species (Newman and Macdonald 2013).

A last point of principle regarding the chapters that follow (in which we focus largely on studies concerning vertebrates): although ultimately motivated to contribute to the solution of practical problems, our approach is always to try better to understand the species involved—just as responsible investors would never risk capital by dabbling in a market place of which they had no understanding, so our studies of ecology and behaviour provide the evidence for action. In some cases, the biological evidence is in the weave and waft of the practical problem (e.g. Chapters 10 and 12 on buzzards *Buteo buteo* and moles), but in others we present foundation chapters (e.g. Chapters 2, 4, and 6 on foxes, badgers, and mink, respectively) as a basis for the practical—and policy-relevant—chapters.

We begin here, as is the tradition, at the beginning; as we increasingly recognize the value of multifunctional landscapes and the (re-)integration of wildlife and agriculture, how did this antagonistic dichotomy ever arise?

1.2 Humans and agriculture

In spite of Palaeolithic technologies leading to ever more advanced tool use, our hunter–gatherer ancestors of the Pleistocene faced precisely the same day-to-day, season-to-season, and year-to-year challenges in food acquisition as any other member of the fauna. As we shall elaborate in Chapter 2 for the fox and Chapter 4 for badgers, the dispersion of resources in space and time dictates socio-spatial geometries in animal societies (Carr and Macdonald 1986; Macdonald and Johnson 2015), where ultimately, if the security with which food can be found in terms of both volume and frequency falls below the species' tolerance for periods of starvation (Newman et al. 2011), individuals (or indeed, for mobile grazers, entire populations) must either move, or die. The 'carrying capacity' of the environment is therefore not simply a measure of food abundance, but also of consistency and predictability of supply (McLeod 1997). So, how to cheat the vagaries of nature? Bend the rules—produce the food yourself.

Mithen (2007) points out that in the entire 250,000 years that Neanderthals (with similar-sized brains to *Homo sapiens*) persisted, they achieved almost no cultural advancement before eventually dying out, but humans underwent a radical change in societal and economic organization at around 12,000 BP. Our ancestors had crossed a crucial threshold from nomadic hunter–gatherer to settled early food production, spurring a dramatic change in social intelligence leading to the formation of societies (Mithen 2007). Dense stands of wheat and barley that were already being used as a food source from the wild increasingly became subject to planned cultivation, with early seed selection for desirable traits[1] (Miller 1992). Recent research (e.g. Zeder 2008, 2011) has revealed that domestication of plants and animals occurred almost at the same time (*c*. 11,500 BP)[2]; archaeological evidence indicates an increasing use of stone, then metal, farming implements, and livestock to pull ploughs, with bones found at dispersed human encampments including cattle, sheep, goats, and pigs, all herded by pastoral nomads during this Neolithic revolution (see Zeder 2008). Unbeknown to our prehistoric ancestors, they were also causing the first anthropogenic damage to the environment (as evidenced at several archaeological sites in south west Asia and north Africa), with soil erosion due to land use (van Andel et al. 1990) and depletion of vegetation by domesticated goats (Bellwood 2001) being the first step into habitat loss. It was also with the origins of agriculture in Africa that human–wildlife conflict started to mirror the same issues that we currently face (Treves and Naughton-Treves 1999). For example, by 2000 BC, Egyptians living by the Nile delta had already recorded crop damage caused by hungry hippopotamuses, while crocodiles fed on livestock and attacked humans (Lamarque et al. 2009). While larger grazers could be fenced out of crop fields, or hunted as an additional food source, smaller rodent species began exploiting human food stores, and the first agricultural pests were born (Stenseth et al. 2003).

Thus, burgeoning human populations had required a solution to their issues of food limitation and it seems to have been the greater demand for food security that 'pushed' societies into agriculture, rather than being 'pulled' by greater ease of production and innovation (Cohen 2009; see also Diamond 2002). But with greater food availability and security came a rapid increase in human population (but see Aiméa et al. 2013 for evidence of pre-agricultural expansions). Gone were the small, egalitarian, closely related bands of hunter–gatherers, which were increasingly replaced by villages, then towns, and ever more hierarchically organized societies which, released from the constant

[1] Rye with cultivated traits dates from *c*. 12,000 BP, recovered in Abu Hureyra, Syria (Hillman et al. 2001).

[2] The first domesticated dog remains, found in a grave excavated in Oberkassel, Germany, date from 14,000 BP (Clutton-Brock 1995), suggesting that domestication had commenced.

struggle to find food, gave rise to civilizations, leading to the development of socio-economic classes, job specialization, and governments (Carneiro 1970; Richerson and Boyd 2001).

This development of an agrarian society was perhaps the first example in human evolutionary history of a major adaptive step forward in the natural selection of our species. Curiously, therefore, we see here that the anthropologist and the ecologist operate in the same theatre of theory, where the same rules of natural selection drive tactical adaptation (Carneiro 1970)—only we humans proved rather good at it, in retrospect almost too good at it, and we now find such extremes of exploitation necessary to sustain our population that we leave little space for nature. It would not be an exaggerated claim that agricultural development has had greater consequences for natural habitats and biodiversity than any other force (Tilman et al. 2011), leading us to our next deadly risk factor, habitat loss.

1.3 Habitat loss

Considering the now chequerboard geometry of lowland England's agricultural landscape (a spatially heterogeneous environment in ecological terms) and temporal roller-coaster (with seasonal and inter-annual switches between crops and management perturbations), it is hard to imagine an ecosystem in which the consequences of patch fragmentation, isolation, and meta-population dynamics could be more vividly relevant (Macdonald et al. 2000a). It can be argued that habitat loss due to agriculture is what determines the needs and feasibility of nature conservation (Macdonald and Johnson 2000), driving catastrophic cascades, and leading to declining ecosystem services (Dobson et al. 2006).

If farms are to be shared with wildlife, that ecological patchwork quilt of variously inhospitable crops must be penetrated by passageways that allow organisms, at whatever scale is relevant to them, to escape the vortices of environmental and demographic runs of bad luck, known as stochasticity, and other evils of small population size (itself a deadly risk factor downstream of our seven). Even in the most strictly protected areas, wildlife may be threatened if the next such refuge cannot be reached (having all their metaphorical eggs in just one basket is risky, however sumptuous it may be, Elmqvist et al. 2003). The fundamental principles of linkage and meta-population function (Macdonald et al. 2000b; Akçakaya et al. 2007) apply across a range of organisms, from the management of field margins for butterflies and moths (Macdonald and Feber 2015: Chapter 8) to the design of corridors through farmland to join up the jaguar *Panthera onca* reserves of Latin America (Rabinowitz and Zeller 2010), lion *Panthera leo* populations in Africa (Morandin et al. 2014), or restoring tiger *Panthera tigris* populations in the Caspian region (Driscoll et al. 2011). Of course, enhancing connectivity between species-rich areas must take into account the differences in mobility between species. The winged and fleet of foot (whether at the scale of moth or roe deer) may be able to jump hostile crop monocultures, but the less agile are imprisoned: without a linking hedgerow what chance has a hazel dormouse of surviving the journey from one coppiced woodland to the next, and without a field margin how does a grass snake seek new hunting grounds?

As it happens, grass margins provide an example of what happens when habitat loss puts conflicting pressures on the habitats that remain. Farmer and conservationist alike take an interest in field margins: the former, nervous they could serve as sources of weeds or other pests (Macdonald and Feber 2015: Chapter 2), or a motorway for unwelcome predators (field margins attract predators and can negatively affect spring survival in released grey partridges *Perdix perdix*; in Macdonald and Feber 2015: Chapter 13), the latter, valuing them as a refuge for nature in a multi-functional landscape (Macdonald and Feber 2015: Chapters 2 and 3) and, no less important, as escape routes out of perilously isolated fragments of habitat. Should nature be allowed, even encouraged, to wend its way through commercial farms, or should it be segregated from them? The options are, at the extremes, characterized as 'land sharing' or 'land sparing' (see Fischer et al. 2008; Phalan et al. 2011; Rey-Benayas and Bullock 2012), the former integrating, the latter segregating farming and wildlife. This was perhaps first famously recognized in Aldo Leopold's 'land ethic' (in his 1949 book, *A Sand County Almanac*), where he wrote that there is a need for a 'new ethic', an 'ethic dealing with humans' relation to land and to the animals and plants which grow upon it'. On the familiar idiom of 'horses for courses', there is no one right answer for all places, nor necessarily even for all species in one place. In the UK, and much of continental Europe the goal is harmonious sharing, fostered through agri-environment schemes which apply modest measures, including restoration and enhancement of features such as ditches and hedgerows, that benefit wildlife (Phalan et al.

2011). The land sparing approach is more characteristic of the USA, where nature is protected in National Parks and reserves, while farmland is managed intensively to maximize production.

Of course, it takes two to tango so, in the case of land sharing, the reciprocal question arises, for a given species, of how fervently the farmer does not want to cohabit with it (and here there is also a regional/cultural dimension to attitudes towards conservation and wildlife, as we have discovered by interviewing farmers) (see Chapters 3, 12, and 14, this volume) and, from the species' perspective, how poorly it is adapted to co-occupancy with people. Even an affluent American apiarist, neighbourly to nature, baulks at a bear in his backyard, fearing it might eat more than his honey (Can et al. 2014) and small-holders in Zimbabwe are unenthused by lions among their cattle, regardless of how important is the species' conservation status (Loveridge et al. 2010). While it might be painless to devote agriculturally useless land to nature, the reality is much more likely to be that, especially in developing countries, somebody could scratch some sort of subsistence from it, such that the land spared for nature limits somebody's options, thus fomenting resentment towards conservation. The profit-and-loss hope is that in spared (i.e. protected) areas, nature generates more value than does farming (Dickman et al. 2011; Barrett et al. 2013).

In Britain, conservation thinking and policy are moving from the isolationist risks of habitat-driven management to landscape-scale approaches (Lawton et al. 2010). In England, more than a third of bird, mammal, reptile, amphibian, and bumblebee species are estimated to function at a scale larger than the typical English farm (average 146 ha, McKenzie et al. 2013). Agri-environment schemes (AES) are offered as one medicine for the ails of nature on farmland. While the extent to which they have been successful in improving outcomes for wildlife is uncertain (Kleijn et al. 2011), first principles (of ecology) dictate that there are likely to be more biodiversity benefits if they are applied at a landscape scale, in order to foster connectivity. This presumption is captured in the coining of new words in the emerging vocabulary of AES, such as 'farmscape' and 'pondscape' (e.g. Merckx et al. 2009; Davies et al. 2009, respectively). Returning to our human risk factor, and with it our human solutions, we describe, in Chapter 14, this volume, how partnerships of non-governmental organizations can persuade contiguous farmers to join agri-environment schemes to restore habitats and create a continuum of conservation practices across a large area, rewarding the taxpayer with more water voles and more moths per buck. Similar virtues of joined-upness emerged from our work on bats and dragonflies in Macdonald and Feber (2015): Chapters 9 and 10, respectively; a host of other researchers in the UK and far beyond have compiled evidence on the potency of the landscape or catchment-level approach. Of course implementing conservation action on this scale requires manpower, and Chapter 15, this volume, describes how the many available hands of 'eco-volunteers' and Citizen Scientists make this work lighter and can reach beyond the field into a landscape-scale conservation approach, while at the same time engaging people with nature (Fig. 1.2).

Figure 1.2 Citizen Scientists can provide valuable help with landscape-scale data collection: we found moths to be a good taxon for engaging members of the public with such projects. Photograph © Yifie Zhang, Earthwatch.

1.4 Chemical inputs

A major driving force behind habitat loss is agrochemical inputs, one of the most important factors involved in the intensification of agriculture globally, and next on our agricultural list of deadly risk factors. The overall amounts of agrochemicals applied to farmland have increased inexorably. In 2010 alone, the UK applied 1,029,000 tonnes of nitrogen fertilizers, 7465 tonnes of herbicides, and 889 tonnes of insecticides (FAOSTAT 2014) to its land. Many studies have demonstrated links between agrochemical inputs, yields, and biodiversity; for example, countries with higher wheat yields have more threatened or recently extinct arable weed species (Storkey et al. 2012), plant species richness is negatively related to nitrogen input at a country-level scale (Kleijn et al. 2009), and bird population declines and range contractions have been greatest in countries with the highest cereal production (Donald et al. 2001, 2006). Recent incentives and agri-environment schemes introduced to address this issue have halted and even reversed declines in some groups but, in general, European farmland biodiversity continues to decline (e.g. EEA 2013), amid heightened concerns over the potentially harmful effects of modern pesticides such as neonicotinoids (Hallmann et al. 2014). In Macdonald and Feber (2015), we showed how pesticide use (insecticides and herbicides) impacts on the plant and invertebrate diversity of arable farmland (Chapters 2, 3, and 6), and how nitrogen and phosphates used as fertilizers can leach into watercourses and cause eutrophication which destroys freshwater habitats, affecting freshwater communities and reducing their scope for dispersal (Chapter 11). Indeed, in Macdonald and Feber (2015: Chapter 10), we explain how it is not only the proximity between ponds that is important for dragonfly and damselfly populations in the landscape, but the viability of each pond is crucial, such that its water is of sufficiently good quality to allow them to complete their life cycle. Chemicals are therefore linked to all the issues of fragmentation discussed in Section 1.3.

The effects of such vast chemical inputs have been many and varied, connecting in some way to all the deadly risk factors in our list and, with pressures on the land for food security and biofuels, increasing globalization, and the impacts of climate change, there seems little chance of their use abating. Chemical inputs can change the course of events because of the ecological inconvenience that everything is linked to everything else. For example, to use examples from this book, herbicides applied to arable fields have reduced the abundance and diversity of farmland plants and, in turn, the invertebrates which depend on them, reducing the insect food needed by grey partridge chicks, leading to the decline of this native gamebird, and to the widespread release of the non-native pheasant (Chapter 9, this volume). The provision of grain to help these pheasants to survive provides a rich food source for another invader, brown rats, widely regarded as pests, due to the agricultural damage that they cause (predominantly through contamination of food stuffs, damage to property and materials, and risks to human and farm animal health). The damage on farms caused by rats was estimated to have cost the UK farming industry £14 to £28 million in 2007 (Buckle 2009).

Using another example from our studies (Chapter 4, this volume), we find that high population densities of badgers in Britain, particularly in the south-west, are facilitated by the modern farmed landscape, where fertilized grasslands provide not only high-quality pasture for cattle (indeed fertilizer use is one factor leading to the year-on-year increase in dairy herd sizes in England, an exacerbating factor in the spread of bTB in cattle), but also worm-rich foraging grounds for badgers. Increased badger numbers increase the potential for conflict with farmers through their crop damage, poultry disturbance, and predation on gamebirds, and of course via their contentious role in the transmission of bovine tuberculosis. Back to linkages, climate change may further exacerbate conflicts wrought by chemical inputs and agricultural intensification (Section 1.8), since, in general terms, the hungrier and more stressed a species, the more likely it is to cause a nuisance, be this the tuberculous badgers visiting milking sheds and cattle troughs (Garnett et al. 2002, 2003), or buzzards predating free-ranging chickens (Stahl et al. 2002). Brown rat numbers in England appear to be increasing (PTES 2008; TMP 2009) and climate change may exacerbate the situation if trends towards a warmer, wetter climate in the UK continue (Buckle 2009). Of course, changes wrought by agrochemical use can cascade back directly into farm economics and have the potential to inflame relations between humans and wildlife. For example, the use of nitrogen fertilizers has led to a widespread change in grassland management, from hay to silage production. However, silage can get contaminated by *Listeria* present in the soil, affecting the livestock that consumes it, and farmers, as we describe in Chapter 12, this volume, frequently see moles as responsible for the contamination of silage and for spoiling their crops (and incomes) with mole hills.

There are few more notorious examples of the unintended consequences of chemicals on wildlife than that of DDT, brought to the public's attention nearly 50 years ago by Rachel Carson's seminal book, *Silent Spring*. DDT and other organochloride insecticides were widely applied to crops in the 1950s and 1960s and accumulated up the food chain. Being less able to metabolize these chemicals, concentrations tended to be higher in birds (Gómez-Ramírez et al. 2014), particularly species which fed on other birds, such as sparrowhawks *Accipiter nisus* (Newton et al. 1993). Organochlorines were responsible for the collapse of this species in Britain in the 1960s—famously by thinning their egg-shells (see summary in Chapter 10, this volume). At around the same time, in the late 1950s, pollution of freshwater habitats by organochlorine pesticides (used for sheep dip and seed dressings) contaminated fish and almost drove the Eurasian otter to extinction in Britain due to the lethal accumulation of dieldrin in their prey (Mason and Macdonald 1993). It was during this time, when the otter was absent from much of England, that American mink began their spread across the country, leading us to our next deadly risk factor, alien invasives.

1.5 Non-native invasive species

Non-native invasive species, those moved beyond their natural geographical haunts (intentionally or as an unexpected result of transporting people and products), pose undesirable biological, practical, and ethical problems, as mind-bending as any of the seven deadly risks. Furthermore, invasives are expensive: problems/damage caused by non-native plants and animals cost the British economy an estimated £1.7 billion/year (Williams et al. 2010) and $35 billion/year to the USA (Jenkins 2013). Although, with knee-jerk naturalistic jingoism, it might seem pretty obvious which species are invasive, deeper thought exposes myopically blurred boundaries. Colonization of new turf is an entirely natural process that underlies the biogeographical kaleidoscope of species' distributions (see climate change, Section 1.8), and equally it is commonplace for one species to be transported to new worlds by another (remember that *a flea has smaller fleas that on him prey; and these have smaller still to bite 'em*[3]).

Further blurriness, summarized for UK mammals by Macdonald and Burnham (2010), and globally by Davis (2009), bedevils the roles of domestic animals brought to the UK by our Neolithic ancestors, or the commensal rodents who have evolved, entirely 'naturally', to hitchhike with people. A painfully agricultural paradox is provided by the hedgehog *Erinaceus europaeus*: on many lowland UK farms, where hedgehogs occur naturally, they now plummet to rarity partly because of agricultural intensification and partly because their spines are an inadequate deterrent against badgers (Young et al. 2006), which have themselves become prolific because of conditions created by farming (and even more so recently, seemingly due to anthropogenic climate change: Chapter 4, this volume). Meanwhile, on the Hebridean islands, where hedgehogs do not occur naturally, they prosper, threatening endangered wading birds (Charadrii) by eating their eggs, and thus have to be killed by conservationists (Jackson 2007).

All these tangled webs complicate beguilingly simple questions such as when is it 'ethical' to control (normally to kill) a non-native species (see Delfosse 2005)? Macdonald and Burnham (2010) attempted a rule of thumb that intervention is justified if (a) the stakeholder suffers loss to the invasive 'pest' and (b) the intervention, normally lethal, reduces that loss cost-effectively. Fulfilment of these conditions is clear when the non-native species damages economic interests, human health, or human well-being, but the justification for control becomes problematic when the value of the invasive is non-monetary but based on rarity: for example, we protect the Chinese water deer *Hydropotes inermis inermis* because the introduced UK population may represent 10% of the rapidly declining global population (Cooke and Farrell 1998), whereas the equally invasive, although less globally rare and more ecologically problematic muntjac deer *Muntiacus reevesi* is excluded from the closed season afforded to (not entirely) native deer. There is fierce debate whether rare and localized populations of black rats *Rattus rattus* in Britain, which have been deadly to humans in the past, should be protected or eradicated (Macdonald and Tattersall 2001).

The wider public may sometimes find rarefied the niceties of distinctions drawn by hard-line conservationists on the basis of non-nativeness. For example, the American grey squirrel *Sciurus carolinensis* was introduced in the UK in the nineteenth century. Its origins probably do not much trouble the people who thrill to see their antics in otherwise lifeless urban parks but, nonetheless, the greys can infect the now idolized (formerly detested by foresters) native red squirrel *Sciurus vulgaris* with poxvirus (Rushton et al. 2000), cause tree damage (Mayle et al. 2009), and may locally depress certain species of woodland birds (Newson et al. 2010). Any or all of these might be good reasons to kill them, but the sole fact of their origins would seem an increasingly flimsy justification for doing so. Contradictions breed like rabbits: rabbits are the most expensive invader in Britain, costing more than £183 million/year in damage to crops (Williams et al. 2010), but while they are hated as pests by farmers, they are celebrated in British culture: who would think of Peter Rabbit as an unwanted invader?

[3] Poem by Johnathan Swift (1733) *On Poetry: a Rhapsody*.

A further question is thus: can non-native origins ever be shed? Ecologists consider a non-native species as 'naturalized' when it is established in the wild and can self-sustain populations without human support (think again of rabbits, and indeed fallow deer *Dama dama*, edible dormice *Glis glis*, and grey squirrels in the UK). Where a non-native has become so embedded in the native community that its presence is essential for the functioning of the ecosystem of which it has become a part, and the ecosystem has been so modified that it would not return to its pre-invasion state if the incomer was removed, the species could be termed an ecological citizen (Macdonald and Burnham 2010). Under these conditions, it may remain entirely legitimate to kill the incomer because it is a pest, but less so to justify this on the grounds of its origins[4].

Self-interest, not surprisingly, can trump ecological righteousness. Those fishermen harvesting American signal crayfish for sale to rustic restaurants may not be much troubled by their provenance, nor even their devastating ecological impacts (Chapter 8, this volume). A rampaging example is the seemingly blind eye turned to the inevitable impact of releasing 35 million chicken-sized birds, and their supplementary food, into the countryside annually in the form of put-and-take pheasant shooting (there are, of course, native, but sadly rare, game birds; see Macdonald and Feber, 2015: Chapter 13). This is perhaps a less perceptible type of farming, but one that has a potentially huge impact on wildlife (in 2010, the autumn biomass of pheasants considerably exceeded that of all the other seed-eating species combined: Chapter 9, this volume).

For centuries, the rural approach to dealing with wildlife viewed as pests, including invasives, has been captured in the pithy dictat that 'the only good one is a dead one'. But in practice things are often not so simple: we show in Chapter 5, this volume, how perturbation following culling can worsen the transmission of bovine tuberculosis in badgers, and in Chapter 8, this volume, we demonstrate how killing American signal crayfish improves the condition of the survivors. But over and above these numerical complications, lethal control can go wrong, and in the absence of a clean kill the topic of animal welfare becomes an important element of conservation science (Treves and Naughton-Treves 2005; McLaren et al. 2007; Box 1.1).

Box 1.1 Welfare—an important arrow in the conservationist's quiver

Much of the work we report in this book, and its sister volume, *Wildlife and Farming: Managing for Nature on Lowland Farms*, was done in or around Wytham Woods. From a European perspective, this was the birthplace of ecology (see Savill et al. 2010), which was delivered by Oxford University's Professor Charles Elton. It is therefore fitting to point out that Elton was among the first ecologists, in *The Ecology of Invasions by Animals and Plants* (1958), to discuss an ethical dimension to conservation. We consider animal welfare as a scientific topic, and another one of the arrows in the conservation biologist's quiver. Not only are welfare considerations among the pros and cons to be integrated into conservation decisions, but conservation outcomes can be improved through the application of welfare research (see McLaren et al. 2007). Thus, in Macdonald and Feber (2015: Chapter 14), we exemplify how water voles kept in laboratory cages tend to be more stressed and dehydrated than voles kept in outdoor enclosures, and this is directly relevant to the efficacy of reintroduction programmes; better welfare leads to better body condition and more successful reintroductions (see Jule et al. 2008). Indeed, more generally, Harrington et al. (2013) review the tendency to neglect this synergy in the wildlife reintroduction literature, offering scope for improving conservation outcomes, and in Chapter 4, this volume, we touch on the use of new techniques measuring leukocyte coping capacity (white blood cell immune activation) to monitor the stress animals experience as a result of ecological interventions.

It has been said that 'death is not a welfare issue' (Webster 1994), in the sense that being dead is of no consequence, but of course the process of dying is squarely in the welfare realm (Yeates 2010). Indeed, the fact that much 'pest control' on farmland has involved killing, not always cleanly, raises two possibilities for improvement (Dawkins 2006). First, lethal techniques can be improved to minimize

continued

[4] Brown hares *Lepus europaeus* are a UK Biodiversity Action Plan (BAP) species but rabbits are persecuted as an agricultural pest, when both have been in Britain since the Romans.

Box 1.1 *Continued*

Box Figure 1.1.1 Mole trapping raises welfare issues. Photograph © Rosalind Shaw.

welfare impacts, and second, more ingenuity can be put into the innovation of potentially more humane non-lethal alternatives. Both approaches are explored in the case studies in this book. In Baker et al. (2012) we expose a shocking variation in the performance of mole traps (Box Fig. 1.1.1) and break-back traps used for mice and for rats; a clear welfare issue that could be remedied by regulation, which is currently non-existent for these types of trap. Moles *Talpa europaea*, and their hills, are blamed for agricultural damage and for transferring bacteria into silage. Traditionally, they were controlled by strychnine poisoning, kill-trapping, and gassing. Strychnine seems to be the most cost-effective method at a landscape scale, but deaths, conveniently out of sight, were through asphyxia due to paralysis of the respiratory muscles. Strychnine was banned in Britain by the Animals (Cruel Poisons) Regulations 1963, but was continued for mole control under an exemption which ended in 2006, and meanwhile inhibited the development of other, welfare-oriented, solutions. The realities of mole behaviour, and the damage it causes, are the topic of a cost–benefit evaluation that revolves heavily around welfare considerations, in Chapter 12, this volume.

Cruelty associated with lethal control is a component of the acceptability of an intervention, along with efficacy at reducing the actual problem (as very distinct from efficacy at killing perceived culprits). For example, after watching badgers dying from underground poisoning with cyanide gas, the Zuckerman (1980) report raised concerns about (but did not oppose) this method of controlling bovine tuberculosis, and in July 1982 a ministerial panel withdrew gassing on cruelty grounds. In Chapter 5, this volume, we explain that trapping badgers to be shot is also not without its welfare issues, potentially leading to injury if not deployed carefully (Woodroffe et al. 2005), while the mis-shooting of free-running badgers has been exposed to result in slow deaths, due to injury (Munro 2014).

An obvious goal would be to devise more humane and more effective ways of solving wildlife problems and, in Chapter 13, this volume, the potential of this quest is exemplified by such tactics as the use of conditioned taste aversion, to protect vulnerable bird eggs from unacceptable predation (Cox et al. 2004) and to dissuade badgers from eating corn (Baker et al. 2005). As with each of the seven deadly risks, local and global lessons sustain each other. For example, in a scaled-up version of foxes reputedly predating on lambs in the UK, we (McManus et al. 2014) found that predation by leopards, caracal, and jackal on sheep in the Western Cape of South Africa could be reduced more effectively (and more cheaply), by a combination of non-lethal techniques, such as the introduction of guardian animals and protection collars on livestock, than by killing the predators.

1.6 Persecution

The subject of killing leads us back to the species with which this chapter opened, the red fox (Fig. 1.3), which provides at least a metaphor, and often a direct answer, to almost every interesting question in human–wildlife conflict. Perceived as a predator of poultry, lambs, and game, the vulpine reputation for mischief stretches back to antiquity, and nothing captures more vividly the love–hate relationship between farming and wildlife than foxhunting (Chapter 3, this volume). From a reputational low point in the Dark Ages, when King Canute classified them as mere beasts of 'chase', second-raters beside the elite beasts of 'venery', they were upwardly mobile by the thirteenth century, when Edward I took time off from hammering the Scots to hunt a royal pack of fox hounds, and their ascent was complete by the eighteenth century when foxhunting was *de rigeur*, and shaped much of what still charms us about the English countryside. So characteristic of human–wildlife conflict, the algebra that hunters wanted them alive so they could kill them, and farmers wanted them dead, was a paradoxical equation that never quite balanced—because each wanted more equality in the outcome than did the other. In 1566, an Act of Parliament set a bounty of one shilling on foxes. The parish of St Bees in Cumberland, for example, reports a tally of 1096 foxes between 1684 and 1775. By the late eighteenth century, as agriculture intensified, and the shooting of gamebirds (themselves, nowadays, a farmed crop) spread, fox-loathing was at full throttle.

Figure 1.3 The red fox is often the source of human/wildlife conflict on farmland. Photograph © Menno Schaefer/Shutterstock.

Darwin's contemporary, philosopher Herbert Spencer, can have had no experience of foxes, or indeed farming and wildlife, when he wrote 'Every man is free to do that which he wills, provided he infringes not the equal freedom of any other man'[5]—that the same fox is perceived as an asset to the huntsman, a blight to farmer and keeper, and a joy to the wildlife watcher, is just the tip of the iceberg of countryside contradictions. Long before the 'equal freedoms' of fur-trappers, animal rightsists, and even photographers further muddied the mix, the royal chronicler Holinshead wrote to Elizabeth I that: 'if foxes were not preserved for the pleasures of gentlemen they would be utterly destroyed many years ago' (quoted in *Silent Fields* in Lovegrove 2008). In the calculus of farmer/fox relationships, culture dominates and carnivores generally figure disproportionately in the canons of human loathing and reverence, and, to know if killing foxes passes the test of rationality, an answer would need to take account, not only of at least partly monetizable effects on pest control, employment, recreation, tourism, and land management, but also variously non-market factors such as animal welfare, tradition, and fieldcraft.

Wildlife conflict with humans, especially when livelihoods and resources are jeopardized, is a global issue, and lessons from foxes eating free-range poultry in the UK (Moberly et al. 2004) are readily transferable around the world. For example, to the jackals *Canis mesomelas*, eating sheep on farms in South Africa: Klare et al. (2010) radio-tracked 15 jackals, and probed 1000 of their droppings, to show that the 5–10 jackals on a 50 km^2 sheep farm, ate about 60 lambs a year, worth $8900, whereas nearby on a 110 km^2 game farm, they ate about 188 wild ungulates per year, worth $6938 in meat value. A lot of wildlife conflict is imagined, but some is real. Much of the biology, and even more of the human sociology, of lion predation on cows in Bostwana (Hemson et al. 2009) and jaguars in Brazil (Marchini and Macdonald 2012) has the same cocktail of ingredients as fox conflict in the UK. In Bhutan, from 274 farmers, 20% lost their livestock to tigers and leopards *Panthera pardus* in a year costing each an estimated 84% of their yearly income ($250 pa) (Wang and Macdonald

[5] *The Principles of Ethics*, 1897.

2006). In Botswana, lions eating cows cost the community $24,385 pa (from 52 cattleposts), and although the profit from lion-related tourism earned $317,200 per year (Hemson et al. 2009), this did not deter farmers from killing the dollar-printing lions to protect their livestock (largely because of an inequality of outcome, i.e. the people carrying the loss were not the people making the profit—another pervasive sociological lesson and market failure within conservation biology; Dickman et al. 2011). Perhaps, in the first century BC, Catullus had in mind farmers and predators when he famously remarked '*Odi et Amo*'—I hate and I love.

How to address this issue of conflict and depredation? Innovative research (such as the use of their scent marks to influence the movements of African wild dogs *Lycaon pictus*; Jordan et al. 2013) or crafty technologies, such as anti-predator collars to dissuade caracal and jackals from eating sheep (McManus et al. 2014), offer much potential (Chapter 13, this volume), but sometimes reinventing a forgotten wheel pays off: for example, reinvigorating mediaeval techniques, we found we could reduce by 70% the predation of sheep by wolves *Canis lupus* in Slovakia, by (re)introducing Slovensky cuvac dogs to guard the sheep (Rigg et al. 2011). Further afield, in Namibia—home to 25% of the world's remaining 12,000 cheetah *Acinonyx jubatus* (Marker et al. 2003)—we sought to reconcile the conflict arising from cheetah attacks to calves on farms and the farmer's retaliation of killing the cheetahs, despite their protected status. We imported Anatolian sheep dogs and gave them to 117 farmers: although 69% of farmers lost stock prior to owning a dog, after a year, 74% reported no livestock losses (Marker et al. 2010). In a closer-to-home example, we calculated that foxes, which eat rabbits, could, after four years of hunting their prey, save the farmer over £100/ha in rabbit damage to wheat crops (Macdonald et al. 2003).

In Britain, wild raptors have often been at the heart of controversy and conflict and, in Chapter 10, this volume, we delve into the knotty problem of raptor management, focusing on the chequered history of the buzzard in Britain. Killing buzzards is no longer socially acceptable, thanks not only to legislation but also to more tolerant attitudes among those at risk of adverse impacts. But such truces are sometimes uneasy and if problems that affect livelihoods are ignored, there is risk of a backlash, with loss of tolerance that could lead to a renewal of illegal killing. As with all the human–wildlife conflicts in this book, the challenge is seeking solutions that are acceptable to all of the stakeholders. This is particularly the case where our next deadly risk factor is concerned—infectious disease.

1.7 Infectious disease

Our sixth agricultural deadly risk leading to human–wildlife conflict is the transmission of infectious disease, which may be between wildlife, humans, and/or livestock (those diseases afflicting British mammals are summarized in Macdonald et al. 2014). Where these diseases are zoonotic, they can risk human health (Kruse et al. 2004; Chomel et al. 2007), for example severe acute respiratory syndrome (SARS) or Ebola (Mackenzie and Jeggo 2013); or cause significant societal disruption, incurring financial cost, such as the 2001 outbreak of foot and mouth disease in the UK (Keeling et al. 2001). Of course, animal infections are not the only ones to impact farmland. Plant pathogens have left an indelible mark on the countryside. Between 1967 and 1990 a virulent strain of Dutch elm disease spread in Britain, wiping out nearly all mature elm trees. Now the fight is on to stop die-back of ash caused by the fungus *Chalara fraxinea*, which since its detection in one nursery in 2012, has been recorded at 376 sites (Forestry Commission 2013) and, more recently, the already precarious situation of British juniper trees is further threatened by the deadly *Phytophthora austrocedrae* fungus.

In the UK, and arguably much further afield, there has probably never been a wildlife disease whose management was based on such a huge volume of research as is the case for the role of badgers (Fig. 1.4) in the epizootiology of bovine tuberculosis in cattle. Paradoxically, while the data are copious and the evidence far-reaching, a solution has remained elusive. This is the topic of Chapter 5, this volume. Precisely because the problem is so intractable, and the apparent balance of outcomes from previous interventions so marginal, it has become ever more important to understand the nuances of badger biology that might be relevant. This has been one motive for our long-term study of badger biology, the topic of Chapter 4—other motives being to understand the impact of climate change on a medium-sized mammal and indeed to explore solutions to crop damage caused by badgers.

To return to our red fox, its persecution and slaughter has been historically justified as a means to control rabies, for which foxes are the main vector of disease in Europe and North America (see Box 2.4 in Chapter 2, this volume). In Chapter 2, this volume, we unravel their behaviour, ascertaining foxes' home ranges to better understand their distributions and contact rates in the landscape, essential to understanding how disease may spread. Our suggestion in the 1980s was, as subsequently applied to badgers, that culling foxes

Figure 1.4 Knowledge of badger behaviour is crucial to a better understanding of bTB transmission and control. Photograph © Andrew Harrington.

creates a perturbation effect which can increase disease transmission, and thus fox vaccination would be a better solution for rabies control.

Zoonotic diseases may, however, lie undetected in the wild, and in Chapter 11, this volume, we reveal how we sampled 600 farm rats, finding, for the first time in Europe, *Coxiella burnetti* and Hanta virus, the causative organisms, respectively, of serious human diseases, Q-fever and Hantaan fever. The same studies revealed how parasite-altered behaviour in the rats encouraged kamikaze tendencies that led them into the jaws of cats. Transmission of disease also occurs between wildlife and livestock further afield, in the USA, free-ranging domestic ruminants and mule deer *Odocoileus hemionus* share pathogens (Roug et al. 2012) and, globally, wild boar *Sus scrofa* are a source for infectious diseases (e.g. swine fever) transmitted to livestock and humans (Meng et al. 2009).

In Chapter 5, this volume, we argue that in the case of badgers and bTB, it seems that, as with red foxes, the perturbation hypothesis (Macdonald et al. 2006) is a major reason why killing badgers has not controlled TB in cattle in the UK, and indeed the principles behind this hypothesis apply to attempts at lethal control of wildlife worldwide (Swinton et al. 1997; Tuyttens et al. 2000). Indeed, there are diverse reasons why culling is often less promising than one might think: in the case of paratuberculosis transmission between cattle and rabbits it is estimated to require the implausible target of killing more than 95% of the rabbits (Davidson et al. 2008). In other cases, unattractive double standards are displayed to all those with the surname 'rat', which seems to deaden most people's concerns for animal welfare (see Box 1.1).

Surprisingly, humans have deliberately introduced disease as a means to control pests, but have unintentionally created tangled webs that have spun beyond our control. Take rabbits, introduced in the UK for meat and fur in the twelfth century. Perhaps part of the reason it took over 800 years for rabbits to explode into pestilence was because for much of that time people ate them avidly, but by the 1950s they were off the post-war menu and rabbits had become an agricultural pest of such magnitude that the myxoma virus was introduced to control them (Sumption and Flowerdew 1985). This germ warfare resulted in 99% rabbit mortality, which brought drastic changes in agricultural habitats by easing the pressure on previously grazed pasture and winter crops, and by restoring woodland regeneration (see Sumption and Flowerdew 1985). However, with the decline in rabbits, many predators which mainly preyed on rabbits suffered declines (e.g. stoats *Mustela ermineu* and buzzards; Sumption and Flowerdew 1985). Interestingly, without the rabbit's grazing pressures grasses grew long, promoting field vole *Microtus agrestis* populations and, in turn, predators that were able to switch their diet from rabbits to voles thrived, for example, foxes, polecats *Mustela putorius*, and kestrels *Falco tinnunculus*. Similarly, as discovered by 'Mick' Southern's pioneering research in Wytham Woods (see Box 12.3 in Macdonald and Feber, 2015: Chapter 12), with the decline in rabbits, tawny owl *Strix aluco* clutch survival also decreased and owls put a heavier predation pressure on wood mice *Apodemus sylvaticus* and field vole populations (Southern 1954). Rabbits, now recovered, have been involved in the decline of voles, and with them the weasel *Mustela nivalis*, for which voles are the primary prey (Yalden 1999). Rabbits are also an important prey of the introduced American mink, and so have facilitated their spread, and as we elaborate in Macdonald and Feber (2015: Chapters 14 and 15), with devastating effects on native water voles.

Anthropogenic land use changes, of which agricultural intensification is just one, drive a range of infectious disease outbreaks and modify the transmission of endemic infections (Patz et al. 2004), but to these can also be added, as part of our backdrop to the stage, a factor which affects all of our previous risk factors in different ways: climate change.

1.8 Climate change

The thing about climate change is not that it is new (it has happened throughout geological time), nor is the problem that nature cannot adapt to it (it has been a perennial driver of natural selection), but rather that it is currently changing at a more rapid rate, and that ever greater variation is occurring (Nouvellet et al. 2013). It therefore joins our list of seven deadly risk factors.

Farming supports a growing global population in excess of 7 billion, taking humankind way beyond the extremes of what natural food productivity could support. Achieving this super-natural feat is, however, only possible in a system where productivity is constantly 'cropped', and this leads to around 30–35% of typical annual global greenhouse gas (GHG) emissions (Foley et al. 2011), contributing to the anthropogenic acceleration of climate change. The impacts of greater extremes of climatic variability bleed into all of our other agricultural deadly risk factors, altering the economic environment for farmers, as well as the attributes of the countryside for people and wildlife. Farmers and landowners are exposed, not only to the influence of predicted more extreme weather events and seasonal and inter-annual variation in conditions on the types of crops that can be grown, but also to the corollaries of these impacts that are likely to have worse consequences in ever more intensively managed landscapes. For example, after heavy rainfall in 2012, UK farmers claimed to be facing £600 million in lost output[6].

[6] National Farmers' Union: 2012 weather 'cost farming £600m' <http://www.bbc.co.uk/>.

Against the deteriorating fabric of habitat loss, the consequences, direct and indirect, of climate change on wildlife will surely be various, and variously serious (e.g. Newman and Macdonald 2013). While, in the long term, species will adapt their genotypes, through natural selection, to new climatic conditions, in the short term their options are either to seek to remain within their comfort zones by moving (subject to habitat fragmentation), or for species with the necessary behavioural adaptability/plasticity, to explore new and different sources of food. For example, in the UK, some southern butterflies and dragonflies are heading north (Hassall et al. 2007); this may be fine for strong flyers and mobile species able to jump the gaps in habitat, but for the plodders it may be catastrophic. Northern species, adapted to cold, may, in the face of unaccustomed warmth, shuffle off the island, and in some cases this could mean off this mortal coil (e.g. some carabid beetles; Morecroft et al. 2009; LWEC 2014).

Changes in primary productivity may influence herbivores (many rodents, rabbits, hares, and deer) potentially exacerbating conflicts with crop production via absolute changes in food availability, or phenological changes influencing food at critical times. Insectivores (shrews, hedgehogs, moles, and bats) are similarly vulnerable to invertebrate loss (and suffer a double whammy where pesticides also contribute to declines in insect communities). Carnivores are, in turn, vulnerable to change in this herbivore prey base and even versatile omnivorous carnivores are influenced by food-availability fluctuations, for example, species for which we have found earthworm prey is important, such as foxes (Chapter 2, this volume), badgers (Chapter 4, this volume), buzzards (Chapter 10, this volume), and moles (Chapter 12, this volume), will suffer when spring droughts curtail this food source at a crucial time for juvenile development.

Sometimes, while certain changes in climate will be advantageous for a species, it will cascade in a series of negative effects. For example, warm winters benefit the fecundity and development of red deer *Cervus elaphus*, however, as we elaborate in Macdonald and Feber (2015: Chapter 12), too many deer can be problematic for forest regeneration and can cause crop damage. Similarly, in Chapter 4, this volume, we show that badgers experience better over-winter survival when conditions are mild, while cold, damp spring conditions cause badger cub mortality and, in Chapter 11, this volume, how brown rats which carry parasites that can cause disease in humans are more prone to capitalize on food indoors during winter colder weather, increasing the risk of rat–human contact and potentially contaminating feed/silage with disease.

Even these few examples illustrate the diversity of ways in which climate variability affects farmland biodiversity and thus the shape of conflicts between humans and wildlife. For example, predicted milder climates may increase nematode parasites in sheep (Morgan and van Dijk 2012), helminth species of parasites on farmed ruminants and wild ungulates (e.g. deer; van Dijk et al. 2010), and tick loads in otters (Sherrard-Smith et al. 2012). In badgers, the mortality rate caused by parasite load in cubs proves higher with high rainfall in spring (Chapter 4, this volume); and some brown rats' parasites are at their peak at the time of the year when they migrate from crops to grain stores (Chapter 11, this volume).

Habitat changes are also a projected corollary of climate change, interacting with habitat management. In particular, changes in the stability of aquatic habitats are predicted, where riverine specialists, such as water shrews, water voles, re-introduced beavers, otters, and invasive mink, might experience difficulties along water courses in response to drought or flood. Under changing climatic conditions intensive arable management is predicted to spread northwards and westwards in Europe, which, for example, could worsen the issues of habitat degradation and loss for internationally important populations of lesser horseshoe bat *Rhinolophus hipposideros* (Newman and Macdonald 2013). In addition, as the climatic conveyor rolls north, in come continental species—we might welcome incoming winter damselflies *Sympecma fusca*, cattle egrets *Bubulcus ibis*, even spoonbills *Platalea leucorodia*, as they settle cosily into British territory. Natural philosophers may muse, as we did earlier, whether there are double standards at work with respect to what is, or is not, an invasive. In any case, farmers are not going to rejoice at the arrival of new insect agricultural pests.

Here is where practice meets policy. The interaction between farming and climate change that affects nature is compounded substantially by habitat connectivity, or lack of it (Donald and Evans 2006). In response, the Lawton report (Lawton et al. 2010) calls, catchily, for bigger, better, more, and more connected wildlife habitat in Britain, and for the policies to bring these changes about. The UK Natural Environment White Paper (Defra 2011) recommends the creation of Nature Improvement Areas to connect sites for wildlife to 'live in and adapt to climate change', reaching a wide area of the landscape, and allowing generalists and non-generalists to build in resilience and flexibility: the more diverse the agricultural ecosystem (e.g. greater biodiversity), the greater its ability to adapt (LWEC 2014).

1.9 Conclusions

So, what have we learnt? First, in the introductory chapter of Macdonald and Feber (2015: Chapter 1), we asked whether farmland was a fit topic for conservation biologists and a place for conservation. Of course, the answer was yes, and not least because the first lesson of this chapter is that everything is joined up. It is joined up because the principles of ecology apply, whether dealing with what we (transiently) perceive as pest or asset, blurring any demarcation between biodiversity conservation, wildlife management, and pest control, and guaranteeing that whatever action is taken in the name of one of these out-moded distinctions will have consequences in the bailiwicks of the others. It is joined up because the amount of land we have taken into farming dictates the amount available for wildlife. It is joined up not only because farmland and nature interact at their confluence, determined by the geometry of fragmentation, but also because tendrils of wildlife habitat physically do that joining up. It is joined up not just by the processes of natural science, but by the processes of social science, and beyond, through inter-disciplinarity to politics.

Rather, although, farmland is farmland because it produces crops (often food), and is thus different, as a matter of vocabulary, from nature reserve, it also delivers landscape, biodiversity, archaeology, culture, and community and—although oddly only rarely mentioned—income for the farmer (Knowler and Bradshaw 2007). These may once have been policy bling, non-essential co-benefits, but today they are ecosystem services for which society is prepared to pay. So, everything is joined up.

And it is precisely this joined-upness that ensures some level of conflict will always permeate the detailed interface between farming and wildlife, and indeed the interface between them both and wider society. The ploughman, stockman, game-keeper, forester, nature enthusiast all see the same farmscape from a slightly different vantage point, each in turn different from those from which politicians, regulators, sociologists, economists, and the like draw their perspectives. Each may also see the landscape at a different scale, remembering the analogy of views from a hot-air balloon at different altitudes, which we drew in Macdonald and Feber (2015: Chapter 1). The important thing is that their varying perspectives can all be informed by the same evidence and understanding, and generating and interpreting those data has been the *raison d'etre* of the research reported in this volume, and its sister. We hope the detail will be helpful and enduring, but we are sure that the principle of evidence-based, inter-disciplinary compromise will be both. There are purist circles in which the word compromise is a dirty one, and certainly we see ourselves as uncompromising in our commitment to fostering nature as part of humanity's future, whether it be in the English farmscape or global wilderness. But the very fact that no corner of the Earth, and by definition no landscape touched by agriculture, is beyond the influence of people makes it simply silly to deny there is, at some scale, give-and-take in every solution to living with nature, just as there is conflict in every attempt to do so.

The seven deadly risk factors are global, and the common denominators of both problems and solutions are global too. So, for 'fox among your hens' read 'lion among your cows' (Maclennan et al. 2009), for fragmentation of woodland spinneys consider the sugar plantations eroding the chimpanzees' *Pan troglodytes* enclaves in Uganda (Zommers et al. 2012), and for the weasel, nemesis of the hedgerows, think about the pumas *Puma concolor* as they move between forest fragments in Chilean farmland (Murphy and Macdonald 2010). Clearly, these pairings differ in crucially important detail, but they are also similar in crucially important principles, and while landscapes may be fragmented, knowledge need not be: lessons learnt can travel globally. For a world where 50% of people live in towns, think of the spiritual uplift of a flock of snow geese, and the hazards of trying to monetize it, remembering the universality of Nature's value beyond markets. And when it comes to markets, we need to remember that AES payments are fiscal instruments to deliver payment for an ecosystem service. Remember, too, that biodiversity hot spots often coincide with poverty worldwide, for which a quick fix through increased production will be self-defeating if it increases vulnerability to pests, diseases, and natural disasters, and impairs ecosystem services.

So, on this journey of war and peace in the countryside, we end, as we began, with a quotation from Tolstoy: *Everyone thinks of changing the world, but no one thinks of changing himself.*

References

Aiméa, C., Laval, G., Patin, E., Verdu, P., Ségurel, L., Chaix, R., Hegay, T., Quintana-Murci, L., Heyer, E., and Austerlitz, F. (2013). Human genetic data reveal contrasting demographic patterns between sedentary and nomadic populations that predate the emergence of farming. *Molecular Biology and Evolution*, **30**, 2736–2737.

Akçakaya, H.R., Mills, G., and Doncaster, C.P. (2007). The role of metapopulations in conservation. In D. Macdonald and K. Service, eds., *Key Topics in Conservation Biology*, 64–84. Blackwell Publishing, Oxford.

Baker, S.E., Ellwood, S.A., Tagarielli, V.L., and Macdonald, D.W. (2012). Mechanical performance of rat, mouse and mole spring traps, and possible implications for welfare performance. *PLOS ONE*, **7**, e39334.

Baker, S.E., Ellwood, S.A., Watkins, R.W., and Macdonald, D.W. (2005). Non-lethal control of wildlife: using chemical repellents as feeding deterrents for the European badger. *Journal of Applied Ecology*, **42**, 921–931.

Barrett, C.B., Bulte, E.H., Ferraro, P., and Wunder, S. (2013). Economic instruments for nature conservation. In D.W. Macdonald and K. Willis, eds., *Key Topics in Conservation Biology 2*, pp. 59–73. Wiley-Blackwell, John Wiley & Sons Ltd., Oxford.

Bellwood, P. (2001). Early agriculturalist population diasporas? Farming, languages, and genes. *Annual Review of Anthropology*, **30**, 181–207.

Buckle, A. (2009). Rat control 3: climate change. *Farmers Weekly*, 17 April. <http://www.fwi.co.uk/>, accessed October 2012.

Can, Ö.E., D'Cruze, N., Garshelis, L., Beecham, J., and Macdonald, W. (2014). Resolving human-bear conflict: a global survey of countries, experts, and key factors. *Conservation Letters*, **7**, 501–513.

Carneiro, R.L. (1970). A theory of the origin of the state: Traditional theories of state origins are considered and rejected in favor of a new ecological hypothesis. *Science*, **169**, 733.

Carr, G.M. and Macdonald, D.W. (1986). The sociality of solitary foragers: a model based on resource dispersion. *Animal Behaviour*, **34**, 1540–1549.

Chomel, B.B., Belotto, A., and Meslin, F.-X. (2007). Wildlife, exotic pets and emerging zoonoses. *Emerging Infectious Diseases*, **13**, 6–11.

Clutton-Brock, J. (1995). Origins of the dog: domestication and early history. In Serpell, J. ed. *The Domestic Dog: Its Evolution, Behaviour and Interactions with People*. Cambridge University Press, Cambridge.

Cohen, M.N. (2009). Introduction: rethinking the origins of agriculture. *Current Anthropology*, **50**, 591–595.

Cooke, A. and Farrell, L. (1998). *Chinese Water Deer*. The Mammal Society and the British Deer Society, London.

Cox, R., Baker, S., Macdonald, D.W., and Berdoy, M. (2004). Protecting egg prey from Carrion Crows: the potential of aversive conditioning. *Applied Animal Behaviour Science*, **87**, 325–342.

Davidson, R.S., Marion, G., White, P.C.L., and Hutchings, M.R. (2008). Use of host population reduction to control wildlife infection: rabbits and paratuberculosis. *Epidemiology and Infection*, **137**, 131–138.

Davies, B., Biggs, J., Williams, P., and Thompson, S. (2009). Making agricultural landscapes more sustainable for freshwater biodiversity: a case study from southern

England. *Aquatic Conservation: Marine and Freshwater Ecosystems*, **19**, 439–447.

Davis, M.A. (2009). *Invasion Biology*. OUP, Oxford.

Dawkins, M.S. (2006). A user's guide to animal welfare science. *Trends in Ecology and Evolution*, **21**, 77–82.

Defra (2011). UK Natural Environment White Paper 2011. https://www.gov.uk/.

Delfosse, E.S. (2005). Risk and ethics in biological control. *Biological Control*, **35**, 319–329.

Diamond, J.M. (1984). 'Normal' extinction of isolated populations. In M.H. Nitecki, ed., *Extinctions*, pp. 191–246. Chicago University Press, Chicago, USA.

Diamond, J. (2002). Evolution, consequences and future of plant and animal domestication. *Nature*, **418**, 700–707.

Dickman, A.J., Macdonald, E.A., and Macdonald, D.W. (2011). A review of financial instruments to pay for predator conservation and encourage human–carnivore coexistence. *Proceedings of the National Academy of Sciences*, **108**, 13937–13944.

Dobson, A., Lodge, D., Alder, J., Cumming, G.S., Keymer, J., McGlade, J., and Xenopoulos, M.A. (2006). Habitat loss, trophic collapse, and the decline of ecosystem services. *Ecology*, **87**, 1915–1924.

Donald, P.F. and Evans, A.D. (2006). Habitat connectivity and matrix restoration: the wider implications of agri-environment schemes. *Journal of Applied Ecology*, **43**, 209–218.

Donald, P.F., Green, R.E., and Heath, M.F. (2001). Agricultural intensification and the collapse of Europe's farmland bird populations. *Proceedings of the Royal Society of London. Series B: Biological Sciences*, **268**, 25–29.

Donald, P.F., Sanderson, F.J., Burfield, I.J., and van Bommel, F.P.J. (2006). Further evidence of continent-wide impacts of agricultural intensification on European farmland birds, 1990–2000. *Agriculture, Ecosystems and Environment*, **116**, 189–196.

Driscoll, C.A., Luo, S.J., Macdonald, D., Dinerstein, E., Chestin, I., Pereladova, O., and O'Brien, S.J. (2011). Restoring tigers to the caspian Region. *Science*, **333**, 822–823.

EEA (2013). *The European Grassland Butterfly Indicator: 1990–2011*. EEA Technical Report no. 11/2013. European Environment Agency, Copenhagen.

Elmqvist, T., Folke, C., Nyström, M., Peterson, G., Bengtsson, J., Walker, B., and Norberg, J. (2003). Response diversity, ecosystem change, and resilience. *Frontiers in Ecology and the Environment*, **1**, 488–494.

Elton, C.S. (1958). *The Ecology of Invasions by Animals and Plants*. Methuen, London.

FAOSTAT (2014). <http://faostat.fao.org/>, accessed June 2014.

Fischer, J., Brosi, B., Daily, G.C., Ehrlich, P.R., Goldman, R., Goldstein, J., and Tallis, H. (2008). Should agricultural policies encourage land sparing or wildlife-friendly farming? *Frontiers in Ecology and the Environment*, **6**, 380–385.

Foley, J.A., Ramankutty, N., Brauman, K.A., et al. (2011). Solutions for a cultivated planet. *Nature*, **478**, 337–342.

Forestry Commission 2013. http://www.forestry.gov.uk/.

Garnett, B.T., Delahay, R.J., and Roper, T.J. (2002). Use of cattle farm resources by badgers (*Meles meles*) and risk of bovine tuberculosis (Mycobacterium bovis) transmission to cattle. *Proceedings of the Royal Society of London. Series B: Biological Sciences*, **269**, 1487–1491.

Garnett, B.T., Roper, T.J., and Delahay, R.J. (2003). Use of cattle troughs by badgers (*Meles meles*): A potential route for the transmission of bovine tuberculosis (*Mycobacterium bovis*) to cattle. *Applied Animal Behaviour Science*, **80**, 1–8.

Gómez-Ramírez, P., Shore, R.F., van den Brink, N.W., et al. (2014). An overview of existing raptor contaminant monitoring activities in Europe. *Environmental International*, **67**, 12–21.

Hallmann, C.A., Foppen, R.P.B., van Turnhout, C.A.M., de Kroon, H., and Jongejans, E. (2014). Declines in insectivorous birds are associated with high neonicotinoid concentrations. *Nature*, **511**, 341–343.

Harrington, L.A., Moehrenschlager, A., Gelling, M., Atkinson, R.P.D., Hughes, J., and Macdonald, D.W. (2013). Conflicting and complementary ethics of animal welfare considerations in reintroductions. *Conservation Biology*, **27**, 486–500.

Harrison, P. (2006). Having dominion: Genesis and the mastery of nature. *Environmental Stewardship*, **56**, 17.

Hassall, C., Thompson, D.J., French, G.C., and Harvey, I.F. (2007). Historical changes in the phenology of British Odonata are related to climate. *Global Change Biology*, **13**, 933–941.

Hemson, G., Maclennan, S., Mills, G. Johnson, P., and Macdonald, D. (2009). Community, lions, livestock and money: A spatial and social analysis of attitudes to wildlife and the conservation value of tourism in a human–carnivore conflict in Botswana. *Biological Conservation*, **142**, 2718–2725.

Hillman, G., Hedges, R., Moore, A., Colledge, S., and Pettitt, P. (2001). New evidence of late glacial cereal cultivation at Abu Hureyra on the Euphrates. *The Holocene*, **11**, 383–393.

Jackson, D.B. (2007). Factors affecting the abundance of introduced hedgehogs (Erinaceus europaeus) to the Hebridean island of South Uist in the absence of natural predators and implications for nesting birds. *Journal of Zoology*, **271**, 210–217.

Jenkins, P. (2013). Invasive animals and wildlife pathogens in the United States: the economic case for more risk assessments and regulation. *Biological Invasions*, **15**, 243–248.

Jordan, N.R., Golabek, K.A., Apps, P.J., Gilfillan, G.D., and McNutt, J.W. (2013). Scent -mark identification and scent-marking behaviour in African wild dogs (*Lycaon pictus*). *Ethology*, **119**, 644–652.

Jule, K.R., Leaver, L.A., and Lea, S.E. (2008). The effects of captive experience on reintroduction survival in carnivores: a review and analysis. *Biological Conservation*, **141**, 355–363.

Keeling, M.J., Woolhouse, M.E., Shaw, D.J., Matthews, L., Chase-Topping, M., Haydon, D.T., and Grenfell, B.T.

(2001). Dynamics of the 2001 UK foot and mouth epidemic: stochastic dispersal in a heterogeneous landscape. *Science*, **294**, 813–817.

Klare, U., Kamler, J.F., Stenkewitz, U., and Macdonald, D.W. (2010). Diet, Prey Selection, and Predation Impact of Black-Backed Jackals in South Africa. *The Journal of Wildlife Management*, **74**, 1030–1041.

Kleijn, D., Kohler, F., Baldi, A., et al. (2009). On the relationship between farmland biodiversity and land-use intensity in Europe. *Proceedings of the Royal Society of London. Series B: Biological Sciences*, **276**, 903–909.

Kleijn, D., Rundlof, M., Scheper, J., Smith, H.G., and Tscharntke, T. (2011). Does conservation on farmland contribute to halting biodiversity decline? *Trends in Ecology and Evolution*, **26**, 474–481.

Knowler, D. and Bradshaw, B. (2007). Farmers' adoption of conservation agriculture: A review and synthesis of recent research. *Food policy*, **32**, 25–48.

Kruse, H., Kirkemo, A.M., and Handeland, K. (2004). Wildlife as source of zoonotic infections. *Emerging Infectious Diseases*, **10**, 2067–2072.

Lamarque, F., Anderson, J., Fergusson, R., Lagrange, M., Osei-Owusu, Y., and Bakker, L. (2009). *Human-wildlife conflict in Africa. Causes, consequences and management strategies*. Food and Agriculture Organization of the United Nations, Rome.

Lawton, J.H., Brotherton, P.N.M., Brown, V.K., et al. (2010). *Making Space for Nature: a review of England's wildlife sites and ecological network*. Report to Defra.

Leopold, A. (1949). *A Sand County Almanac and Sketches Here and There*. Oxford University Press, Oxford.

Lovegrove, R. (2008). *Silent Fields. The long decline of a nation's wildlife*. Oxfod University Press, Oxford.

Loveridge, A.J., Wang, S.W., Frank, L.G., and Seidensticker, J. (2010). People and wild felids: conservation of cats and management of conflicts. In D. Macdonald and A. Loveridge, eds., *The Biology and Conservation of Wild Felids*, pp. 161–196. Oxford University Press, Oxford.

LWEC (2014). *Living with Environmental Change*. <www.lwec.org.uk>.

Macdonald, D.W. (2013). From Ethology to Biodiversity: Case Studies of Wildlife Conservation. *Nova Acta Leopoldina*, NF 111, Nr. 380, 1–46.

Macdonald, D.W., Boitani, L., Dinerstein, E., Fritz, H., and Wrangham, R. (2013). Conserving large mammals: Are they a special case? In D.W. Macdonald and K. Willis, eds., *Key Topics in Conservation Biology 2*, pp. 277–312. Wiley-Blackwell, John Wiley & Sons Ltd., Oxford.

Macdonald, D.W. and Burnham, D. (2010). *The state of Britain's mammals: a focus on invasive species*. WildCRU, Oxford.

Macdonald, D.W. and Feber, R.E., eds. (2015). *Wildlife Conservation on Farmland. Managing for Nature on Lowland Farms*. Oxford University Press, Oxford.

Macdonald, D.W. and Johnson, D.D.P. (2015). Patchwork planet: the resource dispersion hypothesis, society, and the ecology of life. *Journal of Zoology*, **295**, 75–107.

Macdonald, D.W. and Johnson, P.J. (2000). *Farmers and the Custody of the Countryside: Trends in Loss and Conservation of Non-Productive Habitats 1981–1998. Biological Conservation*, **94**, 221–234.

Macdonald, D.W., Moorhouse, T., and Gelling, M. (2014). *The state of Britain's mammals: a focus on disease*. WildCRU for PTES, Oxford.

Macdonald, D.W., Riordan, P., and Mathews, F. (2006). Biological hurdles to the control of TB in cattle: A test of two hypotheses concerning wildlife to explain the failure of control. *Biological Conservation*, **131**, 268–286.

Macdonald, D.W., Feber, R.E., Tattersall, F.H, and Johnson, P.J. (2000a). *Ecological experiments in farmland conservation*. In Hutchings, M.J., John, E.A. and Stewart, A.J.A., eds., The Ecological consequences of environmental heterogeneity. Blackwell Science, Oxford.

Macdonald, D.W., Stewart, P.D., Stopka, P., and Yamaguchi, N. (2000b). Measuring the dynamics of mammalian societies: an ecologist's guide to ethological methods. In L. Boitani and T.K. Fuller, eds., *Research Techniques in Animal Ecology: Controversies and Consequences*, pp. 332–388. Columbia University Press, New York.

Macdonald, D.W., Reynolds, J.C., Carbone, C., Mathews, F. and Johnson, P.J. (2003). The bio-economics of fox control. In F. Tattersall and W. Manley, eds., *Conservation and Conflict – Mammals and Farming in Britain*, pp. 220–236. Westbury Publishing, West Yorkshire.

Macdonald, D.W. and Tattersall, F. (2001). *Britain's Mammals: The Challenge for Conservation*. Mammals Trust UK, London.

Mackenzie, J.S. and Jeggo, M. (2013). Reservoirs and vectors of emerging viruses. *Current Opinion in Virology*, **3**, 170–179.

Maclennan, S.D., Groom, R.J., Macdonald, D.W., and Frank, L.G. (2009). Evaluation of a compensation scheme to bring about pastoralist tolerance of lions. *Biological Conservation*, **142**, 2419–2427.

Marchini, S. and Macdonald, D.W. (2012). Predicting ranchers' intention to kill jaguars: Case studies in Amazonia and Pantanal. *Biological Conservation*, **147**, 213–221.

Marker, L.L., Dickman, A.J., Mills, M.G.L., and Macdonald, D.W. (2003). Aspects of the management of cheetahs, Acinonyx jubatus jubatus, trapped on Namibian farmlands. *Biological Conservation*, **114**, 401–412.

Marker, L., Dickman, A.J., Mills, M.G.L., and Macdonald, D.W. (2010). Cheetahs and ranchers in Namibia: a case study. In D. Macdonald and A. Loveridge, eds., *The Biology and Conservation of Wild Felids*, pp. 353–372. Oxford University Press, Oxford.

Mason, C.F. and Macdonald, S.M. (1993). Impact of organochlorine pesticide residues and PCBs on otters (*Lutra lutra*) in eastern England. *Science of the Total Environment*, **138**, 147–160.

Mayle, B.A., Proudfoot, J., and Poole, J. (2009). Influence of tree size and dominance on incidence of bark stripping by grey squirrels to oak and impact on tree growth. *Forestry*, **82**, 431–444.

McKenzie, A.J., Emery, S.B., Franks, J.R., and Whittingham, M.J. (2013). Landscape-scale conservation: collaborative agri-environment schemes could benefit both biodiversity and ecosystem services, but will farmers be willing to participate? *Journal of Applied Ecology*, **50**, 1274–1280.

McLaren, G., Bonacic, C., and Rowan, A. (2007). Animal welfare and conservation: measuring stress in the wild. In D. Macdonald and K. Service, eds., *Key Topics in Conservation Biology*, pp. 120–133. Blackwell Publishing, Oxford.

McLeod, S.R. (1997). Is the concept of carrying capacity useful in variable environments? *Oikos*, **79**, 529–542.

McManus, J.S., Dickman, A.J., Gaynor, D., Smuts, B.H., and Macdonald, D.W. (2014). Dead or alive? Comparing costs and benefits of lethal and non-lethal human–wildlife conflict mitigation on livestock farms. *Oryx*, First view, 1–9.

Meng, X.J., Lindsay, D.S., and Sriranganathan, N. (2009). Wild boars as sources for infectious diseases in livestock and humans. *Philosophical Transactions of the Royal Society of London. Series B: Biological Sciences*, **364**, 2697–2707.

Merckx, T., Feber, R.E., Dulieu, R.L., et al. (2009). Effect of field margins on moths depends on species mobility: field-based evidence for landscape-scale conservation. *Agriculture, Ecosystems and Environment*, **129**, 302–309.

Miller, N.F. (1992). The origins of plant cultivation in the Near East. In C.W. Cowan and P.J. Watson, *The Origins of Agriculture: An International Perspective*, eds., pp. 39–58. Smithsonian Institution Press, Washington, DC.

Mithen, S. (2007). Did farming arise from a misapplication of social intelligence? *Philosophical Transactions of the Royal Society B: Biological Sciences*, **362**, 705–718.

Moberly, R.L., White, P.C.L., and Harris, S. (2004). Mortality due to fox predation in free-range poultry flocks in Britain. *Veterinary Record*, **155**, 48–52.

Morandin, C., Loveridge, A.J., Segelbacher, G., et al. (2014). Gene flow and immigration: genetic diversity and population structure of lions (Panthera leo) in Hwange National Park, Zimbabwe. *Conservation Genetics*, **15**, 697–706.

Morecroft, M.D., Bealey, C.E., Beaumont, D.A., et al. (2009). The UK Environmental Change Network: Emerging trends in the composition of plant and animal communities and the physical environment. *Biological Conservation*, **142**, 2814–2832.

Morgan, E.R. and van Dijk, J. (2012). Climate and the epidemiology of gastrointestinal nematode infections of sheep in Europe. *Veterinary Parasitology*, **189**, 8–14.

Munro, R. (2014). *Pilot Badger Culls in Somerset and Gloucestershire. Report by the Independent Expert Panel*. Presented to the Secretary of State for Environment, Food and Rural Affairs, The Rt Hon Owen Paterson MP, March 2014.

Murphy, T. and Macdonald, D.W. (2010). Pumas and people: lessons in the landscape from a widely distributed felid. In D.W. Macdonald and A.J. Loveridge, eds., *The Biology and Conservation of Wild Felids*, pp. 431–452, Oxford University Press, Oxford.

Newman, C. and Macdonald, D.W. (2013). *The implications of climate change for terrestrial UK mammals*. Terrestrial biodiversity climate change impacts report card technical paper. Living with environmental change partnership. <http://www.lwec.org.uk/>

Newman, C., Zhou, Y.B., Buesching, C.D., Kaneko, Y., and Macdonald, D.W. (2011). Contrasting sociality in two widespread, generalist, mustelid genera, meles and martes. *Mammal Study*, **36**, 169–188.

Newson, S.E., Leech, D.I., Hewson, C.M., Crick, H.Q.P., and Grice, P.V. (2010). Potential impact of grey squirrels *Sciurus carolinensis* on woodland bird populations in England. *Journal of Ornithology*, **151**, 211–218.

Newton, I., Wyllie, I., and Asher, A. (1993). Long-term trends in organochlorine and birds in Britain. *Environmental Pollution*, **79**, 143–151.

Nouvellet, P., Newman, C., Buesching, C.D., and Macdonald, D.W. (2013). A multi-metric approach to investigate the effects of weather conditions on the demographic of a terrestrial mammal, the European badger (*Meles meles*). *PLOS ONE*, **8**, e68116.

Patz, J.A., Daszak, P., Tabor, G.M., Aguirre, A.A., Pearl, M., Epstein, J., and Bradley, D. J. (2004). Unhealthy landscapes: policy recommendations on land use change and infectious disease emergence. *Environmental health perspectives*, **112**, 1092.

Phalan, B., Onial, M., Balmford, A., and Green, R.E. (2011). Reconciling Food Production and Biodiversity Conservation: Land Sharing and Land Sparing Compared. *Science*, **333**, 1289–1291.

PTES (2008). *Living with mammals - survey update 2008*. <http://www.ptes.org/>, accessed October 2012.

Rabinowitz, A. and Zeller, K.A. (2010). A range-wide model of landscape connectivity and conservation for the jaguar, *Panthera onca*. *Biological Conservation*, **143**, 939–945.

Rey-Benayas, J.M. and Bullock, J.M. (2012). Restoration of biodiversity and ecosystems services on agricultural land. *Ecosystems*, **15**, 883–899.

Richerson, P.J. and Boyd, R. (2001). Institutional evolution in the Holocene: the rise of complex societies. *Proceedings of the British Academy*, **110**, 197–234.

Rigg, R., Findo, S., Wechselberger, M., Gorman, M.L., Sillero-Zubiri, C., and Macdonald, D.W. (2011). Mitigating carnivore-livestock conflict in Europe: lessons from Slovakia. *Oryx*, **45**, 272–280.

Roug, A., Swift, P., Torres, S., Jones, K., and Johnson, C.K. (2012). Serosurveillance for Livestock Pathogens in Free-Ranging Mule Deer (Odocoileus hemionus). *PLOS ONE*, **7**, e50600.

Rushton, S.P., Lurz, P.W.W., Gurnell, J., and Fuller, R. (2000). Modelling the spatial dynamics of parapoxvirus disease in red and grey squirrels: a possible cause of the decline in the red squirrel in the UK? *Journal of Applied Ecology*, **37**, 997–1012.

Savill, P., Perrins, C., Kirby, K., and Fisher, N. eds. (2010). *Wytham Woods: Oxford's Ecological Laboratory*. Oxford University Press, Oxford.

Sherrard-Smith, E., Chadwick, E., and Cable, J. (2012). Abiotic and Biotic Factors Associated with Tick Population Dynamics on a Mammalian Host: Ixodes hexagonus Infesting Otters, Lutra lutra. *PLOS ONE*, **7**, e47131.

Sillero-Zubiri, C., Sukumar, R., and Treves, A. (2007). Living with wildlife: the roots of conflict and the solutions. In

D. Macdonald and K. Service, eds., *Key Topics in Conservation Biology*, pp. 255–272. Blackwell Publishing, Oxford.

Southern, H.N. (1954). Tawny owls and their prey. *Ibis*, **96**, 384–410.

Stahl, P., Ruette, S., and Gros, L. (2002). Predation on free-ranging poultry by mammalian and avian predators: field loss estimates in a French rural area. *Mammal Review*, **32**, 227–234.

Stenseth, N.C., Leirs, H., Skonhoft, A., Davis, S.A., Pech, R.P., Andreassen, H.P., and Wan, X. (2003). Mice, rats, and people: the bio-economics of agricultural rodent pests. *Frontiers in Ecology and the Environment*, **1**, 367–375.

Storkey, J., Meyer, S., Still, K.S., and Leuschner, C. (2012). The impact of agricultural intensification and land-use change on the European arable flora. *Proceedings of the Royal Society of London. Series B: Biological Sciences*, **279**, 1421–1429.

Sumption, K.J. and Flowerdew, J.R. (1985). The ecological effects of the decline in rabbits (*Oryctolagus cuniculus* L.) due to myxomatosis. *Mammal Review*, **15**, 151–186.

Swinton, J., Tuyttens, F., Macdonald, D., Nokes, D.J., Cheeseman, C.L., and Clifton-Hadley, R. (1997). A comparison of fertility control and lethal control of bovine tuberculosis in badgers: the impact of perturbation induced transmission. *Philosophical Transactions of the Royal Society of London. Series B: Biological Sciences*, **352**, 619–631.

Tilman, D., Balzer, C., Hill, J., and Befort, B.L. (2011). Global food demand and the sustainable intensification of agriculture. *Proccedings of the National Academy of Sciences*, **108**, 20260–20264.

Treves, A. and Naughton-Treves, L. (1999). Risk and opportunity for humans coexisting with large carnivores. *Journal of Human Evolution*, **36**, 275–282.

Treves, A. and Naughton-Treves, L. (2005). Evaluating lethal control in the management of human-wildlife conflict. In R. Woodroffe, S. Thirgood, and A. Rabinowitz, eds., *People and Wildlife: Conflict or Coexistence?* pp. 86–106. Cambridge University Press, Cambridge.

TMP (2009). *Tracking Mammals Partnership Update 2009*. <http://jncc.defra.gov.uk/page-5497>, accessed October 2012.

Tuyttens, F.A.M., Delahay, R.J., Macdonald, D.W., Cheeseman, C.L., Long, B., and Donnelly, C.A. (2000). Spatial perturbation caused by a badger (*Meles meles*) culling operation: implications for the function of territoriality and the control of bovine tuberculosis (*Mycobacterium bovis*). *Journal of Animal Ecology*, **69**, 815–828.

Van Andel T., Zangger E., and Demitrack, A. (1990). Land use and soil erosion in prehistoric and historic Greece. *Journal of Field Archaeology*, **17**, 379–396.

Van Dijk, J., Sargison, N.D., Kenyon, F., and Skuce, P.J. (2010). Climate change and infectious disease: helminthological challenges to farmed ruminants in temperate regions. *Animal*, **4**, 377–392.

Wang, S.W. and Macdonald, D.W. (2006). Livestock predation by carnivores in Jigme Singye Wangchuck National Park, Bhutan, *Biological Conservation*, **129**, 558–565.

Webster, J. (1994). *Animal welfare: A cool eye towards Eden*. Blackwell Publishing, Oxford.

Williams, F., Eschen, R., Harris, A., et al. (2010). *The Economic Cost of Invasive Non-Native Species on Great Britain*. Defra commisioned report from CABI 2010.

Woodroffe, R., Bourne, F.J., Cox, D.R., Donnelly, C.A., Gettinby, G., McInerney, J.P., and Morrison, W.I. (2005). Welfare of badgers (*Meles meles*) subjected to culling: patterns of trap-related injury. *Animal Welfare*, **14**, 11–17.

Yalden, D. (1999). *The History of British Mammals*. T. & A.D. Poyser, London, UK.

Yeates, J.W. (2010). Death is a welfare issue. *Journal of Agricultural and Environmental Ethics*, **23**, 229–241.

Young, R.P., Davison, J., Trewby, I.D., Wilson, G.J., Delahay, R.J., and Doncaster, C.P. (2006). Abundance of hedgehogs (*Erinaceus europaeus*) in relation to the density and distribution of badgers (*Meles meles*). *Journal of Zoology*, **269**, 349–356.

Zeder, M.A. (2008). Domestication and early agriculture in the Mediterranean Basin: Origins, diffusion, and impact. *Proceedings of the National Academy of Sciences*, **105**, 11597–11604.

Zeder, M.A. (2011). The origins of agriculture in the Near East. *Current Anthropology*, **52**, 221–235.

Zommers, Z.A., Johnson, P.J., et al. (2012). Biofuels bonanza? Sugarcane production and poverty in villages surrounding Budongo forest Uganda. *Journal of East African Studies*, **6**, 177–195.

Zuckerman O.M. (1980). *Badgers, Cattle and Tuberculosis*. Ministry Of Agriculture, Fishieries And Food. August 1980.

CHAPTER 2

Foxes in the landscape: ecology and sociality

David W. Macdonald, Patrick Doncaster, Malcolm Newdick, Heribert Hofer, Fiona Mathews, and Paul J. Johnson

With foxes we must play the fox.

Thomas Fuller

2.1 Introduction

Work with red foxes *Vulpes vulpes* in the Upper Thames catchment was the opening chapter in the history of all of the work we report between these covers. The red fox (Fig. 2.1) was the starting point because of its remarkable versatility. While Amy-Jane Beer may have been a touch lyrical in describing the species as 'our closest rival for world domination'[1] (Beer 2007), there is much to marvel at in its biology and biogeography. The red fox[2] has the widest geographical range of any member of the Carnivora, spanning the entire northern hemisphere from the Arctic Circle to North Africa, Central America, and the Asiatic Steppes (Macdonald and Reynolds 2004). Within this immense range, it thrives in tundra, desert, and forest, as well as in human-dominated agricultural landscapes, even where the land is intensively managed. In cities too, foxes prosper (Baker and Harris 2004). In Europe, foxes are absent only from the highest montane areas.

This ubiquity is linked with dietary versatility—foxes are opportunistic omnivores, eating everything from earthworms *Lumbricus terrestris* and beetles to mammals, birds and their eggs, and even fruit (Macdonald and Reynolds 2004). In cities, human refuse can supply up to 60% of their food (Baker and Harris 2004). Diet varies, not just regionally in response to the distribution of prey, but locally too: foxes in adjacent home ranges can eat quite different diets (Macdonald 1987). Foxes' diets also track seasonal fluctuations in prey availability (Baker and Harris 2003). Oxford's urban foxes eat fruit in autumn, and small mammals in late winter, while in spring and summer the abundance of rabbits and birds is exploited (Doncaster et al. 1990). Their habit of food caching gives some security during times of shortage (Macdonald 1976, 1977a; Macdonald et al. 1994).

This dietary flexibility is matched by versatility in using space. In urban areas where food is abundant, home ranges can be less than 25 ha (Newdick 1983; Harris and Rayner 1986). Yet in the cereal fields of North Dakota, or on the barren uplands of England, home ranges can be orders of magnitude larger, up to 2000 ha (Voigt and Macdonald 1984; Macdonald 1987). Foxes' use of their ranges changes with seasonal variation in different foodstuffs (Cavallini and Lovari 1991; Cagnacci et al. 2004).

Believing that intra-specific variation provided a calculus for unravelling how ecological circumstances fashioned societies, we began studying foxes precisely because of their capacity to adapt to almost every habitat in the northern hemisphere, with the expectation that comparisons of contrasting ecological circumstances would flush out variation in their societies, and inform understanding of carnivore society generally. Indeed, in the 1970s and 1980s, studies in the embryonic WildCRU by our group, then known as the Oxford 'Foxlot', revealed for foxes a surprising level of sociality and sophisticated communication. This exposed some tantalizing clues concerning intricate linkages between how foxes used the landscape, their sociality, and the spatial configuration of their key resources.

[1] Among mammal species.
[2] Hereafter just 'fox'.

Wildlife Conservation on Farmland. Conflict in the Countryside. Edited by David W. Macdonald and Ruth E. Feber.
© Oxford University Press 2015. Published 2015 by Oxford University Press.

Figure 2.1 Although red foxes were formerly thought to live solitary lives, our studies revealed that, in some habitats, cubs like these may benefit from the co-operative care of non-breeding vixens in their social group. Photograph © A.L. Harrington.

Interest in foxes goes beyond that of intellectual problem solving (bearing in mind the 'so what?'[3] question at the heart of this book, with which the WildCRU team daily challenges itself). In much of their range, including the UK, foxes are a prominent practical issue. As relatively large and widespread predators, they come into conflict with humans (Macdonald 1987); indeed the fox may have the greatest economic and ecological impact of any British predator (Baker and Harris 2003). Foxes are pests for stock farmers and for game-shooting enterprises and, in Chapter 3, this volume, we explore in detail a quintessentially British aspect of how humans interact with pest predators: that is of course foxhunting, particularly the mounted version, which provokes highly polarized opinions.

Notwithstanding the fascination of fox biology, our initial and primary practical reason for wanting to understand their ecology, and its consequences for their then unknown societies, was the conviction that this understanding would inform rabies control. Foxes in Europe are the principal vector of classical rabies[4] (Macdonald 1980a). The variability in fox ecology and behaviour that we had begun to uncover, informed early models, showing how the pattern of contacts among individuals would affect the likely success of control efforts (Bacon and Macdonald 1980; Macdonald and Bacon 1982).

More recent modelling work now informs vaccination policy by predicting the likely optimal bait density, for example, as a function of fox density (Smith et al. 2008). Contact rates prove a vital element in disease epidemiology, and are inevitably affected, not only by the social perturbation caused by the disease itself, but also by efforts to control it (Macdonald and Voigt 1985). For example, one possibility is that the removal of a territory holder creates a vacuum effect, drawing in susceptible foxes to the vacancy, and exposing them to infected neighbours (Macdonald 1995). This concept of *perturbation* has subsequently become central to the management of badgers in connection with controlling bovine tuberculosis in cattle (e.g. Macdonald et al. 2006; see Chapter 5, this volume). This contact epidemiology has been advanced further through predictive social contact network modelling, where Ball et al. (2009) have shown the importance of heterogeneity in real population structures. Foxes are also important intermediate hosts for various forms of echinococcosis over much of their range (not the UK); this is a dangerous human disease caused by cestode tapeworms of the family Taeniidae; the disease is on the increase in humans as foxes invade cities (Eckert and Deplazes 2004). As Macdonald and Laurenson (2006) highlight, animal behaviour is central to the

[3] The 'so what?' question, first raised in 1977 in the introduction to DWM's DPhil thesis, came to be embodied in the mission statement of the WildCRU—original research linked to practical conservation and management.

[4] As distinct from the lyssavirus forms, associated mainly with bats.

transmission of these and many other diseases, and efforts to control disease that do not take proper account of social dynamics are likely to fail.

Our purpose in this chapter is not merely to present findings on fox behaviour and ecology in Oxfordshire as a foundation for revealing the diversity of their interactions with people, but also to use them as a model to inform wider thoughts on human–wildlife conflicts on lowland farms. Hitherto, each of the WildCRU fox studies has been written up separately, often in doctoral theses and ensuing papers. Here we take a different perspective by attempting the first overview, at a landscape scale, of our Oxfordshire-based research. Although these studies encompass only a few km^2, this lowland landscape includes a great deal of ecological variation; consequently fox groups' ranges include differing proportions of city, farmland, and various types of woodland. A synopsis from the larger spatial scale offers the opportunity for quantification with relevance for ecology, ecosystems, and policy.

2.2 Space use and sociality

Foxes are most often seen foraging singly and this formerly gave rise to a reputation as 'loners'. They breed once annually, mating, in Britain, from December to February, with cubs appearing from March to May (Macdonald and Barrett 1995), so they need not associate continuously for breeding. As late as the 1970s, the prevailing orthodoxy was that the fox represented the lower end of a continuum of sociality among canids. Roger Burrows, for example, noted in 1968 that 'The adult fox is a solitary animal except during the breeding season' (Burrows 1968). Foxes were thought to hunt and live alone while wolves were at the other end of the spectrum, hunting co-operatively in packs for large prey. This orthodoxy permeated veterinary thinking in Europe regarding rabies control (Moelge et al. 1974; Bogel et al. 1976), which essentially assumed that foxes moved in the landscape as solitary individuals amongst whom the contact rate of disease transmission was directly proportional to their numbers; assumptions that were reflected in the brutal rural epithet that 'the only good fox is a dead one'.

At the same time, it was becoming clear that neither co-operative hunting nor defence (of resources or young), then the two most compelling *a priori* explanations for group living, were sufficient to explain the formation of groups in mammals. There were plenty of examples where apparently social carnivores fed predominantly on small prey. In mongooses, for example, some species live in groups (Gittleman 1989), including the highly social meerkats (Macdonald 1992; Macdonald 2013). Co-operative defence is also a plausible explanation for the function of these groups. But there are examples, as first pointed out by Kruuk (1976), where groups are formed in the absence of any co-operative foraging or defence; badgers for example, live colonially (see Chapter 4, this volume).

Fragmentary observations began to suggest that the monogamous and solitary stereotype of foxes was an over-simplification; in London, aggregations of foxes had been seen in resting sites during the day, and there were published reports of vixens sharing dens and rearing cubs together (Storm et al. 1976), a phenomenon which appeared to be well known to gamekeepers and fox hunters (Macdonald 1980c). The elusive and nocturnal habits of foxes thwarted early researchers interested in these mysteries. Occasional observations by even the most diligent and skilled fieldworkers were not enough. Developments in technology, particularly in radio-tracking (Macdonald 1978), together with night vision equipment, provided researchers with the means to shed light on fox behaviour. Catching foxes to attach the transmitters was not easy and, once collared, the early equipment was prone to fail[5]. Nevertheless, between 1973 and 1976, 47 foxes were tracked in Oxfordshire in the first phase of the 'Foxlot' project (Macdonald 1977b; Macdonald 1981).

These foxes ranged over habitat comprising a mixture of suburban gardens and mixed farming, typical of the English midlands, which we came to think of as 'rural suburbia'. It became apparent that there were two distinct patterns of space use by the foxes here; some occupied the same area consistently while others travelled erratically over much larger areas. This conformed to a general distinction among carnivores between 'residents' and 'transients'. The itinerant foxes were almost all males, either young males moving through the area, or older males that may have been touring widely in search of oestrous vixens. Examples of this dichotomy have been observed in carnivores ranging from lions to weasels (Macdonald 1981). Itinerant foxes are sometimes attacked by resident foxes on whose territories they trespass (Macdonald 1979b).

[5] After months of trying, the first fox was caught and collared in February of 1973—its transmitter failed almost immediately (Macdonald 1987, *Running With The Fox*, p. 22).

Figure 2.2 Territories of foxes at Boar's Hill (near Oxford), c 1975. Individual territories (numbered) with boundaries (dotted lines). From Macdonald (1977b).

Far from the expectation of solitary, socially stand-offish singletons, the resident foxes in our first study in the rural suburbia of Boar's Hill (north Oxford, see Fig. 2.2) showed stable space use and occupied shared territories (and dens). These foxes were clearly living in groups, although they tended to forage solitarily. The groups were initially identified through close observation in the field. Radio-tracking together with direct observation of these resident foxes led to a new and very revealing discovery: these foxes lived in groups of 4–5 adult individuals (4.4 ± 1.1 mean and SD), sharing a territory averaging about 40 ha. Even the smallest territory, at 19 ha, was shared by a dog fox and four adult vixens. Moreover, these group territories were delimited by borders (see Box 2.1), which tended to follow topographic features such as roads, ditches, and hedgerows, typical in the UK rural/suburban landscape. Furthermore, inter-group overlap was slight.

Typical group compositions comprised one adult dog fox and several vixens. These vixens were almost certainly related to each other; this was clear from long-term tracking of juveniles as they became integrated into a group, and from observations on colour similarities (e.g. white legs) between group members. Furthermore, observational evidence, combined with the occasional post-mortem, indicated that many vixens did not rear cubs (Macdonald 1979a; 1980c). A group observed at Boar's Hill continuously between 1973 and 1974 consisted of three vixens (aged 9, 6, and probably 2 in 1973), of which only the eldest bred. This proved typical of groups in this habitat[6]. There was no evidence for substantial cub mortality, so the barren vixens, about 60% of the total, either did not conceive or else their embryos did not reach full term. When the breeding female and non-breeders were seen together the 'body language' strongly suggested that the breeding female was dominant. The non-breeder would sometimes be seen, back arched and body crouched, with ears flat against the head pointing backwards, clearly signalling subservience. The

[6] Variations on this theme subsequently became apparent for other foxes around the world (e.g. Hersteinsson and Macdonald 1982; Kruchenkova et al. 2009).

Box 2.1 Scent and communication

Foxes urinate very frequently. Some of the sites they use are very predictable—Niko Tinbergen recounted that the same discarded boot on the sands of Ravenglass[7] was regularly anointed with fresh fox urine for three years. To investigate some of the intricacies of how foxes use their urine in this way, we used semi-tame foxes (also used for the preference and caching studies)[8]. Observation of these foxes (Macdonald 1979b) showed that scent played a key part in the social system—a dominant fox encountering a subordinate will signal its status with ostentatious bouts of urination. Dominant foxes also marked group members, frequently anointing subordinates while they were sleeping. But urine was clearly also used more subtly than in these face-to-face encounters. Captive foxes were observed to markedly change their urination behaviour at about five months of age. They began to 'token' urinate, leaving numerous deposits at apparently conspicuous sites. A single fox following a regular route would repeatedly leave scent marks at the same site. Some areas were avoided consistently. Leash-walked vixens reacted differently to their own urine compared with that of aliens, applied experimentally.

These observations were compatible with the interpretation that the area of token marking defined a fox territory, where one function of the marking was concerned with the spatial organization. Droppings were also used as signals. For example, numerous droppings might be deposited on an inedible sheep carcass, and fox faeces are often seen at rabbit warrens. Faeces may persist at a site for much longer than urine. We tested the idea that fox latrines were located at unusually conspicuous sites by systematically surveying fox faeces in Wytham Woods (Hofer and Macdonald, unpublished observations). The majority of fox faeces were found on conspicuous objects or on trails. The foxes appeared to be using the rule of marking a conspicuous object when off a trail, while if they were on a trail, they avoided other faeces. This would maximize information transmission. Such marked locations probably provide beacons for both visual and olfactory anal gland information. The anal gland contains, among a complex cocktail of chemicals, low volatility fatty acids which release odours slowly (Macdonald and Brown 1985).

sub-dominant vixens were often seen visiting earths (the underground shelters) and were suspected to be provisioning the earth of the breeding vixen. Overt aggression between members of the same group was not common. So a picture of not only group living, but of a relatively advanced social system began to emerge. But these observations were inconclusive, because interactions between identifiable group members were seen so infrequently in the wild.

To take a closer look at fox society, in 1973 we set about observing two family groups in specially constructed enclosures of approximately 0.1 ha. These fox groups comprised individuals of known relatedness. To mimic dispersal we removed all sub-adult males in the autumn, and removed some sub-adult females at random. As expected, in almost all years of observation (7 of 8 enclosure-years in total), only one female in each enclosure produced young (Macdonald 1979a). The non-breeding vixens provided food for the cubs and groomed them. When a dominant vixen was injured, the 'helpers' contributed sufficiently to ensure the cubs' survival. They also defended the enclosures when foxes from outside approached the fence. We also saw that dominance and breeding status went hand in glove, because when two vixens 'flipped' their dominance status—after a turbulent transitional period—the vixen that became dominant started to breed, while the formerly dominant vixen became a helper (Macdonald 1987). Many aspects of this system later proved to be characteristic of canid societies in general (Macdonald et al. 2004).

Our subsequent studies of foxes switched from rural suburbia to an urban landscape in Oxfordshire, UK. These studies revealed similar patterns to those we had seen in suburbia insofar as groups defended territories (Doncaster and Macdonald 1991b). Furthermore, the ranges of individual foxes had similar sizes to those in suburbia. The foxes studied in urban Oxford by Newdick (1983) ranged over about 50 ha, while the eleven group ranges studied by Doncaster (1985) averaged 38.8 ha. The smallest average range sizes observed so far for the species were in urban Bristol, at just 27.5 ha (Harris 1981); on Boar's Hill the smallest

[7] In the English Lake District.

[8] In *Running with the Fox,* I (DWM: Macdonald 1987) describe how hand reared foxes were sufficiently tame to observe their behaviour while being walked on long leashes.

group territory, including five adults, was 19.6 ha. The spatial configuration and social structure of Oxford's urban foxes was similar to those in the suburbs; the group territories overlapped rather little while the ranges of individuals within the groups overlapped a great deal (Macdonald and Newdick 1982; Newdick 1983). As for the rural suburbia of Boar's Hill, territories were shared by a social group of vixens, with only a dominant vixen breeding, and with other vixens commonly helping the alpha female with food provisioning and defence of the young foxes (Macdonald 1979a; Doncaster 1985). Fox populations in London also contained non-breeding adult vixens, though at around 50% not as many as in our rural–suburban interface (Harris 1979).

What did appear to differ for Oxford's urban foxes compared with those in more rural habitats was that the ranges were less spatially stable. Application of some novel methods for spatial analysis revealed a pattern of 'drifting territoriality' (Doncaster and Macdonald 1991a). We found that territories were maintained in the urban setting with tessellated borders that were much more fluid than their more rural counterparts. Parts of each territory were progressively vacated while new areas were increasingly used. The abandoned territory areas tended to become encompassed within the territory of neighbouring groups. Our observations of encounters between neighbours, as one range developed an extension and the other retracted, were too few to reveal the behavioural mechanism underlying these reconfigurations. Each range drifted through the landscape quite rapidly, of the order of 30–40 ha per year, quite close to the typical urban range size and ten times the rate in rural suburbia (Doncaster 1985).

What was the reason for this peculiarity of group dynamics in towns? We hypothesized that higher mortality there, the result of greater disturbance and unpredictability in food availability, might promote this drifting. Higher population turnover and social instability may have perturbed the matrilineal dominance and the mechanisms preventing subordinate vixens from breeding (Doncaster and Macdonald 1991a), and hence led to a lower proportion of barren vixens. At the same time, the turnover of individuals could promote unstable interactions at the borders between territories, and hence a tendency for those borders to drift. Malcolm Newdick observed that border transgression increased when a neighbouring individual was killed in a road accident (Newdick 1983). Drift in fox ranges was observed in Bristol too, though only one third of that observed in Oxford (Doncaster and Macdonald 1996)[9]. There were other differences. Group structure was different for the Bristol foxes, where the presence of more than one adult male was more frequent, averaging 1.8, while more than one male was rare in the Oxford groups.

We hypothesized that, to a large extent, these differences between the socio-spatial dynamics of urban and suburban foxes in Oxford and Bristol most plausibly reflected variation in the predictability of food supply and the knock-on effect of unpredictability on mortality. As such, they demonstrated the potential for complex sociality in foxes. However, over much of their biogeographical range, a simpler societal system prevails, and one which is closer to the conventional model of foxes as 'loners'. Beyond the rich pickings of cities and lowland farmland, foxes may occupy much larger ranges. In the fells of Cumbria, individual foxes range over 1000 ha (Macdonald 1981), while in the similar terrain of the Welsh uplands, Gwyn Lloyd found ranges of around 400 ha (Macdonald 1987). Beyond the UK, the range sizes as summarized by Lloyd (1980) and by Macdonald (1987) are everywhere else more than 400 ha. All these foxes appear to occupy territories in the conventional sense, in that they are defended and with stable boundaries. Pairs are thought to predominate in these large territories (Macdonald 1980c).

For example, the monograph by Storm et al. (1976), describing foxes on the plains and wooded areas of north-central USA, reports that the 'typical' breeding group was a pair (while also noting occasional departures from this: two females with a single male and vice versa were observed twice). Other species of fox living in relatively large and homogenous ranges also live in pairs that endure from year to year. This is true for the swift fox *Vulpes velox* (Moehrenschlager and Sovada 2004), the kit fox *Vulpes macrotis* (List and Macdonald 2003; Moehrenschlager et al. 2007), and the Cape fox *Vulpes chama* (Kamler and Macdonald 2014), and very probably many other vulpines (Macdonald and Sillero-Zubiri 2004). The proportion of breeding females is higher in these pairs: all paired females breed in the absence of suppression of subordinates in group-forming populations. Storm et al. (1976) found that almost 95% of the vixens they examined post-mortem showed evidence of recent breeding. Similar high rates were observed in populations in Southern

[9] Drifting ranges have subsequently been observed in a range of other vertebrates (Taber and Macdonald 1992; Doncaster and Macdonald 1996).

Ontario, and in Germany and Sweden (summarized by Macdonald 1987). The difference between these populations and the rural, suburban, and urban populations of Oxfordshire and London is linked to density dependence. This is a phenomenon of great interest in ecology, whereby population growth rates are inhibited as the size of a population approaches the carrying capacity of the environment.

In foxes, various strands of evidence suggest that the full reproductive potential is realized only where mortality is high enough to keep population density below environmental carrying capacity. This happens where humans kill foxes. The population studied by Storm et al. (1976) was heavily hunted. And in Ohio, USA, birth rate averaged 5.6 cubs per adult female per year in a hunted area, compared with 2.6 cubs in an area where hunting pressure was low (Phillips 1970). Part of the explanation is that fewer females breed when the population size is close to the carrying capacity of the environment. Englund (1970) collected a large sample of dead vixens from hunters in Sweden and observed a great deal of variation in fertility, particularly among younger vixens—yearling fecundity varied between 5 and 88%. Reproductive suppression has also been noted in arctic foxes *Vulpes lagopus* (Kruchenkova et al. 2009) and bat eared foxes *Otocyon megalotis* (Maas and Macdonald 2004). In canids, reproductively suppressed females have low levels of oestradiol and progesterone—the mechanism whereby suppression comes about is likely to be via hormonal signals rather than aggression (van Kesteren et al. 2013). Captive male foxes make no attempt to mate with subordinate females and this does not appear to be related to interference by the dominant female (Macdonald 1980c).

Studies by the Game Conservancy Trust[10] have shown that reproductive suppression in UK foxes is associated with lower rates of culling. Heydon and Reynolds (2000) looked at the uteri of dead foxes from three areas of the UK, two where foxes were heavily culled by gamekeepers and other land managers (in East Anglia and upland Wales), and a third where culling was much lighter (typical Midlands agricultural land). Vixens that showed no evidence of breeding (uterine scars are visible in reproductive females) were found only in the lightly culled area. In addition, vixens in this area also sometimes conceived and then lost all embryos *in utero*. In total, 19% of Midland vixens were in one of these two categories. Where resources are scarce, producing young now may not always be the most successful long-term strategy for maximizing fitness, particularly for a social species. It is clear that fox society is flexible, and that part of that flexibility allows a social response to variation in ecological conditions, including the amount of culling (with clear implications for disease management). The obvious question emerging from this is why do foxes in some places live in pairs while elsewhere they live in groups with reproductive and non-reproductive members?

[10] Now the Game & Wildlife Conservation Trust, GWCT.

2.3 The ecological basis of sociality

The unifying predictive mechanism of Darwinian selection dictates that individuals will form groups only when the forces that shape the behavioural 'decisions' of individuals are such that it is in the interests of those individuals to do so. This may lead to the formation of societies within those groups.

Both within and between species, variation in sociality is more closely linked to ecology than to allometry (Geffen et al. 1996). The likely ecological forces shaping differences were reviewed, for carnivores in general, by Creel and Macdonald (1995) and by Macdonald et al. (2004) for wild canids. Variation in home range size is one of the fundamental variables. The size of fox ranges can largely be explained by well-established relationships between metabolic rate, body mass, and food availability. Simply put, when food is scarce, larger areas are required to supply enough food to meet metabolic needs. These relationships, reviewed by Macdonald et al. (2004), have predictive power for a wide range of taxa: mammalian home ranges scale with body mass to the power of 0.75. While larger canids need to eat larger prey due to the relationship between prey size and the metabolic costs of hunting, a discontinuity occurs somewhere near a critical predator body size of 20 kg (Carbone et al. 1999). For some large canids, one factor influencing sociality may be the advantages of co-operation in hunting the large prey necessary to balance their energy requirements—fox masses fall comfortably on the lower side of the 20 kg threshold (Macdonald and Reynolds 2004). Thus much of the variation in fox range sizes is attributable to food availability as measured in the simple metric of mass per unit area; food is plentiful in urban and suburban habitats, but is harder to come by in the marginal uplands of the UK. Hence, larger ranges are needed there to meet metabolic needs.

For smaller carnivores such as foxes, the accumulating body of evidence suggests that variability in food abundance across space or through time prevails over

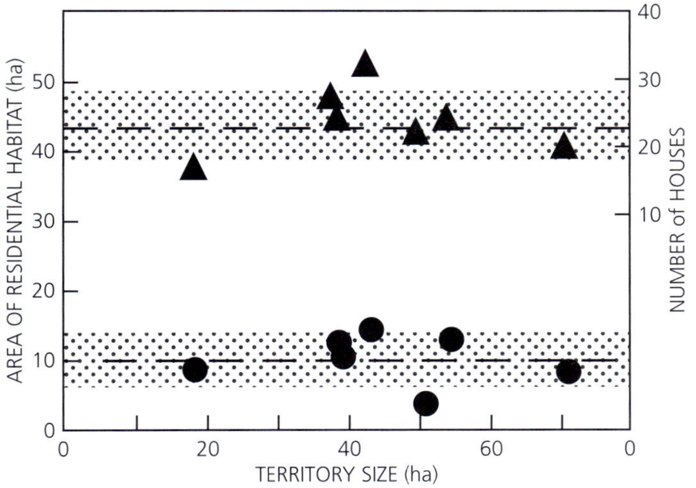

Figure 2.3 Territories (varying between 18.6 and 72.0 ha each) included a rather constant area of human-residential habitat (10.4 ± 3.5 ha) and a similar number of houses (23.7 ± 4.8). Shading indicates 1 SD about the mean. Area of residential habitat and number of houses are correlated (r = 0.68), but in addition to providing household scraps, residential habitat is also rich in earthworms, fruit, passerine birds, and rodents. From Macdonald (1977b).

food availability per se in determining sociality. This idea stems from our early observation on suburban and rural foxes (Macdonald 1977b), that the food supply was not only rich in these habitats, which led to small territory size requirements, but was also patchy. Foxes favoured gardens, where they spent long periods foraging on patches for one of their preferred prey types—earthworms (Macdonald 1980b). While territories of foxes in suburban Oxford varied in size, both the amount of residential habitat and the number of houses within each territory were relatively constant (Fig. 2.3). Even in more urban habitats, foxes still relied on patchily distributed food (Macdonald and Newdick 1982).

If the food supply is evenly[11] distributed in space, however, no specific predictions arise concerning grouping or sociality from variation in availability; the territory size occupied by a social unit should scale in proportion with availability, give or take some adjustment owing to the metabolic cost of patrolling a larger range. That prediction is not borne out by observations on UK foxes, where social structure does appear to be linked to resource abundance. Zabel and Taggart (1989) report a compelling example of a link between resources and sociality *within* a population of foxes on Round Island, Alaska. As the seabirds (kittiwakes *Rissa* spp.) forming the favoured food of foxes crashed over a single year (nesting failed in 1982), the foxes were forced to switch to a less abundant alternative (young auks *Cyclorrhynchus psittacula*) (Zabel and Taggart 1989). Before 1982, kittiwakes formed more than 40% of the items delivered to pups, with auklets at less than 20%. After 1982, kittiwakes declined to approximately 12% while auklets were nearly 40%. This was related to a clear shift in the social system. While polygyny had been predominant (71%) when food was superabundant, all the foxes became monogamous when it was not. Fewer females bred (non-breeding females increased from 53% to 84%). The study period was not long enough for the total number of adult foxes on the island to change (mean N = 29 foxes in the years before the crash in kittiwake numbers, N = 28–30 after), the number of cubs produced decreased by about 75%. One attribute of the seabird distribution which may have been influential was that the kittiwakes nested in colonies on exposed cliffs, while the less favoured auks nested in rock crevasses and scree slopes. The resource-rich state was therefore probably also associated with more spatial aggregation, a situation reminiscent of that for Arctic foxes exploiting seabird colonies (Goltsman et al. 2005).

Some interesting possibilities arise when an important resource is spatially aggregated or temporally variable. A home range that permits a pair to persist might, because of the surplus within a patch, allow conspecifics to live in the same area with secondary

[11] In statistical parlance, a distribution of items in space which is less variable than random, tending to even distribution, is called 'underdispersed'; if the items are more clumped than random the distribution is 'overdispersed'. Items that are randomly distributed in space will correspond with a Poisson distribution. Whether a distribution is even, random, or clumped may also depend on the spatial scale considered.

food security and tolerable cost to the primary occupants. The number of conspecifics that could be tolerated in such a situation is then expected to be related to patch richness for patches defined at an appropriate spatial or spatio-temporal scale (Doncaster and Macdonald 1992; Johnson et al. 2002). This explanation of group living in foxes, which was also developed for badgers by Kruuk and Parish (1987), has been formulated as an hypothesis for explaining group living more generally (Macdonald 1983; Carr and Macdonald 1986).

The general idea, the Resource Dispersion Hypothesis (RDH), is that groups may develop where resources are spatially configured such that the smallest territory which is viable for a pair can sustain additional animals, while also being defensible (Macdonald 1979b, 1980d). The RDH is consistent with the formation of groups where there is no additional benefit to group living from co-operation—a level of sociality that has been termed 'spatial grouping' (Macdonald 1983). The foxes of Boar's Hill (Oxford), while sharing the same territory, were observed to spend rather little of their time interacting during foraging time, totalling only a few minutes in a typical night (Macdonald 1980b). That they are solitary foragers explains why most people see lone foxes, but, as explained above, cohabiting in a shared territory opens the door to vulpine co-operation in the care of cubs, and territorial defence. In overview, it seems that when RDH conditions permit group formation, a template is formed for the evolution of more advanced sociality (Macdonald and Carr 1989).

Since its inception, the RDH has attracted much attention, including criticism for tautology and for failing to make predictions which allow it to be discriminated from other explanations for group living (e.g. Revilla 2003a,b). These points have been addressed by Johnson et al. (2002), Johnson and Macdonald (2003), and Johnson et al. (2003). The debate has arisen at least partly because of misunderstandings about the scope of RDH; it does not, for example, purport to be a universal explanation for group living, nor does it predict that cohesive groups in clearly defined territories invariably form when resources are aggregated. The presence of food-rich patches does not necessarily lead to territory formation—foxes in Saudi Arabia formed loosely knit groups where the harsh desert environment was punctuated by rich feeding sites around farms (Macdonald et al. 1999). In reality, a complex interplay between the costs of group living, such as competition for den sites and dispersal opportunities, interact to determine optimal strategies. This is unsatisfying for theoretical development, and the consensus is that experimentation is required to refine these ideas.

Returning to the specifics of how the RDH affects fox biology in the rural suburbia of the UK, where the landscape is a patchwork of farmland and human settlements of various sizes, aggregation of food resource in this habitat happens to be linked to resource richness, so territories can be small, and probably therefore defensible against other fox groups. The question that arises here is why do some vixens forgo breeding and instead help rear the cubs of the alpha female? The explanation that emerges from RDH concerns food security—it does not pay females to form smaller territories because they will not have reliable access to enough patches to ensure sufficient food in times of shortage; a verbal version of this hypothesis is given in Macdonald (1987), and is shown to be mathematically coherent by Macdonald and Carr (1989) and by Bacon et al. (1991). This remains, though, one of the areas identified by Johnson et al. (2002) where observation is lagging behind theory.

Because some fox groups include mature non-breeding females[12] that forego their own reproductive output but contribute to raising another's offspring, an apparent problem appears for Darwinian orthodoxy. In *The Origin of Species*, Darwin (1859) saw this phenomenon in ants as 'potentially fatal', and discussed it at great length as a 'difficulty sufficient to overthrow my whole theory'[13]. Darwin's own explanations, before any knowledge of genetics, were necessarily inadequate. The modern interpretation of this phenomenon, which is sometimes called 'eusociality' (where some individuals reduce their own lifetime reproductive potential to raise the offspring of others; Nowak et al. 2010), has been to invoke 'kin selection'.

This highly influential concept originated with the great evolutionary biologist Bill Hamilton (Hamilton 1964a,b). He proposed that, in some circumstances, it may benefit animals to co-operate with relatives provided $R > c/b$, where R is relatedness and c/b is the cost–benefit ratio for the co-operation strategy. A recent review (Nowak et al. 2010) argues that 'standard' natural selection is superior. This proposition has provoked considerable debate (Nowak et al. 2011; Abbot et al. 2011). The debate does not affect Nowak et al.'s observation that the formation of spatial groups may be the first stage of eusociality, nor does it alter the fact

[12] But not so in arctic foxes (Goltsman et al. 2005).

[13] In Chapter VII of *The Origin of Species*, 'Instinct'.

that research on the ecological forces shaping early sociality is sparse—and it is precisely these attributes, the formation of spatial groups and a trajectory towards sociality, which RDH addresses.

A further nuance to fox group formation has been advanced by Paolo Cavallini (Cavallini 1996), who proposed that the variability of fox social systems, and their dietary flexibility, is a response to predator pressure in the past: foxes may have evolved in environments with larger predators, such that flexibility was necessary to avoid predation, termed 'the ghost of predators past' (see Blumstein 2006). This idea does not necessarily conflict with RDH as an explanation for the evolution of sociality, and is no easier to test empirically, as it explains the ecology and sociobiology of modern foxes in terms of phylogenetic inertia.

2.4 Landscape and ecology

A revisiting of our extensive data relating to urban foxes and their near neighbours in rural suburbia is timely. Between 1977 and 1983, we conducted a rolling sequence of studies on the behavioural ecology of red foxes in Oxfordshire. Foxes were studied in five areas in and around Oxford. Each encompassed a variety of habitat types, but they could be broadly characterized as 'rural suburbia' where human habitation was mainly large detached houses (Boar's Hill), woodland with a village and agricultural land (Wytham), agricultural (Cumnor), and two areas of Oxford city.

One of the key objectives of our 'Foxlot' research, as detailed above, was to gain a better understanding of how foxes were able to thrive in habitats ranging across the spectrum from completely urban to completely rural. Significantly, as we amassed records for more than 100 radio-tracked foxes—and analysed the contents of their faecal samples—it became clear that ranges almost always comprised a diversity of habitats. For example, the urban ranges included some woodland and meadow, the countryside ranges included human homes. Inter- and intra-specific variations offer a powerful metric for exploring the effect of ecology on aspects of social structure (Macdonald 1983). Studies of sociality, space use, and resource availability are particularly valuable, and the five-year duration and unusually large sample size from our fox study allowed us to ask whether distinct local variations in fox home range size and diet could be attributed to the habitat heterogeneity of this region of the English midland landscape. For example, how did ranges and diets vary between the urban, the more heavily agricultural, and the intermediate sites?

2.4.1 Landscape type and composition, and home range size

Our total collection of 44,404 radio-fixes from 100 foxes allowed us to tackle a range of specific issues surrounding these questions. One fundamental issue is that of home range size, and how it is affected by landscape. Knowledge of fox home range size and the dynamics of range uses would prove important in the event of responding to a rabies outbreak in the UK. We have seen that at a very large spatial scale (comparing the original Boar's Hill study with that in the uplands of Cumbria), ranges are larger in landscapes where food availability is lower and more dispersed. At the smaller spatial scale considered here, range size is likely to reflect a more local variability in food abundance. We estimated home range size and core areas using the 95% core area fixed-kernels method (Worton 1989, implemented in Ranges V software of Kenward and Hodder 1996). We mentioned earlier how Oxfordshire's rural suburban foxes of Boar's Hill lived in groups averaging 4.4 adult individuals sharing a territory of about 40 ha. Inter-individual variation, however, provides a calculus for understanding how ecological circumstances fashioned societies. Taking the total Oxfordshire sample of 100 foxes, average figures conceal considerable home range size differences within Oxfordshire—the smallest range was 4.2 ha, occurring in the village/woodland habitat, while the largest was over three hundred times larger at 1307.9 ha, in one of the Oxford town sites.

The variability was largely attributable to habitat differences between the five study sites, and therefore likely to be accounted for by a landscape effect. Ranges (95% kernel) averaged 185 ha (CI 103.5–266.6) in urban sites, much larger than the mean elsewhere[14] (Table 2.1). This is consistent with foxes perceiving average urban habitat, with its high fraction of paved surfaces, as not being useful—in places where the concrete jungle prevails, ranges need therefore to be larger to accommodate patches of useful habitat, which is likely to be suburban habitat with worm-rich grassland, fruit trees, bird tables, and compost tips that characterized the rural suburbia of Boar's Hill.

Core range sizes as measured by 50% kernels differed between sites[15]. Compared with an average size of 17.6 ha across all the sites (CI 14.3–20.8 ha),

[14] $F_{1,85} = 4.97$, $P = 0.03$ for the urban versus the rest contrast.
[15] $F_{4,77} = 2.34$, $P = 0.07$.

Table 2.1 Range sizes. Fox (95% kernel) range sizes for each study site (ha).

Site	N	Minimum	Maximum	Mean	Std Error	Lower CI	Upper CI
Cumnor	11	23.99	138.51	77.70	9.22	57.15	98.25
Boar's Hill	17	14.50	648.56	125.87	37.46	46.46	205.29
Urban 1	37	10.82	1307.87	194.26	46.02	100.93	287.59
Urban 2	6	12.80	416.00	128.44	59.12	0	280.42
Wytham	16	4.20	132.25	68.07	9.66	47.49	88.66

individual core ranges were largest at the two urban sites[16]: 21.5 ha (CI 8.7–34.3 ha) at the Urban 1 site and 23.2 ha (CI 16.5–29.5 ha) at the Urban 2 site. They were smallest at the sites with woodland plus village (Wytham: 11.5 ha; CI 6.6–15.8 ha) and predominant agriculture (Cumnor: 10.4 ha (4.9–15.9 ha).

We looked in more detail at the effect of landscape structure by exploring whether range size was related to the proportion of different habitat types present in the range; we classified land use into categories using the Institute of Terrestrial Ecology's (ITE 1994) digitized land cover map[17]. The habitat types we recognized were high density urban (urban development), low density urban (suburban/rural development), arable (tilled land), deciduous woodland, grassland (mown/grazed turf; meadow/verge/semi-natural grassland; rough/marsh grass and grass/heath), scrub (scrub/orchard), and 'other' (all other categories). Importantly, each habitat included different types and availability of food. Grassland, for example, is often but not always, rich in worms, with availability related to microclimate (see Box 2.2) while urban areas provide scavenging opportunities, bird tables, and the like. Some scrubland can be rich in rabbits (Macdonald 1980b). We also devised an index for the patch density of each habitat to test for an influence of this aspect of resource dispersion.

Within sites, ranges were smaller where the landscape contained a higher proportion either of grassland or of meadow (Fig. 2.4), the predictive effects of the proportion of grassland and meadow were both statistically significant in GLM models including site[18]. The effect size was strongest for these categories combined—each 1% increase in grassland area was associated with a 2.3% decrease in range area (CI: minus 3.93% to minus 0.5%).

Range composition and the patchiness indices were all inter-correlated. For example, ranges with more grassland were also patchier[19]. So we can conclude that the larger ranges were those where the proportion of grassland was lowest and where patch density was

Box 2.2 Foraging and efficiency

One of the fundamental principles of behavioural ecology has been the assumption that natural selection should act to maximize an organism's foraging efficiency, where efficiency is often understood to be intake rate per unit effort. Modelling work has focused on predicting how an organism's behaviour should be shaped by spatial variation in resource richness in such a way as to achieve this. Situations where resources occur in discrete patches have been particularly useful for testing specific predictions of optimal foraging theory (for example Houston and McNamara 1999). Foxes hunting for worms in grassland provide one such example. At Boar's Hill, worms were particularly abundant in grassland, and foxes configured their ranges to include disproportionate amounts of this grassland (and indeed secured additional worm-rich grass in the form of the extensive lawns associated with detached housing there). Using infra-red binoculars we therefore painstakingly observed how individual foxes foraged for worms (Macdonald 1980b), to reveal how behavioural flexibility allowed foxes to maximize their capture efficiency. We

continued

[16] $F_{1,85} = 2.34$, $P = 0.005$.

[17] We used the IDRISI software version 2.007 (Clark University 1997).

[18] $F_{4,70} = 4.92$, $P = 0.03$ and $F_{4,70} = 5.90$, $P = 0.02$ respectively.

[19] $R = 0.37$, $P < 0.001$.

Box 2.2 *Continued*

found that the use of crude habitat class as a proxy for food availability has to be rather tentative—pasture was not a predictable source of food, and the unpredictability was both spatial and temporal; in some months, worms comprised over 60% of an average fox's calorific intake at Boar's Hill.

Also, some fields were much richer than others and, within fields, richer patches were apparent, which seemed to be related to dunging by horses. The distinction between food abundance and availability is important; worms deep underground are inaccessible. They surfaced in numbers only when it was mild after recent rain[20], in these conditions the ground appeared to seethe with worms.

How did foxes deal with this variability? They foraged more often in worm-rich fields when conditions were right—observations on some individual foxes suggested that wind may have been one cue for worm availability (the foxes foraged in the sheltered lee of spinneys). Having found a rich patch of worms, how long should a fox hunt on it? Foxes conformed to an expectation concerning optimal foraging relating to 'giving-up' time. Having encountered a worm-rich patch in a field, they appeared to hunt worms in that patch until the worm encounter rate fell below a critical value. This critical value was set by the cost of travelling to a new patch, so as to maintain maximal energy budget returns. Consequently, 'giving-up rate' (switching) was higher on nights when worms were more available, and patches were richer.

Analysing a series of 24 observations of foraging success, we used variation in worm availability among the patches available to explore the predictions of the Marginal Value Theorem of Charnov (1976). This prediction was that the final capture rates before leaving a patch should be about the average for the patches available for the session—this maximizes overall intake rate. In Box Fig. 2.2.1 we revisualize Figure 11 of Macdonald 1980a (now plotted on a log scale) and we observe a clear depletion effect—declining success was a cue to find a new patch. The slope of this relationship (on a log scale) was 1.05 (SE = 0.48), indicating a linear effect, and not for example a tendency for the final encounter rate to increase faster than proportional to

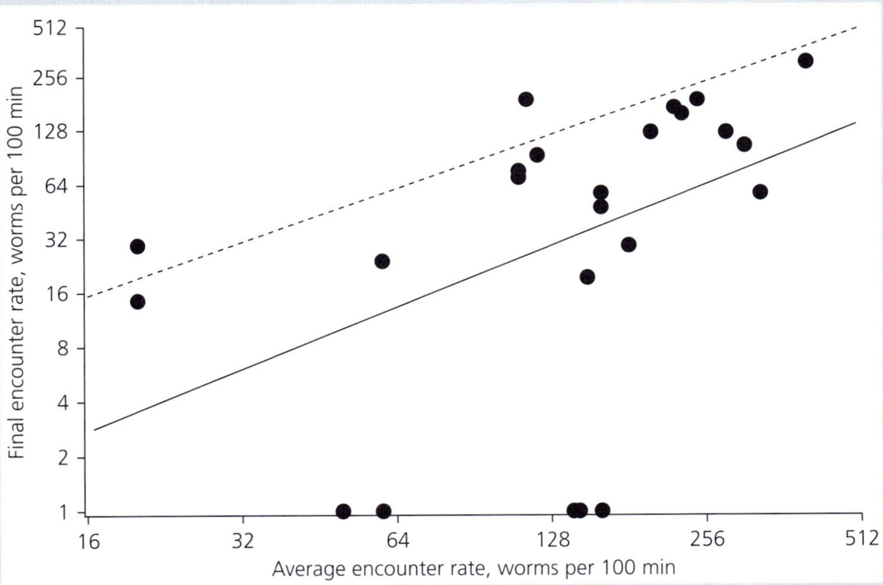

Box Figure 2.2.1 Encounter rate with worms before changing foraging location is significantly lower than the average encounter rate for the night. Dotted line shows expectation of the Charnov (1976) marginal value model. Redrawn from Macdonald 1980a.

continued

[20] Worm abundance declined statistically significantly with time since rain.

Box 2.2 *Continued*

the average rate. This slope is consistent with the 'marginal value' model of Charnov. The intercept in Box Fig. 2.2.1 is however too low. The model predicts quitting at the average rate, while foxes appear to allow the encounter rate to fall somewhat below that value before they 'quit': the line in Box Fig. 2.2.1 lies below that for the theory. That is, relative to the theoretical prediction, the lower intercept indicates that foxes were staying 'too long' in these conditions. Which assumption of the model is likely to be false? One plausible explanation is that foxes 'knew' that rich patches were likely to be scarce, therefore the cost of switching patches is under-estimated by this simple model, which does not incorporate this limit.

On a smaller scale, we saw no evidence that, within patches, foxes changed their search tactics after capturing a worm: detailed scrutiny of the distances moved and angles turned between captures showed no patterns. But there did appear to be distinct types of searching paths, depending on whether conditions on a night were favourable for worm activity: on 'poor' nights, the foraging tracks were much less tortuous than when worm activity was high. Capture rate was significantly higher on the more tortuous trails, suggesting foxes were able to adapt to the very local spatial dispersion and density of surface worms. The overall conclusion was that fox behaviour was sufficiently plastic to adapt to the numerous sources of uncertainty in grassland worm availability.

low, but we cannot disentangle these effects. The observation is consistent with foxes needing to include grassland as part of the range and needing to range further if the grassland is fragmented.

2.4.2 Which habitats do foxes appear to prefer?

To explore this question further, we quantified habitat selection at two levels. First, we can ask how ranges are configured in any study area compared with what we would expect if there were no tendency to avoid, or be attracted to, any habitat type. Second, we can compare how fox locations, as revealed by radio-tracking, are distributed inside the home ranges. Are some habitats over- or under-used compared with their availability? We used compositional analysis (Aebischer et al. 1993) to test for this.

As expected, range habitat profile differed significantly among sites[21]. The mean proportions of urban habitat, woodland, and arable land all differed substantially (Table 2.1). But there was still plenty of scope for foxes to 'choose' different types of ranges at these sites. The nominally urban sites have only 54% of the land class designated 'urban', for example, and 30% of the area of these sites was grassland. The Cumnor site was the most homogenous, with 79% of its area either tilled or grassland.

At the scale of ranges in study sites, foxes tended to avoid scrubland (ranked lowest at all sites in Table 2.2), while grassland was most over-represented within ranges at Boar's Hill and Wytham (where its availability in the study area was lowest). Grassland was a relatively constant proportion, comprising between 28% (Urban 2) and 34% (Boar's Hill) of study areas and fox ranges within them (Table 2.2). We also saw a tendency for foxes to avoid encompassing the 'urban' category in their ranges, which is particularly clear where this type of land use was a dominant feature. In the rural suburbia of Boar's Hill, however, the urban category is relatively favoured: this clearly reveals that the category is inadequately refined to distinguish the very different types of housing that occurred throughout Oxfordshire—with the detached houses and large gardens of Boar's Hill being the very best habitat for foxes, and very different to the more urban housing of the city. (Since our study, human food waste has become much less accessible to foxes in urban and suburban landscapes—the advent of wheely bins has seen to that.)

Were the foxes equally likely to use each of the habitats in their ranges?[22] The answer was clearly 'no',

[21] MANOVA, $F_{10,65} = 8.9$, $P < 0.001$.

[22] We encountered a methodological difficulty in applying compositional analysis for tackling this question, which illustrates an important general point for the assessment of space use at different scales. The amount of habitat of different types was quantified using digital maps, while fox locations were associated with habitats by workers in the field assessing the habitat type by inspection. Hence the resolution with which habitats were identified did not match. Field workers could see that a fox was using patches of scrub in blocks of other habitat types, where scrub was not mapped. Where this occurred we assigned arbitrarily small values (0.1%) of scrub to the range. Significant habitat 'selection' was therefore expected at sites where non-trivial numbers of fixes were observed in scrub.

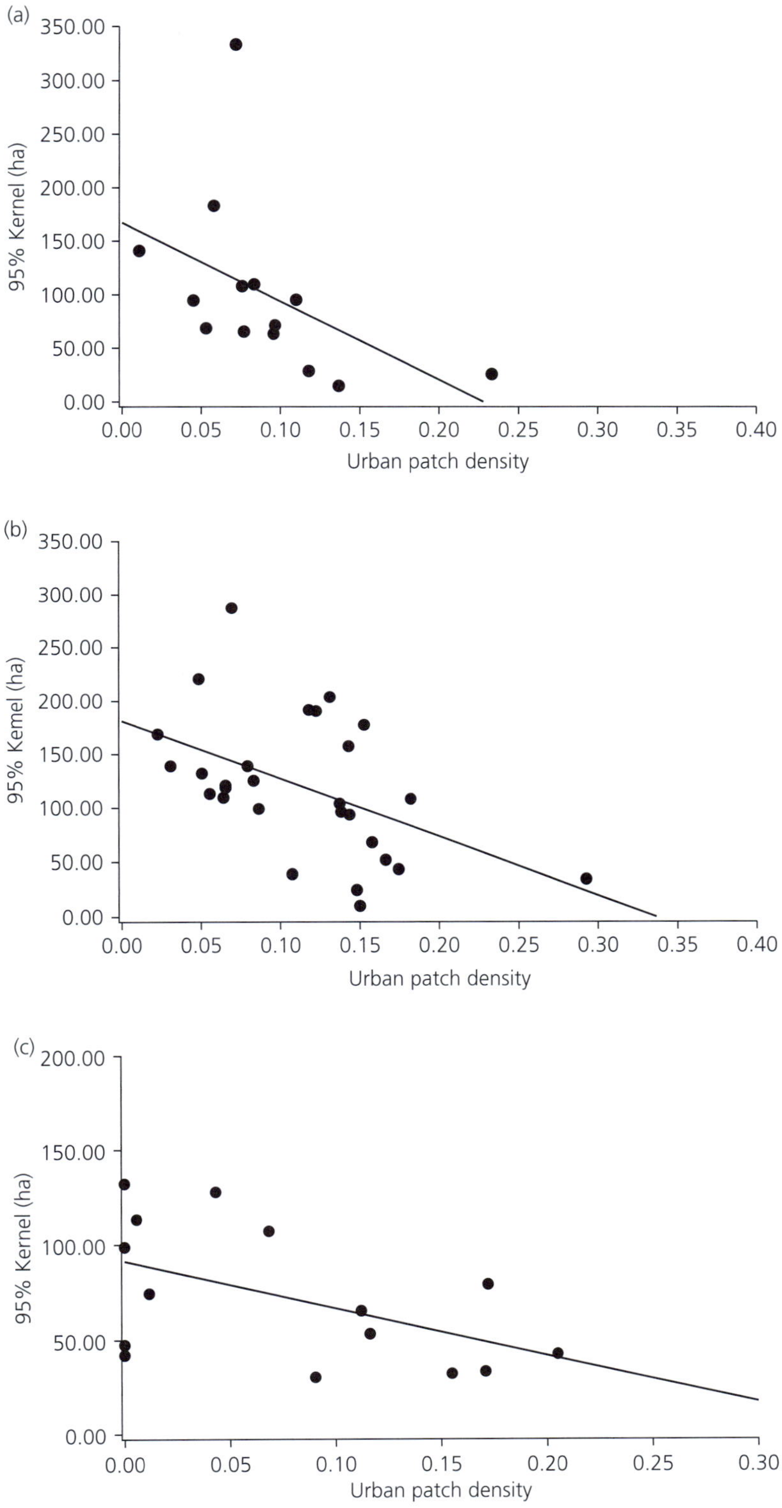

Figure 2.4 Ranges (95% kernels) are larger when patch density is low. (a) Boar's Hill, (b) Urban 2, and (c) Wytham.

Table 2.2 Habitat selection characteristics at different sites. The columns R1 and R2 give the ranked preference for habitat types for ranges in the study site and active fixes in the ranges respectively (5 = most preferred, 0 = avoided). All selection is statistically significant using the compositional analysis of Aebischer et al. 1993), apart from at Wytham and Urban 1 where sample N too low for testing. 5 = maximum selection, 0 = minimum. Refer to Table 2.1 for site descriptions.

Site	N	Variable		Site % area	R1	Mean % range	R2	Mean % active fixes
Cumnor	11	ARABLE	tilled land	40.0	2	46.0	3	33.12
		GRASS	grass/heath	39.0	1	31.0	2	30.85
		URBAN	urban	2.0	4	8.0	0	4.31
		SCRUB	scrub/orchard	8.0	0	0.1	4	2.73
		DECID	deciduous wood	10.0	3	13.0	1	7.25
		other		1.0	5	2.0	5	21.74
Boar's Hill	14	ARABLE	tilled land	20.0	3	27.0	1	11.51
		GRASS	grass/heath	20.0	5	34.0	2	18.62
		URBAN	urban	20.0	1	4.0	4	33.45
		SCRUB	scrub/orchard	9.0	0	0.1	5	7.52
		DECID	deciduous wood	39.0	2	30.0	3	28.47
		other		2.0	4	5.0	0	0.44
Urban 1	30	ARABLE	tilled land	1.0	5	23.0	0	8.74
		GRASS	grass/heath	30.0	2	28.0	4	35.21
		URBAN	urban	54.0	1	34.0	3	33.51
		SCRUB	scrub/orchard	8.0	0	0.3	5	12.18
		DECID	deciduous wood	2.0	4	8.0	1	2.94
		other		2.0	3	7.0	2	7.43
Urban 2	6	ARABLE	tilled land	1.0	-	25.0	-	13.85
		GRASS	grass/heath	30.0	-	34.0	-	40.57
		URBAN	urban	54.0	-	27.0	-	14.58
		SCRUB	scrub/orchard	8.0	-	0.1	-	6.63
		DECID	deciduous wood	2.0	-	11.0	-	9.40
		other		2.0	-	3.0	-	14.97
Wytham	15	ARABLE	tilled land	26.0	4	28.0	1	19.24
		GRASS	grass/heath	28.0	5	32.0	4	35.88
		URBAN	urban	4.0	2	6.0	1	2.84
		SCRUB	scrub/orchard	10.0	0	0.1	*2*	0.09
		DECID	deciduous wood	30.0	3	33.0	*2*	35.10
		other		2.0	1	1.0	*5*	6.85

except at Wytham which had no detectable selection. The column R2 in Table 2.2 tells us how foxes used habitats in their ranges compared with their availability. For example, at Urban 2, an average of 12% of active fixes were observed in scrub (with one individual spending almost half its time in this habitat). Aside from the general tendency to favour scrub, fox use of ranges did not show any consistent bias (Table 2.2). Cumnor foxes spent more time in habitat not falling into the main categories recognized, which were lumped as 'other'; on average 21% of active fixes were in this category, compared with an average of only 2% range area (Table 2.2). At Boar's' Hill, the 'urban' category was used disproportionately; on average more than a third of active fixes were observed there, while it averaged only 4% of range area (Table 2.2). This suggests a dependence on urban gardens—the foxes were excluding the general 'urban' category from their ranges while using those favoured hot spots that were included in their ranges. The Oxford town foxes (both studies) tended to favour grassland in their ranges.

Time spent in any habitat is a relatively coarse indication of whether a habitat is 'important' or not and for what, regardless of the extent to which it is used out of proportion with its availability. Malcolm Newdick observed finer-scale differences (at site Urban 2 in Table 2.1). He observed that urban foxes timed their activity in different habitats to coincide with food availability, or with low human interference (Newdick 1983), and that activity was inversely related to traffic. He also found that the peak in urban garden use occurred at around 3 am, though only some foxes (nine out of 16) showed this form of 'timetabling'. Overall, his urban foxes increased activity in this habitat from 22:00 to 05:00[23], with the steepest increase between 23:00 and 01:00 (Newdick 1983). Urban gardens were used more in humid conditions, and when temperatures were mild[24]. These observations are consistent with foxes adjusting their activity to coincide with worm availability and to avoid encountering people, which may be a characteristic of predators in human influenced landscapes (Valeix et al. 2012). Urban foxes characteristically intersperse their nocturnal activity with frequent inactive periods in winter, though not in summer, with the result that the overall length of activity differs little across the seasons (Doncaster and Macdonald 1997). Other observations on differences in the timing of habitat use have no obvious explanation. Some of the foxes studied by Mike Fenn at Wytham regularly foraged at a farm dump[25]—individuals showed clear consistent differences in the time of day when they did this (Fenn and Macdonald 1995). Some urban foxes in Oxford, observed by both Patrick Doncaster and Malcolm Newdick, behaved as 'commuters', spending the nocturnal hours foraging in one site before travelling some distance to a den during the hours of daylight.

2.4.3 Space use and social effects

Foxes' use of the landscape in relation to that of other foxes tells us something important about their social system. We analysed the interactions of simultaneously tracked animals using grid cells, in order to estimate how often pairs of foxes were more or less in the same place. Dynamic interactions (Doncaster 1990; Doncaster and Macdonald 1997) were evaluated only for individuals that had been tracked on the same nights, and for which separation distances (based on fixes recorded at or close to the same time) were therefore available. We had 19 such pairs of foxes. Expressed simply, dynamic interaction measures the extent to which individuals use the same place at the same time (positive interaction) or the extent to which they avoid each other (negative interaction) (Doncaster 1990). This technique is generally very useful for unravelling the structure of the societies of elusive mammals; for example, we have recently applied a more sophisticated version to lions (Benhamou et al. 2014). The extent to which foxes of different types of relationship tended to spend time together or not tells us something about the use of that habitat.

Based on analyses of 20 dyads, including 19 foxes, did any foxes attract or repel each other? Yes: all cases of negative dynamic interactions (where one fox 'repelled' another) occurred between pairs of females where there was no overlap between core ranges. We identified three fox pairs where core range overlap was minimal, yet the animals still had positive interaction. Where females exhibited a high overlap (≥ 30%) between core areas, however, we observed a positive dynamic interaction (at least 14 minutes per night more than expected if movements had been random). Simply, females in adjoining ranges avoided each other, whereas females in the same range co-located. There were exceptions, however: three fox pairs where core range overlap was minimal, yet the animals still had positive interaction. Two of these pairs (dyads) involved male–male combinations, both exhibiting positive dynamic interaction (it was not possible to tell whether these encounters were agonistic or amicable).

2.4.4 Fox diets and habitat use

Droppings tell us a lot about fox diet. Between 1972 and 1986, we collected 10,452 fox scats (identified by their appearance, position, and characteristically pungent odour)[26] from our Oxford study area. Some of the sites used were also those where we had radio-tracked foxes and could therefore potentially tell us something

[23] $F_{1,9} = 14.4$, $P = 0.005$.

[24] Mantel–Haenszel tests, $\chi^2 = 9.93, 47.9$, $P < 0.01$.

[25] In addition to general farm waste, animal waste from sheep and pigs was dumped there.

[26] These are likely to be a biased sample of scats—scats resulting from low residue diets will be harder to detect.

about how different range compositions relate to the diet of individual foxes.

Microscopy of desiccated faecal samples and use of a standard key (Day 1966) established ten broad food categories (Doncaster et al. 1990): scavenged items (e.g. bone chips, plastic, paper, and string), birds, lagomorphs, small mammals, earthworms (estimated from counting chaetae—the chitinous hairs of annelids—and extrapolating back to probable numbers of worms), other invertebrates, fruits, grass, cereals, and domestic stock (chickens and pets). Variation in the rate of digestion of different food types creates problems when trying to use faeces to measure the relative importance of different dietary items. From several indices (Kamler et al. 2007) we opted to calculate the proportion that each food type contributed to overall energy consumption. First, we estimated the total biomass (fresh weight of food ingested) of each food item using conversion factors derived from feeding trials with a similar canid, captive side-striped jackals *Canis adustus* (Atkinson et al. 2002). Two indices were used to explore the relative importance of each item eaten: the contribution of each dietary item as a proportion of the total ingested, and the contribution of each to the total energy associated with each dropping, using energy values derived from the literature.

Scavenged items, earthworms, rabbits, birds, and small mammals contributed the major part of food biomass at all our sites (Table 2.3). Worms formed up to 44.8% of ingested energy. Wild birds also contributed a large component of the foxes' energy intake (mean 27%, 95% CI: 24–30). In addition to these major foods, foxes took a wide variety of other items. These included fruit, invertebrates, and cereals. Domestic stock (mainly chickens and small pets) formed a further small proportion of the diet: 1.3% (95% CI: 0–2.8) of biomass; 1.2% (95% CI: 0.1–2.3) of energy intake.

More detailed dietary analyses were available for four of our study sites (Boar's Hill, Cumnor, Wytham, and one site in Oxford city), where food types were identified to more specific categories. For example, fresh biomass intake of the category 'Birds' comprised

Table 2.3 Proportionate contribution of major food items in fox diets. Mean % biomass ingested (PBI) and mean % energy ingested (PEI). The means are the mean of the aggregate total for each season (the mean of four numbers). Site compositions as in Table 2.1.

Site	Type	Mean PBI	SE	Mean PEI	SE
Cumnor	Birds	29.13	4.55	14.59	3.79
	Mammals	7.50	2.00	5.41	1.28
	Lagomorphs	24.39	2.90	17.91	3.73
	Scavenge	13.19	1.70	12.63	2.95
	Worms	15.60	7.62	31.5	11.27
Boar's Hill	Birds	24.41	3.56	10.54	2.05
	Mammals	10.08	1.22	7.03	0.77
	Lagomorphs	7.87	0.67	5.19	0.68
	Scavenge	34.31	3.01	28.28	3.49
	Worms	20.75	3.02	44.83	4.18
Urban 1	Birds	26.51	4.40	14.35	2.97
	Mammals	12.93	0.95	11.18	0.67
	Lagomorphs	12.49	0.56	10.04	0.53
	Scavenge	34.92	2.55	35.04	1.37
	Worms	8.15	1.31	21.86	2.98
Urban 2	Birds	22.33	4.97	11.80	2.88
	Mammals	16.55	8.51	14.81	7.91
	Lagomorphs	12.99	3.03	10.82	2.89
	Scavenge	37.06	4.09	37.17	3.36
	Worms	7.15	1.73	19.01	3.59
Wytham	Birds	31.28	6.19	16.23	4.42
	Mammals	8.62	0.64	7.18	1.11
	Lagomorphs	19.46	3.31	14.69	3.12
	Scavenge	24.10	1.37	22.77	1.84
	Worms	13.86	3.82	33.15	6.82

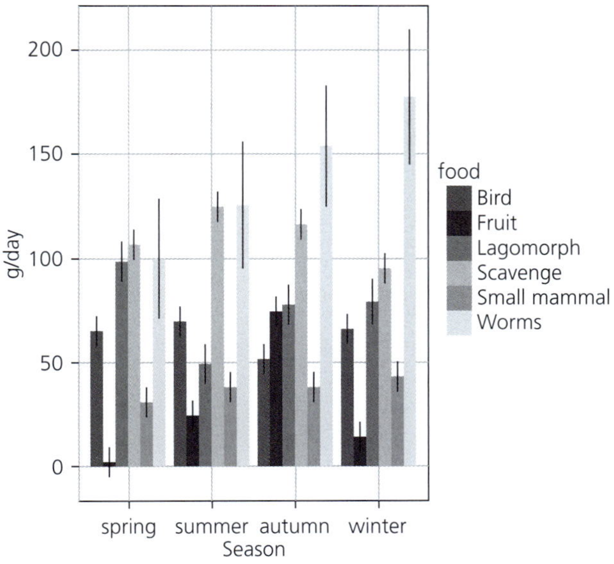

Figure 2.5 Estimated mean daily intake per fox of different food sources by season (error bars are SE).

52% Passeriformes (garden birds and crows), 24% Columbiformes (pigeons), 15% Galliformes (game birds), with the remaining 9% consisting of Gruiformes (coots and moorhens), falcons, and unidentified bird remains. Foxes also consumed an eclectic variety of small mammals: 24% wood mouse *Apodemus sylvaticus*, 20% field vole *Microtus agrestis*, 16% bank vole *Myodes*[27] *glareolus*, 9% brown rat *Rattus norvegicus*, with at least a further 5% of indistinguishable muroid prey. Fruit consumption was largely blackberries (45%), apples (15%), and plums (18%). Some of the Wytham farm foxes, those with access to a large midden, had diets dominated by scavenge (Fenn and Macdonald 1995). For these seven foxes, items scavenged from the midden amounted to between 67% and 81% of the energetic intake.

How do fox diets vary with season? We would expect optimal foraging strategies to differ markedly with seasonal fluctuations in the availability of key dietary components, as is observed in other omnivorous mesocarnivores (Zhou et al. 2011). Baker and Harris (2004) reviewed seasonal patterns in fox diets in Britain. They found that diets contained more birds in summer and that mammals declined in prominence from spring to autumn, but with little consistency between studies. For the Oxfordshire foxes we studied, there was also a clear seasonal pattern, largely due to fluctuation in lagomorph and fruit availability (Fig. 2.5). Peak rabbit availability occurred in the spring, when kits were born. This was reflected in the proportion of rabbits in fox diets. We observed maximum worm intake in winter (Fig. 2.4).

Worm availability also varies markedly between seasons, and is particularly sensitive to prevailing microclimatic conditions. In 1976 when the UK suffered a severe drought, worms were essentially 'off the menu' for a protracted period. Parts of the south west in the UK went 45 days without any rain in July and August[28]. Foxes, of necessity, had worms removed from their diet—a vivid illustration of the impact of changing conditions on food availability.

2.4.5 Links between diet and habitat

Having established that foxes select habitat preferentially from within study sites in defining their home ranges, and having found that they then use these habitat types with a temporal preference, influenced by season, we were able to establish a link between the time foxes spent in different habitat types and the relative amounts that different dietary types contributed to energetic intake.

For example, and reassuringly in the context of habitat proxies for food types, we found that individual foxes whose dietary energy depended more on scavenged food were those that tended to use urban habitat (Fig. 2.6a). The Oxford town foxes used low density

[27] *Clethrionomys* at the time of the study, but since then reassigned to the genus *Myodes*.

[28] <http://en.wikipedia.org/wiki/1976_United_Kingdom_heat_wave>.

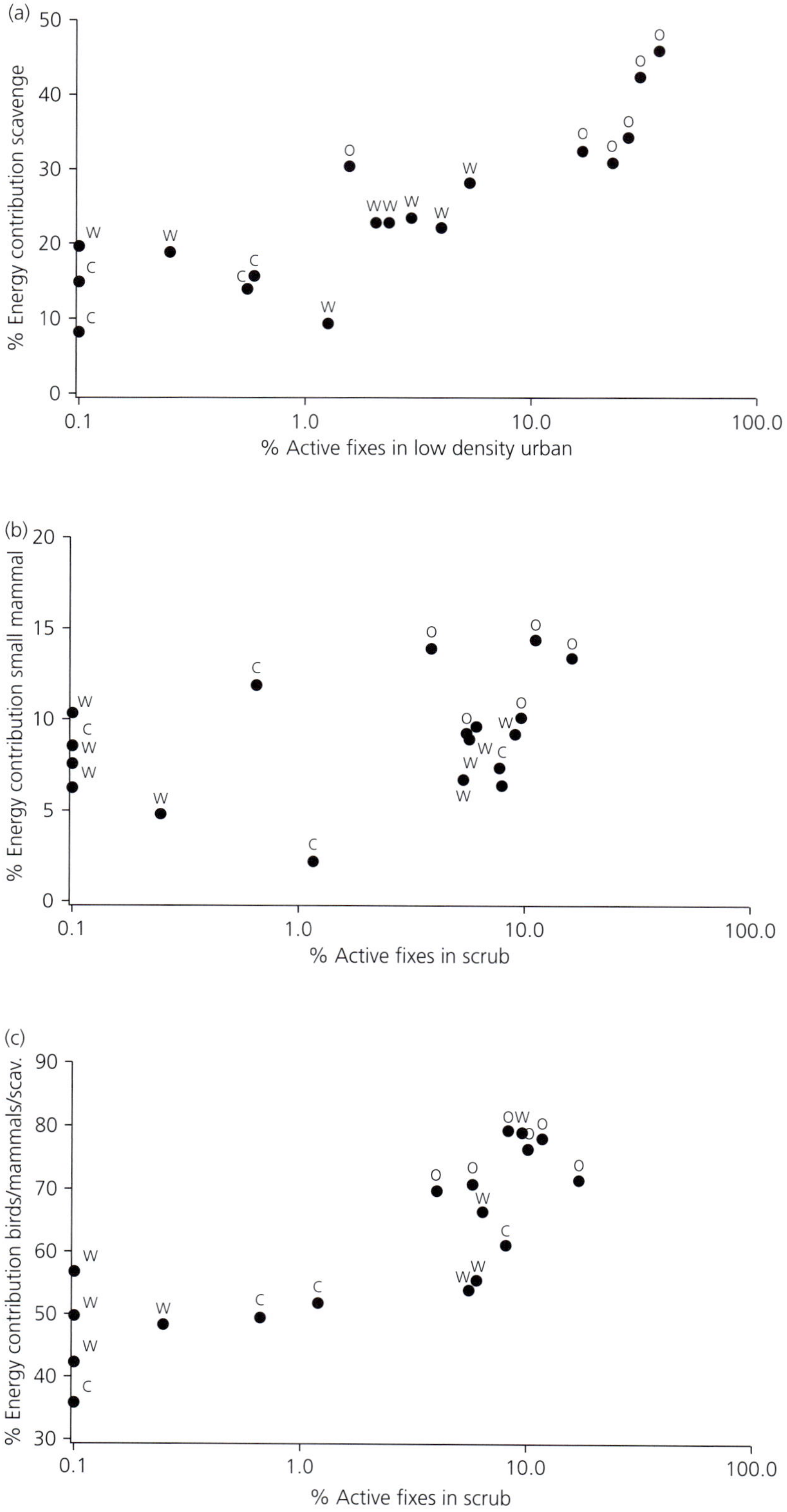

Figure 2.6 Energetic contribution to diet and time spent in habitats. Each point aggregates a site/season combination (O = Oxford (urban), W = Wytham, C = Cumnor).

housing areas the most, and scavenged items contributed a high proportion of total energy intake compared to the other site categories (Doncaster et al. 1990). Similarly, the contribution of small mammals is linked to time spent in scrubland (Fig. 2.6b). Many patches of urban scrubland habitat contained densities of small mammals many times those in neighbouring rural areas (Dickman and Doncaster 1987, 1989).

If we look at the aggregate contribution of birds, small mammals, and rabbits to dietary energy, time in scrubland explains a large part of the variation in dietary energy (Fig. 2.6c)[29]. Time spent in grassland did not influence the dietary contribution of any item, most notably the food type associated with grassland and favoured by foxes, earthworms[30]. Part of the answer is likely to be that habitats are imperfect proxies for food availability (see Box 2.3).

2.5 Conclusion

What emerges from our synthesis of these studies? What can we say about an 'average' fox in this part of Oxfordshire? One of our central findings is that the degree of ecological plasticity in a relatively small area is unparalleled for a carnivore of this size. Our study area encompassed no more than 40 km^2, a fragment of a typical hunt country[31] of over 600 km^2 in the English Midlands (Macdonald and Johnson 1996), and maybe about the size of a plausible rabies control zone (Defra 2011). The landscape is, however, typical of much of lowland UK in its patchwork of farmland, villages, and woodland. Within this patchwork we can learn much from both the consistencies and variability in the fundamental ecology of the 100 or so foxes included in this series of studies. We can see how variation

Box 2.3 Foxes' food preferences

We observed that using the broad habitat classes as proxies for food availability proved problematical: land use is a very rough predictor of resource availability. Furthermore, the 'real' availability of any resource present may depend on numerous factors (which, as we have seen is clearly true of earthworms). One of these is that foxes may have distinct preferences for different foods. By 'preference' we mean, to dispel any anthropomorphic baggage the word may carry, that given a choice, and all else being equal, a fox presented with a choice might not be equally likely to eat the options presented in proportion to their availability. This has some applied significance for administering vaccine or other agents in the field, and apparent 'preference' may be relative, depending on what the available options are.

Information on preference is next to impossible to acquire in the field. To explore food preferences, and the likely impact of those preferences on fox behaviour, we (Macdonald 1977a) designed a series of experiments in semi-natural conditions. Semi-tame hand reared foxes were used, and the experiments were carefully designed to eliminate spatial and other biases. Some observations corroborated what had been indicated indirectly in previous studies. One tame vixen consistently preferred dead field voles *Microtus agrestis* over bank voles *Myodes glareolus* or wood mice *Apodemus sylvaticus*, when these were offered simultaneously in an enclosure.

Choice was not always obvious—the fox sometimes did not eat the apparently preferred option immediately; some were cached, as were some of the remaining options offered. This was pursued by walking the same fox (on a lead) along a repeated route where pairs of *Microtus* and *Myodes* were encountered. This confirmed that *Microtus* was preferred consistently, and when not eaten was more likely to be cached, and more likely to be retrieved and eaten subsequently once cached. When caches were manipulated (by DWM) either by substituting the unfavoured *Myodes* for *Microtus*, or by placing a *Myodes* on top of the cached *Microtus*, the substitute was rejected on rediscovery. Similar experiments demonstrated that foxes avoided shrews (*Sorex*), other insectivores, and carnivores. Recent work exploring the food preferences of foxes in Japan (where *Microtus* is absent), showed red-backed voles (*Myodes*) were preferred to *Apodemus* (Kondo and Shiraki 2012). As those authors comment, the reasons for the preference are unknown. One possibility is that preference is based on properties of the live animal—voles are slower and less agile than mice, and probably therefore return more energy per unit effort *ceteris paribus*.

continued

[29] $r = 0.80, P < 0.001$.
[30] Overall $r = -0.28, P = 0.26, n = 18$.
[31] A hunt country is the area of land hunted by an English fox hunt, see Chapter 3, this volume.

Box 2.3 *Continued*

Preference can change. One vixen readily took dead one-day-old chicks until she was three months old, at which point she rejected them, but later took them to feed her own cubs. Further exploration of this suggested preference depended on season and the presence of cubs. As cubs grew and increasingly became competitors for food, vixens began to favour chicks over preferred rodents. Compared with the preferred *Microtus*, less-preferred food items tended to be cached more often. A separate series of observations and experiments on semi-tame foxes revealed that foxes were rather good at finding their caches, seemingly using both spatial memory and smell (Macdonald 1976); caching seems likely to be an important survival mechanism for foxes in the face of unpredictable food security.

between sites compares with variation within a site for some important aspects of fox behaviour. Our foxes ranged over areas averaging about 140 ha (CI 95–182), towards the lower end of the species range, even within the UK. While, as would be expected from the known dietary breadth of the species, a great diversity of items were observed in their diet: most foxes appeared to subsist on diets dominated by worms, mammals, scavenge, and birds. Given the great interest in the fox as a pest in rural UK (see Chapter 3, this volume), it is noteworthy that, while birds were consistently an important dietary item, a relatively small part of that intake in this particular part of Oxfordshire comprised game birds (*c.* 15% by energy); chickens or other domestic stock were not important dietary items at any site (which does not imply of course that the occasional predation incident is not important to the stock holder). The prominence of lagomorphs (principally rabbits) in fox diet, especially at more rural sites, illustrates the possible benefit of foxes to farmers, where modelling has shown that this rabbit removal service can be economically important under a series of plausible assumptions (Macdonald et al. 2003). The myxomatosis epidemic which drastically reduced UK rabbit numbers in the 1950s illustrated the adaptability of foxes compared with some other predators. Before the outbreak, both foxes and buzzards *Buteo buteo* relied heavily on rabbits; after the outbreak, foxes were able to switch from rabbits to voles and showed no long-term decline, while buzzards apparently were not, and their numbers declined by about 50% (see Chapter 10, this volume).

Our examination of fox interactions suggested that contact within ranges was more frequent than between neighbours, emphasizing the importance of social structure for contact rates (a vital part of fox management in the event of rabies outbreaks). The configuration of home ranges in the landscape showed little consistency among sites, probably reflecting the difficulty of categorizing land use. 'Grassland', for example, will not be qualitatively homogenous or of equal value to foxes at different sites—or even day to day, given the effect of microclimate on worm availability. Hence grassland was avoided at Cumnor (where it was largely grass ley), but included in ranges disproportionately at Boar's Hill (where horse pastures rich in worms predominated) and Wytham farm (where several meadows were richly populated with rabbits) (Table 2.2). In addition, the inclusion of habitats within ranges at a site was not a good indicator of the use of that habitat (in terms of active fixes). At Boar's Hill, foxes did not spend time foraging in grassland in direct proportion to the amount it contributed spatially to their ranges; rather grassland in total was under-used, probably reflecting not only that even worm-rich horse pastures were variously rich in worms from time to time, and also the fact that even at Boar's Hill, some grass leys were effectively useless to foxes for foraging (Macdonald 1980b). Similarly, Cumnor foxes spent more time foraging in the urban components of their ranges (which included some large, detached properties similar to those of Boar's Hill) than the spatial proportion that urban habitat contributed to these ranges would suggest. Habitat differences that are striking and likely to be important for foxes at small spatial scales are difficult to summarize adequately at larger spatial scales.

It is appealing to visualize a 'typical fox', and a broad portrayal of space use and diet is useful. But much of the practical value of our observations lies in the variation among individuals—and individual differences are an important influence on how animal populations react to human activity (Festa-Bianchet and Apollonio 2003). These individual differences, and the difficulty of characterizing a typical fox, are particularly problematic when predicting the spread of disease (in Box 2.4 we consider the implications of our findings

Box 2.4 Foxes and rabies

Rabies is a deadly disease and a considerable risk to humans. It occurs throughout most of the world and we have accounts of the disease in the civilizations of the Nile and Euphrates 2000 years ago (Macdonald 1977c). The virus responsible principally affects the central nervous system, and is usually introduced via a bite wound. There are only two recorded cases where humans who developed symptoms recovered (Bacon and Macdonald 1980). More than 50,000 cases occur worldwide in humans each year, most in developing countries—the disease is virtually absent from the UK, with the most recent human fatality in 2002 as the result of contact with a rabid bat[32] (where the virus is not the same classical species that infects canids in Europe). The UK is currently free of rabies but has been affected through much of recorded history (Macdonald 1980a). The fox is the main wild vector of the disease throughout much of Europe and North America, often comprising 60–85% of recorded cases (Macdonald and Bacon 1982), though the spread of the invasive raccoon dog *Nyctereutes procyonoides* is a significant complication on continental Europe (Smith et al. 2008). As with many wildlife problems, the conventional 'solution' has often been simple: slaughter (Macdonald 1980a). But the more modern approach is to exploit knowledge of the animal's behaviour and ecology to develop a more subtle and socially acceptable solution.

Knowledge of fox densities is required for disease management, and predictable patterns in fox distributions and landscape, linked to food availability, can be used for this (Macdonald et al. 1981). Habitat is also influential for the rate of spread of the disease (Bacon and Macdonald 1981). The ultimate mechanism for the link is how habitat indirectly influences contact rate.

Contact rates between individuals are a vital component in disease dynamics. A good understanding of transmission dynamics is important for control strategies. For example, the rate of transmission must be known in order to determine the level of vaccine coverage necessary to eliminate a disease (Nouvellet et al. 2013). The rate at which foxes acquire rabies is likely to be proportional to the encounter rate between susceptible and infectious foxes (Anderson et al. 1981). The social structure of fox populations will clearly be an important influence on these contact rates. And as we have seen, social organization varies dramatically depending on landscape structure, linked to the availability of food, and how it is dispersed. We showed here that the availability of worm-rich grassland and the habitat patchiness (Fig. 2.3) are predictors of range size. Where food is scarce, as in the uplands of the UK, territories are larger and foxes tend to live in pairs. Contact rates at territory boundaries are likely to be much lower than in rural suburbia, where territories are small and inhabited by groups of foxes. Our observations on females in adjoining territories avoiding each other while those in the same territory co-locate have implications for the rate of disease spread within and between groups. And observations on foxes in the same group in rural suburbia suggest there would be many opportunities for transmission during a four day infectious period (Macdonald and Bacon 1982).

The importance of variation in contact rates resulting from the high level of ecological plasticity in foxes was explored using simulation by Macdonald and Bacon (1982). The crucial aspect of disease transmission was the number of foxes infected by each infectious fox. If this was as high as three, the disease was so severe that the disease drove foxes (and consequently itself) rapidly to extinction. If it was as low as 0.5, the virus soon died out. But at around two, the population was reduced to 20% of that at carrying capacity, followed by a recovery until it was once again able to sustain an outbreak. This cycling behaviour is similar to that observed in Europe, where epizootics of fox-borne rabies follow a three to four year cycle (Bacon and Macdonald 1980).

What does the social system and its variability imply for different disease management options? To cull, to sterilize, or to vaccinate? In Europe, mass campaigns of poisoning, shooting, and gassing had only limited success in controlling the spread of rabies (Macdonald 1987). One problem with fox control is that the disruption to the social system can promote increased contact rates. Territorial behaviour is such that there may be 'surplus' individuals in the population who take up the territories made vacant by culling (Macdonald 1980a). Increased aggression in disputes for territories and mates will likely accelerate disease spread—the 'perturbation' effect (see Chapter 5, this volume). Vaccination, on the other hand, avoids this problem. In Europe, vaccination programmes have eliminated rabies from fox populations across wide areas (Pastoret and Brochier 1998).

The spatial models of Smith and Wilkinson (2003) suggested fox density was likely to be an important influence on

continued

[32] Public Health England <http://www.hpa.org.uk/>, accessed 06/12/13.

Box 2.4 *Continued*

which strategy was optimal. At high fox densities, vaccination by itself was not promising. The models suggested an approximate threshold for 'high' was 0.5 families per km^2, which is within the range implied by territory sizes for all the Oxfordshire sites (our Table 2.1), assuming typical family structures. The existence of a critical density implies an outbreak would be harder to control with increasing fox densities. The landscape typical of rural suburbia would therefore need to be given particular attention during an outbreak, and probably with a combination of strategies involving culling, vaccination, and fertility control (Smith and Wilkinson 2003).

for rabies management). Individual range sizes, for example, were dependent on the patchiness of the habitat; they tended to be larger where the landscape was less patchy, and this was consistent over several sites (Fig. 2.3), in accord with the predictions of the RDH, promoting fox sociality in this type of landscape.

Within sites, differences between foxes in how their active time was allocated to different habitats could be clearly linked to their diet. Our data do not allow us to plot individual fox diet but we would expect variation in the aggregates for season and site to be similar. Time spent in scrubland was particularly influential; foxes active in scrubland had diets that were dominated by birds, small mammals, and scavenge; and Oxford foxes spending more time in urban habitat were likely to have more scavenge in their diet. As was originally observed by Macdonald (1987), fox diets vary substantially over small spatial scales. Individual differences in temporal use of space are also possible—Malcolm Newdick saw some urban foxes with clear activity cycles, he speculated that these foxes were optimizing their foraging success by synchronizing with worm activity while avoiding human disturbance. Other foxes showed no such patterns. Foxes also adjusted their foraging patterns depending on the weather—the greater use of grassland in humid conditions is likely to be related to worm availability.

Our exploration illustrates the complexities of fox ecology in a patchy urban, suburban, and rural landscape. While foxes in this landscape, typical of the English Midlands, show some characteristic features, with family groups occupying relatively small ranges, for example, the variability in space use and diet was considerable among and within sites. One of the most striking aspects of space use and the diets linked to it is the amount of variation between individual foxes apparently having similar options in any landscape. This is exemplified by the Wytham foxes studied by Mike Fenn—those foxes that used a midden ate mostly scavenged items while their near neighbours opted for a rabbit-rich diet. Foxes are able to adjust their behaviour according to local differences in food availability, and depending on weather, time of day, and season. The ecological plasticity of foxes is a considerable complication for land managers with an interest in foxes.

Acknowledgements

Paul Johnson acknowledges the support of the Whitley Trust. Chris Newman made extensive comments on an earlier draft. Ruth Feber and Eva Raebel made numerous constructive editorial suggestions, significantly improving the final version.

References

Abbot, P., Abe, J., Alcock, J., et al. (2011). Inclusive fitness theory and eusociality. *Nature*, **471**, E1–E4.

Aebischer, N.J., Robertson, P.A., and Kenward, R.E. (1993). Compositional analysis of habitat use from animal radio-tracking data. *Ecology*, **74**, 1313–1325.

Anderson, R.M., Jackson, H.C., May, R.M., and Smith, A.M. (1981). Population-dynamics of fox rabies in Europe. *Nature*, **289**, 765–771.

Atkinson, R.P.D., Macdonald, D.W., and Kamizola, R. (2002). Dietary opportunism in side-striped jackals *Canis adustus*. *Journal of Zoology*, **257**, 129–139.

Bacon, P.J., Ball, F., and Blackwell, P. (1991). A model for territory and group formation in a heterogeneous habitat. *Journal of Theoretical Biology*, **148**, 445–468.

Bacon, P.J. and Macdonald, D.W. (1980). To control rabies: vaccinate foxes. *New Scientist*, **87**, 640–645.

Bacon, P.J. and Macdonald, D.W. (1981). Habitat and the spread of rabies. *Nature*, **289**, 634–635.

Baker, P.J. and Harris, S. (2003). A review of the diet of foxes in rural Britain and a preliminary assessment of their impact as predator. In F.H. Tattersall and W.J. Manley, eds., *Conservation and Conflict: Mammals and Farming in Britain*, pp. 120–140. Linnean Society Occasional Publication, Westbury Publishing, Yorkshire.

Baker, P.J. and Harris, S. (2004). Red foxes. The behavioural ecology of red foxes in urban Bristol. In D.W. Macdonald and C. Sillero-Zubiri, eds., *The Biology and Conservation of Wild Canids*, pp. 207–216. Oxford University Press, Oxford, UK.

Ball, F., Sirl, D., and Trapman, P. (2009). Threshold behaviour and final outcome of an epidemic on a random network with household structure. *Advances in Applied Probability*, **41**, 765–796.

Beer, A.-J. (2007). Seeing Red. *Mammals UK*, pp. 10–12.

Benhamou, S. Valeix, M, Chamaillé-Jammes, S., et al. (2014). Movement-based analysis of interactions in African lions. *Animal Behaviour*, **90**, 171–180.

Blumstein, D.T. (2006). The multipredator hypothesis and the evolutionary persistence of antipredator behavior. *Ethology*, **112**, 209–217.

Bogel, K., Moegle, H., Knorpp, F., Arata, A., Dietz, K., and Diethelm, P. (1976). Characteristics of spread of a wildlife rabies epidemic in Europe. *Bulletin of the World Health Organization*, **54**, 433–447.

Burrows, R. (1968). *Wild Fox*. David & Charles, Newton Abbot.

Cagnacci, F., Meriggi, A., and Lovari, S. (2004). Habitat selection by the red fox Vulpes vulpes (L. 1758) in an Alpine area. *Ethology Ecology & Evolution*, **16**, 103–116.

Carbone, C., Mace, G.M., Roberts, S.C., and Macdonald, D.W. (1999). Energetic constraints on the diet of terrestrial carnivores (Addendum). *Nature*, **402**, 442–442.

Carr, G.M. and Macdonald, D.W. (1986). The sociality of solitary foragers: a model based on resource dispersion. *Animal Behaviour*, **34**, 1540–1549.

Cavallini, P. (1996). Variation in the social system of the red fox. *Ethology Ecology & Evolution*, **8**, 323–342.

Cavallini, P. and Lovari, S. (1991). Environmental factors influencing the use of habitat in the red fox, Vulpes-vulpes. *Journal of Zoology*, **223**, 323–339.

Charnov, E.L. (1976). Optimal foraging, marginal value theorem. *Theoretical Population Biology*, **9**, 129–136.

Creel, S.R. and Macdonald, D.W. (1995). Sociality, group size, and reproductive suppression among carnivores. *Advances in the Study of Behaviour*, **24**, 203–257.

Darwin, C. R. (1859). *On the Origin of Species by Means of Natural Selection*. John Murray, London.

Day, M.G. (1966). Identification of hair and feather remains in the gut and faeces of stoats and weasels. *Journal of Zoology*, **18**, 201–217.

Defra (2011). *Rabies Disease Control Strategy*. Crown Copyright, UK.

Dickman, C.R. and Doncaster, C.P. (1987). The ecology of small mammals in urban habitats I. Populations in a patchy environment. *Journal of Animal Ecology*, **56**, 629–640.

Dickman, C.R. and Doncaster, C.P. (1989). The ecology of small mammals in urban habitats II. Demography and dispersal. *Journal of Animal Ecology*, **58**, 119–127.

Doncaster, C.P. (1985). *The spatial organisation of urban foxes (Vulpes vulpes) in Oxford*. Thesis (D Phil), University of Oxford.

Doncaster, C.P. (1990). Nonparametric estimates of interaction from radio-tracking data. *Journal of Theoretical Biology*, **143**, 431–443.

Doncaster, C.P., Dickman, C.R., and Macdonald, D.W. (1990). Feeding ecology of red foxes (*Vulpes vulpes*) in the city of Oxford, England. *Journal of Mammalogy*, **71**, 188–194.

Doncaster, C.P. and Macdonald, D.W. (1991a). Drifting territoriality in the red fox *Vulpes vulpes*. *Journal of Animal Ecology*, **60**, 423–439.

Doncaster, C.P. and Macdonald, D.W. (1991b). Ecology and ranging behaviour of red foxes in the city of Oxford. *Atti Io Simp. Carnivori (Univ. Pavia), Hystrix*, **3**, 11–20.

Doncaster, C.P. and Macdonald, D.W. (1992). Optimum group size for defending heterogeneous distributions of resources: a model applied to red foxes, *Vulpes vulpes*, in Oxford City. *Journal of Theoretical Biology*, **159**, 189–198.

Doncaster, C.P. and Macdonald, D.W. (1996). Intraspecific variation in the movement behaviour of foxes (*Vulpes vulpes*): a reply to White, Saunders and Harris. *Journal of Animal Ecology*, **65**, 126–127.

Doncaster, C.P. and Macdonald, D.W. (1997). Activity patterns and interactions of red foxes (*Vulpes vulpes*) in Oxford city. *Journal of Zoology*, **241**, 73–87.

Eckert, J. and Deplazes, P. (2004). Biological, epidemiological, and clinical aspects of echinococcosis, a zoonosis of increasing concern. *Clinical Microbiology Reviews*, **17**, 107–135.

Englund, J. (1970). Some aspects of reproduction and mortality rates in Swedish foxes (*Vulpes vulpes*) 1961–1963 and 1966–1969. *Viltrevy*, **8**, 1–82.

Fenn, M.G.P. and Macdonald, D.W. (1995). Use of middens by red foxes: risk reverses rhythms of rats. *Journal of Mammalogy*, **76**, 130–136.

Festa-Bianchet, M. and Apollonio, M. (2003). *Animal Behavior and Wildlife Conservation*. Island Press, Washington, DC; London.

Fuller, R.M., Groom, G.B., and Jones, A.R. (1994). The land cover map of great Britain : an automated Classification of Landsat Thematic Mapper data. *Photogrammetric Engineering & Remote Sensing*, **60**, 553–562.

Geffen, E., Gompper, M.E., Gittleman, J.L., Luh, H.K., Macdonald, D.W., and Wayne, R.K. (1996). Size, life history traits, and social organization in the Canidae: a reevaluation. *American Naturalist*, **147**, 140–160.

Gittleman, J.L. (1989). *Carnivore Behavior, Ecology, and Evolution*. Chapman and Hall, London.

Goltsman, M., Kruchenkova, E.P., Sergeev, S., Johnson, P.J., and Macdonald, D.W. (2005). Effects of food availability on dispersal and cub sex ratio in the Mednyi Arctic fox. *Behavioral Ecology and Sociobiology*, **59**, 198–206.

Hamilton, W.D. (1964a). The genetical evolution of social behaviour II. *Journal of Theoretical Biology*, **7**, 17–52.

Hamilton, W.D. (1964b). The genetical evolution of social behaviour I. *Journal of Theoretical Biology*, **7**, 1–16.

Harris, S. (1979). Age-related fertility and productivity in red foxes, Vulpes vulpes, in suburban London. *Journal of Zoology*, **187**, 195–199.

Harris, S. (1981). An estimation of the number of foxes (Vulpes vulpes) in the city of Bristol, and some possible factors affecting their distribution. *Journal of Applied Ecology*, **18**, 455–465.

Harris, S. and Rayner, J.M.V. (1986). Urban fox (Vulpes vulpes) population estimates and habitat requirements in several British cities. *Journal of Animal Ecology*, **55**, 575–591.

Hersteinsson, P. and Macdonald, D.W. (1982). Some comparisons between red and Arctic foxes, *Vulpes vulpes* and *Alopex lagopus*, as revealed by radio tracking. In *Proceedings of the Symposia of the Zoological Society of London*, **49**, pp. 259–289.

Heydon, M.J. and Reynolds, J.C. (2000). Demography of rural foxes (Vulpes vulpes) in relation to cull intensity in three contrasting regions of Britain. *Journal of Zoology*, **251**, 265–276.

Houston, A. and McNamara, J.M. (1999). *Models of Adaptive Behaviour: an Approach Based on State*. Cambridge University Press, Cambridge.

Johnson, D.D.P., Kays, R., Blackwell, P.G., and Macdonald, D.W. (2002). Does the resource dispersion hypothesis explain group living? *Trends in Ecology and Evolution*, **17**, 563–570.

Johnson, D.D.P. and Macdonald, D.W. (2003). Sentenced without trial: reviling and revamping the resource dispersion hypothesis. *Oikos*, **101**, 433–440.

Johnson, D., Macdonald, D.W., Kays, R., and Blackwell, P.G. (2003). Response to Revilla, and Buckley and Ruxton: the resource dispersion hypothesis. *Trends in Ecology and Evolution*, **18**, 381–382.

Kamler, J.F., Ballard, W.B., Wallace, M.C., Gilliland, R.L., and Gipson, P.S. (2007). Dietary overlap of swift foxes and coyotes in northwestern Texas. *American Midland Naturalist*, **158**, 139–146.

Kamler, J.F. and Macdonald, D.W. (2014). Social organization, survival, and dispersal of cape foxes (Vulpes chama) in South Africa. *Mammalian Biology—Zeitschrift für Säugetierkunde*. **79**, 64–70.

Kenward, R.E., and Hodder, K.J. (1996). *Ranges V: an analysis system for biological location data*. Institute of Terrestrial Ecology. Natural Environment Research Council, Cambridge.

Kondo, A. and Shiraki, S. (2012). Preferences for specific food species of the red fox Vulpes vulpes in Abashiri, eastern Hokkaido. *Mammal Study*, **37**, 43–46.

Kruchenkova, E.P., Goltsman, M., Sergeev, S., and Macdonald, D.W. (2009). Is alloparenting helpful for Mednyi Island arctic foxes, Alopex lagopus semenovi? *Naturwissenschaften*, **96**, 457–466.

Kruuk, H. (1976). Functional aspects of social hunting in carnivores. In G. Baerends, C. Baer, and A. Manning, eds., *Function and Evolution in Behaviour*, pp. 119–141. Oxford University Press, Oxford.

Kruuk, H.H. and Parish, T. (1987). Changes in the size of groups and ranges of the European badger (*Meles meles* L) in an area in Scotland. *Journal of Animal Ecology*, **56**, 351–364.

List, R. and Macdonald, D.W. (2003). Home range and habitat use of the kit fox (*Vulpes macrotis*) in a prairie dog (*Cynomys ludovicianus*) complex. *Journal of Zoology*, **259**, 1–5.

Lloyd, H.G. (1980). *The Red Fox*. Batsford, London.

Maas, B. and Macdonald, D.W. (2004). Bat-eared foxes, 'insectivory' and luck: lessons from an extreme canid. In D.W. Macdonald and C. Sillero-Zubiri, eds., *The Biology and Conservation of Wild Canids*, pp. 227–242. Oxford University Press, Oxford, UK.

Macdonald, D.W. (1976). Food caching by red foxes and some other carnivores. *Zeitschrift für Tierpsychologie*, **42**, 170–185.

Macdonald, D.W. (1977a). On food preference in the red fox. *Mammal Review*, **7**, 7–23.

Macdonald, D.W. (1977b). *The behavioural ecology of the red fox, Vulpes vulpes: a study of social organisation and resource exploitation*. Thesis (D Phil), University of Oxford.

Macdonald, D.W. (1977c). The behavioural ecology of the red fox. In C. Kaplan, ed. *Rabies, the facts*, pp. 70–90. Oxford University Press, Oxford.

Macdonald, D.W. (1978). Radio tracking: some applications and limitations. In B. Stonehouse, ed. *Animal Marking: Recognition Marking of Animals in Research*, pp. 92–204. Macmillan, London.

Macdonald, D.W. (1979a). 'Helpers' in fox society. *Nature*, **282**, 69–71.

Macdonald, D.W. (1979b). Some observations and field experiments on the urine marking behaviour of the red fox, *Vulpes vulpes* L. *Zeitschrift für Tierpsychologie*, **51**, 1–22.

Macdonald, D.W. (1980a). *Rabies and Wildlife: a Biologist's Perspective*. Oxford University Press, New York.

Macdonald, D.W. (1980b). The red fox, *Vulpes vulpes*, as a predator upon earthworms, *Lumbricus terrestris*. *Zeitschrift für Tierpsychologie*, **52**, 171–200.

Macdonald, D.W. (1980c). Social factors affecting reproduction behaviour among red foxes (*Vulpes vulpes* L., 1758). In E. Zimen, ed. *The Red Fox: Behaviour and Ecology*, pp. 131–183. The Hague, Boston.

Macdonald, D.W. (1980d). Patterns of scent marking with urine and faeces amongst carnivore communities. In, ed. *Proceedings of the Symposia of the Zoological Society of London*, pp. 107–139.

Macdonald, D.W. (1981). Resource dispersion and the social organisation of the red fox (*Vulpes vulpes*). In J.A. Chapman and D. Pursley, eds., *The First International Worldwide Furbearer Conference*, Frostburg, Maryland, pp. 918–949.

Macdonald, D.W. (1983). The ecology of carnivore social behaviour. *Nature*, **301**, 379–384.

Macdonald, D.W. (1987). *Running with the Fox*. Unwin Hyman, London.

Macdonald, D.W. (1992). *The Velvet Claw*. BBC Books, London.

Macdonald, D.W. (1995). Wildlife rabies: the implications for Britain. Unresolved questions for the control of wildlife rabies: social perturbation and interspecific interactions. In Rabies in a changing world. *Proceedings of the British Small Animal Veterinary Association*, pp. 33–48. Cheltenham.

Macdonald, D.W. (2013). Genus *Suricata*. In J.S. Kingdon and M. Hoffmann, eds., *Mammals of Africa Vol V. Carnivores, Pangolins, Equids and Rhinoceroses*, pp. 346–352. Bloomsbury, London.

Macdonald, D.W. and Bacon, P.J. (1982). Fox society, contact rate and rabies epizootiology. *Comparative Immunology Microbiology and Infectious Diseases*, **5**, 247–256.

Macdonald, D.W. and Barrett, P. (1995). *Collins Field Guide to Mammals of Britain & Europe*. HarperCollins, London.

Macdonald, D.W. and Brown, R.E., eds. (1985). *Social Odours in Mammals*. Oxford University Press, Oxford.

Macdonald, D.W., Brown, L.J., Yerli, S., and Canbolat, A.-F. (1994). The behaviour of red foxes, *Vulpes vulpes*, caching the eggs of loggerhead turtles, *Carreta carreta*. *Journal of Mammalogy*, **75**, 985–988.

Macdonald, D.W., Bunce, R.G.H., and Bacon, P.J. (1981). Fox populations, habitat characterization and rabies control. *Journal of Biogeography*, **8**, 145–151.

Macdonald, D.W. and Carr, G.M. (1989). Food security and the rewards of tolerance. In V. Standen and R.A. Folley, eds., *Comparative Socioecology: the Behavioural Ecology of Humans and other Mammals*, pp. 75–99. Blackwell Scientific Publications, Oxford.

Macdonald, D.W., Courtenay, O., Forbes, S., and Mathews, F. (1999). The red fox (*Vulpes vulpes*) in Saudi Arabia: loose-knit groupings in the absence of territoriality. *Journal of Zoology*, **249**, 383–391.

Macdonald, D.W., Creel, S., and Mills, M.G. (2004). Canid society. In D.W. Macdonald and C. Sillero-Zubiri, eds., *The Biology and Conservation of Wild Canids*, pp. 85–106. Oxford University Press, Oxford, UK.

Macdonald, D.W. and Johnson, P.J. (1996). The impact of sport hunting: a case study. In N. Dunstone and V.L. Taylor, eds., *The Exploitation of Mammal Populations: 1994 Symposium of the Zoological Society of London*, pp. 160–207. Chapman & Hall, London.

Macdonald, D.W. and Laurenson, K. (2006). Infectious disease: inextricable linkages between human and ecosystem health. *Biological Conservation*, **131**(2), 143–150.

Macdonald, D.W. and Newdick, M.T. (1982). The distribution and ecology of foxes, *Vulpes vulpes* (L.), in urban areas. In R. Bornkamm, J.A. Lee, and M.R.D. Seaward, eds., *Urban Ecology*, pp. 123–135. Blackwell Scientific Publications, Oxford.

Macdonald, D.W. and Reynolds, J. (2004). Red fox (*Vulpes vulpes*). In C. Sillero-Zubiri, M. Hoffmann, and D.W. Macdonald, eds. *Canids: Foxes, Wolves, Jackals and Dogs. Status Survey and Conservation Action Plan*, pp. 129–135. IUCN, Gland, Switzerland, and Cambridge, UK.

Macdonald, D.W., Reynolds, J.C., Carbone, C., Mathews, F., and Johnson, P.J. (2003). The bioeconomics of fox control. In F.H. Tattersall and W.J. Manley, eds., *Conservation and Conflict: Mammals and Farming in Britain*, pp. 220–236. Linnean Society Occasional Publication, Westbury Publishing, Yorkshire.

Macdonald, D.W., Riordan, P., and Mathews, F. (2006). Biological hurdles to the control of TB in cattle: a test of two hypotheses concerning wildlife to explain the failure of control. *Biological Conservation*, **131**, 268–286.

Macdonald, D.W. and Sillero-Zubiri, C., eds. (2004). *The Biology and Conservation of Wild Canids*. Oxford University Press, Oxford, UK.

Macdonald, D.W. and Voigt, D.R. (1985). The biological basis of rabies models. In P. J. Bacon, ed. *Population Dynamics of Rabies in Wildlife*, pp. 71–108. Academic Press, London.

Moehrenschlager, A., List, R., and Macdonald, D.W. (2007). Escaping intraguild predation: Mexican kit foxes survive while coyotes and golden eagles kill Canadian swift foxes. *Journal of Mammalogy*, **88**, 1029–1039.

Moehrenschlager, A. and Sovada, M.A. (2004). Swift fox (*Vulpes velox*). In C. Sillero-Zubiri and D.W. Macdonald, eds., *Canid Status and Conservation Action Plan*, pp. 109–116. IUCN Publications, Gland, Switzerland.

Moelge, H., Knorpp, F., Bogel, K., Atrata, F., Dietz, N., and Diethelm, P. (1974). Zur Epidemiologie der Wildtiertollarut. *Zentbl. VetMed. B.*, **21**, 647–659.

Newdick, M.T. (1983). *The behavioural ecology of urban foxes, Vulpes vulpes, in Oxford*. Thesis (D Phil), University of Oxford.

Nouvellet, P., Donnelly, C.A., De Nardi, M., et al. (2013). Rabies and canine distemper virus epidemics in the red fox population of northern Italy (2006–2010). *PLOS ONE*, **8**.

Nowak, M.A., Tarnita, C.E., and Wilson, E.O. (2010). The evolution of eusociality. *Nature*, **466**, 1057–1062.

Nowak, M.A., Tarnita, C.E., and Wilson, E.O. (2011). In defence of inclusive fitness theory. *Nature*, **471**, E9–E10.

Pastoret, P.P. and Brochier, B. (1998). Epidemiology and elimination of rabies in western Europe. *Veterinary Journal*, **156**, 83–90.

Phillips, R.L. (1970). Age ratios of Iowa foxes. *Journal of Wildlife Management*, **34**, 52–56.

Revilla, E. (2003a). Moving beyond the resource dispersion hypothesis. *Trends in Ecology & Evolution*, **18**, 380.

Revilla, E. (2003b). What does the resource dispersion hypothesis explain, if anything? *Oikos*, **101**, 428–432.

Smith, G.C., Thulke, H.-H., Fooks, A.R., et al. (2008). Towards the elimination of rabies in Eurasia: what is the future of rabies control in Europe? In Dodet, B., Miller, T., and Tordo, N., eds., *Developments in Biologicals*, pp. 283–289. Karger, Basel.

Smith, G.C. and Wilkinson, D. (2003). Modeling control of rabies outbreaks in red fox populations to evaluate culling, vaccination, and vaccination combined with fertility control. *Journal of Wildlife Diseases*, **39**, 278–286.

Storm, G.R., Andrews, R.R., Phillips, R.L., Bishop, R.A., Siniff, D.B., and Tester, J.R. (1976). Morphology, reproduction, dispersal and mortality of midwestern red fox populations. *Wildlife Monographs*, **49**, 3–82.

Taber, A.B. and Macdonald, D.W. (1992). Spatial organization and monogamy in the mara *Dolichotis patagonum*. *Journal of Zoology*, **227**, 417–438.

Valeix, M., Hemson, G., Loveridge, A.J., Mills, G., and Macdonald, D.W. (2012). Behavioural adjustments of a large carnivore to access secondary prey in a human-dominated landscape. *Journal of Applied Ecology*, **49**, 73–81.

van Kesteren, F., Paris, M., Macdonald, D.W., et al. (2013). The physiology of cooperative breeding in a rare social canid: sex, suppression and pseudopregnancy in female Ethiopian wolves. *Physiology & Behaviour*, **122**, 39–45.

Voigt, D.R. and Macdonald, D.W. (1984). Variation in the spatial and social behaviour of the red fox, *Vulpes vulpes*. *Acta Zoologica Fennica*, **171**, 261–265.

Worton, B.J. (1989). Kernel methods for estimating the utilization distribution in home range studies. *Ecology*, **70**, 164–168.

Zabel, C.J. and Taggart, S.J. (1989). Shift in red fox, Vulpes-vulpes, mating system associated with El-Nino in the Bering Sea. *Animal Behaviour*, **38**, 830–838.

Zhou, Y.B., Newman, C., Buesching, C.D., et al. (2011). Diet of an opportunistically frugivorous carnivore, *Martes flavigula*, in subtropical forest. *Journal of Mammalogy*, **92**, 611–619.

CHAPTER 3

Foxes in the landscape: hunting, control, and economics

David W. Macdonald and Paul J. Johnson

> Psychologists have shown that in any dispute both sides are convinced that they themselves are reasonable and upright and that their opposite numbers are mulish and dishonest. They can't all be right, at least not all the time.
>
> **Steven Pinker**

3.1 Introduction

The social challenges that can confront wildlife management are nowhere more vividly illustrated than for the red fox *Vulpes vulpes* (simply the 'fox' from here on) in the UK. This seemingly parochial issue captures in microcosm many of the dauntingly incommensurable scientific and social factors that beset conservation decisions globally (Macdonald and Johnson 1996). The image of the fox in the UK is culturally complex. Its reputation as a 'villain' is deeply entrenched; 'theft' of chickens and other stock has enraged as far back as the earliest written accounts. It has consequently been among the most despised of British mammals, and widely killed. In Tudor times anti-vermin legislation was enforced by bounty payments, and the fox appears more consistently than any other species in the written records of these (Lovegrove and Lovegrove 2007). The fox pervades art and literature, including nursery stories, depicted as a creature of guile and resourcefulness. William Caxton's 1481 history of *Reynard the Fox* brings together many different legends: among many misdemeanours, Reynard tricks cats into castrating clergymen, he dupes Bruin the bear, who ends up burned alive, and rapes Hersent the she-wolf. In many European languages the word 'foxy' carries connotations of slyness and deceit.

There is a thread, though, of, sometimes grudging, respect running through its iconography— foxes' perceived intelligence is exemplified by a sixteenth century misericord in Manchester Cathedral, depicting a vixen teaching its cubs to read (Fig. 3.1), and the wily tactics of Reynard in mediaeval bestiaries resonate deeply throughout European culture. In modern times, surveys of the public suggest that foxes are among Britain' s most popular mammals. In a 2008 survey of readers of *BBC Wildlife* magazine, for example, foxes polled third highest behind otters and hedgehogs. In the twentieth century the word 'foxy' has acquired an additional meaning which is unambiguously complimentary (Oxford English Dictionary *U.S. slang*, 'Of a woman: attractive, desirable, pretty; sexy'). Against this complex cultural backdrop, we can consider how attitudes to the treatment of foxes have changed over the years, and ask what this tells us about human–wildlife conflicts more generally.

November 2004 saw a considerable landmark for the relationship between people and foxes in the UK. A vote in the House of Commons made hunting with dogs unlawful in England and Wales[1]. The bill encompassed the hunting of foxes, hares, deer, and mink. An amendment advocated by the then Prime Minister, Tony Blair, which would have allowed a modified and licensed version of the pastime to continue, was voted down.

This was a remarkable political event with implications transcending government's normal concern with a wild mammal, or indeed any environmental issue; a weighty power called the Parliament Act (PA) was used after the House of Lords refused to pass the hunting bill, while the Commons had voted for abolition by a

[1] Effective from 18 February 2005.

Wildlife Conservation on Farmland. Conflict in the Countryside. Edited by David W. Macdonald and Ruth E. Feber.
© Oxford University Press 2015. Published 2015 by Oxford University Press.

Figure 3.1 A sixteenth century misericord in Manchester Cathedral appears to depict a vixen teaching her cubs to read. Photograph by kind permission of the Dean, Manchester Cathedral.

majority of 356 to 166[2]. Since it came into being, the PA has been used only seven times, and as a weapon of last resort in conflicts between the innately conservative Lords and the generally more reforming Commons. Bringing about the ban on hunting with dogs is its most recent use and, by the time this happened, the parliamentary significance of hunting with dogs was not about foxes at all, but more to do with 'who governs Britain' (Mullin and Winstone 2009). Charles Pye-Smith has written an engaging account of the political machinations leading up to the vote, albeit from a transparently pro-hunting perspective (Pye-Smith 2006).

It may seem baffling to onlookers not familiar with the background to foxes in the UK that such a large legal sledgehammer (the 'nuclear' option as Stuart Harrop has called it; Harrop 2006) was brought to bear on such an apparently inconsequential nut. It has been estimated that 700 hours of parliamentary time were expended on the issue. The Hunting Act was the culmination, though not necessarily the end, of a long debate in the UK about hunting with dogs, and how foxes and other problem wildlife species are treated generally. The high parliamentary drama in November 2004 illustrated just how deeply culturally enmeshed were British attitudes to wildlife and the countryside. Tony Blair has since written that the Hunting Bill was one of the measures he most regretted (Blair 2011). The ban turned into one of the most controversial laws of his administration, prompting mass marches in London and even an invasion of Parliament by protesters. In his memoirs Blair (2011) describes the passions aroused as 'primeval', a depth of feeling that took him (and no doubt many others) by surprise. Blair adds: 'If I'd proposed solving the pension problem by compulsory euthanasia for every fifth pensioner I'd have got less trouble for it'. The Prime Minister had set this ball rolling, perhaps unintentionally, by hinting during a television interview that foxhunting might be banned. Once rolling, the ball could not be stopped 'By the end of it, I felt like the damn fox', he said (Blair 2011).

To understand foxhunting in the UK, as Tony Blair admitted he did not, requires us to consider conservation, welfare, economics, and the complex sociology of their inter-relatedness. One of us has been using the term 'interdisciplinarity' in connection with foxhunting since long before it was fashionable in many conservation circles (Macdonald 1987). The issue encapsulates how predators are viewed and valued by different interest groups, and has lessons for conflicts between wildlife and people worldwide.

Ten years after the hunting ban became law, we re-examine the relationships between foxes and people, and how they changed in the period leading to the parliamentary vote of November 2004. Our brief account of the sport's historical roots puts the social complexity of foxes' relationship with man in British agriculture into perspective. The changing legal framework surrounding foxhunting (and the treatment of wild animals in general) reflects how society as a whole balances the interests of farmers, game managers, and sportsmen against the moral acceptability of specific actions. We are able to trace some significant shifts in attitude, and to compare the reported foundation for those attitudes with the evidence for the effects thought to shape them. Our own studies include questionnaire surveys of different interest groups extending over more than 20 years, as well as modelling of the populations of foxes and other pests to gauge their economic impact. We describe how different types of foxhunts and other land managers operated in the landscape. How many foxes did they kill, when, and how? All this adds up to a uniquely detailed case history surrounding a charismatic predator with a complex and fascinating biology (which we explore in some detail in Chapter 2, this volume).

[2] The Parliament Act has its roots in class schism in the UK. The House of Lords is a secondary chamber of government; one of its principal functions has been to modify legislation arising in the principal chamber, the House of Commons. But when Prime Minister Lloyd George proposed a radical land tax in 1911, the Lords, many among them unelected landowners, were able to block its enactment indefinitely. The Parliament Act was intended to prevent constitutional disarray by preventing the Lords from wielding this power. Its creation required the personal support of the monarch. Its most recent use before 2004 was for the 2000 Sexual Offences (Amendment) Act, equalizing male and female age of consent.

3.2 A brief history of foxhunting in the UK

The significance of hunting with dogs can be appreciated only with some understanding of its history: how a hunt functions in the landscape, and in relation to other fox management methods. What is a modern hunt? The organized and formal version does not have a long history compared with hunting of other animals (Marvin 2000a). Mounted hunting as we would recognize it (Fig. 3.2) emerged in the eighteenth century as the Enclosure Acts took common land out of small holdings and converted it to parks and grazing land. The proliferation of wild rabbits at this time may have increased the carrying capacity of the landscape for foxes (Macdonald and Johnson 1996). Having once been considered an inferior beast, foxes became elite quarry. The landscape in the lowlands came to be divided into more or less continuous areas covered by foxhunting clubs (the first was the Charlton Hunt Club, established in 1837). The proliferation of hedges (Rackham 1997) led to jumping on horseback becoming integral to the sport, and the breeding of the hounds was taken very seriously. Foxhunting became one of gentlemen's essential pastimes. Donna Landry comments that the period 1730–1830 saw hunting 'constantly reinvented as a proving ground of English 'manhood'', and 'a test of bravery and an advertisement for British imperial governance' (Landry 2001). It was clearly a sport: foxes were quarry not pests for the hunters, who paid farmers to spare breeding earths. Coverts and spinneys were cultivated to provide shelter for foxes. When numbers were short, they were translocated and released. Foxes were imported from the continent and sold for 10 shillings in the Leadenhall Market in the 1840s (Macdonald 1987). At the same time that they were being preserved for sport, foxes continued to be widely killed as vermin (Lovegrove and Lovegrove 2007). The increasing popularity of shooting in the late eighteenth century created a further conflict of interest with the sport—gamekeepers did not want foxes on their 'shoots'. Nevertheless, foxhunting continued in much the same form until 2004.

What did typical foxhunts do? At the time of the hunting ban, there were 184 hunts registered with the Master of Foxhounds Association (MFHA). The hunting season was in two parts, an autumn or cub hunting phase, from August to October, with few participants, and a hunting 'proper' season from November until May. A mounted field of around 50 followers then met, typically twice a week. The hounds were controlled by a huntsman and his 'whippers in', whose job it was to prevent the hounds from straying. The hunt began with a search for a scent ('draws') in woodland or rough ground. When a scent was found the whole pack pursued it, with varying success depending on conditions. The pursuit lasted anything between a few seconds and half an hour. It ended when the fox was caught and killed by the hounds, or 'went to ground' (escaped underground). Mounted followers of the hunt did not pursue the fox or run ahead of the hounds; indeed to do so would be considered a grave breach of etiquette (Marvin 2000a). Landry (2001) speculates that many participants did not know whether hounds were hunting or not at any time during a day's hunting, and that killing foxes was a secondary outcome of the activity, necessary only to persuade farmers to allow the hunt on their land.

A fox 'gone to ground' could be dug out with the aid of terriers. It has been argued that at this point

Figure 3.2 A typical mounted foxhunt in progress—a hunter and his hounds. Photograph ©Richard Chaff/Shutterstock.

'hunting' ceased and vermin control began (Marvin 2000a). A terrier was released into the earth. In modern times, the terrier was frequently equipped with a radio-collar to aid those digging. The fox might be killed by the terrier(s), but if it was retrieved by the hunters, both law and lore required that it be dispatched humanely.

To explore, among other issues, the prevalence of digging out, a particularly contentious aspect of hunting with dogs, we circulated a questionnaire (in 1980) to all registered MFHA hunts in Britain. Eighty hunts responded, and reported considerable variability (Macdonald and Johnson 1996). For some hunts, no foxes were dug out, while others killed as many as 70% of their annual tally in this way. Part of the explanation for this variability was that some landowners tolerated the hunt only on condition that foxes were dug out. We also circulated a questionnaire to 2288 farmers in 1980, covering a range of issues relevant to wildlife; 887 (38%) were returned. Almost half of farmers reported that they hunted with hounds. Participation varied regionally, as few as 18.2% hunted in Suffolk while 79.2% in the West Country hunted (Macdonald and Johnson 2000).

In the uplands, where the terrain prevents mounted hunting, there were also organized hunts using dogs. 'Footpacks' had an October to March season, unsurprisingly covering a smaller area compared with mounted hunts, and tended to have more hounds. This form of hunting is close to the pedestrian roots of foxhunting (Landry 2001). These hunts followed a fox's scent much in the same way as the mounted hunts. Elsewhere, gunpacks used small packs of hounds, or a team of human beaters to flush foxes into the path of shotguns. Some of these hunts were in 'fox destruction societies' (Lloyd 1980). Both these methods were largely concentrated in the uplands of Wales and the north of England (Lloyd 1980). These 'organized' or communal methods were, in most places, occurring at the same time as shooting and snaring.

3.3 Hunting, the law, and the Burns Inquiry

While foxhunting, as we have described it, continued largely unchanged for around 200 years, attitudes to animals were changing. The historian Keith Thomas describes this attitudinal shift as becoming particularly marked in England in the eighteenth century: there was a developing majority view that people could use animals for food and clothing, but that it was wrong to cause unnecessary suffering (Thomas 1984). Before that, the use of animals for sport was rife throughout Britain. Bulls and bears were baited by being tied to stakes and attacked by dogs; cock fighting was also very popular through all sectors of society. At the Inner Temple on St Steven's day it was customary to bring a fox and a cat into the hall and then to set the hounds upon them. Thomas (1984) describes a wide variety of such informal cruelties, as well as the more organized conventional hunting, and the absence of any apparent moral consideration.

But from about 1740 onwards, protesting voices began to be heard. Numerous philosophical essays and pamphlets were published. Poets were influential; Blake and Shelley were regularly quoted in parliament. Donna Landry finds 'all the ingredients for a modern rejection of bloodsports in the poetry of Wordsworth' (Landry 2001)[3]. By the mid eighteenth century, there was also increasing public debate about the use of animals in medical research (Baker et al. 2012). In 1824, the Society for the Prevention of Cruelty to Animals was formed, and a series of Acts of Parliament were passed against animal cruelty. Thomas (1984) describes how these acts sought to prevent cruelty to horses and cattle (in 1822), dogs (in 1835 and 1849), also banned baiting and cock fighting (in 1835 and 1849 respectively). We don't know what caused this shift in attitudes. The doctrine of stewardship may have increasingly supplanted the idea that animals were created for the use of man, an idea (dominium) traceable back to Aristotle (Pinker 2011). With that stewardship came an obligation for the treatment of animals to be more sensitive.

In the countryside, the pursuit of wild animals for sport became harder to justify. The fox's undisputed impact on poultry and stock made it a justifiable quarry. But foxhunting did not escape opprobrium indefinitely. Thomas (1984) points to an influential article in 1869 by historian E.A. Freeman, attacking foxhunting, as starting the modern agitation. Thomas remarks that this is surprisingly late given the widespread knowledge that foxhunters, far from trying to keep fox numbers down, were carefully conserving them; fox cubs were preserved, foxes were imported from adjacent counties, and 'bagged' foxes were brought in for hunt days. As Jonathan Reynolds remarks in his introduction to a modern re-edition of T.F. Dale's 1912 monograph on the fox (Reynolds et al. 2010), other

[3] In 'Home at Grasmere', Landry comments that Wordsworth visualizes a 'bloodless future' in his dejection 'By sorrow for the hunted beast who here/Had yielded up his breath'.

forms of fox control were seen to be so effective that foxhunting for sport could not be sustained. The hunts tried to protect their sport by paying people to tolerate them (we can see in this practice a forerunner of modern compensation schemes for living with predators outside the UK).

For much of the twentieth century, legislation for animal welfare in the UK excluded wild animals (Harrop 2005). The Protection of Animals Act (passed in 1911) was restricted to domestic animals and captive wild animals (this covered the inspection of traps). Private prosecutions intended to enlarge the scope of the term 'captive' failed; Harrop attributes this to the influence of the hunting lobby[4]. The next landmark was the Scott Henderson report of 1951 (HMSO 1951). The principal conclusion of this committee was that: 'Fox hunting makes a very important contribution to the control of foxes, and involves less cruelty than most other methods of controlling them' (the other methods were shooting, gassing, trapping, and poisoning). It was recommended that foxhunting be allowed to continue. This was based on an explicitly utilitarian function, and the committee implied that hunting as a sport required such a justification: 'any field sport which has a reasonable measure of support and is a traditional activity *and which has some utilitarian value* [our emphasis] should not be interfered with'. They went on to add that interference on the grounds of 'cruelty' required the cruelty involved to be 'excessive' or 'unreasonable'. The Pests Act of 1954 made some provisions, for example on the type and placement of traps, which were motivated by animal welfare (Baker et al. 2012). The continuing change in attitudes to wild animals in the twentieth century is illustrated by some accounts of country sports in popular works of natural history. Modern readers would probably be surprised by the apparent sympathy for badger digging and baiting expressed, as late as 1958, in Ernest Neal's monograph on the badger (Neal 1958), for example. By 1975, that sympathy had evaporated (Neal 1975).

Little changed between the Scott-Henderson report and the 1990s, when a series of bills opposing hunting appeared. The Wild Mammals (Protection) Act was passed in 1996. This gave, for the first time, limited protection to all wild mammals. The act excluded hunting from its protection, with the proviso that it permitted: 'the killing in a reasonably swift and humane manner of any such mammal if he shows it had been injured or taken in the course of either lawful shooting, hunting, coursing or pest control activity'.

So, the weighing of any practical benefit of hunting against its moral acceptability clearly swung (pragmatically or slipperily according to your tastes) on two words: the rather nebulous 'reasonably' and the hugely controversial 'humane'. This was not enough to satisfy campaigners. Continuing controversy led to the Burns Inquiry—there was a clear need for all the relevant facts to be brought together and critically appraised. The Home Secretary appointed Lord Burns to chair a committee to inquire into all the facts surrounding hunting with dogs (the report was published in June 2000: Burns et al. 2000). The remit of the committee was to report on the rural economy, on pest control, and on the social and cultural life of the countryside. The issue of cruelty was central. On this, the inquiry provided some clear signals for reform. Two of these were also conspicuous in our analysis of 1996 (Macdonald and Johnson 1996). Burns indicated that the 'digging out' aspect of hunting (described in Section 3.2) was of particular welfare concern. He also looked at the possibility that drag hunting[5] might offer opportunities to replace the various social functions of the hunt (see Section 3.4.2). For a variety of reasons, he was not enthusiastic (as many hunters are not when this possibility is mooted)[6]. A compromise was in any case rejected, and the Hunting Act became law.

What exactly has been made illegal? The Act bans the hunting of mammals with one or more dogs. So much is clear. A complex series of exemptions and conditions, however, are not so clear. The exemptions are intended to separate the forms of hunting which use dogs but which are not the intended target of the legislation (Harrop 2005). The conditions for legal hunting with dogs include that the dogs should not go below ground, that no more than two dogs are used, and that reasonable steps are taken to shoot dead the quarry animal. This list appears to be devised to target foxhunting for sport, while exempting the use of dogs for other types of hunting. Harrop describes the process as 'tortuous'. The difficulty may be inevitable in trying to proscribe an action based on its motive. If foxes are killed for more than one reason, and it is clear that they are, then asking if the motive is sport or not may be meaningless. Perhaps recognizing that motivation is

[4] In one test case, a court ruled that a stranded whale was not 'captive' and that a defendant inflicting a large wound could not therefore be convicted of cruelty.

[5] A form of hunting where hounds follow a scent trail.

[6] The French essayist of the sixteenth century, Montaigne, observed that a hunt without killing was like intercourse without orgasm (in Thomas 1984, p. 146).

the crucial issue, 'The Case for Repeal' (Alliance 2008), a document prepared by the Countryside Alliance to summarize the case against the Act, makes no mention of sport and observes: 'The Act limits the options available to farmers and land managers for controlling foxes'.

3.4 Quantifying the evidence

What was the evidence synthesized by the Burns Inquiry which led to the conclusions in its report? If we (perhaps optimistically) assume that those recommendations had some influence on Members of Parliament, this was the evidence that also led indirectly to the ban on hunting with dogs. In the latter part of the inquiry, Lord Burns appointed one of us (DWM) as scientific advisor to the committee. A team led by DWM produced a report (Macdonald et al. 2000) that brought together all the scientific evidence on the methods used for fox killing, and on the effectiveness of those methods. The report dealt with all the species within Lord Burns' remit (foxes, deer, hare, and mink). Here, we consider only the evidence for foxes; particularly the social context.

3.4.1 Fox control: why and how?

In an earlier analysis (Macdonald and Johnson 1996), we drew an analogy between foxhunts and natural predators in their home ranges. The foxhunts were compared with a natural ranging predator while the hunt countries were thought of as the home range they patrolled in search of foxes. We thought this analogy might illuminate the 'behaviour' of the hunt. The Reverend Andrew Linzey (of the Oxford Centre for Animal Ethics) argued, when kindly commenting on a first draft of that work, that we had neglected a fundamental distinction between the actions of moral and nonmoral agents, and that in doing so we had implied that what takes place in foxhunting is *ethically* comparable to what takes place in nature. The error that the Reverend Linzey was alluding to is the *naturalistic fallacy* (Honderich 2005): this is the belief that because something is natural it is also in some way *good*[7]. All such comparisons are intended by us to be morally neutral. Lord Burns' remit explicitly excluded moral or ethical issues (Burns Report, p. 7); as he commented in his foreword to our submission to the inquiry (Macdonald et al. 2000), science cannot answer all the questions, but it has an important part to play. In places, though, moral and scientific arguments appear to have been conflated, and judgements about morality are likely to be swayed by the scientific evidence. The Veterinary Association for Wildlife Management (together with UK parliament 'Middle Way') produced a pamphlet comparing hunting by wild predators with that of hounds pursuing foxes (Colville 2007). The similarities between hunting of foxes by hounds and wild predators are clearly alluded to throughout in defence of hunting, and are linked with outcomes that are, at least theoretically, within the scope of science to quantify. For example, one of the conclusions in that review was that 'hunting comes nearest of all control methods to preventing the erosion of the original "wild" gene pool[8] of the quarry species, in effect maintaining biodiversity'.

The numbers of foxes killed by different methods, and the reported motivation for that killing, are clearly quantifiable aspects of fox management. Most of the primary evidence concerning foxes in the UK countryside, and how humans control them, is provided by the work of the Game & Wildlife Conservation Trust (GWCT)[9]. In particular, the 'three-region' study comparing fox management in three disparate parts of the UK, and predator removal experiments, has been illuminating (Heydon and Reynolds 2000a). In the following paragraphs we summarize some results from this study, and those from questionnaire surveys carried out by the WildCRU, which also explored patterns in attitudes and motivation for fox control (Macdonald 1984; Baker and Macdonald 2000; Macdonald and Johnson 2000).

Motivation for fox control varied regionally and predictably with prevailing land use (Fig. 3.3)—in mid Wales, 94% of farmers in the GWCT study cited protection of stock as a reason for killing foxes, while in predominantly arable Norfolk only 24% did so (Heydon and Reynolds 2000a). The WildCRU study of the early 1980s (Macdonald 1984) also recorded domestic stock protection as the most commonly cited reason for killing foxes (70% of respondents). The most cited reason in Norfolk in the GWCT study, and cited by 44.5% of farmers in the WildCRU study (Macdonald and Johnson 1996), was protection of game. Unsurprisingly, most gamekeepers

[7] This confusion has in fact been rife through a wide diversity of investigation into human behaviour; if an aspect of human behaviour is argued to be largely genetic in origin, for example, this is seen by some as legitimizing that behaviour (Pinker 2003).

[8] The author was considering genetically determined antipredator behaviour at this point.

[9] Formerly the Game Conservation Trust.

considered the fox to be a 'serious' pest (evidence to the Burns inquiry from the National Gamekeepers Association, cited in Macdonald et al. 2000, p. 25).

These perspectives are rational. Predator removal experiments carried out by the GWCT show clear impacts on game bird populations, upland waders, and brown hares (Tapper et al. 1996; Fletcher et al. 2010; Reynolds et al. 2010). In three years of heavy control of common generalist predators, including fox, partridge numbers increased 75% each year (see also Macdonald and Feber 2015: Chapter 13). Many pheasants (most of them pen-reared) are also eaten by foxes, though the impact is less well quantified. Draycott et al. (2008) report that predator control (of mainly foxes and corvids) doubled pheasant nest survival rates. Stock loss by farmers is less easy to attribute to foxes. Surveys of farmers, synthesized in Macdonald et al. (2000), suggest that most of sheep farmers believe they lose lambs to foxes, but that the numbers believed to be lost are low, probably fewer than 2%. In their 'three-region' study, Heydon and Reynolds (2000a) found that 24–61% of farmers, depending on region, thought they had lost lambs to foxes. The maximum values reported by any farmer were between 5 and 15% depending on region. These estimates are of course of losses in the presence of fox culling, and the impact of foxes on sheep farming without fox control is poorly known. Foxes, as is well known, also take poultry, and between half and three-quarters of farmers with small outdoor flocks (< 200 birds) report losses to foxes, averaging 15–50% of the flock in a year (Heydon and Reynolds 2000a). Large-scale free range flocks occur predominantly in regions where fox control is heavy. The increase in outdoor rearing of pigs has also left piglets open to fox predation—this is said by pig producers to be economically significant (Macdonald et al. 2000).

Retaliatory or preventative fox killing in reaction to these threats is widespread. The data on this all predate the ban on some methods using dogs, and vary depending on how the question to the landowner was phrased. The presence of hunting on land without the landowner's encouragement also complicates interpretation of the pattern. One study suggests about a third of farmers participated in fox control themselves (Macdonald and Johnson 1996), and about 80% of farmers surveyed in that study believed some fox control occurred on their land (where 'control' included the operation of an organized hunt). Heydon and Reynolds' (2000a) study suggested more widespread control; all their respondents on large (> 200 ha) farms in Wales and the West Midlands, and 91% in Norfolk, said some culling was practised on their land. The majority of farmers on smaller farms also reported fox control (the lowest rate was 71% in Norfolk).

Heydon and Reynolds (2000a) also reported on the contribution made by different methods to the total regional fox cull, before the hunting ban (Fig. 3.4). Rifles dominated in the East Midlands and Norfolk (contributing 64% and 53% of fox culls respectively), while in Wales, driving to standing guns and hunting on foot with hounds collectively contributed 50% of the cull. Snaring was relatively less important, contributing a maximum of 13% (in the East Midlands). A recent study reports that snares are used on 6% of landholdings in England and Wales, with about half of snare users being gamekeepers (Anon 2011b).

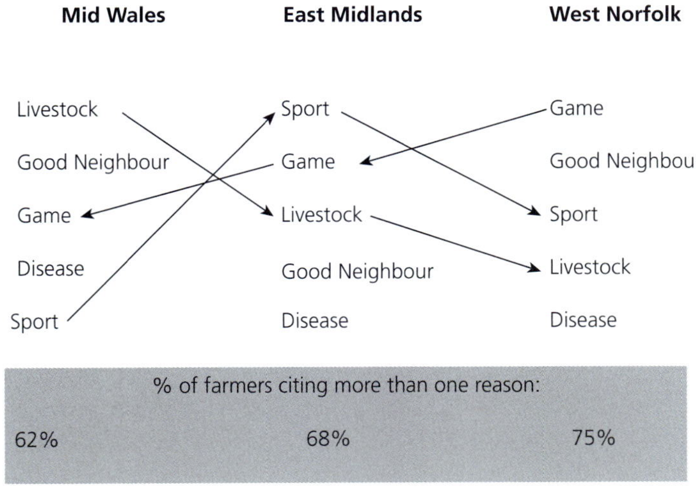

Figure 3.3 Farmers' reported regional differences in reasons for killing foxes, ranked by most cited (top). Drawn from data in Table 3 in Heydon and Reynolds (2000a).

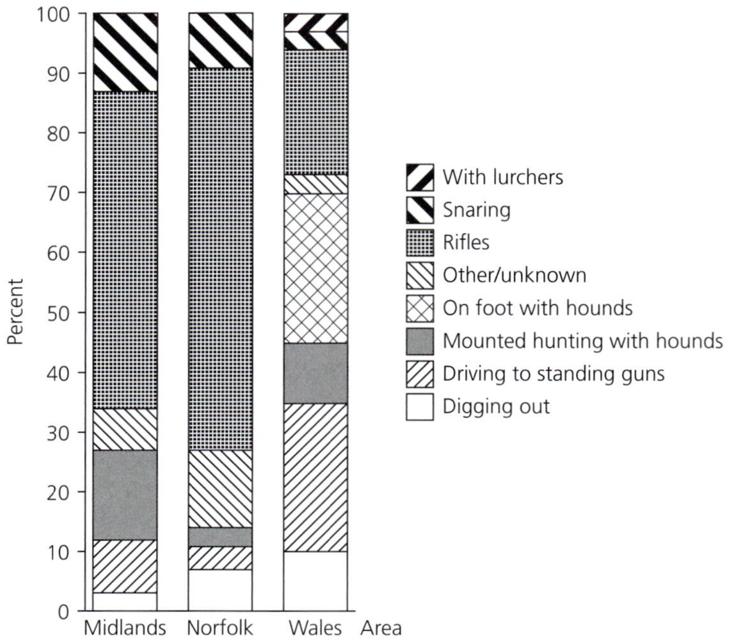

Figure 3.4 Regional variation in fox mortalities to different methods of control. Drawn from data in Table 3 in Heydon and Reynolds (2000a).

Mounted hunts were estimated to account for about 15% of the cull in the East Midlands, and only 3% in west Norfolk. Our Wiltshire study estimated that about 5% of all foxes killed on farms were accounted for by the local hunt (Baker and Macdonald 2000). In the Welsh uplands, footpacks and gunpacks accounted for a large proportion (61%) of a heavy regional cull (Heydon and Reynolds 2000a). The total number of foxes culled by the different methods was one of the elements considered by the Burns Inquiry when the net welfare effect of a ban was estimated. Fox kills by MFHA registered hunts is carefully documented: these records revealed that about 14,000–15,000 foxes were killed annually (Macdonald and Johnson 1996). The hunt kills are correlated with predictions of fox numbers based on land use (Macdonald et al. 1981), suggesting these data had some potential as an index of fox numbers—a 'tally ho' index. There was an upward trend in foxes killed by MFHA hunts between 1960 and the early 1990s (Fig. 3.5). The total killed by shooting or snaring, however, is not known with any precision, but the surveys of the GWCT (Fig. 3.4) indicate that these numbers are considerably higher.

One finding which resonates with carnivore management elsewhere, is that fox control is not necessarily motivated by personal loss. A perception that there were 'too many' foxes was a significant predictor of whether a farmer controlled foxes or not, in addition to whether the farmer believed they suffered some loss to foxes (77% of farmers recording 'significant' damage reported controlling foxes, compared with only 13% who did not; Macdonald and Johnson 2003). The belief that there were 'too many' foxes was widespread, even among farmers not affected by foxes. Farmers may kill foxes as part of a 'good neighbour' policy (Heydon and Reynolds 2000a), where regional rather than local control can be part of the aims (Heydon and Reynolds 2000b). Reported damage on the farm did not affect the likelihood that a farmer participated in mounted hunting (Macdonald and Johnson 1996). Similarly, Baker and Macdonald (2000) observed that the perceived pest status of foxes influenced the number of foxes taken by shooting, but not by mounted hunting (Fig. 3.6).

What effect did this culling have? The Norfolk fox population was suppressed below carrying capacity by game management culling (Reynolds 2000), as was the mid-Wales population by hunting methods involving hounds and terriers. The number of foxes killed in the uplands suggests a similar effect. In the mid 1990s, our analysis of some Welsh upland pack records indicates that farmer fox control societies killed more than one fox per km^2 annually, using a combination of dogs and guns, and in a terrain where fox densities are expected to be at their lowest in the UK (see Chapter 2, this volume). Over much of lowland Britain, however, fox numbers may be close to the carrying

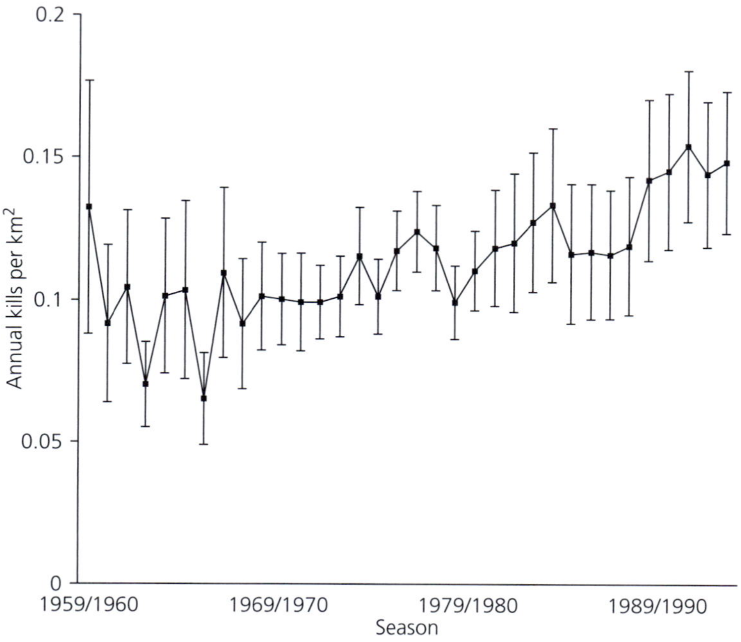

Figure 3.5 Trends in Master of Fox Hounds Association fox kills 1960–1993. From Macdonald and Johnson (1996), reproduced with permission from Elsevier.

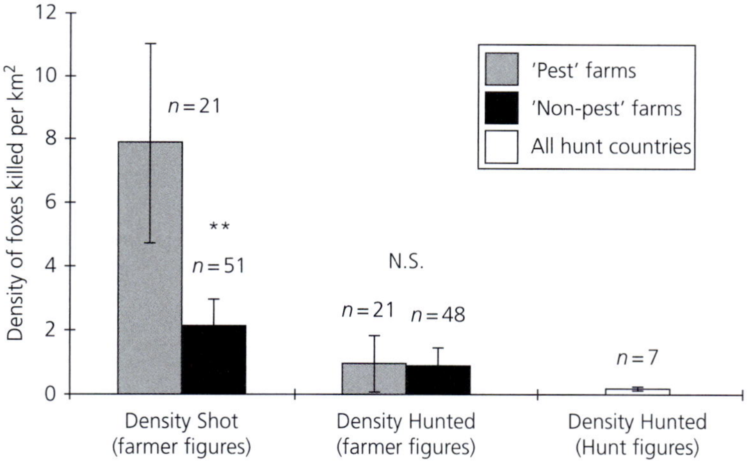

Figure 3.6 The density of foxes reported shot, and killed by hunting, on farms where the fox was, and was not, considered a pest (source: farmers' questionnaire) and the density of foxes killed by the Hunt (source: Hunt Masters' diaries). From Baker and Macdonald (2000).

capacity of the environment, and modelling suggested that mounted hunting alone would not be practically useful for protecting game or stock over a wide area (Macdonald et al. 2000; Rushton et al. 2006). Diaries of mounted hunts recorded the taking of around one fox per 10 km^2 per annum. Simple population modelling told us this was too few to have much impact on the population (Macdonald and Johnson 1996). Surveys by the GWCT (Heydon et al. 2000) estimated that post reproduction numbers in the midlands peaked at 2.62 per km^2 declining to 1.17 per km^2 in spring (close to our own estimates based only on group sizes and ranging behaviour: Macdonald and Johnson 1996).

These exercises modelled the effect at its observed rate—mounted hunts could kill substantially more foxes if they hunted for more days, or dug out more foxes. Fox kills were in proportion with hunting effort (Macdonald et al. 2000). Similarly, if all foxes run to ground were dug out and killed, efficiency could increase—we estimate that the number killed could more than triple (Macdonald et al. 2000). It is also clear that mounted foxhunting was rarely if ever the only method of fox control in place at any location. So asking what its 'solo' effect was may not be an important question. The more important question is about how it has contributed to an overall cull. This is a lot more difficult to establish, as we would need to know how many foxes killed by the hunt were killed *in addition to* those killed by other methods.

3.4.2 Costing fox control, and the hunt as an economic and social actor

Reported motives for killing foxes, together with the impact of that killing, provide a basis for an economic assessment of the practice. If the removal of the foxes prevents the loss of stock, for example, and the costs of that removal are less than the value of the stock lost, the management is cost-efficient. This neglects the value of hunting as a sport. When we first discussed this aspect of our work with the Burns Committee, one if its members (John Marsh) pointed out that neglecting the sporting aspect was more accountancy than economics in its wider sense, which implies a much wider perspective, and the costing of what economists call 'non exchange' attributes: those attributes which cannot appear in a market. For hunting this includes its value as a sport. Most of the hunt's followers are not active participants performing a function necessary for the fox to be captured. They have been likened to a moving 'gallery' of spectators, with an affinity to a particular hunt as that of football supporters to a football club (Scruton 1998).

Non-exchange values such as sport can, in theory, be estimated by using methods such as willingness-to-pay surveys. No such study has been carried out on the sporting value of hunting with hounds. Such studies are generally problematic in being based on hypothetical questions, where no real costs are at stake. A willingness-to-pay study of hunting with hounds would be further beset by bias concerning the perceived possible effect of the study by its participants. But given the emphasis on a utilitarian justification for hunting in the earlier inquiry of 1951, and in the cases advanced in defence of the practice, an accountancy exploration is justified: how much is spent removing those foxes, and what damage to stock is prevented by doing so? A full account of this exploration is presented in Macdonald et al. (2003), but broadly the conclusions are as follows.

First, what do the elements of a mounted hunt cost? The MFHA cite a range of 4–18 'hunt horses' (compared with the median field of 50 reported by Macdonald and Johnson (1996). We used a baseline of ten to cost a hunt. Taking into account the costs of buying these and stabling them, as well as other costs (hunt staff, riding equipment, transport, etc.), we estimated that somewhere in the region of £97,000 (at 2012 prices, all other amounts given here are similarly index-linked for comparability) was needed to run a typical hunt (in the Midlands). An independent estimate of a mean of £71,300 for all hunts was submitted to Lord Burns (Burns et al. 2000). A typical Midlands hunt killed around 100 foxes per season, giving a cost of around £970 per fox removed (index-linked to 2012). The second half of the equation is more difficult: what damage might thereby have been averted by those with stock within such a hunt country? Assuming a linear relationship between number of foxes and reported damage to stock, and that about a quarter of farmers are susceptible to fox damage (Heydon and Reynolds 2000b), we estimated that about £4100 (2012 price) was saved by removing these foxes. ('Damage' here was defined simply as the amount of money each farmer said that foxes cost them annually, either as lost stock or the amount spent preventing loss, or both.). Our rather back-of-envelope calculation, based on a series of approximations, is, then, that this form of fox control is not cost efficient.

If we attempt a similar exercise for a very different hunt in the uplands, a foot-pack not using horses, we arrive at a different conclusion. We looked closely at the Afonwy Hunt, which operates in the region of mid Wales studied by Heydon and Reynolds (Heydon and Reynolds 2000a,b), and for which accounts were available. We estimated each fox culled then cost *c.* £50 (£71 at 2012 prices). Taking into account the market price of lambs at the time, and again making some assumptions about the relationship between the number of foxes culled and the value of the stock they would have killed, fox control had a net benefit, and was therefore cost efficient. There are further implications: the evidence of the GWCT's three-region study was that, unlike in the Midlands, fox numbers were well below carrying capacity in this region (Reynolds 2000). So a complete cessation of hunting could lead to an increase in fox numbers and increased stock losses.

A third accountancy exercise illustrates a different point. Our surveys of landowners provided evidence that foxes were sometimes killed on land where they posed no threat to stock (because there was no stock), but also where they were possibly of some benefit to the landowner by consuming rabbits. Simulation modelling of fox and rabbit populations (Macdonald et al. 2003) showed that foxes could consume enough rabbits to make a substantial economic contribution to arable farming. Figure 3.7 illustrates how the numbers of rabbits could be influenced by foxes given one plausible set of assumptions on rabbit survivorship, and the number consumed by foxes. The effect on rabbit numbers and the prevention of damage to crops translated into savings of up to approximately £40 per hectare per year (at 2000 crop prices, equivalent to £57 in 2012).

These examples do not fully explore all the ramifications of the fox's impact in the countryside, but they illustrate that for a farmer, the economic pros and cons of controlling foxes depend on a balance of issues. That balance varies regionally, and also as the market values of crops and stock change from year to year. It is of course unlikely that individual farmers will base their day-to-day management decisions at any time on these considerations, and the wider picture is more complex.

Among the wider issues is the impact that different fox management might have on habitat conservation. In the past, foxhunting farmers removed less hedgerow, and more frequently reported encouraging other non-productive habitats, such as parkland trees and shelter belts, than did non-hunting farmers (Macdonald and Johnson 2000, 2003). For example, hunting farmers reported removing about 40% less hedgerow than non-hunting farmers in the 1970s (2.5 m/ha, SE = 0.48, compared with 5.8 m/ha, SE = 1.00). By the 1990s, when much less hedgerow was being removed, no such difference was detectable. But trends in habitat creation at that time were linked to participation in field sports—more farmers who hunted and shot (55.4%), or just hunted (39.7%) reported creating woodland than did farmers with no field sports interest (34.8%) (Macdonald and Johnson 2000). Similarly, Oldfield et al. (2003) found that landowners participating in field sports planted more new woodland and hedgerows than those who did not. Some hunts managed woodland, and some bought woodland to secure access (Macdonald et al. 2000). Lord Burns concluded nevertheless that the balance of evidence was that benefits to the environment were local, and that hunting was an increasingly minor motivation for habitat management compared with general market and policy trends (Burns et al. 2000: p. 18). The contribution to the rural economy was estimated to comprise between 6000 and 8000 full-time equivalent (FTE) jobs, 700 of which were directly employed by hunts. The rest were judged to depend on foxhunting indirectly.

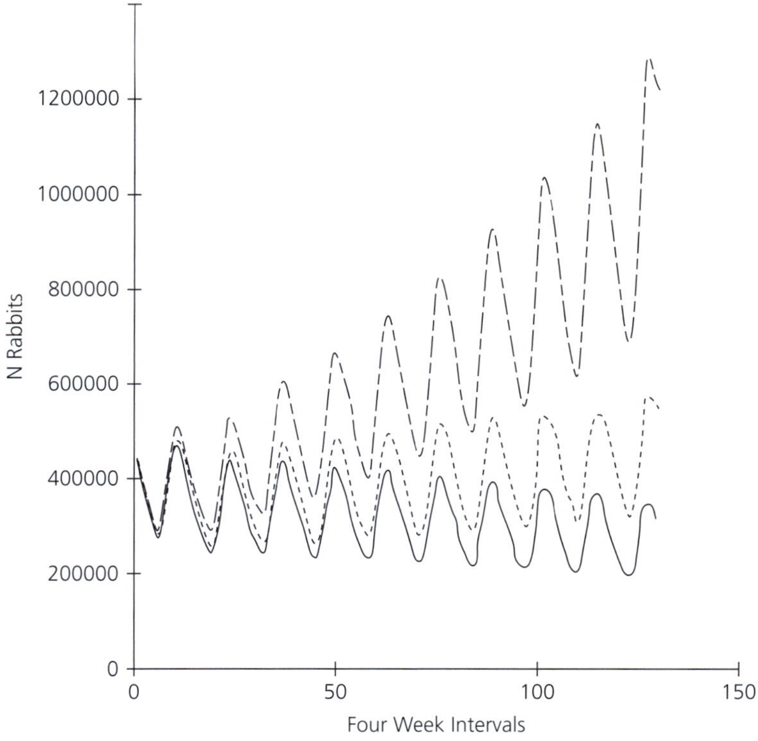

Figure 3.7 Modelling rabbit numbers with (long dashes) and without (solid line) foxes and with fertility control (short dashes), given some plausible assumptions concerning the density of both foxes and rabbits. See Macdonald et al. 2003 for the precise quantitative assumptions.

Hunting is also valued as a social activity, which is difficult to cost, and to disentangle from its value for sporting recreation[10], and its cultural significance. Its participants often report resentment of what they perceive to be uninformed interference. The anthropologist Garry Marvin has discussed foxhunting as a complex 'cultural performance' enacted within the countryside (Marvin 2000b), with some similarities with bullfighting in its ritualization. He argues that the hounds, the participants most under the control of human agency, are the central part of the hunt, not the fox. The death of the fox is viewed as an end-point in the performance, not the point of the activity.

Foxes are valued, even by farmers who control them. This was, before the hunting ban, predictably more marked among hunting farmers. Our surveys of farmers (Macdonald and Johnson 1996) reported more hunting farmers approved of fox conservation (28.9%) than did non-hunting farmers (14.7%). Even among those farmers reporting fox damage on their farm, 17.6% approved of fox conservation. Together with the numbers agreeing with the statement that foxes were 'too numerous' (65%), the evidence is that fox conservation is supported by a substantial minority of farmers. Farmers regularly report a perception of 'balance' and the idea that there is an appropriate number of foxes, somewhat below that observed, seems common.

3.4.3 Acceptability

Measuring the welfare implications of fox management is clearly more contentious than counting dead foxes or estimating the cost of damage attributable to living foxes. It takes us into the realm of animal psychology and controversial judgements about animal suffering. There is no doubt, however, that this is central to the perceived acceptability of a practice, and is at the heart of the social challenges with which fox control is confronted. Supposed 'cruelty' was a conspicuous part of the debate about hunting before the enactment of the current legislation[11]. A legal judgment of 'cruelty' based on unnecessary suffering allows for subjective judgement; as Marian Dawkins points out, the word 'cruelty' includes as part of its definition the idea that an action is morally bad (Dawkins 2012). And whether an action is 'cruel' or not depends on the reason for that action.

Acceptability was addressed in our synthesis (Macdonald et al. 2000) principally by looking at the answers to a series of questions to different interest groups about aspects of the treatment of foxes. We do not mean to imply that if a practice is reported to be 'acceptable' then that practice is legitimized in some way[12]. The patterns in these answers may tell us something about the effect of human experience of wildlife on these judgements. Two WildCRU surveys in the 1980s quantified the difference between farmers and urban populations (Macdonald and Newdick 1982; Macdonald 1984). Only 12% of urban people (N = 2100) approved of foxes being killed for sport with hounds, compared with 68% of farmers (N = 760). Fox control, in general, to protect stock, revealed less polarization; protection of stock was cited as an acceptable reason to control foxes by 49% of urban participants compared with 68% of farmers. Among urban respondents, exposure to the countryside had a lasting effect: 53% of those brought up in the countryside favoured fox control in the countryside, compared with 46% of those brought up in the city (Macdonald and Johnson 1996). When asked if the presence of 'too many' foxes justified control, background was less influential, 22.6% and 19.5% respectively agreed (a non-statistically significant difference).

Farmers consistently reported that shooting was a more humane method of fox control compared with the alternatives—49% of the Wiltshire farmers

[10] The Burns Committee commented 'We do not underestimate the importance for those who take part of the opportunities for social interaction provided by hunting'. At least for some hunts, there is anecdotal but convincing evidence that Lord Burns was underestimating this: '[Lizzie, not her real name, of the Portman Hunt], who can't resist the adrenalin rush says: "Sex is so easy on the hunt. You're close to death and drunk – I neck half a bottle of sloe gin every time I go out – so sex comes easily. Plus it's incredibly glamorous. Almost any man in hunting clothes looks good on a horse"'. The Observer magazine (16/10/11).

[11] Welfare ethics, is increasingly prominent whether the quarry are leopards, lions, or foxes. Lindsey et al. (2006) commented on the increasing resistance to killing of animals for sport among urban populations of the developed world. As for foxes, where welfare is considered, the method of culling then comes under scrutiny, and the use of dogs attracts particular opprobrium. The use of dogs for hunting in the USA has generated a debate with striking parallels with UK foxhunting, surrounding utility and welfare; in 2012, the hunting of bears and bobcats with dogs was banned—opponents argued that hunting was not humane while its supporters argued that animals killed instead by shooting were more likely to be wounded. As for the UK foxhounds, the fate of the 22,000 hunting hounds after a ban was prominent in the controversy.

[12] Our earlier critic, the Reverend Linzey, commented 'common approval is neither here nor there when it comes to morality'.

surveyed in 1995 considered it to be both effective and humane (Baker and Macdonald 2000). Snaring was thought to be humane by fewest farmers, only 1.4% of Wiltshire farmers in 1995[13]. Experience of fox damage was influential; higher proportions of farmers who reported that they had experienced damage by foxes said all the methods were more humane, compared with those who had not.

When judging under what circumstances it is acceptable to use lethal force to kill foxes or other predators, conservationists themselves occasionally have to make awkward decisions. The acceptability of killing foxes in the UK to protect terns (rare native species) is judged differently compared with that of gamekeepers killing foxes to protect pheasants (a superabundant introduced alien). This judgement is often based on the conservation status of the prey species[14]. In Norfolk, fox predation of coastal birds has been an increasing problem and reserve-owning conservation bodies carry out commissioned fox culling (Reynolds 1998). This has usually been a reluctant and controversial option (Macdonald et al. 2000).

Are there any hard facts on the relative welfare merits of different methods of killing foxes? Few data on the physiological effects of foxhunting were available to the Burns Inquiry to assess the humaneness of foxhunting. These led to some tentative conclusions. A single study which attracted disproportionate attention showed (hardly surprisingly) increased heart and metabolic rates of foxes chased in captivity (Kreeger cited in Macdonald et al. 2000). The significance of elevated heart and respiratory rates during the chase was contentious. It was concluded by Lord Burns, however, in an unusual phrase, the meaning of which has been much debated, that both the chase phases, and the practice of 'digging out', 'compromised the welfare of' the foxes. Fragmentary observations and post-mortem data suggested that, while the oft-repeated claim that the fox was despatched by a single bite from the lead hound was false, nonetheless the time to insensibility after the chase was 'short'. These assessments tended to match those of farmers affected by foxes, rather fewer of whom (34.8%) judged terrier work humane, compared with the use of hounds (59.3%).

Wounding by shooting was particularly contentious. Evidence from a variety of sources assessed by Bateson and Bradshaw (1999) suggested a wounding rate for trained stalkers, shooting red deer, of around 2% of those shot that escape wounded. More may be wounded and dispatched after further shots. The relevance of these observations for foxes is debatable. Several strands of evidence (summarized here from Jonathan Reynolds' account in Macdonald et al. 2000) suggest that the risk of wounding is greater for foxes. With a rifle, the target required to effect a rapid kill is small; at 100 m the main body of a fox subtends an angle of about 0.23 degrees, implying that a hand movement of the order of 3 mm sweeps the whole body length of the fox (if presented side-on). The target areas for a clean kill are much smaller than the fox's body length, of the order of 4 cm or so in the head region, and 7 cm in the thoracic region (Bucknell 2008). But because most soft-nosed friable bullets are the norm, a hit outside the 'target' area may cause such extensive wounding as to be rapidly fatal. How often is hard to estimate. There are too many unknowns for the acceptability of rifle shooting to be meaningfully compared with alternative cull methods.

The evidence is no clearer for shotguns: they may be effective at short range, but this diminishes rapidly with distance and the dispersal of the shot. An experimental study of wounding rates using targets and assessment of the area hit (Fox et al. 2005) suggests serious wounding rates could be close to 50%. The authors acknowledged that the 'real' range of shooting conditions is unknown, and detail of their protocols and conclusions have been debated (Baker and Harris 2005, 2006; GCT 2006). Wounding frequency of foxes shot in the wild is known only from a single study in rural Denmark: Bertsden (1999), cited by Fox et al. (2005) reported that 25% of culled foxes carried shotgun pellets from previous injuries.

It seems we cannot make an informed judgement on the welfare of shooting with either guns or rifles. This is a considerable knowledge gap. The relative welfare merits of shooting versus hunting are particularly important if shooting were to increase in response to the ban—as has been mooted (e.g. by Pye-Smith 2006). There is currently no evidence on this either way.

[13] Snaring meets the standard set by the Agreement on International Humane Trapping Standards (Anon. 2011b)

[14] The two partridge species in the UK provide an interesting example of two closely related species where management is carried out with rather different decision-making frameworks. The red-legged partridge is, like the pheasant, invasive and widely shot for sport. The grey partridge is also a gamebird, but is native; it is moreover a BAP species which is being actively reintroduced to parts of the UK (Rantanen et al. 2010; Macdonald and Feber 2015: Chapter 13). The issue of native or invasive has a significant impact, and can also create different priorities for the predator species—control of the red fox in Australia, where it is invasive, is more likely to be considered acceptable by conservationists than it is within its natural range.

3.5 Beyond the hunting ban

The 'beyond' here can be interpreted at two levels. Parochially, we can ask what impact has the illegalization of foxhunting had on the UK landscape. But, and perhaps of more lasting interest, we can look 'beyond' foxhunting in the UK to see how it illuminates general characteristics of sport hunting worldwide, particularly where dogs are used. One of the prominent questions discussed while the ban was being considered was what would be its impact on foxes and on rural society? Can we now answer this question? The short answer is 'no'. At the time of writing, the effect of the ban on attitudes and actions with respect to foxes in the agricultural landscape, and more widely, is unclear, despite the passing of several years. Its enactment has not provided the before–after experiment in social perturbation that once seemed possible. Some of the more apocalyptic predictions of the ban's opponents have clearly not come to pass. All 184 MFHA hunts are still active in the field, participants claim to be exercising horses and dogs, sometimes following artificial trails. There has therefore been little or no unemployment or other secondary economic impacts of the ban, nor any slaughter of horses or hounds. National Hunt racing has been unaffected. The social function of the hunts continues as before. The police have acknowledged the difficulty in policing the Act, particularly with respect to identifying the conditions whereby hunting using dogs remains legal. There have, though, been prosecutions. Approximately 280 successful prosecutions had taken place up to August 2013[15]. Defra (2012) reported that many of these convictions were for offences such as rabbiting without the landowner's permission.

Few of these convictions appear to have involved MFHA hunts. Three members of the Sussex based Crawley and Horsham Hunt were found guilty of Hunting Act offences at Haywards Heath Magistrates court in May 2012: a judge was unconvinced that film of hounds chasing a fox was of an 'accidental' event which was incidental to lawful trail hunting[16]. In December 2012 the RSPCA brought a successful prosecution of individuals of the Heythrop Hunt, based on video evidence of illegal hunting (BBC 2012). In August 2013, four members of the Middleton Hunt, a registered hunt based in north Yorkshire, pleaded guilty to offences under Section 1 of the Act at York Magistrates Court, again based on video evidence, supplied by the League Against Cruel Sports. What are social attitudes to these convictions? We do not know whether convictions like this are reputationally damaging, or if they are worn as badges of honour in what is perceived to be a battle against unjust legislation. Supporters of the ban report a reduction in publicity for hunt meets, but little evidence for widespread law breaking (Anon. 2011a).

Another predicted consequence of a ban was that the 'culture of tolerance', whereby farmers and other land managers were restrained in the interests of the hunt, would be disrupted. It was suggested that the removal of this restraint would lead to an increase in the use of other cull methods, and with the perverse risk for the ban's supporters that the net effect on welfare could be negative. Again, there is no evidence on this either way and nor is there is any evidence for any impact on fox numbers nationally: both The National Game Bag Census of the GWCT and the mammal data collected as part of the Breeding Bird Survey of the British Trust for Ornithology show that fox numbers were stable between *c.* 1995 and 2010[17,18].

The illegality of hunting may satisfy some of those who supported a ban. Others are more actively seeking enforcement. There is suspicion that beyond the sight of any onlookers, it is 'business as usual' for many hunts. Little is known on how the foxhunters have changed their behaviour to stay within the law. The issue is not dead: at the time of writing the majority Conservative Party in the UK coalition government has promised a free vote on the repeal of the hunting act should they be returned to government. Are hunters keeping the hunting infrastructure 'ticking over' in the hope that the ban will be reversed, or are they seeking a more permanent long-term transformation? And is the 'culture of tolerance' a genuine phenomenon allowing foxes to persist where they might otherwise be exterminated, but only while farmers and other land managers believe a repeal of the ban is possible?

Taking a wider perspective, how does our overview of the social, economic, and welfare aspects of foxhunting inform predator conflict issues more generally? For many landowners, negative attitudes to predators are rational, and the work of the GWCT has shown this for UK landowners depending on pheasant and partridge shooting where foxes are allowed to thrive. At the same time, predator control concerns much more than the balance sheet for agricultural production,

[15] <http://huntingact.org/?q=node/18>.

[16] <http://www.huntingact.org/ (09/07/2012)>.

[17] <http://www.gwct.org.uk/>.

[18] Aebischer et al. (2003) argued that claims that the suspension of hunting provoked by the foot-and-mouth epidemic of 2000 was informative for assessing the effect of hunting on fox numbers were seriously flawed.

and conservation concerns more than rarity. In cities, some people value the presence of foxes in their gardens while others pay pest control professionals to remove them. We suspect that urban attitudes are volatile. An opinion poll in August 2010 reported that almost a fifth of respondents were 'afraid of wildlife', following unconfirmed reports of a fox attack on two nine-month-old children in June of that year (The Telegraph, London, 02/10/2010). In 2013, another fox attack on an infant attracted much publicity (<http://www.bbc.co.uk/news/uk-21399709>). While difficult to quantify, the reaction in correspondence and phone-ins suggested public perception of the risk of such incidents was disproportionately high.

In both continental Europe and the USA, where wolf numbers are recovering, public attitudes to predators are divided by a rural/urban schism, similar to that for foxes in the UK: 'the farther away people live from wolves, the more they like them' writes one correspondent of the recent emphasis on wolf conservation (Anon. 2012). But rural attitudes are not homogenous either. Our surveys of UK farmers revealed support for fox conservation, even among non-hunting farmers and by farmers suffering fox damage (our observation that attitudes of people living in cities but brought up in the countryside were more similar to the rural view than the urban, also suggests that rural attitudes are less volatile). Landowners seem to have a perception of a tolerable number of foxes. So even if they regularly report that there are 'too many' foxes, they do not wish foxes to be absent from the landscape and are, in that sense, in favour of fox conservation. Some people have a more clearly identifiable rational interest in the presence of foxes: hunting farmers were, unsurprisingly, more likely to support fox conservation, and during the Burns Inquiry the social aspect of hunting was prominent among its defenders. Parallels elsewhere include jaguar hunting in Brazil, which is often linked with threat to stock (or even to humans), but is also thought by some participants to be a cultural activity worthy of preservation (Cavalcanti et al. 2010)[19]. As for foxes in the UK, where there is peer pressure to kill foxes as part of a 'good neighbour' policy, jaguar killing is endorsed by one's peers.

The UK debate about foxhunting is essentially a prolonged national argument concerning the question 'under what conditions can killing "pests" be tolerated?' When, for the purposes of the Burns Inquiry, we addressed a more limited question about the necessary conditions for control to be considered rational, we suggested that the answers to three questions needed to be 'yes'. These were: first, does the stakeholder suffer loss to the pest? second, does the action taken reduce that loss; and finally, is the action cost effective? Mounted hunting may not meet this standard, and nor does fox control by any means for some arable farmers who might benefit from fox presence. The limitation of this perspective is that it is based on simplistic accountancy rendered in pounds, shillings, and pence, and is also limited to the judgement of a single stakeholder. A verdict on rationality might be different if we account for reputational damage to a farmer who opts out of the 'good neighbour' policy, or for the value of hunting as a sport. Non-market goods provided by predators, including sport can, at least in theory, be valued. And as we have seen for foxes in the UK, evaluation by society as a whole depends on divergent assessments of wider issues; the value of the predator for tourists (including users of the countryside in the UK), and how acceptable the practice is for animal welfare, for example. Foxes in the UK do not need preservation, but on a local scale, their conservation may be an issue requiring decisions about priorities. If the killing of foxes on reserves to preserve rare birds is more acceptable than the killing of foxes to protect pheasants, the decision is based on a ranking of the conservation merits of terns versus pheasants.

Where predator killing by landowners is motivated principally by threats to stock, avoiding conflict by better or novel stock husbandry can be promoted as an option, whether the problem is foxes or lions (see examples in Macdonald and Loveridge 2010). But where it is not, behavioural change requires an interweaving of education, incentivization, and regulation (Macdonald 2013). Killing of lions by herdsmen in Botswana illustrates a link between financial incentives and education (Hemson et al. 2009). Local tourism benefits from the presence of lions, but herdsmen who do not see any of the income from tourism and therefore continue, rationally, to kill the lions that take their livestock. But herdsmen who were aware of tourism were more likely to value lions (Hemson et al. 2009), in the same way that some arable farmers in the UK might be swayed by knowledge of the amount of money foxes could save them. In Scotland, fish farmers tolerate losses to ospreys *Pandion haliaetus* because wildlife enthusiasts will pay to watch these birds fish. Macdonald et al. (2010) and Dickman et al. (2011) show how economic instruments are used to conserve endangered felids.

The framework for the behavioural economics of predator killing can be changed by longer-term

[19] Dogs are important for jaguar hunting in the Pantanal.

education. In the Pantanal in Brazil, persuading jaguars not to eat cattle is largely futile. Instead, children have been targeted via activities in school (Marchini 2010); the hope is that they will learn to value jaguars, both influencing their parents and changing later societal judgement about the acceptability of culling. There is reason to believe that predator tolerance can be increased by education, even if depredation is not reduced (Marker et al. 2003). And education can be delivered effectively beyond the classroom. In Zimbabwe, WildCRU's lion theatre group has performed for thousands of children (Telford 2006). Attitudes to fox culling, particularly among urban populations have almost certainly been shaped by foxes' depiction in popular culture.

The judgment of government in the wake of the Burns report was, ultimately, that regulation was appropriate to ban hunting with dogs. It is too early to know if this regulation can be judged to be effective, and opinions on how to make that judgement will be diverse. We can draw a parallel with medical ethics, where, when a change in behaviour is desired (such as less smoking), coercion is considered to be both less desirable and less effective compared with persuasion[20]. Methods for changing human behaviour increasingly exploit the irrational aspects of human decision making that have been uncovered by psychologists and economists. Conservation of predators in conflict with humans may have lessons to learn there—'nudge' type policy tools lying between persuasion and coercion are being explored by policy makers for reasons of public health (Dolan et al. 2013). It may transpire that the blanket ban on hunting with dogs was too simplistic a measure to tackle the complexities of foxes and foxhunting in UK farmland. Changing attitudes and actions to any wildlife problem requires a thorough understanding of what is driving human decision making (St John et al. 2013).

Acknowledgements

The authors would like to thank Jonathan Reynolds and Amy Dickman for very constructive comments on earlier drafts of this chapter. They also acknowledge the contribution of Jonathan Reynolds, Fran Tattersall, Chris Carbone, Steve Rushton, Mark Shirley, and Jochen Langbein to the work carried out for our contribution to the Burns Inquiry, which we draw on here. Paul Johnson acknowledges the support of the Whitley Trust. Ruth Feber and Eva Raebel provided meticulous and valuable editorial input.

[20] Professor Richard Ashcroft, presentation 'Behaviour change: moving the debate on from coercion', Oxford Martin School Oxford 31/01/2012.

References

Aebischer, N., Baker, S.E., Johnson, P.J., Macdonald, D.W., and Reynolds, J.C. (2003). Hunting and fox numbers in the United Kingdom. *Nature*, **423**, 400.

Alliance, C. (2008). *The Case for Repeal*. Countryside Alliance, p. 16.

Anon. (2011a). *The 2010/2011 Hunting Season*. Goldaming.

Anon. (2011b). *Determining the extent of use and humaneness of snares in England and Wales*. Game and Wildlife Conservation Trust.

Anon. (2012). Call of the Wild. *The Economist*, 22 December.

Baker, S.E., Ellwood, S.A., Tagarielli, V.L., and Macdonald, D.W. (2012). Mechanical performance of rat, mouse and mole spring traps, and possible implications for welfare performance. *PLOS ONE*, **7**.

Baker, P.J. and Harris, S. (2005). Shooting in the dark. *Animal Welfare*, **14**, 275–276.

Baker, P.J. and Harris, S. (2006). Still shooting in the dark. *Animal Welfare*, **15**, 89–90.

Baker, S.E. and Macdonald, D.W. (2000). Foxes and foxhunting on farms in Wiltshire: a case study. *Journal of Rural Studies*, **16**, 185–201.

Bateson, P. and Bradshaw, E.L. (1999). How often do stalkers wound red deer? *Deer*, **11**, 180181.

BBC (2012). <http://news.bbc.co.uk/>. Accessed online 21/12/2012.

Bertsden, J. (1999). *Shooting of Game/Crippling/Wounding*. The Ministry of Environment and Energy, Denmark.

Blair, T. (2011). *A Journey*. Arrow, London.

Bucknell, R.W. (2008). *Foxing with Lamp and Rifle*. Foxearth, Chelmsford.

Burns, T., Edwards, V., Marsh, J., Soulsby, L., and Winter, M. (2000). *Report of the Committee of Inquiry into Hunting With Dogs in England and Wales*, CM 4763. The Stationary Office, Norwich.

Cavalcanti, S.M.C., Marchini, S., Zimmermann, A., Gese, E.M., and Macdonald, D.W. (2010). Jaguars, livestock, and people in Brazil: realities and perceptions behind the conflict. In D.W. Macdonald and A.J. Loveridge, eds., *The Biology and Conservation of Wild Felids*, pp. 383–402. Oxford University Press, Oxford.

Colville, K. (2007). *The Natural Chase*. In All Parliamentary Middle Way Group/Veterinary Association for Wildlife Management. Askew Design.

Committee of Inquiry into Hunting With Dogs in England and Wales. (2000). *Report of the Committee of Inquiry into Hunting With Dogs in England and Wales*. Stationery Office, London.

Dawkins, M.S. (2012). *Why Animals Matter: Animal Consciousness, Animal Welfare, and Human Well-being*. Oxford University Press, Oxford.

Defra (2012). *Hunting with Dogs*. <http://www.defra.gov.uk/rural/countryside/hunting/>. Accessed online 20/12/2012.

Dickman, A.J., Macdonald, E.A., and Macdonald, D.W. (2011). A review of financial instruments to pay for predator conservation and encourage human–carnivore coexistence. *PNAS*, **108**, 13937–13944.

Dolan, P., Hailsworth, M., Halpern, D., King, D., and Vlaev, I. (2013). *MINDSPACE*. The Cabinet Office. <http://www.instituteforgovernment.org.uk/our-work/better-policy-making/mindspace-behavioural-economics>. Accessed online 18/07/2013.

Draycott, R.A.H., Hoodless, A.N., Woodburn, M.I.A., and Sage, R.B. (2008). Nest predation of common pheasants Phasianus colchicus. *Ibis*, **150**, 37–44.

Fletcher, K., Aebischer, N.J., Baines, D., Foster, R., and Hoodless, A.N. (2010). Changes in breeding success and abundance of ground-nesting moorland birds in relation to the experimental deployment of legal predator control. *Journal of Applied Ecology*, **47**, 263–272.

Fox, N.C., Blay, N., Greenwood, A.G., Wise, D., and Potapov, E. (2005). Wounding rates in shooting foxes (Vulpes vulpes). *Animal Welfare*, **14**, 93–102.

GCT (2006). *Shooting as a potential tool in badger population control*. Defra.

Harrop, S.R. (2005). The Hunting Act 2004—cruelty, countryside, conservation, culture or a class act. *Environmental Law Review*, **7**, 201–207.

Harrop, S.R. (2006). Parliamentary process, opposition in the House of Lords and the Hunting Act 2004—Regina (Jackson and others) v Attorney General *Environmental Law Review*, **8**, 299–305.

Hemson, G., Maclennan, S.D., Mills, G., Johnson, P.J., and Macdonald, D.W. (2009). Community, lions, livestock and money: A spatial and social analysis of attitudes to wildlife and the conservation value of tourism in a human-carnivore conflict in Botswana. *Biological Conservation*, **142**, 2718.

Heydon, M.J. and Reynolds, J.C. (2000a). Fox (Vulpes vulpes) management in three contrasting regions of Britain, in relation to agricultural and sporting interests. *Journal of Zoology*, **251**, 237–252.

Heydon, M.J. and Reynolds, J.C. (2000b). Demography of rural foxes (Vulpes vulpes) in relation to cull intensity in three contrasting regions of Britain. *Journal of Zoology*, **251**, 265–276.

Heydon, M.J., Reynolds, J.C., and Short, M.J. (2000). Variation in abundance of foxes (Vulpes vulpes) between three regions of rural Britain, in relation to landscape and other variables. *Journal of Zoology*, **251**, 253–264.

HMSO (1951). *Report of the Committee on Cruelty to Wild Animals*.

Honderich, T. (2005). *The Oxford Companion to Philosophy*. Oxford University Press, Oxford.

Landry, D. (2001). *The Invention of the Countryside: Hunting, Walking, and Ecology in English Literature, 1671–1831*. Palgrave, Basingstoke.

Lindsey, P.A., Alexander, R., Frank, L.G., Mathieson, A., and Romanach, S.S. (2006). Potential of trophy hunting to create incentives for wildlife conservation in Africa where alternative wildlife-based land uses may not be viable. *Animal Conservation*, **9**, 283–291.

Lloyd, H.G. (1980). *The Red Fox*. Batsford, London.

Lovegrove, R. and Lovegrove, R. (2007). *Silent Fields: the Long Decline of a Nation's Wildlife*. Oxford University Press, Oxford.

Macdonald, D.W. (1984). A questionnaire survey of farmers' opinions and actions towards wildlife on farmlands. In D. Jenkins, ed. *ITE Symposium No. 13: Agriculture and the environment*, pp. 171–177.

Macdonald, D.W. (1987). *Running with the Fox*. Unwin Hyman, London.

Macdonald, D. W. (2013). From ethology to biodiversity: case studies of wildlife conservation. *Nova Acta Leopoldina*, **111**(380), 111–156.

Macdonald, D.W., Bunce, R.G.H., and Bacon, P.J. (1981). Fox populations, habitat characterization and rabies control. *Journal of Biogeography*, **8**, 145–151.

Macdonald, D.W. and Feber, R.E., eds. (2015). *Wildlife Conservation on Farmland. Managing for Nature on Lowland Farms*. Oxford University Press, Oxford.

Macdonald, D.W. and Johnson, P.J. (1996). The impact of sport hunting: a case study. In N. Dunstone and V.L. Taylor, eds., *The Exploitation of Mammal Populations: 1994 Symposium of the Zoological Society of London*, pp. 160–207. Chapman & Hall, London.

Macdonald, D.W. and Johnson, P.J. (2000). Farmers and the custody of the countryside: trends in loss and conservation of non-productive habitats 1981–1998. *Biological Conservation*, **94**, 221–234.

Macdonald, D.W. and Johnson, P.J. (2003). Farmers as conservation custodians: links between perception and practice. In F.H. Tattersall and W.J. Manley, eds., *Conservation and Conflict: Mammals and Farming in Britain*, pp. 2–16. Linnean Society Occasional Publication, Westbury Publishing, Yorkshire.

Macdonald, D.W. and Loveridge, A.J., eds. (2010). *The Biology and Conservation of Wild Felids*. Oxford University Press, Oxford.

Macdonald, D.W., Loveridge, A.J., and Rabinowitz, A. (2010). Felid futures: crossing disciplines, borders and generations. In D.W. Macdonald and A.J. Loveridge, eds., *The Biology and Conservation of Wild Felids*, Oxford University Press, Oxford.

Macdonald, D.W. and Newdick, M.T. (1982). The distribution and ecology of foxes, *Vulpes vulpes* (L.), in urban areas. In R. Bornkamm, J.A. Lee, and M.R.D. Seaward, eds., *Urban Ecology*, pp. 123–135. Blackwell Scientific Publications, Oxford.

Macdonald, D.W., Reynolds, J.C., Carbone, C., Mathews, F., and Johnson, P.J. (2003). The bioeconomics of fox control. In F.H. Tattersall and W.J. Manley, eds., *Conservation and Conflict: Mammals and Farming in Britain*, pp. 220–236. Linnean Society Occasional Publication, Westbury Publishing, Yorkshire.

Macdonald, D.W., Tattersall, F.H., Johnson, P.J., et al. (2000). *Managing British Mammals: Case Studies from the Hunting Debate*. WildCRU, Oxford.

Marchini, S. (2010). *Human dimensions of the conflicts between people and jaguars (Panthera onca) in Brazil*. Thesis, University of Oxford. DPhil. Access 2010.

Marker, L.L., Mills, M.G.L., and Macdonald, D.W. (2003). Factors influencing perceptions and tolerance toward cheetahs on Namibian farmlands. *Conservation Biology*, **17**, 1290–1298.

Marvin, G. (2000a). The problem of foxes. In J. Knight, ed. *Natural Enemies. People-Wildlife Conflicts in Anthropological Perspective*, pp. 189–211. Routledge.

Marvin, G. (2000b). Natural instincts and cultural passions. *Performance Research*, **5**, 108–115.

Mullin, C. and Winstone, R. (2009). *A View from the Foothills*. Profile, London.

Neal, E.G. (1958). *The Badger*. Penguin Books, Harmondsworth.

Neal, E.G. (1975). *The Badger*. Collins, London.

Oldfield, T.E.E., Smith, R.J., Harrop, S.R., and Leader-Williams, N. (2003). Field sports and conservation in the United Kingdom. *Nature*, **423**, 531–533.

Pinker, S. (2003). *The Blank Slate: the Modern Denial of Human Nature*. Penguin, London.

Pinker, S. (2011). *The Better Angels of our Nature: the Decline of Violence in History and its Causes*. Allen Lane, London.

Pye-Smith, C. (2006). *Rural rites hunting and the politics of prejudice*. Parliamentary Middle Way Group.

Rackham, O. (1997). *The History of the Countryside*. Phoenix Giant, London.

Rantanen, E.M., Buner, F., Riordan, P., Sotherton, N., and Macdonald, D.W. (2010). Habitat preferences and survival in wildlife reintroductions: an ecological trap in reintroduced grey partridges. *Journal of Applied Ecology*, **47**, 1357–1364.

Reynolds, J.R. (1998). Foxes and wildlife tourism:. *The Game Conservancy Review*, **29**, 86–91.

Reynolds, J.C. (2000). *Fox Control in the Countryside*. Game Conservancy Trust, Fordingbridge.

Reynolds, J.C., Stoate, C., Brockless, M.H., Aebischer, N.J., and Tapper, S.C. (2010). The consequences of predator control for brown hares (Lepus europaeus) on UK farmland. *European Journal of Wildlife Research*, **56**, 541–549.

Rushton, S.P., Shirley, M.D.F., Macdonald, D.W., and Reynolds, J.C. (2006). Effects of culling fox populations at the landscape scale: a spatially explicit population modelling approach. *Journal of Wildlife Management*, **70**, 1102–1110.

Scruton, R. (1998). *On Hunting*. Yellow Jersey, London.

St John, F.A.V., Keane, A.M., and Milner-Gulland, E.J. (2013). Effective conservation depends upon understanding human behaviour. In D.W. Macdonald and K.J. Willis, eds., *Key Topics in Conservation Biology 2*, pp. 344–361. Wiley-Blackwell, John Wiley & Sons Ltd., Oxford.

Tapper, S.C., Potts, G.R., and Brockless, M.H. (1996). The effect of an experimental reduction in predation pressure on the breeding success and population density of grey partridges Perdix perdix. *Journal of Applied Ecology*, **33**, 965–978.

Telford, J. (2006). *Study of the impact of the Ingonyama Theatre Group on Conservation awareness in the Hwange area*. WildCRU.

Thomas, K. (1984). *Man and the Natural World: Changing Attitudes in England, 1500–1800*. Penguin, London.

CHAPTER 4

Badgers in the rural landscape—conservation paragon or farmland pariah? Lessons from the Wytham Badger Project

David W. Macdonald, Chris Newman, and Christina D. Buesching

> Society exists only as a mental concept: in the real world there are only individuals.
>
> **Oscar Wilde, 1854–1900**

4.1 Introduction

Understanding badger behaviour and ecology is not only fundamentally interesting, but also contributes to resolving the conflicting imperatives of conservation or control. Reconciling the badger's ecological and political fortunes is a challenge so anguished that it has riven public opinion and faced the British government with an agonizing dilemma.

As the symbol of the UK's Wildlife Trusts, the badger *Meles meles* is emblematic of the British countryside (Fig. 4.1), but their vulnerability to reckless disturbance, or even deliberate persecution, has led to them being one of the most protected species in Britain since 1973 (Badger Act 1973; Schedule 5 of the Wildlife and Countryside Act 1981; Protection of Badgers Act 1992). Nevertheless, up to 10,000 badgers per year were still being killed annually through illegal digging of setts and badger baiting in the 1970s and 1980s (Cresswell et al. 1990). Even with the added protection afforded by the 1992[1] legislation, the illegal persecution of badgers continues[2].

[1] The Badger Act 1992 lists as offences: (1) Taking, injuring, or killing badgers; (2) Cruelty; (3) Interfering with badger setts; (4) Selling and possession of live badgers; (5) Marking and ringing (except under licence).

[2] e.g. Mooney, J. (2009). Dogs seized in police raids on badger baiters. *Sunday Times*, 22 February, retrieved from <http://www.thesundaytimes.co.uk>; Farming News 2012. Rise in badger persecution follows cull postponement. *Farming News*, 9 November, retrieved from <http://www.farming.co.uk>; BBC News 2013. Badger baiting raids: three men arrested and dogs seized. *BBC News*, 7 March, retrieved from <http://www.bbc.co.uk>.

While clearly warranting protection, badger conservation is complicated by their conflicts with people over crop damage and transmission of bovine tuberculosis to cattle (see Chapter 5, this volume). Here, we provide the biological framework within which to consider how these issues impact this farmland mammal. Using science not only to enthral but to inform, thereby underpinning policy: these have ultimately been the aims of our now 27-year study of badgers in Wytham Woods.

To avoid reinventing the wheel, it is prudent to check the literature—Giraldus Cambrensis reported to King Henry II in 1187 (Topographica Hibernica; see Wright 1913) that:

> There is an animal called the badger. It bites and is unclean. It makes holes underground by scraping and digging with its feet. Some badgers are born to be servants of others. They lie on their backs and pile on their bellies soil, which the others have dug out. Then they hold a piece of wood in their mouth and clutch the soil with their four feet. The others sink their teeth into the wood and drag them backwards out of their holes, to the astonishment of anyone who sees it.

So, key facts about badgers include that they bite, they carry parasites, and they dig and live underground. Giraldus also highlighted their propensity for group living (a point missed by Beatrix Potter in *The Tale of Mr. Tod* (1912), who observed that 'badger was solitary, grumpy and had no friends'). Indeed, as we shall elaborate, an extraordinary facet of badger biology is that,

Wildlife Conservation on Farmland. Conflict in the Countryside. Edited by David W. Macdonald and Ruth E. Feber.
© Oxford University Press 2015. Published 2015 by Oxford University Press.

Figure 4.1 David Macdonald and Chris Newman release a badger at its sett. Photograph ©A.L. Harrington.

at high densities, their social organization switches from the largely solitary, intra-sexual territoriality typical of the family Mustelidae (Powell 1979; Sandell 1989; Newman et al. 2011), to conspicuous group living. As we shall see, the formation of these social groups, or 'clans', is seemingly linked to the dispersion and abundance of food. Thus while they forage alone for their assorted omnivorous diet (Kruuk 1978a; Kilshaw et al. 2009), badgers living at high densities (in excess of ten badgers/km^2) on lowland British farmland (and Ireland—Sleeman et al. 2009) live socially, albeit in a society that is intriguingly different to that of any other carnivore, not least because co-operation among badgers, while strictly reciprocal, is fleeting (Macdonald et al. 2000; Stopka and Johnson 2000).

The picture of badger society familiar in the lowland UK (the focus of this chapter) is therefore but one point on a broad continuum of intra-specific variation. National population densities vary across Europe, from averages as low as < 1 badger/km^2 in Poland (Kowalczyk et al. 2000), Portugal (Rosalino et al. 2004), and Spain (Revilla and Palomares 2002), through to 1.39 badgers/km^2 in Great Britain (21.3% of the total species population; Wilson et al. 1997) and 2.75 badgers/km^2 in Northern Ireland (Feore and Montgomery 1999). Some of the highest regional densities are to be found on the farmland of southern England, and can exceed 15 badgers/km^2 (e.g. Rogers et al. 1997; Macdonald and Newman 2002; for overviews see Griffiths and Thomas 1993; Roper 2010).

Group sizes vary with population density (but non-linearly; see Roper 2010), from little more than pair-living (Kruuk 1989; Rodriguez et al. 1996; Revilla and Palomares 1999, 2002; Kowalczyk et al. 2000, 2003a,b; Tanaka et al. 2002; Revilla 2003a,b) to, in the most extreme examples, 25 individuals (Cheeseman et al. 1987; da Silva et al. 1993; Buesching et al. 2003). Where badgers form social groups these are composed mainly through individuals remaining in their natal group as they mature (Kruuk and Parish 1982; Cheeseman et al. 1987; Woodroffe et al. 1995; Macdonald et al. 2008). Groups centre on a communal den, or 'sett' (Thornton 1988), with group perimeters (notionally 'territorial boundaries') marked by shared defecation sites, termed latrines (Kruuk 1978b; Delahay et al. 2000; Stewart et al. 2001; Kilshaw et al. 2009).

The parallel to our own society is thought-provoking: from low-density hunter–gatherer systems, to high-density cities—where agriculture transformed our own carrying capacity. In 1798, political economist and demographer Thomas Malthus foretold that: 'The power of population is so superior to the power of the earth to produce subsistence for man, that premature death must in some shape or other visit the human race.' Malthus's pessimism was premature (but perhaps only by a few centuries), not foreseeing improvements in health care and food production technologies, but his logic nonetheless spurs two lines of questioning, as pertinent to our study of badgers as to human society. First, what limits the carrying capacity of the environment, how does this regulate populations extrinsically and what can lead to changes and enhancement of this carrying capacity? And, second, how does a society adjust to greater opportunity? Here, we summarize our answers to these over-arching questions, at scales from molecular genetics to population socio-demographics—challenging aims because badgers are difficult to study, being timid, nocturnal, and all looking similar to each other.

4.2 The Wytham Badger Project

Although studies of Wytham's badgers go back to the late 1960s (Hancox 1973), our personal involvement began in 1972 when one of us (DWM) trudged through Wytham Woods in the inspirational footsteps of Hans Kruuk[3], who pioneered the study of carnivore societies, as a member of nobel laureate Niko Tinbergen's Animal Behaviour Research group at Oxford. In this chapter, however, we draw principally on data that stem chiefly from 1987 onwards, when we began systematic individual monitoring of badgers in Wytham; a 424 ha mixed forest, 5 km north west of Oxford, UK (GPS ref: 51:46:26N; 1:19:19W—for details see Macdonald and Newman 2002; Macdonald et al. 2004b; Savill et al. 2010).

Since 1987, we have sought to live-trap the entire population three to four times per annum—but, knowing this is unrealistic, our compromise is to trap as large a proportion as we can. To date, we have amassed over 9500 captures of 1548 individuals, recording up to 60 details of badger morphometry, reproductive condition, parasite burden, tooth-wear, fighting injuries, and health indices per capture. In addition, we also collect a wealth of samples (genetic, scent gland, faeces, urine, etc.), enabling the detailed studies we describe below.

On first capture, each badger is given a unique identifying tattoo on the inner thigh and 975 of 1410 individuals (69%) trapped between 1987 and 2010 (Annavi et al. 2014a) were first marked as cubs. We follow each individual from 'womb to tomb', to reveal survival patterns, dispersal rates, and how they respond to changes in food supply, environmental conditions, and disease (Macdonald and Newman 2002). Throughout, we have ensured the highest welfare standards, refining trapping protocols continuously (McLaren et al. 2005; Thornton et al. 2005; Montes et al. 2011). Indeed, the project stimulated development of a novel 'leukocyte coping capacity' technique for measuring stress responses in wildlife ecology studies (McLaren et al. 2003; Montes et al. 2004), now applied, for instance, in our water vole studies (see Macdonald and Feber (2015): Chapter 14).

4.3 Defining social groups, breeding groups, ecological groups, and their corollaries

Given that, over the species' range, badger densities vary so extensively, and given that differences in density are clearly associated with different social organization, our starting point is to ask which extrinsic conditions set environmental carrying capacity? How do these regulate badger population dynamics? And, how do badgers respond to changes in carrying capacity—particularly in the light of habitat management and climate change? In essence: *Why do badgers live in groups?*

Axiomatically, group living evolves when the net benefits of association with other conspecifics in the group exceed the costs (Alexander 1974). The selective advantage of group living is often obvious in terms of increased vigilance, or foraging capacity, or reduced reproductive costs. But living in groups can also involve costs, such as worsened risk of infectious disease, or heightened competition. Individuals must therefore weigh the costs and benefits of sociality; but when the benefits predominate, animals will tend to congregate. This is reviewed for carnivores in Creel and Macdonald (1995) (see also Cahan et al. 2002; Macdonald et al. 2004a, 2010b), but Kruuk (1978b) was the first to highlight that badger society presents a paradox: although at higher-densities they live in conspicuously sociable groups, they display minimal co-operation (Macdonald 1983; Kruuk and Macdonald 1985; Woodroffe and Macdonald 1993).

We have devoted over 25 years to resolving Kruuk's paradox. Nevertheless, the sociological benefits to group living (e.g. Creel 2003; Stevens and Hauser 2004) remain elusive for badgers (see Macdonald 1983). As we will elaborate, we have uncovered signs of only nascent sociality, including occasional allo-maternal care (Dugdale et al. 2010, but see Fell et al. 2006), allo-grooming (Stewart and Macdonald 2003; Johnson et al. 2004), and communal sett maintenance (Stewart et al. 1999), with associated benefits for neo-natal cub survival (Kaneko et al. 2010). Social interactions are moderated through scent, where allo-scent marking leads to a shared group odour (Buesching et al. 2002b,c, 2003).

Further heightening the paradox, some 50% of badger litters born to social groups in our study area

[3] Hans Kruuk supervised the D. Phil. of DWM, completed in 1977. In turn, DWM's first badger doctoral student in Wytham, Heribert Hofer, completed his thesis in 1986.

are fathered by males from other groups (Dugdale et al. 2007; Annavi et al. 2014a), exposing that the social group is a fuzzy concept here, being neither a discrete breeding unit nor an exclusive clique.

4.4 Resource dispersion and the ecology of society

The diversity between and, especially, within, species of the Carnivora provides powerful algebra for solving the equation of how selective pressures combine to fashion sociality. The simplest ecological model is that increases in prey abundance (noting that only a proportion of food present may actually be available) equate to increases in habitat quality per unit area and, thus, higher carnivore population densities and, pro rata, smaller individual ranging areas (Macdonald 1981; Gittleman and Harvey 1982; Reiss 1988).

Territory holding is often an adaptation to defend a limiting resource, usually food (e.g. Brown 1964, 1982; Davies and Houston 1984). Since territoriality involves costs (e.g. defence against conspecifics) as well as benefits, theory predicts that territories will generally not be larger than the minimum size required to satisfy the requirements of all the occupants (insofar as further enlargement would bring extra costs without net benefits).

When clumped resources are dispersed widely, however, the extent of the home range necessary to ensure resource (typically food) security becomes indefensible—leading to overlap between individual ranges. This proposition is formalized by the Resource Dispersion Hypothesis (RDH)[4]: that groups may develop where resources are dispersed such that the smallest economically defensible territory for a pair can also sustain additional animals (Macdonald 1983). Under these circumstances, secondary animals can cohabit with the primary occupants at little or no cost (Carr and Macdonald 1986; Blackwell and Macdonald 2000; Johnson and Macdonald 2003). The 'cost' to these secondary animals has been debated (Woodroffe and Macdonald 1993; Revilla 2003a,b); it need not be cost-free (cf. Roper 2010), but must be both bearable and less expensive than other sociological options (see Carr and Macdonald 1986; Newman et al. 2011).

Patch richness determines the number of individuals viable in the group and patch dispersion determines the group range; due to this independence larger groups need not necessarily maintain larger territories (e.g. Macdonald 1983; Carr and Macdonald 1986). Group living, however, inevitably leads to competition between group members for access to limited resources (e.g. mates, food, and breeding sites), and variation in the availability and distribution of these resources might therefore influence the rewards of tolerance and marginal values of group membership (Macdonald and Carr 1989; Kokko and Rankin 2006; Silk 2007; Cameron et al. 2011; Huchard and Cowlishaw 2011; see also Annavi 2012).

The RDH explains how spatial congruity can arise from selective pressures in the absence of any sociological benefit of co-operation, providing grist for the evolutionary mill (Carr and Macdonald 1986). As a consequence, purely ecological conditions can facilitate cohabitation of a territory, resulting in group living in a minimalist form, termed 'spatial groups'.

It transpires that badgers provide the perfect exemplar of an RDH driven society; indeed they have underscored the development of this socio-ecological framework. Having researched carnivore social behaviour in Africa (Kruuk 1972), Hans Kruuk's interest in Wytham badgers was initially spurred by the different type of less co-operative sociality that they exhibit (Kruuk 1978b). With respect to testing the RDH framework, Kruuk conceived that, in the case of badgers in the UK, this smallest economically defensible territory would be defined by the availability, dispersion, and richness of patches of the badgers' primary food source, the earthworm *Lumbricus terrestris*. Earthworms surface, and thus become available to lone foraging badgers (Noonan et al. 2014), in response to optimal microclimatic conditions (Bouché 1977; Kruuk 1978a; Macdonald 1983), a response likely to be affected by climate variability. As a consequence, patches variously rich in earthworms become available in different habitats at different times (da Silva et al. 1993). When worm (and other seasonal food resource) patches are irregularly shaped, widely dispersed, and vary a great deal in nightly productivity, the minimum territory required by an individual badger will be highly contorted and unsustainably expensive to defend. Kruuk's original principle was thus that, because each badger requires several blocks of worm-rich habitat to provide food reliably, it was the economics of resource dispersion that militated for badger group formation. Across Britain, Kruuk and Parish (1982) went on to identify that more scattered distributions of blocks of worm-rich habitat correlate well with badger territory size.

[4] Although this framework is sufficiently flexible to take what is, for some, an annoying variety of forms (e.g. Revilla 2003a,b, but see Johnson et al. 2002b; Johnson and Macdonald 2003).

Resource dispersion, as a determinant of badger socio-spatial variations, also explains patterns in badger society across their biogeographical range (Johnson et al. 2002a). Research from low-density badger populations (e.g. our collaborations in Portugal: Rosalino et al. 2004; and Japan: Kaneko et al. 2014) reveals that where fruit, cereals, invertebrates, amphibians and mammals contribute more towards primary food resources than do earthworms, less gregarious social systems arise (reviewed by Roper 1994, 2010).

4.5 Changes in population density: what determines carrying capacity for badger populations?

Intriguingly, badger numbers in Wytham have almost trebled since our study began, catapulting these badgers from their archetypal social system into high density and group living in an evolutionary heartbeat; a steep gradient for natural selection to operate upon. These local data, corroborated by national and international trends, prompted us to consider what might have influenced the site's carrying capacity so dramatically? To these ends the resource conditions of the landscape of lowland British farmland (worm-rich pastures supplemented by arable crop fields in a mosaic of copses and hedgerows) proves highly pertinent.

Back in 1987 we caught 93 adults and 30 cubs in Wytham. This rose to 235 adults and 67 cubs by 1996 (Macdonald and Newman 2002), since when the population has stabilized at around 200 + adults and *c.* 45 cubs per year (44.55 ± 5.37; range = 16 to 86; period 2000–2010; Annavi et al. 2014a). Equivalent to over 40 badgers/km^2 across the population's stable 6 km^2 foraging range (Macdonald et al. 2002, 2008), this is the highest European badger density ever recorded (Macdonald et al. 2009). Importantly, our trapping regime does not entail pre-baiting; thus, the 250 g of bait provided in traps on only three nights of each season (much of which is eaten by corvids or squirrels), is trivial in population terms (see Macdonald et al. 2009) and does not explain the population increases (contrary to Roper 2010).

Moreover, the Wytham badger population is not alone in exhibiting manifest increases in recent decades. Based on extrapolation from sett surveys, Wilson et al. (1997), estimated that the number of badgers in England and Wales increased by 77% between 1988 and 1997. An update of this national survey, commissioned by Defra, and again using setts as a proxy for badger social groups, has reported an estimated 71,600 (66,400–76,900) social groups in England and Wales (Judge et al. 2014), giving a mean density of 0.485/km^2 (95% confidence interval 0.449–0.521). Twenty-five years on from the first national survey (Wilson et al. 1997), this equates to an 88% (70–105%) increase in the number of social groups inferred across England and Wales; a rate of increase of 2.6% (2.2–2.9%) per annum. Clearly this continuing pattern of population increase is of concern to farmers in conflict with badgers, and warrants explanation in the context of attempts to control bovine TB (see Chapter 5, this volume).

4.5.1 How are badger populations regulated?

Most populations are subject to complex regulatory mechanisms (Murdoch 1970, 1994; Royama 1992; Krebs 1994; Turchin 1995), based fundamentally on age-specific natality, mortality, fecundity, age at first and last reproduction, and immigration and emigration; understanding the foundation to these rates is central to ecology and to species management. Using models of population growth, termed stage-classified matrix models (i.e. parameterization of a life-table data matrix model; see Caswell 1989a; Oli and Zinner 2001) we have been able to summarize metrics of age-specific survival and reproduction for the Wytham badger population, at a level of detail achieved only infrequently for large mammals in the wild (see Macdonald et al. 2009). Using this approach, we have estimated a mean annual growth rate (λ) of 1.065, with a mean generation time of 5.8 years.

While changes in adult survival and fecundity have the greatest potential to influence population growth rate, fecundity in badgers is constrained by litter size, which is relatively small compared to other carnivores (Creel and Creel 1991; Creel and Macdonald 1995), with a mean of 1.62 ± 0.62 SD (Macdonald and Newman 2002) to more recently 1.46 ± 0.65 (Macdonald et al. 2009; Annavi et al. 2014a) with 93% of litters typically including fewer than three cubs (range 1–5; Annavi et al. 2014a). Age at sexual maturity does not prove especially influential to the demographics of this population; at age two, and never sooner, 21% of females and 36% of males first reproduce successfully (Dugdale et al. 2011a). The ratio of fecundity divided by age at first reproduction characterizes a species' life-history dynamics (Oli 2004). For Wytham's badgers, this value is 0.134 (Macdonald et al. 2009), and typifies a 'slow type' life history, governed more by environmental carrying capacity than by reproductive output; that is, offspring are relatively 'expensive' (Pianka 1970; Promislow and Harvey 1990; Bielby et al. 2007).

Organisms must adapt, evolve, and proliferate constantly, not merely to gain reproductive advantage, but also simply to survive while pitted against ever-evolving opposing organisms in an ever-changing environment—the so-called 'Red Queen hypothesis' (Smith 1976). Slow life-history dynamics limit this capacity, exposing such organisms to the vicissitude of nature's 'Court Jester' of abiotic factors (Benton 2009). Crucially then, it is survival rate, for both adults and juveniles (i.e. recruitment into the adult population) that proves the principal vulnerability-driving change in badger population regulation (Macdonald et al. 2009), modelled in this population with means of 0.67 (SE: 0.03) and 0.83 (SE: 0.01) for juveniles and adults respectively. As a result, adult survival accounted for the greatest component of the total change in λ (64%), with juvenile survival following (30%), and fertility accounting for just 5% (Macdonald et al. 2009).

4.5.2 Land use and landscape effects on food availability

Pre-historically, the old growth forests in the UK would have more closely resembled those habitats still intact in the eastern part of Europe, which support a much lower density of badgers (e.g. Kowalczyk et al. 2000). With the advent of agriculture in Britain, these forests were largely cut down (Green 1990; Macdonald and Feber 2015: Chapter 1) and the resultant mosaic is a landscape of farmland and woodland, together with more recent increases in rough grazing for cattle and sheep, coupled with arable fields. It is this very landscape, combined with the temperate and moist British climate, which favours earthworm availability and, in turn, fuels high density, group-living badger populations.

Building on the work by Kruuk and Parish (1982), which established that habitat affects local badger density, group, and territory sizes, our research (da Silva et al. 1993) has found that variation among territories in Wytham in the availability of food-rich rough pasture habitats was linked to the reproductive rates and body weights of their occupants. Furthermore, we found evidence that conversion of arable fields to pasture, prompted by the Common Agricultural Policy (1962) altered badger food supply and, as a consequence, smaller and richer badger territories could fit into the landscape (as we elaborate later in Section 4.8.1; Macdonald et al. 2004c). Thus the group territory size reduced from 0.9 km^2 in 1974 to 0.3 km^2 in 1987, with associated changes in territorial configurations.

In 1988, the mean number of badgers per group was 11.3 (range 6–26), of which, on average, 8.4 were adults (range 4–19). Although, under optimal weather conditions, pasture yields the greatest availability of earthworms, not all badger territories in Wytham encompass pasture (and those that do, do so to varying extents), whereas all include some deciduous woodland, which provides a good and less variable source of earthworms, along with a wider range of other foods (such as seasonal berries, nuts, and forest beetles). As a result, although we found that while group size was unrelated to territory size, the number of cubs born per territory, and the number of breeding females therein did increase with the proportion of territory composed of deciduous woodland, as did the average body weight of male badgers.

4.5.3 Density dependent regulation

While increasing population density is usually accompanied by reduced individual quality and breeding success, the increase in badger numbers we observed through the 1990s occurred without compromise to either average weight or fecundity. Indeed, mean badger condition increased by 15% from 1994–1997, compared with 1990–1993, which, as we will explain, was associated with changing socio-spatial dynamics (Macdonald et al. 2004c, 2009). Our detailed analysis of badger demography exposed that a population growth rate of nil would occur in this Wytham population with 147 female individuals (the offspring-bearing sex is modelled) present—hence 294 badgers if an even sex ratio is projected (95% confidence interval: 124–170) (Macdonald et al. 2009)—almost exactly the current population (that is, *c*. 270–300). This cessation of population growth marks the point of sustainable carrying capacity. Indeed, we are able to estimate maximum growth rate, λ_{max}, unconstrained by carrying capacity (K) but based upon observed reproductive values, for this population. This λ_{max} value of 1.30 (95% confidence interval: 1.13–1.52) contrasts with the actual observed λ of 1.065, revealing the extent to which K governs population dynamics. Additionally, at the group level, we also identified a measurable cost to living in Wytham's larger social groups, which can have twice the average group size of ten residents. From 3528 captures made from 1990–1997, we calculated that between groups with the smallest and largest memberships, body condition ranged over ±9%, body weight ±12%, and fecundity ±20% (Macdonald et al. 2002). These effects were most prominent for individuals in breeding condition. Group size (i.e. mean number of individuals residing within a group range) also had a negative impact on annual variation in fecundity; initially we observed that in groups where only one female bred, these pregnancies were typically successful, whereas when 4–5 females per group attempted to breed, success rate dropped to around 30% (Woodroffe and Macdonald 1995a). Exploring this further with our recent genetic data (Annavi et al. 2014a) revealed an individual breeding success of 72% when 1–3 females were present in a group, 46% with 4–5 females, 31% with 6–10 females, and 21% with 11–17 females present. Overall, however, relative social group density explains only about 6.7% of total variability in body condition (Macdonald et al. 2002), for which the primary driver turns out to be seasonality in food availability (Noonan et al. 2014).

Since the early 1990s, when badger numbers in Wytham increased significantly, land use in Wytham has not changed greatly, and nor have territory sizes (Macdonald and Newman 2002; Macdonald et al. 2009). This leads us to consider another set of variables that affect earthworms: those associated with the weather (Macdonald and Newman 2002; Macdonald et al. 2010b; Newman and Macdonald 2013; Nouvellet et al. 2013).

4.6 Climatic conditions and the effects of weather variability

Aside from disease (discussed in Section 4.7), in terms of 'top-down' population regulation, with no contemporary predators in the UK, badgers are literally *under the weather*. In Wytham, we have found climatic conditions to be the major regulator of population dynamics (Macdonald et al. 2009, 2010b; Nouvellet et al. 2013), with the increase in badger numbers through the 1990s (Macdonald and Newman 2002) being strongly associated with a trend towards warmer weather (Fig. 4.2) and particularly to the lessening of winter frosts (Macdonald and Newman 2002).

4.6.1 Shifting constraints: easier winters, harder summers, and weather variability

Individual fitness is linked to an optimal range of environmental conditions (Chase and Leibold 2003) that define a species' climatic niche (Parker and Maynard-Smith 1990; Kerr 2007), termed a 'Goldilocks zone', in deference to Robert Southly's fabled novel in which bears could tolerate porridge that was 'neither too hot, nor too cold, but just right' (Nouvellet et al. 2013). Simply, for badgers, frost-free winters are good and drought summers are bad. For example, Wytham badgers are around 20% heavier in the warmest rather than the coldest Januaries (Fig. 4.3: ranging from averages of 9–11.5 kg for males and 8.1–10.1 kg for females), during the latter of which they remain underground and deplete fat reserves in a winter torpor (Newman et al. 2011; Noonan et al. 2014). At a biogeographical scale, warmer winters have also been associated with a range expansion and population increase among Finnish badgers (Bevanger and Lindstrom 1995; Kauhala 1995).

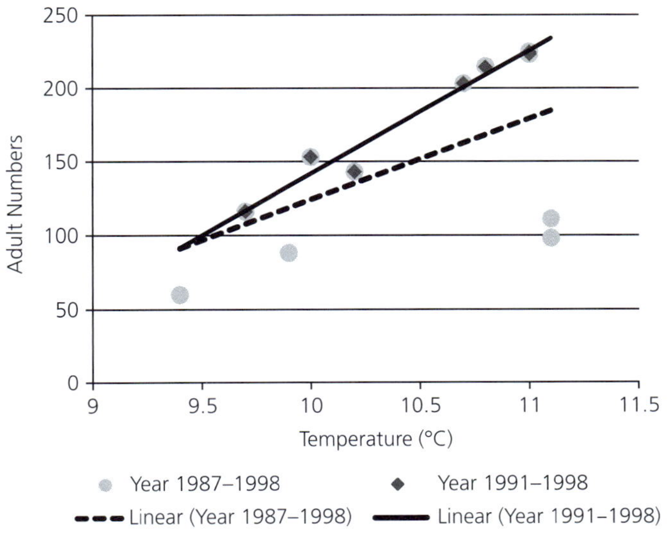

Figure 4.2 Adult badger population size at Wytham Woods in relation to mean annual air temperature. Adult numbers increased with increases in temperature over 1987–1998 years, where each year was milder than the last. Modified from data in Macdonald and Newman (2002) and Newman (2000).

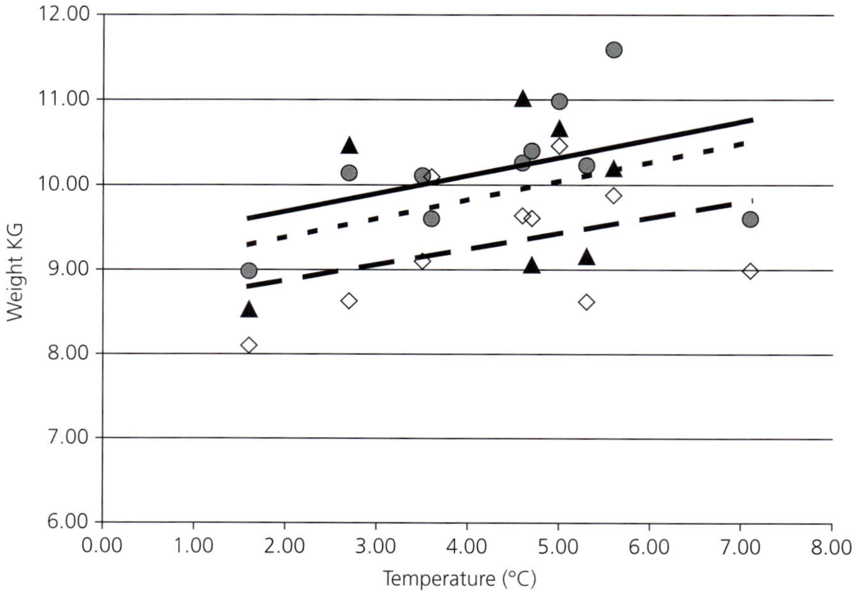

Figure 4.3 Milder January temperatures (1987–1998) correlate with heavier badger body-weight (circles/solid line = males; diamond/dashed line = females; triangles/dotted line = pregnant). Modified from data in Macdonald and Newman (2002) and Newman (2000).

In Wytham, we found not only 'Goldilocks zone' conditions to be important, but also that the 'predictability' of seasonably normative conditions favoured optimal badger survival dynamics. Having exposed that mean temperature, μT, increased with year, equivalent to a rise of 2 °C since 1881 (Nouvellet et al. 2013), we used multi-model inference procedures, based on 1125 life histories (from 1987 to 2008), to decompose population responses to mean trends, yearly amplitude of change, and residual variation in temperature and rainfall for the Oxford area, using daily records. This revealed that annual survival (mean adult: 81%, SE 0.01; mean juvenile: 67%, SE 0.03; Macdonald et al. 2009, 2010b) and recruitment rates both responded to changing mean annual temperature and rainfall trends (Macdonald et al. 2010b), as well as to weather variability (Nouvellet et al. 2013). We infer that predictable conditions allow badgers to prospect risk most effectively and optimize their annual routines, thus defining the badgers' climatic niche (Nouvellet et al. 2013).

Variation in mean rainfall was by far the most influential predictor in our analysis (Fig. 4.4). Most notably, while adult survival rates were higher in years that were wetter than normal (mean annual; residual standard deviation around daily predicted values), cub survival and recruitment rates exhibited a quadratic response, being lower in years that were both wetter and drier than normal; also benefiting during years with lower amplitude of temperature change between winter and summer. Both cub and adult survival rates also exhibited a range of tolerance for residual standard deviation around daily predicted temperature values, beyond which survival rates declined (Fig. 4.4).

Our interpretation is that, while dry weather is 'bad' for foraging badgers, too much rainfall is also bad—and persistently wet conditions in the spring expose vulnerable cubs to parasitic diseases and greater hypothermic stress, through soaking (Webb and King 1984). On average, 25% of cubs die in their first year (Macdonald and Newman 2002), so summer droughts or floods can have a significant influence on badger population dynamics (Woodroffe and Macdonald 1995b; Macdonald and Newman 2002; Macdonald et al. 2010b).

Furthermore, our latest genetic work (Annavi 2012; Annavi et al. 2014b) has exposed that cubs with heterozygous[5] fathers (but not mothers) had greater first year survival probability than cubs with less heterozygous fathers, but crucially *only in wet years* (Annavi et al. 2014b). Female badgers can thus seek to enhance their reproductive fitness by mating with heterozygous

[5] Heterozygosity being a measure of the genetic variation in a population; with respect to individuals having different alleles at one or more corresponding chromosomal loci.

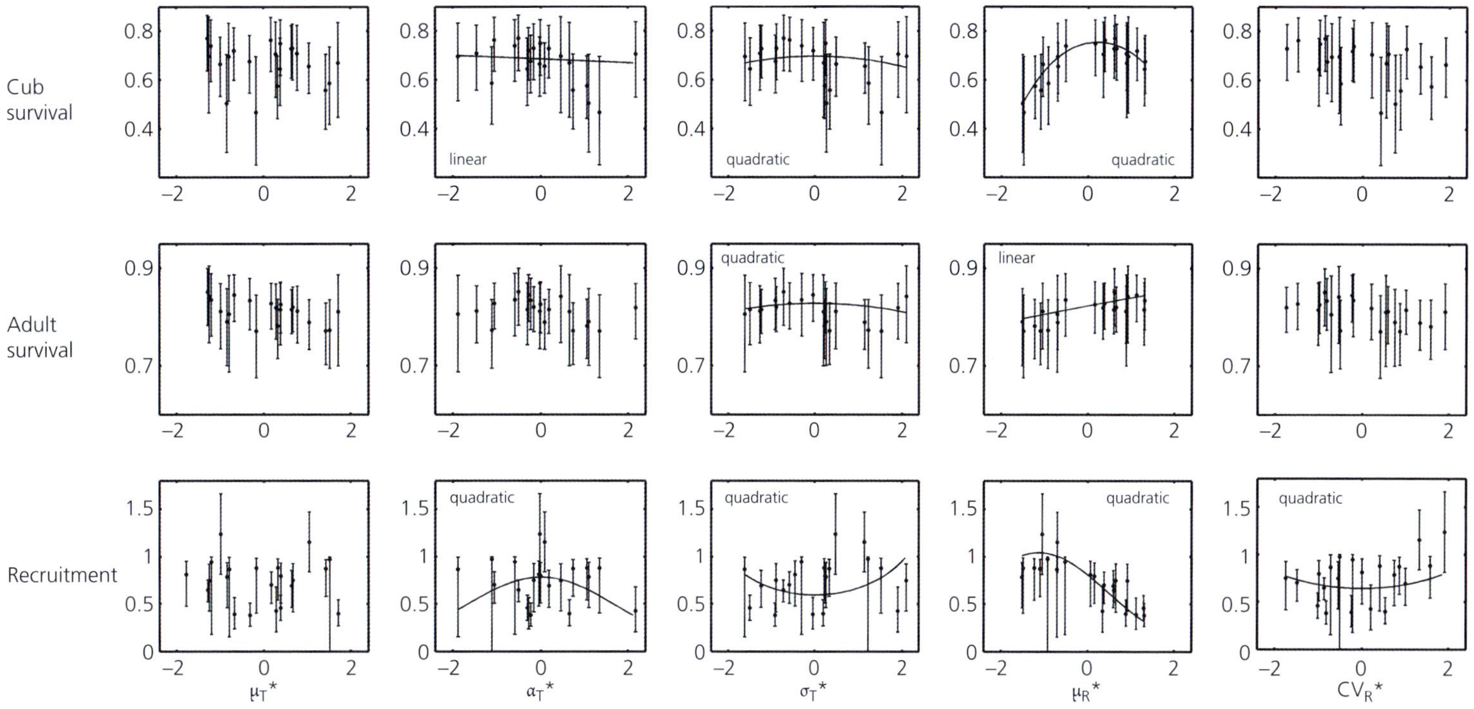

Figure 4.4 Survival rate estimates for cubs and adults, and recruitment as a function of the climate metrics: Mean (average) temperature for that year: μ_T; Amplitude of seasonal temperature changes: α_T—representing the maximum temperature deviation (in summer or winter) from the average yearly temperature; Residual standard deviation around the daily predicted temperature values, σ_T—providing a measure of the yearly temperature variability; Daily mean rainfall, μ_T, and; Coefficient of rainfall variation CV_R. Estimates of survival rates and recruitment together with their 95% confidence intervals (error bars) are based on model averaging. Solid curve represents statistically significant link between life history parameters and climate metrics. For each significant relationship, we indicate whether the linear or quadratic (or both) component(s) was (were) significant. Importantly, a significant linear component does not imply a straight line in the shown representation, as the relationship is defined as linear within a logistic transformation (for survival rates) or within a log transformation (for recruitment). From Nouvellet et al. (2013).

males—but these benefits only manifest under advantageous conditions, when the subtleties of differential survival can be expressed (Harrison et al. 2011; Annavi et al. 2014b); food shortages resulting from summer drought are an indiscriminate cause of mortality, no matter what genetic advantages an individual might possess.

The effects of weather are also compounded by disease; endoparasitic infections exacerbating the malnutrition experienced by cubs (as discussed in Section 4.7; Anwar et al. 2000; Newman et al. 2001; Macdonald et al. 2010b; Nouvellet et al. 2011, 2013; Sin et al. 2014). In 1990 and 1996 we found severe summer droughts combined with parasitoses to cause over 60% cub mortality. In addition we estimate, from ultrasound, that on average *c*. 36% of cubs die underground before they are ever caught by our trapping regime and enter onto our records (Macdonald and Newman 2002).

Matching species' mechanistic responses to fine-scale environmental variability, while imperative, proves challenging and, of course, if the weather is disadvantageous badgers have the option to remain underground. The decision to do so on any given night must, however, weigh the likely energy costs of activity under inclement conditions (mating, scent marking, etc.) against the possible gains from foraging, both in relation to the badgers' contemporaneous body-fat reserves.

Using the latest accelerometery technology (Ellwood et al. 2007; Shepard et al. 2008), our current research (Noonan et al. 2014) is establishing that the tendency for badgers to be active outside the sett differs significantly between individuals. During the relatively wet autumn of 2012, badgers with the longest *per noctem* activity subsequently exhibited higher Body Condition Indices (BCI) when recaptured. In contrast, under drier 2010 autumn conditions, badgers in good BCI engaged in less *per noctem* activity, while badgers with poor BCI were the most active. Furthermore, when compared on the same calendar dates, to control for night length, duration of mean badger nightly activity was longer (9.5 hrs ± 3.3 SE) in 2010 than in 2012 (8.3 hrs ± 1.9 SE). We thus emphasize how a 'one size fits all' approach should not be applied to ecological responses.

4.6.2 Climatic interactions with Road Traffic Accidents (RTAs)

Using data from over 700 badger RTAs, gathered largely in south west England between 1972 and 1999, we also discovered that weather conditions affect the rate of RTAs; in Wytham, RTAs account for around 44% of annual badger mortality (Macdonald and Newman 2002). Nationally, around 50,000 badgers were killed on British roads each year in the 1980s, equivalent to 30–66% of adult and post-emergence cub mortality (Skinner et al. 1991; Clarke et al. 1998; no more recent national surveys).

Crucially, we found from regional data, gathered across southern England, that RTA rates doubled for every 1 °C increase in mean daily temperature in February (Fig. 4.5: Macdonald et al. 2010b). Since the peak

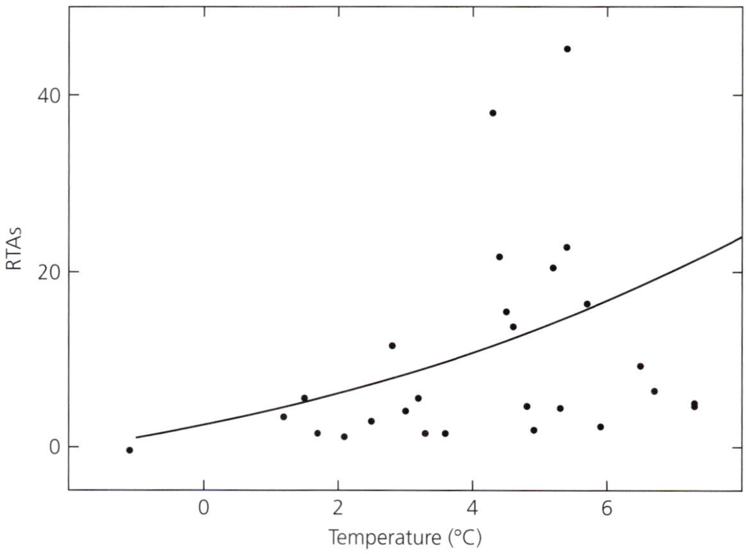

Figure 4.5 Road traffic accidents (RTAs) occurring in February in relation to temperature, based on over 700 badger RTAs gathered largely in south west England between 1972 and 1999. Drawn from data in Macdonald et al. (2010b).

of badger mating occurs in February (Davies et al. 1987), warmer weather may lead to wider-ranging courtships and more libidinous distractions from road safety. If warming trends continue (Root et al. 2006; Macdonald et al. 2010b; Newman and Macdonald 2013) these effects, compounded by population fragmentation caused by traffic and expanding road networks, have the capacity to elevate badger mortality rates significantly.

Changes in seasonal weather conditions may also prove influential for other species with high rates of RTA: WildCRU studies, ranging from otters *Lutra lutra* in the UK (Philcox et al. 1999) to raccoon dogs *Nyctereutes procyonoides viverrinus* in Japan (Saeki and Macdonald 2004), have revealed similar RTA peaks in early and late winter, as crepuscular emergence times coincide with evening and morning rush hours, but are yet to examine interactions with weather conditions.

4.7 Parasitic disease and immunity

Having noted that disease proves a major determinant of survival in this badger population, we proceeded to explore the interplay between the pathogenicity of infections and nutritional status (Anderson and May 1979); that is, poor nutrition (e.g. stated habitat and climatic effects on environmental carrying capacity and food competition) amplifies the morbidity caused by disease.

Although their susceptibility to bovine tuberculosis is the most notorious (see Chapter 5, this volume), badgers suffer a variety of infections. For example, the coccidial gut parasites *Isospora melis* and highly pathogenic *Eimeria melis* occur commonly among badgers (Anwar et al. 2000). *E. melis* causes diarrhoeal enteritis, epithelial sloughing, steatorrhoea, and malaise, consequently, infected cubs lose fluid and suffer malabsorption and anorexia (Newman et al. 2001). In Wytham, from 1993–1995, we found (Newman et al. 2001) that cubs sampled during their first year had an *E. melis* prevalence of 66.4% (n = 159) for (cf. 8.5% in adults; n = 242), with a 100% rate of infection when sampled immediately after their first emergence from the sett in May (n = 41). Infectious intensities reached over 20,000 parasitic oocysts/g of faeces during this critical phase of maturation. When these infections coincide with shortages of food and water, typical of dry spring weather, cub cohort mortality rates peak at over 90%, where mortality is due to malnutrition (Newman et al. 2001; Nouvellet et al. 2013), whereas in years of plenty, over 50% of the cohort will typically survive (Macdonald and Newman 2002). Cubs surviving infection, however, typically suffer stunted growth: the most heavily infected (surviving) males attain adult body lengths typically 5 cm shorter (7% of 70.5 cm), and females 3.5 cm (5% of 670 cm) shorter, than the least severely infected (Newman et al. 2001). This is because male cubs grow faster than females and thus suffer a greater impairment due to coccidiosis, as a trade-off between somatic investment and skeletal development (Newman et al. 2001).

Two blood parasites, *Babesia missirolii* and *Trypanosoma pestanai*, also occur commonly in Wytham's badgers (Macdonald et al. 1999) and in other populations (Lizundia et al. 2011). Our genetic PCR analysis of 207 badgers undertaken in 2009 detected *T. pestanai* in 31% of individuals (42% of males; 27% of females), again with a tendency for juveniles to be most susceptible to infection (40% incidence), but without evident pathology. Our study discovered that the host-specific badger flea *Paraceras melis* is the primary vector transmitting *T. pestanai* (Lizundia et al. 2011), which has implications for grooming behaviour, as we discuss in Section 4.10.3.

Curiously, we also discovered that juvenile infections also affect the black pattern on the roof of badgers' mouths (termed palatine maculations). Patterns tend to develop more asymmetrically by adulthood for badgers with the highest intensity of coccidiosis (*E. melis*), relative to less severely infected contemporaries. This type of 'Fluctuating Asymmetry' (Parsons 1992) is recognized as a measure of the inability of individuals to buffer their development from environmental stresses (Nouvellet et al. 2011).

Ultimately, the insults caused by malnutrition, disease, and injury, as well as social stress, affect animals at the cellular level. This causes the production of damaging oxygen-free radicals through metabolic processes that can have harmful consequences for a diverse range of molecules such as DNA, lipids, proteins, and sugar. The next phase in our ongoing investigation was therefore to investigate the badgers' immune response and capacity to survive this 'oxidative stress'. Negative effects of oxidative stress are mitigated by protective molecules and enzymes, termed 'antioxidants', which neutralize free radicals or arrest their production (McGraw et al. 2010). Antioxidant systems are costly to implement (Montes et al. 2011), and so the development of these defences is delayed in most taxa (Khan and Black 2003; Fontagné et al. 2008), instead investing energy into growth during maturation (termed heterochrony—MacKinney and MacNamara 1991).

Our work (against a paucity of studies on wild mammals) has uncovered the capacity for badger cubs to engage in the unexpectedly precocious development of antioxidant capacity. We found that, in 2011 (with a dry spring), the mean antioxidant capacity of 16-week-old cubs was equivalent to that of prime-age adults (1–5 years old; Bilham et al. 2013). Compared to individuals aged ≥ 6 years, cubs exhibited a significantly higher mean total plasma antioxidant capacity. In this year, however, the cub cohort size of 12 was well below the average of 44.55 ± 5.37 (range 16–86: years 2000–2010), where plausibly, only those capable of resolving oxidative stress survived until we were able to capture (weaned) cubs in early June. Subsequent years of study have begun to suggest the occurrence of antioxidant phenotypes. We posit that the cub phenotype better able to survive antioxidant stress (generated by coccidial infection and catabolic metabolism; see Bilham et al. 2013; also Levitis 2011) may achieve this as a developmental trade-off, where other aspects of their maturation (especially development of subcaudal scent secretion; Bilham et al. 2013) appear to be compromised. In a benign year, those not as invested in expensive antioxidant defences may have the developmental advantage (balancing selection reminiscent of the heterozygosity finding we described above). We have also established that food deprivation per se leads to oxidative stress directly: emaciated badgers of all ages mount significantly lower antioxidant responses, exposing individuals to physiological damage (Montes et al. 2011).

4.7.1 Acquired immunity and mate choice: the Major Histocompatibility Complex (MHC)

The Red Queen hypothesis (first mentioned in Section 4.5.1) specifically proposes that it is sexual reproduction that provides the genetic diversity necessary to counter the rapid adaptations of pathogens during the co-evolutionary 'arms race' (van Valen 1973). Furthermore, according to the Hamilton–Zuk hypothesis (Hamilton and Zuk 1982; Hamilton et al. 1990), individuals are predicted to choose those mates that will maximize pathogen resistance in an individual's progeny, as part of this arms race.

In terms of the adaptive immune system, the Major Histocompatibility Complex (MHC) is a diverse gene family that encodes cell surface glycoproteins that bind and present antigens to T-cells, which belong to a group of white blood cells known as lymphocytes, and play a central role in cell-mediated immunity (Swain 1983). A variety of species is already known to be able to discriminate the MHC profile of potential mates and thus optimize the immune systems of their offspring (Penn and Potts 1999; Penn et al. 2002; Milinski et al. 2005). This can be detected from visual (e.g. Dunn et al. 2013) or olfactory (Johnston et al. 1997) cues (see Hamilton and Zuk 1982). Elaborating on initial work to characterize class I and class II genes (Sin et al. 2012b,c), we were prompted to investigate: (a) if MHC confers resistance across a representative range of pathogens in badgers, and (b) whether this influences mate choices.

Two MHC alleles (alternative forms of a gene—one member of a pair) were found to confer qualitative and/or quantitative resistance to certain pathogens, while a further three alleles were associated with pathogen susceptibility (Sin et al. 2014). MHC heterozygotes, however, did not exhibit a lower pathogen burden or infection status, with respect to the thirteen pathogens we examined. Indeed, contrary to the MHC-disassortative mating hypothesis (e.g. Juola and Dearborn 2012), which predicts general preference for MHC-dissimilar mates (generating offspring heterozygosity), we found that badgers were more likely to breed with MHC-*similar* mates (Sin et al. in press). This overall effect was found both when badgers mate within their own group or with individuals from neighbouring groups. In several other species, MHC dissimilarity and heterozygosity have proven advantageous, however MHC diversity can cause a loss of T-cells, the lowering of the immune capacity in favour of diversity (Lawlor et al. 1990; Nowak et al. 1992), which also increases the risk of auto-immune diseases (Kaufman 1999; Puurtinen et al. 2005). Importantly, the most significant pathogen driving natural selection in this badger population, *Eimeria melis*, is a coccidian, which are primarily under innate, not acquired, immune control.

As to whether badgers can detect one another's MHC genetics, and therefore gain information on immunocompetence, the microflora hypothesis (Yamazaki et al. 1990; Lanyon et al. 2007) proposes that MHC genes control odour indirectly by influencing the growth of microbiota commensal on sweat and scent gland secretion, which contribute to variation in individual odour. As we shall go on to elaborate (Section 4.10.1), badgers have a unique subcaudal gland encoding individual specific parameters (Buesching et al. 2002b,c) and we have evidenced that microbiota play a key role in metabolizing these primary gland products, to yield characteristic scent profiles (Sin et al. 2012a). We are now investigating whether it is plausible that they use this scent profile to make 'informed choices' concerning with whom they mate.

4.8 Responsive and adaptive behaviours facilitating group living and promoting co-operation.

So far, we have focused our attention on the mechanisms of population regulation, exposing a system where high badger density is associated with group living, and where extrinsic factors, such as resource dispersion and habitat, climate, and parasites, regulate badger population dynamics. Group living and the need to deal with the stresses of high density have consequences for individuals, and we now go on to consider how behavioural adaptations underscore the other major theme of our research: badger sociology.

4.8.1 Socio-spatial re-organization and sett dispersion

Dens are important resources for a variety of carnivores, and particularly for badgers (see Kaneko et al. 2010), despite the work involved in excavating them (Roper 1992; Stewart et al. 1999; Macdonald et al. 2004c; Rosalino et al. 2005; Loureiro et al. 2007).

Although the subterranean ecotope has been exploited by myriad mammalian species, from small rodents and eutherians to the largest burrowers, aardvarks *Orycteropus afer*, burrows have particular importance for carnivores, connected to providing a secure neo-natal environment for the extended periods of maternal/parental care necessary to raise altricial young (see Noonan et al. 2014). Either by extension of philopatry and/or through delayed dispersal, dens can also impose communal aggregation upon individuals; den use is associated with 'social-group' formation in a number of carnivores, for example, meerkats *Suricata suricatta*, European badgers, dwarf mongooses *Helogale parvula*, banded mongooses *Mungos mungo*, and bat-eared foxes *Otocyon megalotis*. Despite these recognized roles and advantages of burrow use, due to the inaccessibility of this subterranean environment, little is known about the behavioural aspects of burrow use; thus 'inactivity' is often assumed, but without substantive evidence.

To redress this lack of knowledge we have recently begun using collars fitted with Magnetic Induction (MI) transmitters to monitor underground badger activity, with 10 cm accuracy in three dimensions (Noonan et al. 2015). Our preliminary analysis of the movements of four individuals has already revealed that on average badgers spent 12.3 ± 1.69 SD hrs in the sett *per diem*, 6.4 ± 1.0 SD during daylight hours, and 5.8 ± 0.88 SD during the night-time. Interestingly, these tagged badgers spent on average no more than 2.2 ± 3.1 SD consecutive hours away from the sett and returned to the sett an average of 2.3 ± 1.45 SD times throughout the night, with resting periods during these visits lasting a mean of 0.77 ± 0.91 SD hrs. Furthermore, when returning to the sett during nocturnal periods, badgers tended to use peripheral chambers (i.e. chambers located near the outer circumference of the sett), whereas central chambers were used for longer during diurnal periods. Badgers used 1.55 ± 0.61 SD chambers *per diem* (n = 49, range: 1–3) and, while each badger used a unique subset of the available chambers ($\chi^2 = 115.4$, df = 7, $p < 0.0001$), badgers used more than one chamber *per diem* nearly 50% of the time. A chamber of primary fidelity could be identified for each badger, on average favoured for 2.7 ± 1.7 SD consecutive days (n = 16, range: 1–7 days), before moving to a different chamber. Subterranean ecotope conditions exacerbate infectivity where den sharing has been implicated in the spread of pathogens and parasites (see Section 4.7).

These setts provide the foundations for badger occupation within the landscape, each acting as a hub within the group's active foraging range. To monitor the spatial organization of Wytham's badgers we have, for many years, used a technique called 'baitmarking' (Stewart et al. 1997a; Macdonald et al. 2004c; Kilshaw et al. 2009)—originally invented by Skoog (1970) and refined in Wytham by Hans Kruuk (Kruuk 1978a). Approximately bi-annually, we deliver a sticky mixture of peanuts, syrup, and uniquely coloured inert plastic beads to each badger sett, daily over a two-week period. This readily consumed confection is part-digested by the badgers, but the coloured marker beads are passed in their faeces, which are (conveniently, for our purposes) deposited in pits at well-used latrine sites, apparently marking the periphery of each group's range.

We interpret these surveys to show how groups have persistently compartmentalized the Wytham Estate into a series of discrete tessellating group ranges, giving the appearance of a high level of social integration and organization (Vicente et al. 2007; Hewitt et al. 2009). Despite their spacing, however, there is scant evidence that any group ranges thus defined are actively 'defended', and thus we use the term 'territories' cautiously (Woodroffe et al. 1995; Macdonald et al. 2004d; Fig. 4.6). In Wytham, ever greater division into smaller ranges (and thus more latrine sites) over time has accompanied increasing population density. While the 6 km^2 foraging range used by the Wytham badgers has remained fairly constant since the first baitmarking

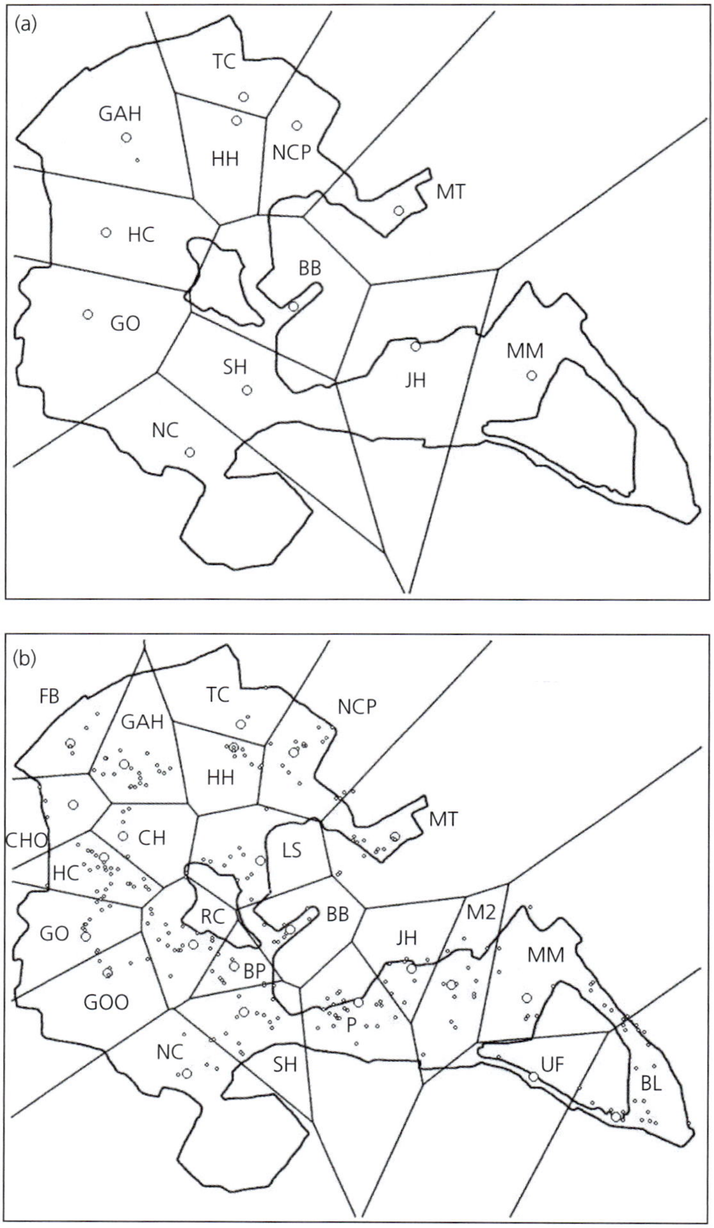

Figure 4.6 Setts (dots) and territories (dividing lines) (modified from Macdonald et al. 2004c). (a) 1978; (b) currently.

survey of 1974, the number of territories had increased from 12 (Kruuk 1978a, 1989) to 20 by 1993 (Macdonald and Newman 2002), with a further increase to 23 by 2002 (Macdonald et al. 2004c) perpetuated to the time of writing this chapter (Fig. 4.6a: 1974, b: currently). As Plato observed (Plato's Laws 360 B.C.E), 'Yet it is clear that if a number of sons are born and the land is correspondingly divided there will inevitably come to be many poor men.'

Doncaster and Woodroffe (1993) proposed that suitable sett sites might be limiting and therefore influence the social structure of badgers (see also Blackwell and Macdonald 2000). In contravention, new sett sites at Wytham appear to have been readily available (Macdonald et al. 2004c). While the number of group ranges has reached a plateau, the number of burrows that badgers have dug has proliferated in relation to increasing population size (see Macdonald et al. 2004c).

In 2004, we recorded 279 active sett sites (defined minimally as a 'badger hole') within Wytham's 4.24 km^2 of woodland (Macdonald et al. 2004c), giving a sett density of 0.71 setts/ha; comprising 1130 entrances, 906 (80%) of which were active. Major setts (> 5 entrances) numbered 23 in 1987, when the trapping regime began, with 67 outliers (Doncaster and Woodroffe 1993), whereas we currently have 63 major setts and 216 outliers. These new setts were not measurably inferior to established ones (in terms of the weight or reproductive success of residents), thus site limitation does not seem to have been a limiting factor in social configurations (see Blackwell and Macdonald 2000). Setts have, however, been found to be limiting in some low-density badger population areas in Mediterranean habitat (e.g. Loureiro et al. 2007).

So, on what criteria do badgers select sett sites? Overall, in Wytham they favoured sites with sandy, well-drained soils, situated on convex and moderately inclined north-west facing slopes, at moderate elevation for the site (65–165 m). We found body weight to be, on average, 7% lower at setts in clay soils (Macdonald et al. 2004c), possibly because these locations are in ranges that provide less food, or that burrows in clay require occupants to expend more energy, due to cooler temperatures in less optimal/shallower burrow systems (Kaneko et al. 2010). Alternatively, intrinsically inferior badgers may get pushed to the edge of society.

Regionally and nationally, sett densities and characteristics also vary with habitat type (Macdonald et al. 1996). Tunnels typically do not exceed depths of 2 m (Kaneko et al. 2010; Noonan et al. 2015), but can form intricate dendritic systems extending over 100 m from one entrance to that most distant (Roper 1992). The 1130 tunnels in Wytham likely represent over 158 tonnes of spoil! Management of these digging activities is a practical problem for farmers/landowners.

4.8.2 More places to reside—less social stress?

Social stress is broadly recognized to be a bad thing, leading to anxiety and even aggression in human society (Aseltine et al. 2000). We therefore expect such stressors to influence animals in the wild in similar ways, both behaviourally and endocrinologically (Blanchard et al. 2001). For example, the neuropeptide oxytocin, the affection hormone, inhibits infanticide in mice (McCarthy 1990), and is associated with maternal behaviours and pair bonding, and the alleviation of anxiety (Lee et al. 2009). Conversely, inhibition or deficit of oxytocin is associated with sociopathy in humans (Conlisk 2011). Notably, oxytocin must be inhibited for delayed implantation, which is a badger hallmark trait, potentially to the detriment of an affectionate nature (Douglas et al. 1998). To these ends, Cresswell et al. (1992) reported that the presence of annex setts (i.e. smaller setts, usually within 150 m of a main sett, not necessarily in use all the time) correlated with increased productivity in younger sows, and suggested that additional sett sites enable younger sows and their cubs to avoid the aggression of older and 'more dominant' individuals (though dominance can be difficult to establish for badgers; Macdonald et al. 2002; Hewitt et al. 2009).

In particular, we have noted from increasing numbers in the Wytham population that crowding proves stressful; behaviourally, the increase in socio-spatial dispersion that accompanied population growth reduced the symptoms of crowding stress, such as impaired body condition and bite-wounding and infanticide (e.g. Cresswell et al. 1992; Tuyttens et al. 2000; Macdonald et al. 2002). Furthermore, cubs raised in the more stable environs of main setts proved more likely to survive than those born at outliers (Kaneko et al. 2010).

4.8.3 Movement without borders

Although badgers living in groups may benefit from an increased probability of intercepting intruders into their group range (Kruuk 1989), perplexingly we have found scant evidence of active territorial defence (Stewart et al. 2001; Tinnesand et al. in press)—badgers from different social groups meet frequently at boundary latrines, and pass by each other amicably into their neighbour's range (Paget and Middleton 1974; Christian 1995; Dyo et al. 2010). From the alternative perspective, our MI tracking has recorded one badger (of four collared) staying away from its sett of residence for three successive nights, presumably visiting a group elsewhere (Noonan et al. 2015).

High rates of extra-group paternity (Section 15.9.1) further reveal that, in Wytham, badger groups are neither spatially, nor genetically segregated (Dugdale et al. 2008; Annavi et al. 2014a,b). A human socio-political analogy might be the zone created by the Schengen Agreement (1985), eliminating border controls and allowing free travel between member European States—and yet borders still actually delimit these separate territories. In badger society, perimeter paths tend to link a series of latrines to create group range interfaces, which are frequently reinforced by both neighbouring groups with fresh faeces (Stewart et al. 2002): neighbouring groups place a similar quantity of faeces at shared latrines (supporting the faeces matching hypothesis; Stewart et al. 2001). This suggests that faecal volume may provide a signal of the encounter likelihood and/or foraging pressure for badgers along a particular border.

When visiting a latrine, a badger typically stays around 23 seconds and squats an average 1.6 times, with no significant differences between sexes or age classes (Stewart et al. 2001). Sniffing occupies 54% of the badgers' summed time budget at latrines, crouching (squat mark urination and defecation combined) comprises around 16% of latrine time, while digging and scuffing only 5%. The remaining 25% of time spent at latrines is distributed among behaviour not specifically associated with signalling (e.g. 4% of latrine time is spent feeding, self-grooming, and engaging in social behaviour; 7% in an alert posture).

The conventional 'active territorial defence' hypothesis (Maher and Lott 1995), would interpret latrine use as a stand-off in terms of each group's resource holding potential, signalling encounter likelihood across the boundary. By contrast, to account better for the actual role of latrines in badger spatial organization in the absence of overt defence, and in accord with frequent extra-group mating, we conceived the 'Passive Range Exclusion Hypothesis', or PREH (Stewart et al. 1997a; see also Kilshaw et al. 2009; Buesching et al. subm.). This proposes that badgers setting out to forage from their setts will consume their available food supply as they go, creating a gradient as they move away from the sett, leaving a depleted zone behind them. Ultimately, badgers foraging in this way will reach the advancing front of their neighbours' explorations. Should these badgers pass each other and forage beyond this contact zone they will enter into the zone of diminishing return, and thus this border presents an isopleth (equal contour) of resource depletion between groups. Foraging beyond the group boundary is thus not necessarily an adversarial transgression; though to approach a neighbouring sett directly can sometimes spark an aggressive response (Macdonald et al. 2004b; Hewitt et al. 2009). By either hypothesis, these matched effects of faecal volume and sett proximity provide a simple mechanism for communication among neighbours that can generate the emergent complexity of coordinated group ranges without central control (Tinnesand et al. in press).

We have also shown that, in Wytham, badgers do not simply defecate at the section of boundary closest to where they happened to be active, but rather they distribute faeces to achieve comprehensive coverage of the border (Kilshaw et al. 2009). Badgers apparently either partition the collective activity of marking their territories, or are all attracted to the same site to communicate through scent signals (Tinnesand et al. in press)—akin to a badgerine version of 'Facebook'—with the result that entire group boundaries are marked regularly and consistently.

Recognizing that latrines communicate olfactory information (Stewart et al. 2002), a potentially parsimonious hypothesis on which we are working currently is the extent to which it is the network of paths linking latrines that is the true focus of the badgers' interest. Paths are linked in a dendritic network often with latrines marking junctions, and thus potentially acting as navigational cues (see Stopka and Macdonald 2003). Under this scenario latrines function as information nodes, providing 'scent signposts' along this path network—'notes to self' or 'notes to others' on previous nights' foraging success, resource depletion, and possibly who else is active in the area; ally or enemy. By analogy, we use signposts to navigate our road networks, rather than building roads to link signposts!

4.8.4 Philopatry, dispersal, and the advantages of trespassing

Dispersal and movement patterns of badgers are important because social contacts spread disease, and populations that are destabilized (for example, due to forestry/habitat intervention) are known to harbour more disease due to the immune-compromise caused by perturbation. This is a particular issue with regard to bTB transmission on farmland (see Chapter 5, this volume).

Between 1987 and 2003, we investigated the dispersal patterns of 267 badgers over 5255 trapping events (Macdonald et al. 2008; note, this provides a maximum of 4/365th of nightly locations per year, and so gives us a robust minimum movement metric) and found that permanent dispersal was infrequent, generally occurring between adjacent groups and without any tendency to respond to population density (Woodroffe et al. 1995; Macdonald et al. 2008). The mean dispersal distance of 530 m (SD = 393 m, SE = 55 m) in Wytham approximates the average distance between neighbouring group setts. About a third (96/267) of individuals sampled was never trapped outside of their natal social group (Macdonald et al. 2008). A further 39.9% were captured in at least two groups during their lifetimes, 16.0% in three groups, 5.2% in four groups, and 2.9% in more than four groups. Only 19.1% (51) of these 267 animals actually satisfied our definition of dispersal[6], of these, 28 were males and 23 were females. Overall, males tended to be captured at more social groups than were females.

With respect to motivation, dispersing males showed a propensity to move to larger groups, and to groups with a preponderance of females (Tinnesand et al. in press). When we radio-tracked movement patterns between 1987 and 1992 (Woodroffe et al. 1995), we found the tendency for females to disperse and cover larger distances exceeded that of males: 11 dispersing males all relocated to adjoining territories whereas, of 14 dispersing females, seven transcended their neighbours' territory to join groups one-social unit further away (Woodroffe et al. 1995). In the broader data set, however, between 1987 and 2003 (Macdonald et al. 2008), we could detect no overall gender bias in dispersal (median dispersal distances: males = 400 m; females = 414 m), although four out of five dispersing events over 1 km were by males (Macdonald et al. 2008). Interestingly, events involving two individuals dispersing to the same group simultaneously were recorded only for females (n = 3 occasions), where the former resident females were usurped, supporting earlier observations by Woodroffe et al. (1995).

Aside from permanent dispersal, we also observed frequent transitory exchange between groups (see Fig. 5.11 in Chapter 5, this volume)—interesting insofar that it is these transitory movements that are likely to underscore inter-group mating and disease transmission. Of the 5255 trapping events investigated (1987–2003), 863 (16.4%) involved individuals caught at a social group differing from that of the previous trapping occasion, but on a temporary basis, and not fulfilling our definition of dispersal. Overall, 19.8% of the population, at any given time, was found in a social group other than their principal, or natal, affiliation (Macdonald et al. 2008). Among these 1040 events, movements between social groups were most common in autumn (17.1%) and least common in winter (10.9%). Short-term visits by males were most frequent in autumn and spring (18.4 and 18.2% respectively), while those for females peaked in summer (15.9%). Males also tended to be captured at approximately twice as many social groups as were females. As with permanent dispersal, males tended to visit groups that included more females, while females made visits to groups whose members included fewer females (i.e. less breeding competition). As above, however, this is again only a robust minimum record of movement, based on usually no more than four trapping records per year, and so sett visits, and particularly inter-group range transgression, seem likely to be far more commonplace than had been the conventional wisdom.

Dispersal did not affect the likelihood of reproduction by females: 54.6% (12, n = 22) of females were observed to be pregnant post-dispersal, compared with 61.5% (16, n = 26) of females that did not disperse (Macdonald et al. 2008). We also found (Macdonald et al. 2008) that, when adjusted for age and number of captures, badgers that had a history of movement between social groups were significantly more likely to be scarred from fighting.

Given the limitations of examining sett affiliation patterns from our trapping data, which strongly infer that a lot more movement might be going on at a finer-grained temporal scale, we have recently implemented a new tracking technology using radio-frequency identification (RFID) collars to keep continuous tabs on around one quarter of the population (Dyo et al. 2010). Preliminary results are leading us to investigate whether patterns and rates of territorial transgression, and individual differences in tendencies to cross borders, resonate with pedigree records of extra-group mating (Domingo-Roura et al. 2003; Dugdale et al. 2007; Annavi et al. 2014a).

4.9 Mating systems: trespassers *will be rewarded*

Badger mating systems turn out to be both complex and interesting with respect to natural selection. Female badgers typically give birth to their litters in around mid February, then immediately have a post-partum oestrus, which results in a mating peak. Gestation lasts four days pre-implantation, but is then suspended until implantation at around the winter solstice (Woodroffe 1995), resuming for around 50 days thereafter. The prolonged period between conception and gestation is due to delayed implantation (Thom et al. 2004), or embryonic diapause, where a fertilized egg (blasotcyst) suspends development temporarily and remains free in the uterus lumen rather than implanting directly (Macdonald 1993). During this time, females may continue to ovulate and can conceive while they are already pregnant, a phenomenon termed superfoetation; females can also conceive with different males within a single oestrus, termed

[6] A badger was flagged as 'resident' within a social group if the two most recent captures—as well as at least one of two previous captures before that—were made within the same social group.

superfecundation (Yamaguchi et al. 2006); both of these can lead to multiple paternity within litters (Evans et al. 1989; Dugdale et al. 2007).

From our genetic analyses, mean litter size in the Wytham population was 1.4 ± 0.06 (range = 1–4, mode = 1, n = 427 litters: Dugdale et al. 2007)—close to the 1.62 we (Macdonald and Newman 2002) had calculated previously by dividing cubs present at each group by the number of lactating females (see Dugdale et al. 2011b). Litter size varies with food supply, from one cub per lactating female in the driest, most food scarce years, to 2.5 cubs in the wettest and most food abundant years (Macdonald and Newman 2002).

In order to establish parentage, we genotyped 915 of 1080 individuals, caught between 1987 and 2005 (see Dugdale et al. 2007: as DNA samples permitted), subsequently extending this sample to 1020 of 1247 badgers, caught between 1988 and 2010 (i.e. to include re-analysing some same individuals with a larger marker set and different pedigree analysis software; see Annavi et al. 2014a). This revealed the extent to which this badger population is polygynandrous: we found that males and females both breed promiscuously, and multiple females within groups produce young (Dugdale et al. 2007; Annavi et al. 2014a). Of 143 multiparous litters we investigated between 1987 and 2005 (Dugdale et al. 2007), assigned with 80% confidence (44.8% of a total of 320 litters), 64 (45%) had multiple paternity.

Our most recent analysis (1988–2010: Annavi et al. 2014a,b), however, used a different pedigree program (called MasterBayes) and 35 micro-satellite markers, and found that, of the 378 litters for which both parents were assigned, 36.8% (139; n = 378) were multiparous (117 twins, 20 triplets, two quadruplets), and of these, 27% (38; n = 139) had multiple paternity, although no more than two fathers were detected. Of the 502 candidate mothers and 612 candidate fathers marked between 1987 and 2010 (Annavi et al. 2014a), only 45.42% females (228) and 32.37% males (201) recruited into the adult breeding population actually bred successfully. Overall, reproductive success declined with the age of both male and female parents (Dugdale et al. 2011b). We also observed multiple breeders per group, with a mean of 1.9 ± 0.2 mothers and fathers per social-group-year (Dugdale et al. 2008).

Compared with other mammals, we deduce that reproduction may be costly for badgers, since both gestation and peak mating periods occur in the winter (Woodroffe and Macdonald 1995b) when the animals are less active and feed less (Newman et al. 2011; Noonan et al. 2014). Although gestation does not appear to influence body condition, by the end of lactation (May), we found that breeding females have suffered a marked weight loss (Woodroffe 1995), which is regained by the autumn. Lactating two-year olds experience a higher age-specific mortality than their contemporaries that did not breed, without any apparent difference among older females.

In terms of individual male reproductive strategy, we have observed two endocrinological phenotypes (Buesching et al. 2009)—some extend testicular activity into the autumn, others conserving reproductive effort with mid-summer testes ascent (see also Woodroffe et al. 1997). All animals over the age of six adopted the conservative strategy, indicating that males either switch phenotypes with age, or that extended testicular activity results in lower life expectancy, further evidencing the costs of male reproduction. Additionally, we (Woodroffe and Macdonald 1995b) observed that sustaining testicular activity later into autumn correlated with more bite wounds and anaemia.

We have also noted differences in the sex ratio of litters and annual cub cohorts. From a long-term mean sex ratio of 50:50 (mean female = 50.95 ± 7.80) (Macdonald and Newman 2002; Dugdale et al. 2003), we observed a male bias coincident with significantly larger cub cohorts (up to almost 60% male in the most productive years; 43% male in the least productive years) and relatively early annual implantation dates (before Christmas—56% male; mid-January 41% male). Females can adapt their implantation date somewhat to prevailing conditions; Woodroffe (1995) discovering from ultrasound scanning that implantation commenced at around 21 December (17 December for earliest individual) for females with higher body condition, and as late as 31 December (11 January for latest individual) for lower body condition. Generally for mammals, male offspring are more costly to raise successfully to reproductive quality (Trivers and Willard 1973), exemplified here by the fact that female badgers appear to optimize cub productivity in response to the level of resources that they can invest, such that when females are fatter, the sex ratio is biased towards sons (Dugdale et al. 2003).

4.9.1 Extra-group paternity

The extent of extra-group paternity among badgers, emerging from our work, reveals a very different picture of their society from the insular 'family groups' portrayed in the traditional ecological knowledge of this species (e.g. Neal 1948).

In Wytham, 48% (n = 655 cubs) of assigned paternities (Annavi et al. 2014a,b) resulted from extra-group matings, with 85% of these (n = 315 cubs) attributed to neighbouring group males. As a consequence, 47% (n = 378) of litters included at least one extra-group offspring (EGO) (Annavi 2012). Whatever the function of the social group, it is clearly not an exclusive breeding unit! Males securing extra-group paternity (EGP) increase their fitness where this leads to siring additional offspring (e.g. Isvaran and Clutton-Brock 2007). The selective advantage of EGP for females is less obvious (see Annavi et al. 2014a), prompting us to tackle the following questions about mate choices.

4.9.1.1 Do females seek mating opportunities with more heterozygous extra-group males?

Individuals may benefit, in terms of offspring fitness (termed *heterosis*— ostensibly 'hybrid vigour') by choosing more heterozygous mates ('Good-gene-as-heterozygosity' hypothesis; Brown 1997). Heterozygosity is associated with a variety of fitness related traits, such as survival probability (Charpentier et al. 2008), reproductive success (Slate et al. 2000), and disease resistance (Acevedo-Whitehouse et al. 2005). In Wytham, however, we found that neither the average nor the maximum of within-group heterozygosity of candidate fathers affected the number/proportion of offspring fathered by extra-group males within a litter (Annavi et al. 2014b). Overall, both sexes exhibited only about 30% fidelity to within-group matings, versus *c.* 40% always mating outside their group, with the rest showing a mixed mating tactic. EGP in badgers therefore does not seem to have evolved in order to seek out mates with higher levels of heterozygosity.

So how do females decide whether to select the (often more than one) father(s) of their cubs from within or beyond their social group? Perhaps it simply reflects the relative numbers of candidate fathers within and beyond the group? Partly, yes, the more candidate fathers there were at home, the lower the rate of EGP; whereas the more there were in the neighbouring groups, the higher the rate of EGPs. Insofar as we expect groups formed by the delayed dispersal of offspring to be relatives, females might also select extra-group mates to minimize the risk of inbreeding.

4.9.1.2 Does Extra-Group Paternity (EGP) mitigate inbreeding in badgers?

Inbred badgers had a lower lifetime breeding success than non-inbred badgers (Annavi et al. 2014b). Inbreeding depression (reduced fitness due to breeding with kin) is costly (Pusey and Wolf 1996), so EGP might favour offspring fitness (e.g. Dugdale et al. 2007; Sillero-Zubiri et al. 1996). In Wytham, badger groups consist of closely related individuals (Dugdale et al. 2008), thus the potential for inbreeding is high. We found that with greater average pair-wise relatedness between mothers and within-group candidate fathers (ranging 0.125, 0.25, 0.375, 0.5), the proportion of EGPs per social group per year was greater (respectively 0.22; 0.5; 0.65; 0.8; Annavi et al. 2014b).

4.9.1.3 Do Extra-Group Offspring (EGO) perform better than Within-Group Offspring (WGO)?

The fitness benefits hypothesis states that females may derive inclusive fitness benefits if EGO out-perform WGO, in terms of survival and reproduction. In seeking EGPs, females might therefore be seeking advantageous genetic compositions for their cubs, in terms of disease resistance (but see Section 4.7.1), reproductive potential, and longevity. In short, did EGO perform better than WGO? Not entirely—EGP proved more costly for badgers in some regards. EGO had *c.* 34% lower first-year survival probability than did WGO. EGO also lived on average 1.3 years less than WGO (Annavi 2012). WGO females were, however, assigned significantly more litters over their lifetime than EGO (2.4 versus 1.6) and produced significantly more offspring over their lifetimes (3.4 versus 2.1). WGO females also exhibited marginally higher lifetime reproductive success (40.5% of EGO, 42.7% of WGO; as did males: 35.2% of EGO, 39.3% of WGO; cohorts 1988–2008). Nonetheless, males with extra-group fathers went on to father more offspring in their lifetime than did males who had in-group fathers (3.4 versus 2.3), in favour of the fitness benefits hypothesis.

In terms of lifetime number of litters, we found that, although mothers did worse in terms of producing both offspring and grand-offspring if they produced EGO, males benefited in terms of grand-descendents if they produced EGO (Annavi et al. 2014b). Thus we observe an antagonistic sex effect, such that genes arising through EGP are beneficial when expressed in males, but detrimental when expressed in females. This raises the question of whether the fitness benefits of producing extra-group sons outweighs the costs of producing extra-group daughters—a question we cannot yet answer. Consequently, if sons inherit their father's traits, females may increase their indirect reproductive success by choosing males that will ultimately produce a greater number of descendents, provided that females can detect these 'sexy' phenotypic traits through cues (i.e. Sexy-son Hypothesis; Fisher 1930) or under circumstances where these males produce more competitive sperm (Annavi 2012).

4.9.2 Mating strategy—adaptive to group living, or responsive to group living?

Plausibly, EGP may confer fundamental sociological advantages to the mating generation, such as fulfilling the females' requirements for extra copulations to induce ovulation (Yamaguchi et al. 2006), or confusing paternity both within the home group and within the neighbouring group, to safeguard against non-paternal male infanticide (Lups and Roper 1990; Cresswell et al. 1992; Agrell et al. 1998). This backcloth enables us to consider a key question in the evolution of badger promiscuity (Annavi et al. 2014a,b): *Do badgers simply mate at random?*

Species in the mustelid family generally, but not invariably, exhibit intra-sexual territoriality, where individuals maintain territories only with respect to members of the same sex (Powell 1979; Erlinge and Sandell 1986; Yamaguchi and Macdonald 2003), with male spacing affected by female spacing, and access to mates, as well as by the distribution of available food resources (Helldin 1999; Zalewski 2007). This is broadly the system adopted by badgers in low-density populations in continental Europe (see Johnson et al. 2000; Newman et al. 2011). Under these circumstances, females may not have evolved to be 'choosey', but when they occur at high densities in large groups this may manifest as indiscriminate promiscuity in badgers. This mating tactic, under these circumstances, however, may be anachronistic and maladaptive, and as we mentioned in Section 4.9.1.1, we observe that while around 40% of individuals engaged in promiscuous EGP, 30% were never promiscuous outside of their group, with the rest showing a mixed mating tactic (Annavi 2012).

4.10 Behaviours governing badger social dynamics

Although we have found little evidence of the adaptive significance of group living in badgers, other than that it appears to be facilitated by the dispersion of resources created by lowland agriculture, badgers nonetheless display abundant social behaviour, albeit of a narrow repertoire.

Badgers are hard to tell apart so, in the early 1990s, we pioneered infra-red video surveillance techniques (Stewart et al. 1997b), combined with fur-clipping[7] for individual identification (Stewart and Macdonald 1997). This revealed that, while all members of a social group dig to refurbish their main sett, this activity is undertaken preponderantly by one or a few mature males (Stewart et al. 1999) and by non-breeding individuals (Dugdale et al. 2010).

These males, and those that that we discussed in Section 4.9 that retain reproductive condition for the longest period into the autumn—all less than six years of age (Buesching et al. 2010) and often immigrants (Woodroffe and Macdonald 1995b)—might classically have been assumed to be socially dominant. High testosterone levels, however, which are necessary for spermatogenesis and thus reproduction, come at a cost (see Section 4.9): they are associated with higher energetic requirements and thus increased metabolic rates, resulting in significantly lower body condition, and leading to a lowered immune response, and thus higher parasite loads (Buesching et al. 2010). Evidence of a linear social hierarchy in either a feeding or a mating context has proven equivocal, and the benefits of high social rank among badgers remains unclear (Macdonald et al. 2002, 2004a; although see Hewitt et al. 2009).

There are, nonetheless, asymmetries in adult badger society: although mutual grooming is highly reciprocal when it occurs (see Section 4.10.3: Johnson et al. 2004; Macdonald et al. 2000); some individuals within a social group both allo-groom and allo-scent mark certain individuals more than others (Buesching et al. 2003), and indeed the inter-individual flows of these two behaviours are highly correlated. This results in sub-groups of individuals which interact significantly more with each other than with other members of the social group. Rank is not related to age, nor does it appear to affect the likelihood of an individual being wounded, or to its ability to access food (Macdonald et al. 2004b), or an individual's breeding status. Overall, we found that hierarchies were dominated primarily by breeding females—especially in the context of breeding competition, with females ranking higher than males (Hewitt et al. 2009).

4.10.1 Scent marking and olfactory communication

We have found that when social interaction does occur, it follows a reciprocal 'tit-for-tat' pattern indicative of rather unsophisticated sociality (Macdonald et al. 2000; Buesching et al. 2003; Stewart and Macdonald 2003; Johnson et al. 2004). Vocalization plays a role in badger communication (Wong et al. 1999), but visual signals are limited. In subterranean or nocturnal darkness it is

[7] Removal of dark guard hair to expose white under-fur to create an easily recognisable pattern, diagnostic of ID.

the persistent message encoded by scent that badgers rely upon primarily (Buesching and Macdonald 2001). Understanding the way badgers communicate with scent, both inter-individually and in terms of navigating their environment is highly pertinent to understanding the social dynamics of their society.

We have already described the use of faeces to leave a scent mark at latrines. There, and elsewhere, badgers also mark using their subcaudal gland: a unique pocket, or lumen under the tail, above the anus, divided into two halves by a septum and lined with exocrine glands which produces a fatty secretion, which they smear onto one another (*allo-marking*; Buesching et al. 2003) and onto features in their environment (e.g. the sett, paths, and feeding sites: *object-marking*; Buesching and Macdonald 2004) in a brief squatting action, to convey olfactory information about their individuality, fitness, and group membership (Buesching et al. 2002a,b,c). Our research has shown that secretion volume (ranging up to a maximum of 3 ml) and colour (from white, through beige to brown) vary with its chemical composition (Buesching et al. 2002a): males tend to produce paler and more substantial volumes of secretions than do females, although females produce more, and paler secretions while they are rearing cubs (Buesching et al. 2002a).

Cubs start to produce their first secretion when they are about 4–6 months of age (Buesching et al. 2002a). Cubs still too young to produce their own secretion (Buesching et al. 2002a) perform 'scent theft', during which they crawl under the belly of adult group members squeezing out between their hind legs, thus practically forcing the adult to scent mark their backs (Fell et al. 2006). Notably, the olfactory coding of the secretion is analogue rather than digital, and thus no single component characterizes the secretion as 'male' or 'female', but the ratios of a number of different components in the chemical mixture encode all-important information. While primary gland products provide the olfactory substrate, we are investigating currently how it is that the bacterial flora living in the subcaudal pouch, which metabolize these primary gland products, lead to a characteristic odour profile (Sin et al. 2012a).

In terms of social dynamics, adult and reproducing females are the preferred targets of allo-marking for other non-reproducing females, whereas adult males in reproductive condition are marked by non-reproducing males three times more often than any other class of badgers (Buesching et al. 2003). Sufficient allo-marking is required to maintain the group smell, but too much would result in the loss of individual-specific information (Buesching and Macdonald 2004).

During the mating season, males with descended testes advertise their reproductive status by object-marking 15 times more often than males that are not in reproductive condition (ascended testes), scent-marking on average once every five minutes throughout the night; whereas reproducing females object-mark almost ten times as often as any other individuals during the cub-rearing season in spring (approximately once every 20 minutes; Buesching and Macdonald 2004)—potentially to signal a claim on resource necessary for raising cubs successfully (e.g. Mertl-Millhollen 2006). As reproducing adults deposit most scent marks, object-marking is likely to serve not only to signal that the individual is present in that group range, but also as an individual fitness advertisement signal (Buesching and Macdonald 2001, 2004; Buesching et al. 2003).

4.10.2 Limited evidence of allo-parental care

Allo-parental behaviour (i.e. cubs being tended by individuals other than their parents) is a common feature of group-living carnivore societies (Creel and Creel 1991; Woodroffe and Vincent 1994). While this presents a significant advantage to group living, its occurrence among badgers remains tentative (Woodroffe and Macdonald 2000). Several litters of cubs can be raised in one sett at the same time and, from genetic and video-surveillance evidence, might occasionally suckle cubs that are not their own (Dugdale et al. 2010). Nevertheless, the availability of more potential helpers in larger groups, or in groups with more than one breeding female, does not increase cub survival rates, and thus results in long-term fitness costs (Dugdale et al. 2010).

4.10.3 Mutual grooming and social interaction: do badgers groom for pleasure or for purpose?

We have mentioned that aside from benefits under certain ecological circumstances, group living can also bring costs, where the evolutionary stability of the group-living strategy depends on the ratio of these costs to benefits. One of these costs is sharing communal dens (Stewart et al. 1999; Kaneko et al. 2010), which provides favourable conditions for the proliferation and transmission of ectoparasites and any diseases they may vector (e.g. trypanosomiosis, as discussed in Section 4.7). Badgers typically carry lice *Trichodectes melis*, host specific fleas *Paraceras melis*, ticks (predominantly *Ixodes hexagonus*), and mite infestations (particularly *Cheyletiella parasitovorax*) (Newman et al.

2004). Although swapping sleeping chambers may help (Noonan et al. 2014), fleas can live for up to 89 days away from their host, although lice only survive for three days independently (Cox et al. 1999).

Grooming is a potent defence against ectoparasites (Macdonald et al. 2000), and the badger's tactic of rapidly grooming a site, changing sides, and enlisting conspecifics to reach inaccessible spots counters the flea's escape behaviour (Cox et al. 1999; Stewart and Macdonald 2003). Although badgers spend a considerable proportion of their time grooming each other (Stewart and Macdonald 2003), this behaviour is usually directed to areas of their bodies that are hard to reach by themselves (Johnson et al. 2004). Allo-grooming involves substantial reciprocity and grooming partners do not give one another any credit, i.e. they stop grooming within one second of each other. This type of interaction fulfils the requirements of the iterated prisoner's dilemma, i.e. a tit-for-tat strategy (Macdonald et al. 2000). All hosts that come in contact with each other end up with similar parasitic loads, and allo-grooming can lower these parasite burdens among group-members (the 'Ideal Flea Distribution Hypothesis'; Johnson et al. 2004).

4.10.4 Mutual tolerance and agonistic behaviour

Earlier we touched on how living in close proximity can inflate 'social stress' (Section 4.8.2) (e.g. Sapolsky 2011). Given our increasing realization that badger society at high density appears ever more to be simply the low-density system crammed into an awkward accommodation—how do badgers ameliorate the potential for mutual aggression, remembering their disproportionately dangerous bite (207 newtons of force; Christiansen and Wroe 2007)? Pocock (1911) was the first to propose that badgers' characteristic black and white facial stripes act as an aposematic warning of this capacity to put up a fight, a trait shared with other small, but fierce species (Newman et al. 2005). Certainly foxes *Vulpes vulpes* in Wytham are subordinate to badgers (Macdonald et al. 2004a; see Chapter 2, this volume).

We found that, in Wytham, bite wounding rates increased with population size, from on average 7% when the population was < 100 individuals in 1990, to in excess of 20% by 1999, when the population rose above 200 individuals (Macdonald et al. 2004b). Although 24.9% of 261 male badgers we examined between 1990 and 1999 had severe wounds present (21% on the rump; 13% on the head and neck), only 18% of 237 females were similarly wounded (9% on the rump; 12% on the head and neck). The rate of bite wounding for males also increased with the number of badgers living in adjoining territories (Macdonald et al. 2004c; Dugdale et al. 2007). Cubs tend to bear wounds much less frequently; of 1076 observations made by Delahay et al. (2006) in the south west of England, bite wounding ranged from 1% in male cubs to 2% in female cubs, with 53% of these being to the head. Indeed, when we have observed cubs fall victim to infanticide at Wytham, it is just this sort of head injury that appears to be the cause of death. These conflicts leave Wytham's badgers bearing scars, particularly on their rumps, with 44% of males, and 34% of females thus scarred (Macdonald et al. 2004b). We also found that lifetime scar acquisition increased with (seasonally adjusted) badger weight. Fresh bites, however, most frequently afflict individuals in very poor condition (Delahay et al. 2006), from which we infer that hunger may motivate more reckless competition for food, or that heavier animals were potentially assertive in deterring conspecifics (Macdonald et al. 2004b).

Although no seasonality in bite wounding was evident in Wytham Woods, most wounding has been associated with the late winter mating season at other sites (Delahay et al. 2006). Behaviourally, the increased socio-spatial dispersion we observed (Macdonald et al. 2004c—Section 4.8.1) as density increased (Macdonald et al. 2002) might also be significant in reducing social tension, biting, and infanticide. Between 1987 and 2004, 83% (n = 18) of the cubs born at four of our most intensively monitored main setts survived, compared with only 27% (n = 11) at outliers within the same territories (Kaneko et al. 2010). It has also been our observation that, through the 2000s, since the Wytham population stabilized at around 200 adults plus *c.* 45 cubs per year, we have observed that the frequency and severity of biting has decreased to around one third of pre-sett proliferation levels.

4.11 Conclusions

Badgers—*conservation paragon or farming pariah?*

The ecology of badgers in Britain, and the issues they can pose in the countryside, can often be traced back to aspects of agricultural and forest management. In this sense, any problems arising from their activities are man-made. As though a 'perfect storm', the badgers' capacity to thrive because of worm-rich foraging grounds and a more varied seasonal food supply has been enhanced very substantially by a reduction in native UK forests and conversion to farmland over the last millennium, where pasture proves especially earthworm productive (da Silva et al. 1993). Indeed, 'copses' characteristic of the British landscape, often provide just enough protection to allow badgers to dig concealed setts while still having prime access to the open land around. The rural realties are that:

(i) Badgers dig setts, often in unwanted and inconvenient places on farmland.
(ii) Badgers also create extensive latrines, filled with faeces.
(iii) Their transgressions, damaging crops or disturbing poultry and game birds, are undoubtedly a nuisance.
(iv) They have a contentious role in the transmission of bovine tuberculosis, the subject of Chapter 5, this volume.

To compound this suite, climate change, and especially the reduction in winter frosts, has encouraged numbers to burgeon. Even if winter warming continues, however, the linear relationship with badger numbers will not continue, because once a substantial number of nights are above around 4 °C, further benefits are less tenable. Indeed other constraints such as summer drought and unseasonable weather variability appear now to be afflicting badger demographics in other complex ways. Furthermore, any substantial deviation in environmental carrying capacity, resulting from more variable and less seasonally predictable weather conditions, could stress high-density populations and force badgers into greater conflicts with land managers and farmers.

From the conservation ecology perspective, badgers clearly provide an intriguing model for the evolution of the nascent stages in sociality (Woodroffe and Macdonald 1993; Newman et al. 2011). Alloparental behaviour is exceptional (Woodroffe 1993; Woodroffe and Macdonald 2000; Fell et al. 2006) and dominance hierarchies highly context dependent (Macdonald et al. 2002; Hewitt et al. 2009), but other behaviours, such as allo-grooming, plural mating, and allo-marking indicate that badgers have adapted to take some advantage from group living, where their socio-ecology lags their actual ecological circumstances.

Knowledge of the biology of species that challenge us is imperative, where the mixed attributes of a thriving British badger population present very tangible challenges. As Ruben Blades eloquently proposed: 'I think we risk becoming the best informed society that has ever died of ignorance'!

Acknowledgements

Many people have contributed enormously to our study—we warmly and gratefully acknowledge in our device of referring to 'our study' and things 'we have found' that this phrase reflects the efforts of a great many involved through the years. Hannah Dugdale's involvement heralded a turning point in our genetic analyses, as did Pierre Nouvellet's insights into the impacts of climate change. Other DPhil students contributing to various aspects of our understanding of badger biology (in chronological order) include Jack da Silva, Heribert Hofer, Rosie Woodroffe, Paul Stewart, Xavier Domingo-Roura, Sandra Baker, Inigo Montes, and most recently Geetha Annavi, Simon (Yung Wa) Sin, Kirstin Bilham, and Mike Noonan (who we thank for comments on this chapter), along with countless interns and volunteers. We are grateful to Ruth Feber and Eva Raebel for their input to the chapter.

References

Acevedo-Whitehouse, K., Vicente, J., Gortazar, C., Hofle, U., Fernandez-de-Mera, I.G., and Amos, W. (2005). Genetic resistance to *Bovine tuberculosis* in the Iberian wild boar. *Molecular Ecology*, **14**, 3209–3217.

Agrell, J., Wolff, J.O., and Ylönen, H. (1998). Counter-strategies to infanticide in mammals: costs and consequences. *Oikos*, **83**, 507–517.

Alexander, J.D. (1974). The evolution of social behaviour. *Annual Review of Ecology and Systematics*, **5**, 325–383.

Anderson, R.M. and May, R.M. (1979). Population biology of infectious diseases: Part I. *Nature*, **280**, 361–367.

Annavi, G. (2012). *Genetic, socio-ecological and fitness correlates of extra-group paternity in the European badger (Meles meles)*. D.Phil Thesis. University of Oxford.

Annavi, G., Newman, C., Buesching, C.D., Burke, T., Macdonald, D.W., and Dugdale, H.L. (2014b). Heterozygosity–fitness correlations in a wild mammal population: single locus, paternal and environmental effects. *Ecology and Evolution*, **4**, 2594–2609.

Annavi, G., Newman, C., Dugdale, H.L., Buesching, C.D., Sin, Y.W., Burke, T., and Macdonald, D.W. (2014a). Neighbouring-group composition and within-group relatedness drive extra-group paternity rate in the European badger (Meles meles). *Journal of Evolutionary Biology*, **20**(10), 2191–2203.

Anwar, M.A., Newman, C., Macdonald, D.W., Woolhouse, M.E.J., and Kelly, D.W. (2000). Coccidiosis in the European badger (*Meles meles*) from England, an epidemiological study. *Parasitology*, **120**, 255–260.

Aseltine Jr., R.H., Gore, S., and Gordon, J. (2000). Life stress, anger and anxiety, and delinquency: an empirical test of general strain theory. *Journal of Health and Social Behavior*, **41**, 256–275.

Benton, M.J. (2009). The Red Queen and the Court Jester: species diversity and the role of biotic and abiotic factors through time. Science, **323**(5915), 728–732.

Bevanger, K. and Lindstrom, E.K. (1995). Distributional history of the European badger *Meles meles* in Scandinavia during the 20th century. *Annales Zoologicae Fennici*, **32**, 5–9.

Bielby, J., Mace, G.M., Bininda-Emonds, **O.R.P.**, et al.(2007). The fast-slow continuum in mammalian life history: an empirical re-evaluation. *The American Naturalist*, **169**, 748–757.

Bilham, K., Sin, Y.W., Newman, C., Buesching, C.D., and Macdonald, D.W. (2013). An example of life history antecedence in the European badger (*Meles meles*): rapid development of juvenile antioxidant capacity, from plasma vitamin E analogue. *Ethology Ecology and Evolution*, **25**, 330–350.

Blackwell, P.G. and Macdonald, D.W. (2000). Shapes and sizes of badger territories. *Oikos*, **89**, 392–398.

Blanchard, R.J., McKittrick, C.R., and Blanchard, D.C. (2001). Animal models of social stress: effects on behavior and brain neurochemical systems. *Physiology and Behavior*, **73**(3), 261–271.

Bouché, M.B. (1977). Stratégies lombriciennes. In U. Lohm and T. Persson, eds., Soil organisms as components of ecosystems. *Ecological Bulletins*, *(Stockholm)*, **25**, 122–286.

Brown, J.L. (1964). The evolution of diversity in avian territorial systems. *Wilson Bulletin*, **76**, 160–169.

Brown, J.L. (1982). Optimal group size in territorial animals. *Journal of Theoretical Biology*, **95**, 793–810.

Brown, J.L. (1997). A theory of mate choice based on heterozygosity. *Behavioral Ecology*, **8**, 60–65.

Buesching, C.D., Clarke J.R., Ellwood, S.E., King, C., Newman, C., and Macdonald, D.W. (2010). The mammals of Wytham Woods. In P.S. Savill, C.M. Perrins, K.J. Kirby, and N. Fisher, eds., *Wytham Woods: Oxford's Ecological Laboratory*, pp. 173–196. Oxford University Press, Oxford, UK.

Buesching, C.D., Heistermann, M., and Macdonald, D.W. (2009). Seasonal and inter-individual variation in testosterone levels in badgers *Meles meles*: evidence for the existence of two endocrinological phenotypes. *Journal of Comparative Physiology A: Neuroethology, Sensory, Neural, and Behavioral Physiology*, **195**, 865–871.

Buesching, C.D. and Macdonald, D.W. (2001). Scent-marking behaviour of the European badger (*Meles meles*): resource defence or individual advertisement. *Chemical signals in vertebrates*, **9**, 321–327.

Buesching, C.D. and Macdonald, D.W. (2004). Variations in scent-marking behaviour of European badgers *Meles meles* in the vicinity of their setts. *Acta Theriologica*, **49**, 235–246.

Buesching, C.D., Newman, C., and Macdonald, D.W. (2002a). Variations in colour and volume of the subcaudal gland secretion of badgers (*Meles meles*) in relation to sex, season and individual-specific parameters. *Mammalian Biology-Zeitschrift für Säugetierkunde*, **67**, 147–156.

Buesching, C.D., Stopka, P., and Macdonald, D.W. (2003). The social function of allo-marking behaviour in the European badger (*Meles meles*). *Behaviour* **140**, 965–980.

Buesching, C.D., Waterhouse, J.P., and Macdonald, D.W. (2002b). Gas chromatographic analysis of the subcaudal gland secretion of the European badger (*Meles meles*) Part I: Chemical differences related to individual-specific parameters. *Journal of Chemical Ecology*, **28**, 41–56.

Buesching, CD., Waterhouse, J.P., and Macdonald, D.W. (2002c).Gas chromatographic analysis of the subcaudal gland secretion of the European badger (*Meles meles*) Part II: Time-related variation in the individual-specific composition. *Journal of Chemical Ecology*, **28**, 57–69.

Buesching, C.D., Newman, C., Service, K., Riordan, P. and Macdonald D.W. (subm.) Between constraint and redundancy: Latrine marking patterns in European badgers (Meles meles) with respect to population density and range size.

Cahan, S.H., Blumstein, D.T., Sundström, L., Liebig, J., and Griffin, A. (2002). Social trajectories and the evolution of social behavior. *Oikos*, **96**, 206–216.

Cameron, C., Berteaux, D., and Dufresne, F. (2011). Spatial variation in food availability predicts extrapair paternity in the arctic fox. *Behavioral Ecology*, **22**, 1364–1373.

Carr, G.M. and Macdonald, D.W. (1986). The sociality of solitary foragers: a model based on resource dispersion. *Animal Behaviour*, **34**, 1540–1549.

Caswell, H. (1989a). *Matrix Population Models: Construction, Analysis and Interpretation*. Sinauer.

Charpentier, M.J.E., Williams, C.V., and Drea, C.M. (2008). Inbreeding depression in ringtailed lemurs (*Lemur catta*): genetic diversity predicts parasitism, immunocompetence, and survivorship. *Conservation Genetics*, **9**, 1605–1615.

Chase, J.M. and Leibold, M.A. (2003). *Ecological Niches: Linking Classical and Contemporary Approaches*. University of Chicago Press, Chicago.

Cheeseman, C.L., Wilesmith, J.W., Ryan, J., and Mallinson, P.J. (1987). Badger population dynamics in a high-density area. *Symposia of the Zoological Society of London*, **58**, 279–294.

Christian, S.F. (1995). Observations of extra-group mating and mate-defence behaviour in badgers, *Meles meles*. *Journal of Zoology*, **237**, 668–670.

Christiansen, P. and Wroe, S. (2007). Bite forces and evolutionary adaptations to feeding ecology in carnivores. *Ecology*, **88**, 347–358.

Clarke, G.P., White, P.C.L., and Harris, S. (1998). Effects of roads on badger *Meles meles* populations in south-west England. *Biological Conservation*, **86**, 117–124.

Common Agricultural Policy: Council Regulation (EEC) No 25 of 4 April 1962 on the financing of the common agricultural policy. http://www.cvce.eu/content/publication/2005/7/12/bd078708-573f-4b2f-a38e-e0c44913b030/publishable_en.pdf.

Conlisk, J. (2011). Professor Zak's empirical studies on trust and oxytocin. *Journal of Economic Behavior and Organization*, **78**, 160–166.

Cox, R., Stewart, P.D., and Macdonald, D.W. (1999). The ectoparasites of the European badger, *Meles meles*, and the behavior of the host-specific flea, *Paraceras melis*. *Journal of Insect Behavior*, **12**, 245–265.

Creel, S. (2003). Why co-operate? Game theory and kin selection. *Trends in Ecology and Evolution*, **8**, 71–72.

Creel, S.R., and Creel, N.M. (1991). Energetics, reproductive suppression and obligate communal breeding in carnivores. *Behavioral Ecology and Sociobiology*, **28**, 263–270.

Creel, S., and Macdonald, D. (1995). Sociality, group size, and reproductive suppression among carnivores. *Advances in the Study of Behavior*, **24**, 203–257.

Cresswell, W.J., Harris, S., Cheeseman, C.L., and Mallinson, P.J. (1992). To breed or not to breed: an analysis of the social and density-dependent constraints on the fecundity of female badgers (*Meles meles*). *Philosophical Transactions of the Royal Society of London*, **338**, 393–407.

Cresswell, P., Harris, S., and Jeffries, D.J. (1990). *The History, Distribution, Status and Habitat Requirements of the Badger in Britain*. Nature Conservancy Council, Peterborough.

da Silva, J., Woodroffe, R., and Macdonald, D.W. (1993). Habitat, food availability and group territoriality in the European badger, *Meles meles*. *Oecologia*, **95**, 558–564.

Davies, N.B. and Houston, A.I. (1984). Territory economics. In J.R. Krebs and N.B. Davies, eds., *Behavioural Ecology: An Evolutionary Approach*, pp. 148–169. Blackwell, Oxford.

Davies, J.M., Roper, T.J., and Shepherdson, D.J. (1987). Seasonal distribution of road kills in the European badger (*Meles meles*). *Journal of Zoology*, **211**, 525–529.

Delahay, R.J., Brown, J.A., Mallinson, P.J., et al. (2000). The use of marked bait in studies of the territorial organization of the European Badger (*Meles males*). *Mammal Review*, **30**, 73–87.

Delahay, R.J., Walker, N.J., Forrester, G.J., et al. (2006). Demographic correlates of bite wounding in Eurasian badgers, *Meles meles* L., in stable and perturbed populations. *Animal Behaviour*, **7**, 1047–1055.

Domingo-Roura, X., Macdonald, D.W., Roy, M.S., et al. (2003). Confirmation of low genetic diversity and multiple breeding females in a social group of Eurasian badgers from microsatellite and field data. *Molecular Ecology*, **12**, 533–539.

Doncaster, C.P., and Woodroffe, R. (1993). Den site can determine shape and size of badger territories: implications for group-living. *Oikos*, **66**, 88–93.

Douglas, D.A., Houde, A., Song, J.H., Farookhi, R., Concannon, P.W., and Murphy, B.D. (1998). Luteotropic hormone receptors in the ovary of the mink (Mustela vison) during delayed implantation and early-postimplantation gestation. *Biology of Reproduction*, **59**, 571–578.

Dugdale, H.L., Davison, D., Baker, S.E., Ellwood, S.A., Newman, C., Buesching, C.D., and Macdonald, D.W. (2011b). Female teat size is a reliable indicator of annual breeding success in European badgers: Genetic validation. *Mammalian Biology-Zeitschrift für Säugetierkunde*, **76**, 716–721.

Dugdale H.L., Elwood S.A., and Macdonald D.W. (2010). Alloparental care and long-term costs of mothers tolerating other members of the group in a plurally breeding mammal. *Animal Behaviour*, **84**, 721–735.

Dugdale, H.L., Macdonald, D.W., and Newman, C. (2003). Offspring sex ratio variation in the European badger, *Meles meles*. *Ecology*, **84**, 40–45.

Dugdale, H.L., Macdonald, D.W., Pope, L.C., and Burke, T. (2007). Polygynandry, extra-group paternity and multiple-paternity litters in European badger (*Meles meles*) social groups. *Molecular Ecology*, **16**, 5294–5306.

Dugdale, H.L., Macdonald, D.W., Pope, L.C., Johnson, P.J., and Burke, T. (2008). Reproductive skew and relatedness in social groups of European badgers, *Meles meles*. *Molecular Ecology*, **17**, 1815–1827.

Dugdale, H.L., Pope, L.C., Newman, C., Macdonald, D.W., and Burke, T. (2011a). Age-specific breeding success in a wild mammalian population: selection, constraint, restraint and senescence. *Molecular Ecology*, **20**, 3261–3274.

Dunn, P.O., Bollmer, J.L., Freeeman-Gallant, C.R., and Whittingham, L.A. (2013). MHC variation is related to a sexually selected ornament, survival, and parasite resistance in common yellowthroats. *Evolution*, **67**, 679–687.

Dyo, V., Ellwood, S.A., Macdonald, D.W., et al. (2010). Evolution and sustainability of a wildlife monitoring sensor network. In *Proceedings of the 8th ACM Conference on Embedded Networked Sensor Systems*, pp. 127–140. ACM.

Ellwood, S.A., Wilson, R.P., and Addison, A.C. (2007). Technology in conservation: a boon but with small print. In: Macdonald, D.W. and Service, K., eds., *Key Topics in Conservation Biology*, pp. 105–119. Blackwell Publishing, Oxford.

Erlinge, S. and Sandell, M. (1986). Seasonal changes in the social organization of male stoats, *Mustela erminea*: an effect of shifts between two decisive resources. *Oikos*, **47**, 57–62.

European Union, *Convention Implementing the Schengen Agreement of 14 June 1985 between the Governments of the States of the Benelux Economic Union, the Federal Republic of Germany and the French Republic, on the Gradual Abolition of Checks at their Common Borders ("Schengen Implementation Agreement")*, 19 June 1990, available at: http://www.refworld.org/docid/3ae6b38a20.html [accessed 26 February 2015].

Evans, P.G.H., Macdonald, D.W., and Cheeseman, C.L. (1989). Social structure of the Eurasian badger (*Meles meles*): genetic evidence. *Journal of Zoology*, **218**, 587–595.

Fell, R.J., Buesching, C.D., and Macdonald, D.W. (2006). The social integration of European badger (*Meles meles*) cubs into their natal group. *Behaviour*, **143**, 683–700.

Feore, S. and Montgomery, W.I. (1999). Habitat effects on the spatial ecology of the European badger (*Meles meles*). *Journal of Zoology*, **247**, 537–549.

Fisher, R.A. (1930). *The Genetical Theory of Natural Selection*. Oxford University Press, Oxford.

Fontagné, S., Lataillade, E., Breque, J., and Kaushik, S. (2008). Lipid peroxidative stress and antioxidant defence status during ontogeny of rainbow trout (*Oncorhynchus mykiss*). *British Journal of Nutrition*, **100**, 102–111.

Gittleman, J.L. and Harvey, P.H. (1982). Carnivore home-range size, metabolic needs and ecology. *Behavioural Ecology and Sociobiology*, **10**, 57–63.

Green, B.H. (1990). Agricultural intensification and the loss of habitat, species and amenity in British grasslands: a review of historical change and assessment of future prospects. *Grass and Forage Science*, **45**, 365–372.

Griffiths, H.I. and Thomas, D.H. (1993). The status of the badger (*Meles meles*) (L. 1758) (Carnivore, Mustelidae) in Europe. *Mammal Review*, **23**, 17–58.

Hamilton, W.D., Axelrod, R., and Tanese, R. (1990). Sexual reproduction as an adaptation to resist parasites (a review). *Proceedings of the National Academy of Sciences of the USA*, **87**, 3566–3573.

Hamilton, W.D. and Zuk, M. (1982). Heritable true fitness and bright birds: a role for parasites? *Science*, **218**, 384–387.

Hancox, M.K. (1973). *Studies on the ecology of the Eurasian badger* (*Meles meles L.*). D.Phil Thesis. University of Oxford.

Harrison, X.A., Bearhop, S., Inger, R., et al. (2011). Heterozygosity-fitness correlations in a migratory bird: an analysis of inbreeding and single-locus effects. *Molecular Ecology*, **20**, 4786–4795.

Helldin, J.O. (1999): Diet, body condition, and reproduction of Eurasian pine marten (*Martes martes*) during cyclic microtine density. *Ecography*, **22**, 324–336.

Hewitt, S.E., Macdonald, D.W., and Dugdale, H.L. (2009). Context-dependent linear dominance hierarchies in social groups of European badgers Meles meles. *Animal Behaviour*, **77**, 161–169.

Huchard, E. and Cowlishaw, G. (2011). Female–female aggression around mating: an extra cost of sociality in a multimale primate society. *Behavioral Ecology*, **22**, 1003–1011.

Isvaran, K. and Clutton-Brock, T.H. (2007). Ecological correlates of extra-group paternity in mammals. *Proceedings of the Royal Society of London B: Biological Science*, **274**, 219–224.

Johnson, D.D.P., Jetz, W., and Macdonald, D.W. (2002a). Environmental correlates of badger social spacing across Europe. *Journal of Biogeography*, **29**, 411–425.

Johnson, D.D., Kays, R., Blackwell, P.G., and Macdonald, D.W. (2002b). Does the resource dispersion hypothesis explain group living? *Trends in Ecology and Evolution*, **17**, 563–570.

Johnson, D.D., and Macdonald, D.W. (2003). Sentenced without trial: reviling and revamping the resource dispersion hypothesis. *Oikos*, **101**, 433–440.

Johnson, D.D., Macdonald, D.W., and Dickman, A.J. (2000). An analysis and review of models of the sociobiology of the Mustelidae. *Mammal Review*, **30**, 171–196.

Johnson, D.D.P., Stopka, P., and Macdonald, D.W. (2004). Ideal flea constraints on group living: unwanted public goods and the emergence of cooperation. *Behavioral Ecology*, **15**, 181–186.

Johnston, R.E., Sorokin, E.S., and Ferkin, M.H. (1997). Scent counter-marking by male meadow voles: females prefer the top-scent male. *Ethology*, **103**, 443–453.

Judge, J., Wilson, G.J., Macarthur, R., Delahay, R.J., and McDonald, R.A. (2014). Density and abundance of badger social groups in England and Wales in 2011–2013. *Scientific Reports*, **4**.

Juola, F.A. and Dearborn, D.C. (2012). Sequence-based evidence for major histocompatibility complex-disassortative mating in a colonial seabird. *Proceedings of the Royal Society of London. Series B*, **279**, 153–162.

Kaneko, Y., Kanda, E., Tashima, S., Masuda, R., Newman, C., and Macdonald, D.W. (2014). The socio-spatial dynamics of the Japanese badger (*Meles anakuma*). *Journal of Mammalogy*, **95**, 290–300.

Kaneko, Y., Newman, C., Buesching, C.D., and Macdonald, D.W. (2010). Variations in badger (*Meles meles*) sett microclimate: differential cub survival between main and subsidiary setts, with implications for artificial sett construction. *International Journal of Ecology*, **2010**, ID 859586, 10 pages.

Kaufman, J. (1999). Co-evolving genes in MHC haplotypes: the 'rule' for non-mammalian vertebrates? *Immunogenetics*, **50**, 228–236.

Kauhala, K. (1995). Mayran levinneisyys ja runsaus Suomessa. *Suomen Riista*, **41**, 85–94.

Kerr, B. (2007). Niche construction and cognitive evolution. *Biological Theory*, **2**, 250–262.

Khan, J.Y. and Black, S.M. (2003). Developmental changes in murine brain antioxidant enzymes. *Pediatric Research*, **54**, 77–82.

Kilshaw, K., Newman, C., Buesching, C.D., Bunyan, J., and Macdonald, D.W. (2009). Investigating coordinated latrine use by European badgers *Meles meles*: Potential consequences for territory defence. *Journal of Mammalogy*, **90**, 1188–1198.

Kokko, H. and Rankin, D.J. (2006). Lonely hearts or sex in the city? Density-dependent effects in mating systems. *Proceedings of the Royal Society of London. Series B*, **361**, 319–334.

Kowalczyk, R. Bunevich, A.N., and Jedrzwjewska, B. (2000). Badger density and distribution of setts in Bialowiezqa Primeval Forest (Poland and Belarus) compared to other Eurasian populations. *Acta Theriologica*, **45**, 395–408.

Kowalczyk, R., Jedrzwjewska, B., and Zalewski, A. (2003a). Annual and circadian activity patterns of badgers *Meles meles* in Bialowieza Primeval Forest (E Poland) compared to other Palaearctic populations. *Journal of Biogeography*, **30**, 463–472.

Kowalczyk, R., Zalewski, A., Jedrzwjewska, B., and Jedrzwjewska, W. (2003b). Spatial organization and demography of badgers *Meles meles* in Bialowieza Forest (Poland) and the influence of earthworms on badger densities in Europe. *Canadian Journal of Zoology*, **81**, 74–87.

Krebs, C.J. (1994). *Ecology*. Harper-Collins, New York.

Kruuk, H. (1972). *The Spotted Hyena: A Study of Predation and Social Behavior*. University of Chicago Press.

Kruuk, H. (1978a). Foraging and spatial organisation of the European badger, *Meles meles* L. *Behavioral Ecology and Sociobiology*, **4**, 75–89.

Kruuk, H. (1978b). Spatial organization and territorial behaviour of the European badger. *Journal of Zoology*, **184**, 1–19.

Kruuk, H. (1989). The *Social Badger: Ecology and Behaviour of a Group-living Carnivore (Meles meles)*. Oxford University Press.

Kruuk, H. and Macdonald, D.W. (1985). Group territories of carnivores: empires and enclaves. In R.M. Sibly and R.H. Smith, eds., *Behavioural Ecology. Symposia of the British Ecological Society*, **25**, 521–536. Blackwell Scientific, Oxford.

Kruuk, H. and Parish, T. (1982). Factors affecting population density, group size and territory size of the European badger, *Meles meles*. *Journal of Zoology*, **196**, 31–39.

Lanyon, C.V., Rushton, S.P., O'Donnell, A.G., Goodfellow, M., Ward, A.C., Petrie, M., Jensen, S.P., Gosling, L.M., and Penn, D.J. (2007). Murine scent mark microbial communities are genetically determined. *FEMS Microbiology Ecology*, **59**, 576–583.

Lawlor, D.A., Zemmour, J., Ennis, P.D., and Parham, P. (1990). Evolution of class-I MHC genes and proteins: from natural selection to thymic selection. *Annual Review of Immunology*, **8**, 23–63.

Lee, H.J., Macbeth, A.H., Pagani, J.H., and Young, W.S. (2009). Oxytocin: the great facilitator of life. *Progressive Neurobiology*, **88**, 127–151.

Levitis, D.A. (2011). Before senescence: the evolutionary demography of ontogenesis. *Proceedings of the Royal Society B-Biological Sciences*, **278**, 801–809.

Lizundia, R., Newman, C., Buesching, C.D., et al. (2011). Evidence for a role of the host-specific flea (*Paraceras melis*) in the transmission of *Trypanosoma (Megatrypanum) pestanai* to the European badger. *PLOS ONE*, **6**, e16977.

Loureiro, F., Rosalino, L.M., Macdonald, D.W., and Santos-Reis, M. (2007). Use of multiple den sites by Eurasian badgers, *Meles meles*, in a Mediterranean habitat. *Zoological science*, **24**, 978–985.

Lups, P. and Roper, T. (1990). Cannibalism in a female badger (*Meles meles*): infanticide or predation? *Journal of Zoology*, **221**, 314–315.

Macdonald, D.W. (1981). Resource dispersion and the social organization of the red fox (Vulpes vulpes). In *Worldwide Furbearer Conference Proceedings, 1981* (Vol. 2, pp. 918–949). University of Maryland Press.

Macdonald, D.W. (1983). The ecology of carnivore social behaviour. *Nature*, **301**, 379–384.

Macdonald, D.W. (1993). *The Velvet Claw: a Natural History of the Carnivores*, p. 210. Parkwest, Miami.

Macdonald D.W., Anwar M., Newman C., Woodroffe R., and Johnson P.J. (1999). Inter-annual differences in the age-related prevalences of *Babesia* and *Trypanosoma* parasites of European badgers (*Meles meles*). *Journal of Zoology*, **247**, 65–70.

Macdonald, D.W., Buesching, C.D., Stopka, P., Henderson, J., Ellwood, S.A., and Baker, S.E. (2004a). Encounters between two sympatric carnivores: red foxes (Vulpes vulpes) and European badgers (*Meles meles*). *Journal of Zoology*, **263**, 385–392.

Macdonald, D.W. and Carr, G.M. (1989). Food security and the rewards of tolerance. In V. Standen and R. Foley, eds., *Comparative Socioecology: The Behavioural Ecology of Humans and Animals*, pp. 75–99. Blackwell Scientific Publications, Oxford.

Macdonald, D.W., Creel, S., and Mills, M.G.L. (2004d). Society. In D.W. Macdonald and C. Sillero-Zubiri, J. Loveridge, eds., *The Biology and Conservation of wild canids*, pp.825–106. Oxford University Press, Oxford, UK.

Macdonald, D.W. and Feber, R.E., eds. (2015). *Wildlife Conservation on Farmland. Managing for Nature on Lowland Farms*. Oxford University Press, Oxford.

Macdonald, D.W., Harmsen, B. J., Johnson, P. J., and Newman, C. (2004b). Increasing frequency of bite wounds with increasing population density in Eurasian badgers, *Meles meles*. *Animal Behaviour*, **67**, 745–751.

Macdonald, D.W., Mitchelmore, F., and Bacon, P.J. (1996). Predicting badger sett numbers: evaluating methods in East Sussex. *Journal of Biogeography*, **23**, 649–655.

Macdonald, D.W., Mosser, A. and Loveridge, A.J. (2010a). Field Society. In D.W. Macdonald and A.J. Loveridge, eds., *The Biology and Conservation of wild felids*, pp. 125–160. Oxford University Press, Oxford, UK.

Macdonald, D.W. and Newman, C. (2002). Badger *Meles meles* population dynamics in Oxfordshire, UK, numbers, density and cohort life histories, and a possible role of climate change in population growth. *Journal of Zoology*, **256**, 121–138.

Macdonald, D.W., Newman, C., Buesching, C.D., and Johnson, P.J. (2008). Male-biased movement in a high density population of the Eurasian badger (*Meles meles*). *Journal of Mammalogy*, **89**, 1077–1086.

Macdonald, D.W., Newman, C., Buesching, C.D., and Nouvellet, P. (2010b). Are badgers '*Under The Weather*'? Direct and indirect impacts of climate variation on European badger (*Meles meles*) population dynamics. *Global Change Biology*, **16**, 2913–2922.

Macdonald, D.W., Newman, C., Dean, J., Buesching, C.D., and Johnson, P.J. (2004c). The distribution of Eurasian badger *Meles meles* setts in a high-density area: field observations contradict the Sett Dispersion Hypothesis. *Oikos*, **106**, 295–307.

Macdonald, D.W., Newman, C., Nouvellet, P.M., and Buesching, C.D. (2009). An analysis of Eurasian badger (*Meles meles*) population dynamics: implications for regulatory mechanisms. *Journal of Mammalogy*, **90**, 1392–1403.

Macdonald, D.W., Newman, C., Stewart, P.D., and Domingo-Roura, X. (2002). Density-dependent regulation of body weight and condition in badgers (*Meles meles*) from Wytham Woods, Oxfordshire. *Ecology*, **83**, 2056–2061.

Macdonald, D.W., Stewart, P.D., Stopka, P., and Yamaguchi, N. (2000). Measuring the dynamics of mammalian societies: an ecologist's guide to ethological methods. In L. Boitani and T.K. Fuller, eds., *Research Techniques in Animal Ecology: Controversies and Consequences*, pp. 332–388. Columbia University Press, New York.

MacKinney, M.L., and MacNamara, K.J. (1991). *Heterochrony: the Evolution of Ontogeny*. Plenum Press.

Maher, C.R., and Lott, D.F. (1995). Definitions of territoriality used in the study of variation in vertebrate spacing systems. *Animal Behaviour*, **49**, 1581–1597.

McCarthy, M.M. (1990). Oxytocin inhibits infanticide in female house mice *Mus domesticus*. *Hormones and Behavior*, **24**, 365–375.

McGraw, K.J., Cohen, A.A., Costantini, D., and Hõrak, P. (2010). The ecological significance of antioxidants and oxidative stress: a marriage between mechanistic and functional perspectives. *Functional Ecology*, **24**, 947–949.

McLaren, G.W., Macdonald, D.W., Georgiou, C., Mathews, F., Newman, C., and Mian, R. (2003). Leukocyte coping capacity: a novel technique for measuring the stress response in vertebrates. *Experimental Physiology*, **88**, 541–546.

McLaren, G.W., Thornton, P.D., Newman, C., Buesching, C.D., Baker, S.E., Mathews, F., and Macdonald, D.W. (2005). The use and assessment of ketamine-medetomidine-butorphanol combinations for field anaesthesia in wild European badgers (*Meles meles*). *Veterinary Anaesthesia and Analgesia*, **32**, 367–372.

Mertl-Millhollen, A.S. (2006). Scent marking as resource defense by female Lemur catta. *American Journal of Primatology*, **68**, 605–621.

Milinski, M., Griffiths, S., Wegner, K.M., Reusch, T.B.H., Haas-Assenbaum, A., and Boehm, T. (2005). Mate choice decisions of stickleback females predictably modified by MHC 190 peptide ligands. *Proceedings of the National Academy of Sciences of the United States of America*, **102**, 4414–4418.

Montes, I., McLaren, G.W., Macdonald, D.W., and Mian, R. (2004). The effect of transport stress on neutrophil activation in wild badgers (*Meles meles*). *Animal Welfare*, **13**, 355–359.

Montes, I., Newman, C., Mian, R., and Macdonald, D.W. (2011). Radical health: ecological corollaries of body condition, transport stress and season on plasma antioxidant capacity in the European badger. *Journal of Zoology*, **284**, 114–123.

Mooney, J. (2009). Dogs seized in police raids on badger baiters. *Sunday Times*, 22 February, retrieved from <http://www.thesundaytimes.co.uk>.

Murdoch, W.W. (1970). Population regulation and population inertia. *Ecology*, **51**, 497–502.

Murdoch, W.W. (1994). Population regulation in theory and practice. *Ecology*, **75**, 271–287.

Neal, E.G. (1948). *The Badger* (Vol. 1). Collins.

Newman, C. (2000). The demography and parasitology of the Wytham Woods badger population. *D.Phil Thesis, Oxford*.

Newman, C., Buesching, C.D., and Macdonald, D.W. (2004). First report of Cheyletiella parasitovorax infestation in the Eurasian badger (*Meles meles*). *Veterinary Record*, **155**, 180–181.

Newman, C., Buesching, C.D., and Wolff, J.O. (2005). The function of facial masks in "midguild" carnivores. *Oikos*, **10**, 623–633.

Newman, C. and Macdonald, D.W. (2013). *The implications of climate change for terrestrial UK mammals*. Terrestrial biodiversity climate change impacts report card technical paper. Living with environmental change partnership. <http://www.lwec.org.uk/>.

Newman, C., Macdonald, D.W., and Anwar, A. (2001). Coccidiosis in the European badger, *Meles meles* in Wytham Woods: infection and consequences for growth and survival. *Parasitology*, **123**, 133–142.

Newman, C., Zhou, Y-B., Buesching, C.D., Kaneko, Y., and Macdonald, D.W. (2011). Contrasting sociality in two widespread, generalist, mustelid genera, *Meles* and *Martes*. *Mammal Study*, **36**, 169–188.

Noonan, M.J., Markham, A., Newman, C., Trigoni, N., Buesching, C.D., Ellwood, A.A., and Macdonald, D.W. (2014). Climate and the individual: inter-annual variation in the autumnal activity of the European badger. *PLOS ONE*, **9**(1), e83156.

Noonan, M.J., Markham, A., Newman, C., Trigoni, N., Buesching, C.D., Ellwood, A.A., and Macdonald, D.W. (2015). A new Magneto-Inductive tracking technique to uncover subterranean activity: What do animals do underground?. *Methods in Ecology and Evolution*. doi: 10.1111/2041-210X.12348.

Nouvellet, P., Buesching, C.D., Dugdale, H.L., Newman, C., and Macdonald, D.W. (2011). Mouthing off about developmental stress: individuality of palate marking in the European badger and its relationship with juvenile parasitoses. *Journal of Zoology*, **283**, 52–62.

Nouvellet, P.M., Newman, C., Buesching, C.D., and Macdonald, D.W. (2013). A multi-metric approach to investigate the effects of weather conditions on the demographic of a terrestrial mammal, the European badger (*Meles meles*). *PLOS ONE*, **8**(7): e68116.

Nowak, M.A., Tarczy-Hornoch, K., and Austyn, J.M. (1992). The optimal number of major histocompatibility complex molecules in an individual. *Proceedings of the National Academy of Sciences of the United States of America*, **89**, 10896–10899.

Oli, M.K. (2004). The fast–slow continuum and mammalian life-history patterns: an empirical evaluation. *Basic and Applied Ecology*, **5**, 449–463.

Oli, M.K. and Zinner, B. (2001). Partial life-cycle analysis: A model for birth pulse populations. *Ecology*, **82**, 1180–1190.

Paget, R.J. and Middleton, L.F. (1974). Some observations on the sexual activities of badgers in Yorkshire in months December-April. *Journal of Zoology*, **173**, 256–260.

Parker G.A. and Maynard-Smith, J. (1990). Optimality theory in evolutionary biology. *Nature*, **348**, 27–33.

Parsons, P.A. (1992). Fluctuating asymmetry: a biological monitor of environmental and genomic stress. *Heredity*, **6**, 361–364.

Penn, D.J., Damjanovich, K., and Potts, W.K. (2002). MHC heterozygosity confers a selective advantage against multiple-strain infections. *Proceedings of the National Academy of Sciences USA*, **99**, 11260–11264.

Penn, D.J. and Potts, W.K. (1999). The evolution of mating preferences and major histocompatibility complex genes. *American Naturalist*, **153**, 145–164.

Philcox, C.K., Grogan, A.L., and Macdonald, D.W. (1999). Patterns of otter *Lutra lutra* road mortality in Britain. *Journal of Applied Ecology*, **36**, 748–762.

Pianka, E.R. (1970). On r- and K- selection. *The American Naturalist*, **104**, 592–597.

Pocock, R.I. (1911). LXXXVII—Some probable and possible instances of warning characteristics amongst insectivorous and carnivorous mammals. *The Annals and Magazine of Natural History*, **8**, 750–757.

Potter, B. (1912). *The Tale of Mr. Tod*. Frederick Warne and Co.

Powell, R.A. (1979). Mustelid spacing patterns: variations on a theme by *Mustela*. *Zeitschrift für Tierpsychologie*, **50**, 153–165.

Promislow, D.E., and Harvey, P.H. (1990). Living fast and dying young: a comparative analysis of life-history variation among mammals. *Journal of Zoology*, **220**, 417–437.

Pusey, A., and Wolf, M. (1996). Inbreeding avoidance in animals. *Trends in Ecology and Evolution*, 11(5), 201–206.

Puurtinen, M., Ketola, T., and Katiaho, J.S. (2005). Genetic compatibility and sexual selection. *Trends in Ecology and Evolution*, **20**, 157–158.

Reiss, M. (1988). Scaling home range size: body size, metabolic needs and ecology. *Trends in Ecology and Evolution*, **3**, 85–86.

Revilla, E. (2003a). Moving beyond the resource dispersion hypothesis. *Trends in Ecology and Evolution*, **18**, 380.

Revilla, E. (2003b). What does the resource dispersion hypothesis explain, if anything? *Oikos*, **101**, 428–432.

Revilla, E. and Palomares, F. (1999). Changes in the behaviour of a male Eurasian badger: evidence in favour of the anti-kleptogamy hypothesis? *Acta Theriologica*, **44**, 471–476.

Revilla, E. and Palomares, F. (2002). Spatial organization, group living and ecological correlates in low-density population of Eurasian badgers, *Meles meles*. *Journal of Animal Ecology*, **71**, 497–512.

Rodríguez, A., Martin, R., and Delibes, M. (1996). Space use and activity in a Mediterranean population of badgers *Meles meles*. *Acta Theriologica*, **41**, 59–72.

Rogers, L.M., Cheeseman, C.L., Mallinson, P.J., and Clifton-Hadley, R.S. (1997). Demography of a high density badger (*Meles meles*) population in the west of England. *Journal of Zoology*, **242**, 705–728.

Root, T.L., Liverman, D., and Newman, C. (2006). Managing biodiversity in the light of climate change: current biological effects and future impacts. In D. Macdonald and K. Service, eds., *Key Topics in Conservation Biology*, pp. 85–104. Blackwell Publishing, Oxford.

Roper, T. J. (1992). Badger *Meles meles* setts–architecture, internal environment and function. *Mammal Review*, **22**, 43–53.

Roper, T.J. (1994). The European badger *Meles meles*: food specialist or generalist? *Journal of Zoology*, **234**, 437–452.

Roper, T. (2010). *Badger*. Harper Collins Publishers, London, United Kingdom.

Rosalino, L.M., Loureiro, F., Macdonald, D.W., and Santos-Reis, M. (2005). Dietary shifts of the badger (*Meles meles*) in Mediterranean woodlands: an opportunistic forager with seasonal specialisms. *Mammalian Biology-Zeitschrift für Säugetierkunde*, **70**, 12–23.

Rosalino, L.M., Macdonald, D.W., and Santos-Reis, M. (2004). Spatial structure and land cover use in a low density Mediterranean population of Eurasian badgers. *Canadian Journal of Zoology*, **82**, 1493–1502.

Royama, T. (1992). *Analytical population Dynamics* (Vol. 10). Springer.

Saeki, M. and Macdonald, D.W. (2004). The effects of traffic on the raccoon dog (*nyctereutes procyonoides viverrinus*) and other mammals in Japan. *Biological Conservation*, **118**, 559–571.

Sandell, M. (1989). The mating tactics and spacing patterns of solitary carnivores. In J.L. Gittleman, ed. *Carnivore Behaviour, Ecology, and Evolution*, pp. 164–182. Cornell University Press, New York, USA.

Sapolsky, R.M. (2011). Physiological and pathophysiological implications of social stress in mammals. In: *Handbook of Physiology, The Endocrine System, Coping with the Environment: Neural and Endocrine Mechanisms*, pp. 517–532. John Wiley & Sons, Inc., New Jersey.

Savill, P., Perrins, C., Kirby, K., and Fisher, N. (2010). *Wytham Woods: Oxford's Ecological Laboratory*. Oxford University Press.

Shepard, E.L.C., Wilson, R.P., Quintana, F., et al. (2008) Identification of animal movement patterns using tri-axial accelerometry. *Endangered Species Research*, **10**, 47–60.

Silk, J.B. (2007). Social components of fitness in primate groups. *Science*, **317**, 1347–1351.

Sillero-Zubiri, C., Gottelli, D., and Macdonald, D.W. (1996). Male philopatry, extra-pack copulations and inbreeding avoidance in Ethiopian wolves (*Canis simensis*). *Behavioral Ecology and Sociobiology*, **38**, 331–340.

Sin, Y.W., Annavi, G., Dugdale, H., Newman, C., Burke, T., and Macdonald, D.W. (in press). MHC-similar mate preference in European badgers (*Meles meles*). *Molecular Ecology*.

Sin, Y.W., Buesching, C.D., Burke, T., and Macdonald, D.W. (2012a). Molecular characterization of the microbial communities in the subcaudal gland secretion of the European badger (*Meles meles*). *FEMS Microbiology Ecology*, **81**, 648–659.

Sin, W.A., Dugdale, H.L., Newman, C., Burke, T., and Macdonald, D.W. (2014). Pathogen burden, co-infection and major histocompatibility complex variability in the European badger (*Meles meles*). *Molecular Ecology*, **23**(20), 5072–5088.

Sin, Y.W., Dugdale, H.L., Newman, C., Macdonald, D.W., and Burke, T. (2012b). MHC class II genes in the European badger (*Meles meles*): characterization, patterns of variation, and transcription analysis. *Immunogenetics*, **64**, 313–327.

Sin, Y.W., Dugdale, H.L., Newman, C., Macdonald, D.W., and Burke, T. (2012c). Evolution of MHC class I genes in the European badger (*Meles meles*). *Ecology and Evolution*, **3**, 285.

Skinner, C., Skinner, P., and Harris, S. (1991). The past history and recent decline of badgers *Meles meles* in Essex—an analysis of some of the contributory factors. *Mammal Review*, **21**, 67–80.

Skoog, P. (1970). *The Food of the Swedish Badger, Meles meles L.* Svenska jägareförbundet. 101–850–719 (*Last edited on* 2002/02/27 17:35:44 US/Mountain).

Slate, J., Kruuk, L.E.B., Marshall, T.C., Pemberton, J.M., and Clutton-Brock, T.H. (2000). Inbreeding depression influences lifetime breeding success in a wild population of red deer (*Cervus elaphus*). *Proceedings of the Royal Society of London Series B: Biological Sciences*, **267**, 1657–1662.

Sleeman, D.P., Davenport, J. Clegg, T.A., et al. (2009). How many Eurasian badgers *Meles meles* L. are there in the Republic of Ireland? *European Journal of Wildlife Research*, **55**, 333–344.

Smith, J. M. (1976). A comment on the Red Queen. *American Naturalist*, **110**, 325–330.

Stevens, J.R. and Hauser, M.D. (2004). Why be nice? Psychological constraints on the evolution of cooperation. *Trends in Cognitive Sciences*, **8**, 60–65.

Stewart, P.D., Anderson, C., and Macdonald, D.W. (1997a). A mechanism for passive range exclusion: evidence from the European badger (*Meles meles*). *Journal of Theoretical Biology*, **184**, 279–289.

Stewart, P.D., Bonesi, L., and Macdonald, D.W. (1999). Individual differences in den maintenance effort in a communally dwelling mammal: the Eurasian badger. *Animal Behaviour*, **57**, 153–161.

Stewart, P.D., Ellwood, S.A., and Macdonald, D.W. (1997b). Remote video-surveillance of wildlife–an introduction from experience with the European badger *Meles meles*. *Mammal Review*, **27**, 185–204.

Stewart, P.D. and Macdonald, D.W. (1997). Age, sex, and condition as predictors of moult and the efficacy of a novel fur-clip technique for individual marking of the European badger (*Meles meles*). *Journal of Zoology*, **241**, 543–550.

Stewart, P.D. and Macdonald, D.W. (2003). Badgers and badger fleas: strategies and counter-strategies. *Ethology*, **109**, 751–764.

Stewart, P.D., Macdonald, D.W., Newman, C., and Cheeseman, C. (2001). Boundary faeces matching in the European badger (*Meles meles*): a mechanism for range exclusion? *Journal of Zoology*, **255**, 191–198.

Stewart, P.D., Macdonald, D.W., Newman, C., and Tattersall, F.H. (2002). Behavioural mechanisms of information transmission and reception by badgers, *Meles meles*, at latrines. *Animal Behaviour*, **63**, 999–1007.

Stopka, P. and Johnson D.D.P. (2000). Badger (*Meles meles*) as a model species for the development of ecological and behavioural research. *Lynx*, **31**, 125–131.

Stopka, P. and Macdonald, D.W. (2003). Way-marking behaviour: an aid to spatial navigation in the wood mouse (*Apodemus sylvaticus*). *BMC Ecology*, **3**, 3.

Swain, S.L. (1983). T cell subsets and the recognition of MHC class. *Immunological Reviews*, **74**, 129–142.

Tanaka, H., Yamanaka, Y., and Endo, K. (2002). Spatial distribution and sett use by the Japanese badger, *Meles meles anakuma*. *Mammal Study*, **27**, 15–22.

Thom, M.D., Johnson, D.D., and Macdonald, D.W. (2004). The evolution and maintenance of delayed implantation in the Mustelidae (Mammalia: Carnivora). *Evolution*, **58**(1), 175–183.

Thornton, P.S. (1988). Density and distribution of badgers in S.W. England—a predictive model. *Mammal Review*, **18**, 11–23.

Thornton, P.D., Newman, C., Johnson, P.J., Buesching, C.D., Baker, S.E., Slater, D., Johnson, D.P., and Macdonald, D.W. (2005). Preliminary comparison of four anaesthetic techniques in badgers (*Meles meles*). *Veterinary Anaesthesia and Analgesia*, **32**, 40–47.

Tinnesand, H.V., Buesching, C.D., Noonan, M.J., Newman, C., Zedrosser, A., Rosell, F., and Macdonald, D.W. (in press). Will tresapassers be prosecuted or assessed according to their merits? A consilient interpretation of territoriality in a group-living carnivore, the European badger (*Meles meles*). *PLOS-1*.

Trivers, R.L, and Willard, D.E. (1973). Natural selection of the parental ability to vary sex ratio. *Science*, **179**, 90–92.

Turchin, P. (1995). Population regulation: old arguments and a new synthesis. In N. Cappuccino and P.W. Price, eds., *Population Dynamics: Novel Approaches and Synthesis*. Academic Press, San Diego, California.

Tuyttens, F.A.M., Delahay, R.J., Macdonald, D.W., Cheeseman, C.L., Long, B., and Donnelly, C.A. (2000). Spatial perturbation caused by a badger (*Meles meles*) culling operation: implications for the function of territoriality and the control of bovine tuberculosis (*Mycobacterium bovis*). *Journal of Animal Ecology*, **69**, 815–828.

van Valen, L. (1973). A new evolutionary law. *Evolutionary Theory*, **1**, 1–30.

Vicente, J., Delahay, R.J., Walker, N.J., and Cheeseman, C.L. (2007). Social organization and movement influence the incidence of bovine tuberculosis in an undisturbed high-density badger Meles meles population. *Journal of Animal Ecology*, **76**, 348–360.

Webb, D.R. and King, J.R. (1984). Effects of wetting on insulation of bird and mammal coats. *Journal of Thermal Biology*, **9**, 189–191.

Wilson, G., Harris, S., and McLaren, G. (1997). *Changes in the British Badger Population, 1988–1997*. People's Trust for Endangered Species, London.

Wong, J., Stewart, P.D., and Macdonald, D.W. (1999). Vocal repertoire in the European badger (*Meles meles*): structure, context, and function. *Journal of Mammalogy*, **80**, 570–588.

Woodroffe, R. (1993). Alloparental behaviour in the European badger. *Animal Behaviour*, **46**, 413–415.

Woodroffe, R. (1995). Body condition affects implantation date in the European badger, *Meles meles*. *Journal of Zoology*, **236**, 183–188.

Woodroffe, R. and Macdonald, D.W. (1993). Badger sociality—models of spatial grouping. *Symposia of the Zoological Society of London*, **65**, 145–169.

Woodroffe, R. and Macdonald, D.W. (1995a). Female/female competition in European badgers *Meles meles*: effects on breeding success. *Journal of Animal Ecology*, **64**, 12–20.

Woodroffe, R. and Macdonald, D.W. (1995b). Costs of breeding status in the European badger, *Meles meles*. *Journal of Zoology*, **235**, 237–245.

Woodroffe, R. and Macdonald, D.W. (2000). Helpers provide no detectable benefits in the European badger (*Meles meles*). *Journal of Zoology*, **250**, 113–119.

Woodroffe, R., Macdonald, D.W., and Cheeseman, C.L. (1997). Endocrine correlates of contrasting male mating strategies in the European badger (*Meles meles*). *Journal of Zoology*, **241**, 291–300.

Woodroffe R., Macdonald, D.W., and da Silva, J. (1995). Dispersal and philopatry in the European badger (*Meles meles*). Journal of Zoology, **237**, 227–239.

Woodroffe, R., and Vincent, A. (1994). Mother's little helpers: patterns of male care in mammals. *Trends in Ecology and Evolution*, **9**, 294–297.

Wright, T. (1913) *The Historical Works of Giraldus Cambrensis*. London.

Yamaguchi, N., Dugdale, H.L., and Macdonald, D.W. (2006). Female receptivity, embryonic diapause, and superfetation in the European badger (*Meles Meles*): implications for the reproductive tactics of males and females. *The Quarterly Review of Biology*, **81**, 33–48.

Yamaguchi, N. and Macdonald, D.W. (2003). The burden of co-occupancy: intraspecific resource competition and spacing patterns in American mink, *Mustela vison*. *Journal of Mammalogy*, **84**, 1341–1355.

Yamazaki, K., Beauchamp, G.K., Imai, Y., et al. (1990). Odor types determined by the major histocompatibility complex in germ-free mice. *Proceedings of the National Academy of Sciences USA*, **87**, 8413–8416.

Zalewski, A. (2007). Does size dimorphism reduce competition between sexes? The diet of male and female pine martens at local and wider geographical scales. *Acta theriologica*, **52**, 237–250.

CHAPTER 5

Badgers and bovine tuberculosis: beyond perturbation to life cycle analysis

David W. Macdonald, Rosie Woodroffe, and Philip Riordan

Science is organized common sense where many a beautiful theory was killed by an ugly fact.

Thomas H. Huxley

Introduction

By the time you read this, the current crop of topical questions about controlling bovine tuberculosis (bTB)—can farmers shoot badgers *Meles meles* humanely and safely? can badgers be counted and monitored accurately and precisely?—will have been superseded by a new crop. Ephemeral detail is not the meat of an enduring chapter but behind details lie principles that will outlive any transitory nuance of policy. Our purpose here, against the background of badger biology, summarized in Chapter 4, this volume, is to expose some of these principles that will be continuingly relevant to whatever policy imperative unfolds.

There can be no doubt that the control[1] or eradication[2] of bTB in the UK is hindered by infection within badger populations. In this chapter, we will first summarize sufficient of the history and epizootiology of bTB to interpret the role that badgers play in maintaining infection, and to understand attempts that have been made to mitigate this role. Within this framework, we will then synthesize our own insights on how culling badgers changes the behaviour of surviving animals, and the consequences for the transmission of infection, both within badger populations and from badgers to cattle.

Then, having considered how these findings bear on current proposals to manage bTB through the culling of more badgers, we turn finally to the constellation of wider factors that might together offer a balanced perspective of the options. This is analogous to a life cycle analysis, such as that we describe in Macdonald and Feber (2015): Chapter 7, where we evaluate the environmental impacts of wider farm policy.

5.1 bTB: disease and epidemiology

Robert Koch discovered the pathogen causing tuberculosis, *Mycobacterium*, in 1882, and developed the first tuberculin 'skin' test[3] used in diagnosis in 1890 (Gradmann 2001).

In 1907, the interim report of the Royal Commission on Tuberculosis demonstrated that *M. bovis* was a zoonotic disease (transmitted from cattle to man), and its control has been a challenge ever since. Between 1912 and 1937, cattle herds infected with *Mycobacterium bovis* were responsible for the death of *c*. 65,000 people in England and Wales, due to human TB contracted from consuming raw infected milk (Wilson 1943; Good and Duignan 2011). Initial control was limited to clinical examination of herds and bacteriological examination of milk, and slaughtering of cows infected with *M. bovis* was voluntary. By 1934, the Committee on Cattle Disease, the Gowland Hopkins Committee, reported

[1] Control: reduction of disease incidence to a locally acceptable level as a result of deliberate efforts.

[2] Eradication: reduction to zero incidence of infection in a defined geographical area as a result of deliberate efforts.

[3] A small amount of the tuberculin (*M. bovis* or same family of bacteria) is injected into skin and if the animal is infected, the skin shows an allergic reaction.

Wildlife Conservation on Farmland. Conflict in the Countryside. Edited by David W. Macdonald and Ruth E. Feber.
© Oxford University Press 2015. Published 2015 by Oxford University Press.

that 40% of cows in dairy herds were infected (Economic Advisory Council 1934) and, during the 1930s, bTB was responsible for about 2500 human deaths per year in the UK[4] (Defra 2013a).

Control of bTB in the cattle industry from that time has focused on the testing and slaughter of infected cattle. In 1935, the government launched the first national Attested Herds Scheme programme for milk production[5], which involved voluntary regular skin testing and compulsory slaughter (with compensation), and encouraged farmers to take up the scheme with bonus payments (e.g. one penny per gallon of milk from an attested herd, replaced by £1 per animal in 1938; Defra 2011a). The programme faltered during the war years, but in 1950 the government introduced a national compulsory TB Area Eradication Plan, using the tuberculin skin test and immediate slaughter of reactor cattle[6], for which farmers were compensated. The eradication scheme began in counties that already had a high percentage of attested herds from the voluntary scheme and, by October 1960, the whole of the country was attested. The number of test reactors was reduced from 15,000 in 1961 to 569 in 1982, reaching an historical minimum in 1979 when less than 0.49% of herds (0.02% of all cattle) tested had a reactor (Defra 2011a; Macdonald and Burnham 2011). However, the disease still existed in pockets in the south west of England and, by the mid 1980s, herd breakdowns started to rise in this area. In a decade, the proportion of total herds with reactors in the South West increased from 0.75% in 1986 to 2.61% in 1996 (versus 0.10% to 0.51% for the same period in the rest of England; Krebs 1997), where reactors have remained three times higher than in the rest of Britain (e.g. Fig. 5.1).

In 2013, the total annual herd testing area covered 60,000 km^2 of England and the entirety of Wales (Godfray et al. 2013). During 2011, 11.5% of herds in England were under movement restrictions at some point due to bTB (23.6% of herds in the south west of England; Defra 2013a). Indeed, the compulsory slaughter of reactor cattle or their direct contacts have increased in Britain from fewer than 7000 in the late twentieth century to 32,620 in 2013 (Defra and AHVLA 2014)

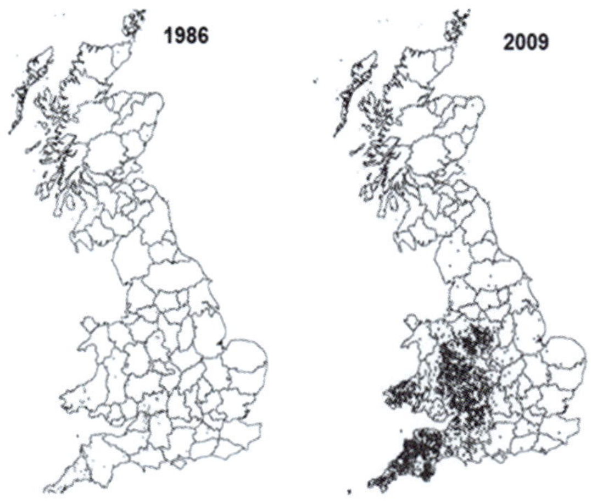

Figure 5.1 Bovine TB cases in England, showing isolated cases of herd breakdowns in 1986 and more widespread patterns in the South West in 2009 (maps from Defra <http://www.defra.gov.uk/>, ©Crown Copyright).

(Fig. 5.2), with a (provisional) 2013 incidence rate of 4.5% of new herds infected[7] (Defra and AHVLA 2014).

Pasteurization of milk and regular testing and slaughter of infected cattle mean that bTB is now no longer a major threat to humans in Britain: in the UK, from 5200 culture-confirmed cases of human TB in 2012, only 0.7% were transmitted by *M. bovis*, compared to 97% by *M. tuberculosis* (Public Health England 2013); but bovine tuberculosis remains a serious health and welfare threat to the cattle industry in Britain and Ireland, and tackling the disease comes at considerable financial cost. Over the last decade, English and Welsh governments have spent an estimated £0.5 billion on testing, compensation, and research (Godfray et al. 2013). Control of bTB alone is estimated to be costing the English taxpayer £100 million in 2014, with testing and compensation being the largest costs (Defra 2014a). The average cost is estimated at £34,000 per incident[8] (Defra 2013a), from which about two thirds are paid by the Government as compensation for animals compulsorily slaughtered and costs of testing. However, the remainder falls upon the individual farmer as on-farm costs of testing, disruption to business through movement restrictions, and consequential losses (Defra 2011a).

[4] 6% of total UK deaths from TB were due to *M. bovis*, compared to current rates of 0.7% (Public Health England 2013).

[5] Administered by the Ministry of Agriculture and Fisheries in England and Wales, and the Department for Agriculture for Scotland.

[6] Reactor cattle are cattle that had a positive test result—an allergic reaction to the skin test injection—and are therefore considered to be infected with bTB (<http://www.defra.gov.uk/>).

[7] New herd incidents with Officially TB-free status withdrawn (OTFW), divided by the number of tests on Officially TB-free herds (OTF).

[8] The average cost of a herd having TB-free status suspended but not withdrawn is estimated at £9400 (Defra 2011a).

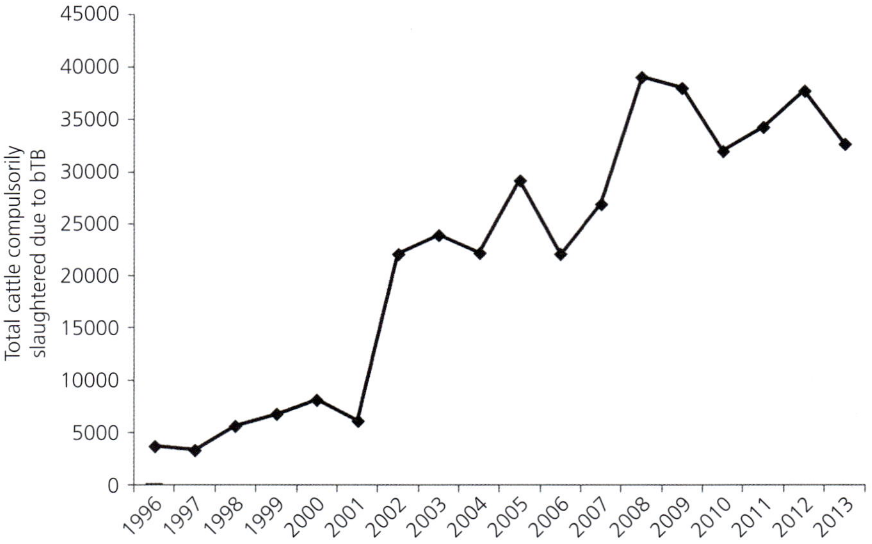

Figure 5.2 Total cattle compulsorily slaughtered between 1996–2013 due to bTB, including reactors or contacts (2011–2013 provisional data). The 2001 foot-and-mouth outbreak disrupted the national bTB testing programme, which led to a fall in the numbers of TB reactors detected in 2001, and a sharp increase in 2002 as tuberculin testing resumed. Data extracted from Defra and AHVLA (2014). ©Crown Copyright.

The latest plan, launched in April 2014, is to eliminate bovine TB from England by 2038 (Defra 2014a). The plan includes operating two cattle-TB testing regimes within three risk areas (high, edge, and low risk) and herds are to be tested annually in endemic 'high risk' areas (including counties in the South West, West, East Sussex, and Wales) and at the edges of these endemic areas (Nottinghamshire, Leicestershire, Northamptonshire, Berkshire, Hampshire, and parts of Cheshire, Derbyshire, Warwickshire, Oxfordshire, and East Sussex), while herds elsewhere in England are considered to be at 'low risk' and will be tested every four years[9] (Defra 2014a).

5.2 Badgers and bTB

Why have things deteriorated so badly since 1986? Possible factors include a relaxation of movement controls in the 1970s, a subsequent increase in cattle movements, a trend towards larger herds, and a hiatus in testing during the foot-and-mouth disease outbreak in 2001, which restricted access to the countryside and significantly increased the number of cattle that tested positive, and were therefore slaughtered, after the ban was lifted (see Fig. 5.2).

The control of bTB in cattle in Britain has been thwarted partly by the presence of a wild reservoir of infection in European badgers. In the UK, bTB was first detected in a badger in 1971, on a cattle farm undergoing a prolonged episode of the disease, and it may be that badgers originally contracted bTB from cattle (Macdonald and Burnham 2011). *M. bovis* can potentially spread between cattle and wildlife through direct contact, as well as indirectly via urine, faeces, and sputum contaminating the shared environment. Bacteria in the environment could remain viable for months (Young et al. 2005; Courtenay et al. 2006). Among badgers, infection can also spread through bite wounds (Jenkins et al. 2012). Whether badgers are a maintenance host[10] for the disease is still unknown, but mortality rates of badgers excreting *M. bovis* are higher than those of test-negative individuals (0.667 excreting versus 0.315 test-negative for males; 0.480

[9] Cattle that react to the skin test are removed for slaughter and the owner is compensated. The Officially Tuberculosis-Free (OTF) status of the remainder of the herd is suspended or withdrawn (if disease is confirmed) and it is placed under movement restrictions until all remaining cattle in the herd have cleared two further tests taken 60 days apart. Movement restrictions are then lifted but further tests are required 6 and 12 months later. If the herd remains clear, testing reverts to the routine test frequency set for the area (Defra 2012a).

[10] Maintenance hosts can maintain infection without external re-infection from another species, and are a source of disease to other species (becoming disease vectors).

versus 0.243 for females; Wilkinson et al. 2000). Other infected mammal species are rare in the UK (Mathews et al. 2006a; Delahay et al. 2007), although some pockets of prevalence have been detected in farmed and park deer[11] (Delahay et al. 2007). Estimates of the average infection prevalence in badgers vary; for example, post-mortems conducted on the initial proactive culls of the Randomised Badger Culling Trial (RBCT: see Section 5.4) revealed prevalence values ranging from 2–38% between areas, with infection more likely in adults than in cubs, and in males rather than females[12]. These figures need to be interpreted carefully. On the one hand, high estimates come from the period immediately after the 2001 foot-and-mouth disease epidemic when badgers appeared to have been experiencing significant transmission from cattle. On the other hand, all of these estimates are likely to be under-estimates because they are based on rapid post-mortems which have limited sensitivity, detecting roughly half the infections revealed by detailed post-mortem (Crawshaw et al. 2008).

Wales and the South West and the Midlands in England show the highest incidence of cattle TB (Fig. 5.1), together with high numbers of badgers, while low levels of bTB are present in the eastern and northern regions of England, and Scotland is officially bTB free[13]. In those clusters of high incidence, TB is endemic and the same strains of *M. bovis* usually infect cattle and badgers (Smith et al. 2003), with infections in both species being spatially associated at a scale of 1 to 2 km (Woodroffe et al. 2005). Isolates of *M. bovis* have been genotyped by spacer oligonucleotide typing ('spoligotyping'; Kamerbeek et al. 1997) allowing for allocation of each isolate to one of the small number of readily identifiable *M. bovis* clones that occur in Britain (Smith et al. 2003). However, Goodchild et al.'s (2012) comparisons between genotype diversity from *M. bovis* isolates from badgers and cattle revealed that transportation of cattle plays a greater part in the spatial distribution of *M. bovis* than does the movement of badgers, and it has been suggested that 25–70% of breakdowns are of non-badger origin (NE 2011a). In a separate analysis, Donnelly and Nouvellet (2013) concluded that roughly 6%—and a maximum of 25%—of herd incidents in high risk areas were caused directly by transmission from badgers to cattle, with the remainder resulting from onward spread within the cattle population. The rates of transmission between cattle and badgers, or even within cattle, are unknown. Indeed, the patterns of breakdown observed in low incidence areas correlate with cattle movements from high incidence areas and, in 'officially tuberculosis free areas' such as Scotland, nearly all breakdowns are attributable to cattle movements (Godfray et al. 2013).

At the national level, between 1972 and 1978, Wilesmith (1983) reported a higher incidence of TB in cattle in 10 km × 10 km grid squares thought to support higher densities of badger setts (see also Griffin et al. 1993). However, subsequent studies have consistently failed to detect associations between badger density and cattle TB risk, whether through case control studies (Johnston et al. 2005, 2011), correlations across RBCT areas (Donnelly and Hone 2010), or transects through TB 'hotspot' areas (Mathews et al. 2006b). Indeed, there is evidence that the prevalence of infection within badger populations is lower at high population densities (Woodroffe et al. 2009a). The national level correlation may occur because badgers thrive in mixed pasture and woodland landscapes where most cattle farming occurs, but in short, the links between badger population density, prevalence of bTB amongst them, and the incidence of bTB in cattle are less clear now than formerly thought.

5.3 History of badger control measures in the UK

Since it had been established in 1971 that badgers could be infected with bTB, attention turned towards complementing the cattle test-and-slaughter scheme with various strategies for badger culling, prompting the then UK Ministry of Agriculture, Fisheries and Food (MAFF), to implement various 'badger removal operations' (BROs; Tuyttens et al. 2000a). These were (1) gassing (1975–1982)—the removal of infected social groups within an average 10 km^2 area around the farm with the breakdown, by gassing setts (which was discontinued when it proved to be inhumane); (2) clean ring

[11] *M. bovis* can also infect goats, pigs, sheep, camelids (e.g. farmed alpacas and llamas), as well as domestic mammals. These species could cause infections in wildlife species and companion animals which could affect humans, and although they are not a major source of infection for cattle, there could be rare and occasional localized transmission (Godfray et al. 2013).

[12] Post-mortems of badgers killed by road traffic accidents in the 1970s to 1990 give rates of infection prevalence of up to 25% within cattle high-incidence areas, although the result may be biased if infected badgers are more likely to be killed by road accidents (Godfray et al. 2013).

[13] Scotland gained Officially Tuberculosis Free (OTF) status in 2009 by the European Standing Committee for Food Chain and Animal Health. This status is only awarded after meeting certain criteria, for example, for six consecutive years, the percentage of bovine herds infected has been < 0.01% (British Veterinary Association 2014).

(1982–1986)—samples of carcasses were taken from all setts used by social groups within the breakdown area and, where an infected individual was detected, the entire group was cage-trapped and shot (this extended to the surrounding area until a 'clean ring' of uninfected groups was found (the mean removal area was 9 km^2); (3) the so-called 'interim' strategy (1986–1998)—badgers were trapped and killed, whether infected or not, on the land which had been used by reactor cattle (the mean area was 1 km^2); and (4) the live test trial (1994–1996)—the aim was to remove all badgers from setts where one or more infected badgers had been detected by blood test, within all land potentially used by the badger, and extending up to cattle herds that had been at risk from these badgers (with a mean area of 9 km^2) (Krebs 1997; Smith et al. 2001). The objective in each case was to remove badgers in the vicinity of TB-affected farms in the hope of avoiding re-infection of cattle. Although, in principle, culling might have been expected to succeed (by lowering host density therefore reducing contact between infected and uninfected badgers)[14], TB rates in cattle rose steadily from the mid 1980s. In 1997 an independent review of the problem, chaired by Professor (now Lord) John Krebs, recommended a randomized controlled field trial (Section 5.4) to quantify the extent to which badger culling might reduce TB in cattle (Krebs 1997).

At this point it is necessary to introduce the perturbation hypothesis, which postulates that

> killing individuals may affect the survivors in ways (behavioural, physiological, immunological) that cause a disproportionate, and perhaps counter-productive, effect
>
> Macdonald et al. 2006

This idea has wide implications for wildlife management (Tuyttens and Macdonald 2000; Choisy and Rohani 2006; Loveridge et al. 2007). In an early model (Swinton et al. 1997), we showed that perturbation had the potential to increase bTB transmission, rather than reduce it, if removal of badgers was incomplete. As Macdonald et al. (2006) summarized, *this could occur when the complete or partial emptying of territories might change spatial organisation, alter frequency and tenor of social interactions and change demography and reproductive biology. Each of these factors might affect the probability of transmission (and thus the reproductive rate of the disease) within badgers and between badgers and cattle. In addition, the stress associated with perturbation may cause immunosuppression, both increasing the rate at which bacilli are shed (because disease is further progressed), and increasing the risk that susceptible individuals become infected.* In short, a possible consequence of killing some badgers was the increased ranging behaviour and susceptibility of the survivors, so that transmission of disease to cattle was unintentionally, worsened.

Writing now, in 2014, we can interpret the unfolding understanding of perturbation with the benefit of hindsight. The idea originated in the ineffectiveness of killing foxes as a means of controlling rabies (Macdonald 1980), in comparison to vaccination (Bacon and Macdonald 1980; Macdonald and Bacon 1982), illustrating the peril of ignoring social behaviour in disease control (Macdonald 1995). Two caveats are worth mentioning before delving further into perturbation. First, it doubtless occurs to a greater or lesser extent in all badger populations, as individuals die of various causes (and these may sometimes impact a group as heavily as do some examples of culling: e.g. in our studies in Wytham Woods (Chapter 4, this volume), there is a 22.5% annual mortality rate in adults, so it can easily happen that members of a particular social group suffer 50% mortality due to road traffic accidents in a given year). So, in the context of attempted control of bTB, perturbation refers to the additional consequences of killing larger numbers of badgers, over and above the reverberating consequences of background mortality. Second, perturbation affects both badgers outside the culled zone, and badgers that escape culling within that zone. It might strike the reader as odd that any badgers should be left alive in the cull zone. This can happen, not only because culling teams do not have access to all of the land within culling areas, but also because badgers can be difficult to catch. Likewise, immigration of badgers into the culled areas to some extent depopulates adjoining land, meaning that the population impacts of culling extend beyond the area actually culled. The key point about the perturbation hypothesis is the proposal that killing badgers causes a counter-productive effect such that each surviving individual is more likely to pass the infection on, making it less likely that culling will drive the disease beneath the threshold of persistence.

Concerns that killing badgers, or rather killing insufficient numbers of them, might not be an efficacious way to control bTB in cattle have a long history (e.g. Overend 1980; Macdonald 1984), but now we can summarize evidence for perturbation at two scales, (a) the

[14] Under Kendell's threshold theorem, the reproduction rate of an infectious disease (R_0) might be expected to decline beneath the threshold of persistence ($R_0 = 1$) if numbers of infectious individuals are reduced.

large-scale (~100 km^2) epizootiological outcomes for bTB in cattle of killing badgers, and (b) the fine-scale (individual badgers and badger groups) behavioural responses that shed light on some mechanisms that explain perturbation. The former stem largely from the Randomised Badger Culling Trial (RBCT), which ran from 1998 to 2007, whereas for the latter we will summarize our field studies at North Nibley (which began in 1995, before the RBCT) and in another experiment, Triplet E (begun in 2000), which was embedded within the RBCT.

5.4 The Randomised Badger Culling Trial (RBCT)

The overall aim of the Randomised Badger Culling Trial (RBCT) was to evaluate the impact of badger culling on bTB in cattle. The RBCT took place in 30 replicate areas of 100 km^2, each with a high incidence of herd breakdowns over the three previous years. The 30 areas were grouped into ten sets of three, and areas within each triplet were subject to a randomized intervention strategy: proactive, reactive, or no culling. Proactive culling areas were subjected to culling on all accessible land, repeated roughly annually. In reactive culling areas, badgers were culled on a single occasion on and around farms where a recent confirmed bTB outbreak in cattle had occurred. Areas without culling acted as experimental controls. The trial cost *c.* £50 million between 1998 and 2007 (the last badger was killed on 31 October 2005).

5.4.1 Proactive culling

In proactive culling areas the aim was to remove, at the outset, as large a proportion as possible of badgers resident in the entire proactive trial area and to maintain the population suppressed throughout the trial period by regular follow-up culling operations (Bourne et al. 2007). In proactive areas, results of the RBCT showed that: (i) while culling was occurring, the number of confirmed herd breakdowns within the culling area was 23% lower than that inside non-culling areas; but (ii) herd breakdowns increased by 25% in the 2 km-wide perimeter surrounding the core culling area (where culling had perturbed badger populations; Woodroffe et al. 2006b) (Table 5.1a). Continued monitoring after the cessation of culling revealed that the beneficial effects of culling within culled areas increased after culling ended, and persisted for several years, albeit diminishing over time (Jenkins et al. 2010 and associated updates).

5.4.2 Reactive culling

In the reactive areas, badgers were culled locally on and near farmland where recent outbreaks of TB had occurred in cattle, with the aim of removing all social groups of badgers having access to the breakdown farm. Reactive culling was stopped by Ministers in November 2003 when interim results showed an increase of 22% in the incidence of new confirmed TB cattle herd breakdowns (95% CI: 2.5% to 45% increase; $p = 0.025$) (Donnelly et al. 2003; Bourne et al. 2007). This harmful effect disappeared after culling ended (Bourne et al. 2007). Vial and Donnelly (2012) subsequently found that greater numbers of badgers reactively culled in the vicinity of cattle herds were associated with significantly increased bTB risk. Furthermore, Karolemeas et al. (2012) showed that for breakdowns that started during the RBCT culling period, reactive culling was associated with increased odds of prolongation (odds ratio of 1.7 within the core areas; 95% credible interval [CI] 1.1–2.4), while this effect was not present after the culling stopped.

5.4.3 Conclusions of the RBCT study

The Independent Scientific Group (ISG, of which RW was a member), who designed and oversaw the study, concluded that 'badger culling cannot meaningfully contribute to the future control of cattle TB in Britain', that 'some policies under consideration are likely to make matters worse rather than better', and that 'weaknesses in cattle testing regimes mean that cattle themselves contribute significantly to the persistence and spread of disease in all areas where TB occurs' (Bourne et al. 2007).

5.5 Evidence of a perturbation effect

Results from two large-scale studies on badger sociality, the North Nibley experiment and the Triplet E Experiment (TEE), suggested mechanisms that explain the perturbation revealed by the RBCT. This work was undertaken collaboratively between WildCRU and colleagues at the then Central Science Laboratory (CSL; subsequently FERA and now the Animal Health and Veterinary Laboratories Agency, AHVLA).

5.5.1 Study sites

The North Nibley study (1995–2000), in Gloucestershire, involved comparing badgers' biology and the status of bTB before and after parts of the study

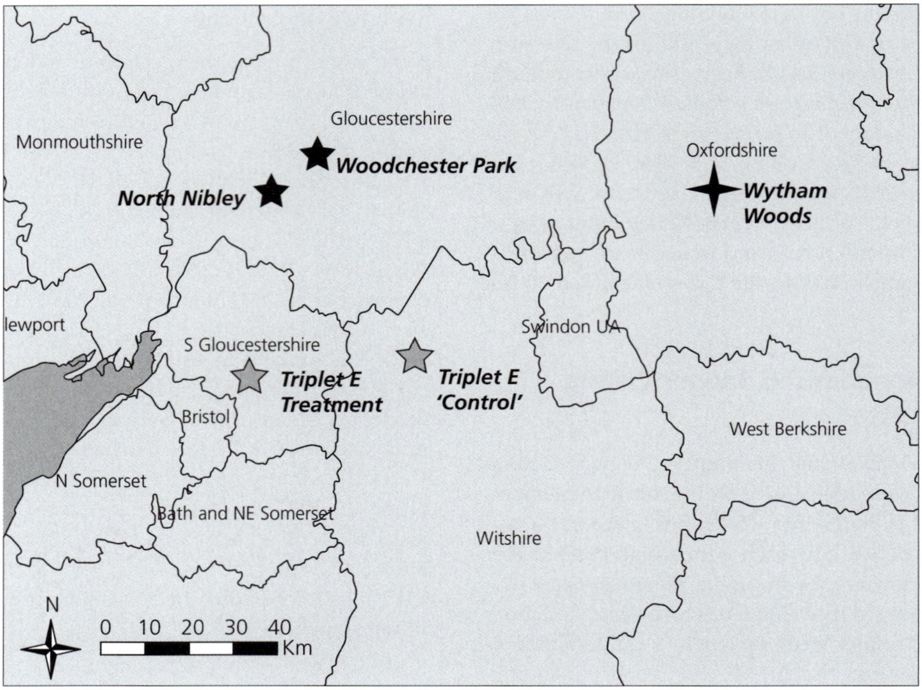

Figure 5.3 Location of study sites: North Nibley, Woodchester Park, Wytham Woods, and Triplet E Experiment (treatment (RBCT reactive culling) and control (RBCT survey-only)). ©Crown Copyright.

area were subjected to a UK Ministry of Agriculture, Fisheries and Food badger removal operation. In the TEE (2000–2004), we undertook the same comparisons but we also included a non-culling control area (Fig. 5.3). Both studies were compared with two long-term studies of relatively undisturbed populations: at Woodchester Park, Gloucestershire, where badger culling had ceased in 1978, and at Wytham Woods, Oxfordshire (Delahay et al. 2000; Macdonald et al. 2006; Chapter 4, this volume) where no culling had been conducted (Fig. 5.3).

Figure 5.4 depicts the badger population trends at North Nibley (TB-infected/culled) in comparison with those at Woodchester Park (TB-infected/no

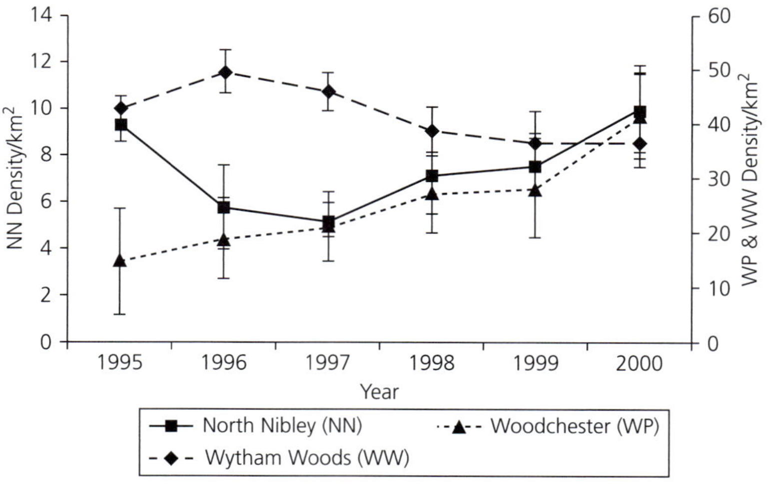

Figure 5.4 Estimated annual population densities of badgers (km^{-2}) at North Nibley (expressed on the left ordinate), Woodchester Park, and Wytham Woods (expressed on the right ordinate). Error bars indicate 95% confidence interval. From Macdonald et al. (2006), reproduced with permission from Elsevier.

intervention) where the population was increasing, and Wytham Woods (TB-free/no intervention) where it was falling back slightly following a period of expansion, which we interpret as a response to climate change (see Chapter 4, this volume). At North Nibley and in the TEE, badger social groups were categorized as either 'Removed' (where culling took place), 'Neighbouring' (groups adjacent to Removed groups), or 'Other' (groups at least one social group range away from any Removed group), while groups within the survey-only population at TEE were labelled as 'Control'.

North Nibley study (1995–2000)

The objective of the North Nibley study was to compare the behaviour of badgers (and their neighbours) from setts at which culling had or had not been attempted. The study (Swinton et al. 1997; Tuyttens et al. 1999, 2000a,b,c; Macdonald et al. 2006) started in 1995 when a tuberculin test identified a bTB outbreak in cattle in the area. Badgers were identified, by government officials, as the likely source. In the 17 km^2 study area, 12 badger-culling interventions (live test and interim strategies) had taken place over the previous 20 years (most recently in 1994), and hence it, like many badger populations in SW England, had already been serially disturbed. Between January and September 1995 before culling started, our team identified 42 setts, marked 62 individuals, radio-tracked 14, and matched 558 latrines to their social groups by bait-marking[15]. In September 1995, officials started live-trapping of badgers in a 6.5 km^2 area around the farm and tested their TB status by ELISA live-test for antibodies to *M. bovis*, while we monitored infection status by culture of samples of faeces, urine, tracheal aspirate, and pus from wounds from 309 individuals (868 sampling events).

The diagnostic power of ELISA and culture tests is limited due to false negatives, but six groups, using eight setts, were identified as infected, and within the 6.5 km^2 area, 27 badgers were killed by government officials in 1995 (live test) and two in 1996 (interim strategy). Estimates within the 17 km^2 area suggested that the badger population was reduced from 145 animals in 1995 to 118 in 1996 (a reduction of 27 badgers, in consonance with the 29 individuals culled). Monitoring of the population continued until 2000.

[15] Small, indigestible coloured plastic chips mixed with peanuts and syrup are given to the badgers, with each social group receiving a different colour. Latrine surveys for these chips then reveal the outer limits of each group's range.

Triplet E Experiment (2000–2004)

The aim of the Triplet E Experiment (TEE: Macdonald et al. 2006; Riordan et al. 2011) was, likewise, to compare behaviour and patterns of bTB infection in the badger population before and after culling. Unlike the Nibley study, the TEE included a control non-culling area. Since the RBCT was, at the time, comparing proactive and reactive with no-culling control areas, the TEE experiment was set up within one of the RBCT triplets. In the reactive culling trial area of the RBCT 'Triplet E', we set up a 37.3 km^2 study area, along with a comparative 18.9 km^2 study area within the no-culling ('survey-only') trial area of the same triplet. Badgers were bait-marked and live-trapped in both study areas. Field teams made 663 badger captures in the culling ($n = 481$; 71 active setts) and non-culling ($n = 182$; 38 active setts) areas, involving a total of 423 individuals, from which 466 samples for bTB culture were collected and 40 individuals tested positive. The work started in 2000, was suspended due to the 2001 epidemic of foot-and-mouth disease, and recommenced in October 2001 until December 2003. In the reactive culling study area, culling operations were undertaken by Defra: three reactive culls took place in 2002–2003, with 77 badgers trap-shot[16], estimated to be 34–43% of the individuals known to be resident in the target setts. Radio-collars were also fitted to 50 (treatment) and 20 (control) adult badgers.

5.5.2 Evidence of culling-induced effects on badger behavioural ecology (immigrants and survivors)

In the TEE treatment and control areas, inter-group movement[17] (IGM, which might represent inter-group dispersal) was rare (12 observations out of 663 trapping events), but 75% occurred in the treatment area post-culling, and was more likely to involve badger movements into Removed groups ($\chi^2 = 7.20$; $p < 0.01$), both from other culled groups or non-culled neighbouring groups (Riordan et al. 2011). This coincides with results on social perturbation from Woodchester

[16] Culling November 2002: 38 badgers from 14 social groups; January 2003: six badgers from four social groups; August 2003: 33 badgers from 13 social groups (Riordan et al. 2011).

[17] Inter-group movement was defined for the purposes of the study as occurring when a badger was trapped at a sett belonging to a different group than the one where it was captured previously, or when the badger was located at a sett belonging to a different group by night-time radio-tracking or daytime positioning.

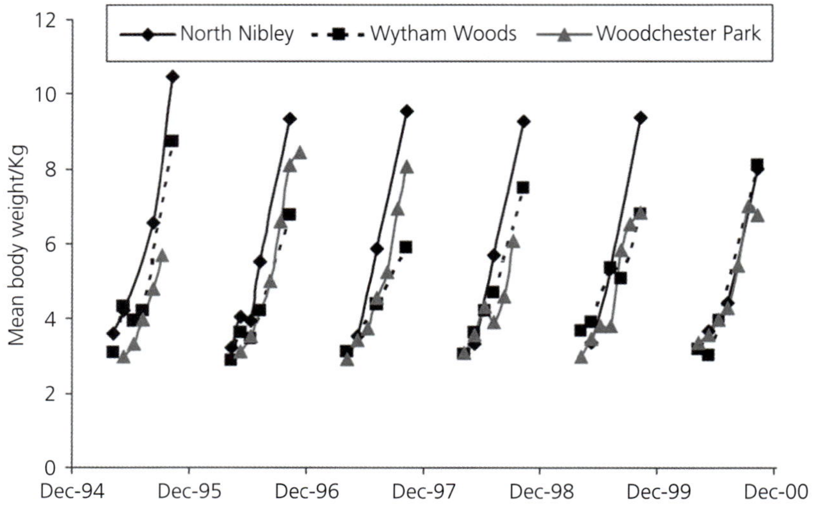

Figure 5.5 Growth curves for badger cubs at North Nibley, Wytham Woods, and Woodchester Park over the course of each year in which they were born. From Macdonald et al. (2006), reproduced with permission from Elsevier.

Park, Gloucestershire (and see Cheeseman et al. 1988, 1993), where by means of bait marking (Chapter 4, this volume), it seemed that culling resulted in a vacuum effect, drawing neighbouring individuals into culled areas (Carter et al. 2007). Trapping records within North Nibley for between 1995 and 1997 also suggested that badgers tended to move more between social groups, especially after a culling operation (Tuyttens et al. 2000c), although the effect was weak. Stronger evidence emerged from the RBCT where, as proactive culls occurred, an increasing proportion of captures were in the periphery of the culling areas (Woodroffe et al. 2006a) and an increasing proportion had genotypes indicating they were likely immigrants (Pope et al. 2007).

Findings from North Nibley suggested that females were the first to re-colonize vacated areas, and that all were cubs, yearlings, or young adults (previously known elsewhere) when they moved in. Tuyttens et al. (2000c) suggests that this is consistent with the idea that young females are being reproductively suppressed, and their dispersal might be explained as an attempt to move to unoccupied land that will allow them to increase productivity, without aggression from dominant, older females (Chapter 4, this volume). However, results from TEE showed no sex-biased dispersal. Badgers at North Nibley were generally younger, heavier, and in better condition than in contrasting undisturbed and high density populations (i.e. Wytham Woods and Woodchester Park), presumably because they were relieved of density-dependent constraints relative to environmental carrying capacity (Fig. 5.5) (Tuyttens et al. 2000c).

Within the TEE, 70 badgers were fitted with radio-transmitters[18] and 8311 fixes collected to plot the home ranges of individual badgers in both treatment and control areas (Riordan et al. 2011). Home-range sizes in the treatment areas tended to decrease through the winter culling phases from an average of 21.4 ha pre-culling (sd = 22.8 ha) to 9.4 ha post-culling (sd = 12.7 ha), but the size of summer (June, July, August) range areas increased during inter-culling phases[19], from 28.3 ha (sd = 11.3 ha) to 40.6 ha (sd = 27.6 ha). The effect of pre- to post-culling phases on the proportional overlap between badger (summer) home ranges varied significantly between social group types (Removed, Neighbouring, Other, and Control) (Cull × Group Type: $F_{2,6} = 5.48$; $p = 0.04$): Inter-range overlap increased following culling, from an average of 7.5 to 13.0 ha in Neighbouring groups, and from 8.2 to 10.8 Ha in Removed groups, but overlap did not change significantly in Control areas.

[18] 173 MHz band, attached to a leather collar (BioTrack Ltd, Wareham, Dorset, UK), which remained active for an average of 230 days (sd = 54 days).

[19] Defra performed three badger culling operations within the treatment area: (i) November 2002 (38 badgers removed from 14 social groups), (ii) January 2003 (six badgers removed from four social groups), and (iii) August 2003 (33 badgers removed from 13 social groups) (Riordan et al. 2011).

The number of home ranges with which each individual badger overlapped (summer) increased for Neighbouring and Other groups (from an average of 0.8 to 2.7 groups), but declined (from 0.8 to 0.2) in Removed groups (Riordan et al. 2011). These results reflected those based on bait-marking in the TEE (Macdonald et al. 2006) (Fig. 5.6a,b,c). Similarly, at Nibley, one year after the culling, badgers used latrines 35% further away from their setts, a distance that only decreased by 10% after two years (Tuyttens et al. 2000b). Indeed, at Nibley we found that social group ranges enlarged by 68% after the culling (Fig. 5.7a,b), and the amount of proportional overlap between groups increased for all years after the cull for which we have data (Fig. 5.7c), and though there was some fluctuation between the different group categories between years, the overall trend was for persistent spatial disruption to the Nibley population.

We mentioned above that, among the mechanisms that might contribute to a perturbation effect, sociological disturbance might cause sufficient stress to induce immuno-suppression and hence heighten an animal's susceptibility to infection (Gallagher and Clifton-Hadley 2000; Montes et al. 2011). Indeed, bite wounds may provide a proxy for social stress (they increased significantly during a period when the undisturbed badger population at Wytham doubled in size; Macdonald et al. 2004a; Chapter 4, this volume).

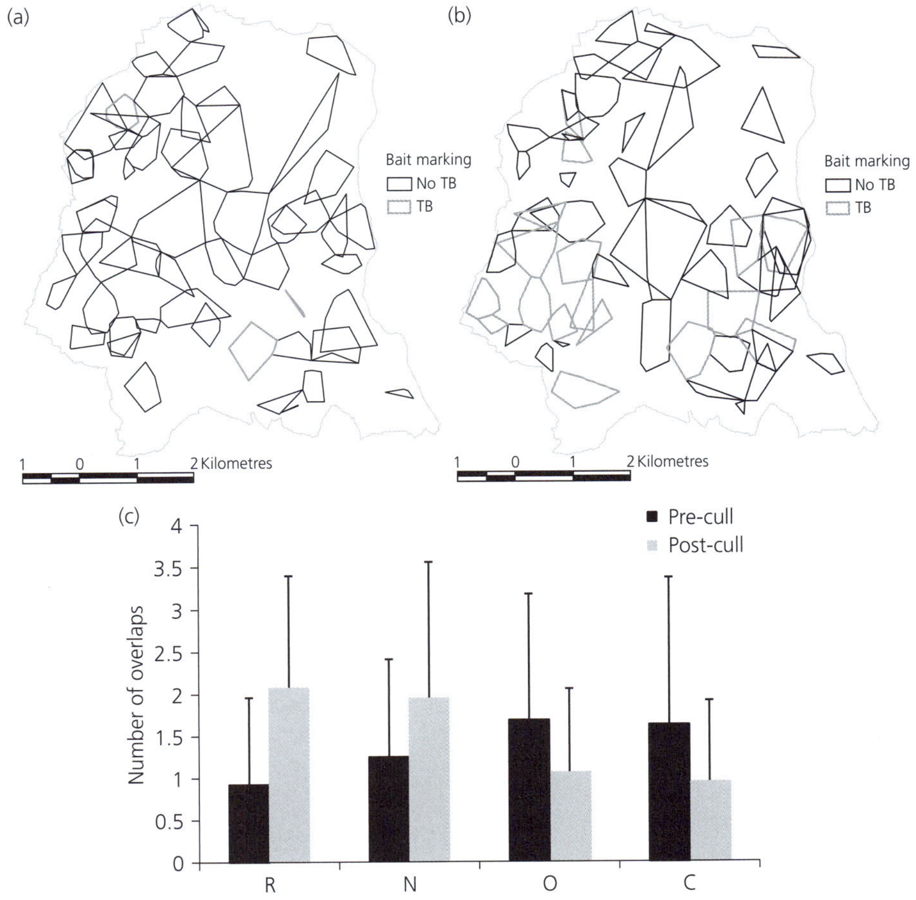

Figure 5.6 Socio-spatial organization of the TEE population and range overlaps. Territorial boundaries of badger social groups were delineated by bait-marking for: (a) 53 groups in the treatment area before culling (Spring 2002) and (b) 55 groups after culling (Spring 2003). Thicker outlines show groups having members identified as positive for *M. bovis*; (c) Proportional overlap between social group territories of each group type: Removed (R), Neighbouring (N), Other (O), and Control (C). Proportional overlap changed following culling (RM-GLM [Cull × Group Type]: $F_{3,64} = 2.97$; $p = 0.038$), with both Removed and Neighbouring groups increasing their overlap, and Other and Control groups decreasing. Reproduced with permission from Elsevier.

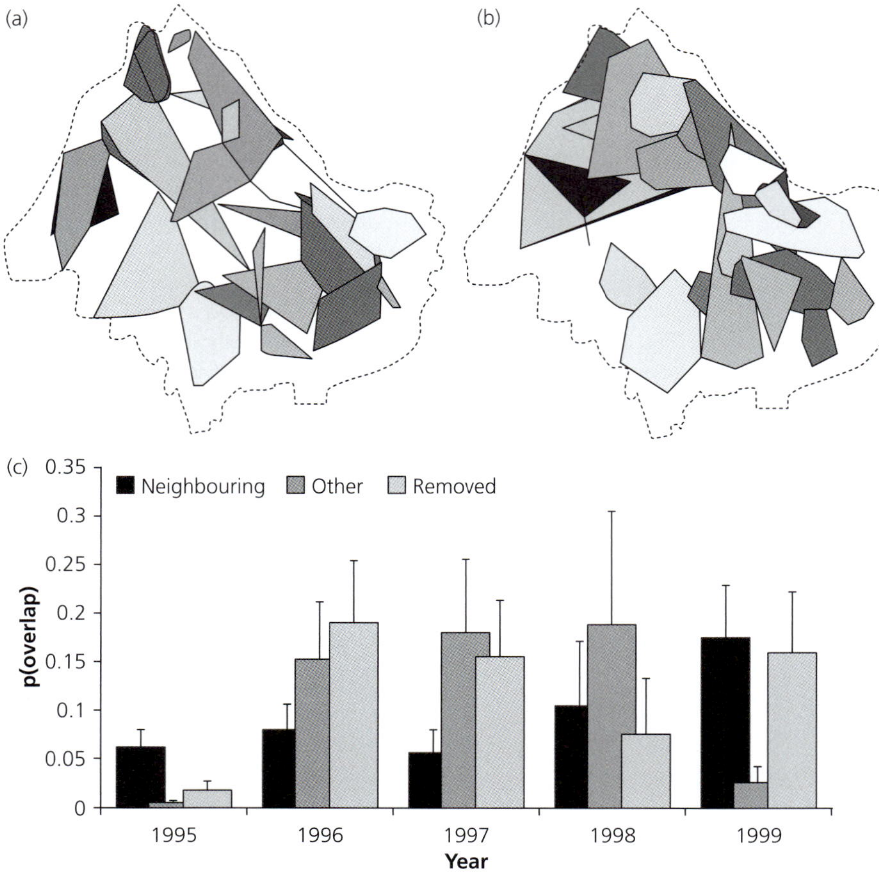

Figure 5.7 Socio-spatial organization of the North Nibley population and range overlaps. Outer limits of each group range as determined by bait-marking: (a) in 1995, before the main cull, showing a relatively well separated population with most groups forming exclusive ranges. Where significant overlaps do occur, these mostly show social groups utilizing more than one sett; and (b) following the cull in 1996, the situation shows an appreciable breakdown of the social group delineations; (c) Proportion of overlaps across the study area within the three categories (GLM: $F_{2,15} = 4.81$; $p < 0.001$) of Nibley badger social groups: 'removed' at which culling took place in 1995, neighbouring groups adjacent to removed groups, and 'other' groups, which are at least one social group range away from any removed group. Adapted from Macdonald et al. (2006), reproduced with permission from Elsevier.

The overall rate of male badgers receiving fresh bite wounds in the TEE was 22.6%, which was high when compared to the rate of 24.9% at Wytham Woods, where the population was at 8–9 times the density of the TEE populations (Macdonald et al. 2004a; Macdonald et al. 2006). Macdonald et al. (2006) suggested that the higher bite rates within the TEE populations could be related to the long history of culling in the TEE study areas, since the proportion of badgers with fresh bite wounds increased significantly following culling in Neighbouring and Removed social groups, following the winter 2002/2003 reactive cull. Bite-wounded animals are more likely to be infected with *M. bovis*; evidence suggests that infected animals are more likely to be bitten, and that bite-wounding transmits infection (Jenkins et al. 2012).

5.5.3 Evidence of culling-induced effects on prevalence and distribution of bTB in badgers

We monitored the infection status of badgers at both North Nibley and TEE. Prior to the culling (in September 1995) at North Nibley, 14.5% (9/62) of badgers tested positive for bTB using ELISA tests (but only two to bacterial culture, culture prevalence 3.2%; Macdonald et al. 2006). Between 1996 and 1999, after culling,

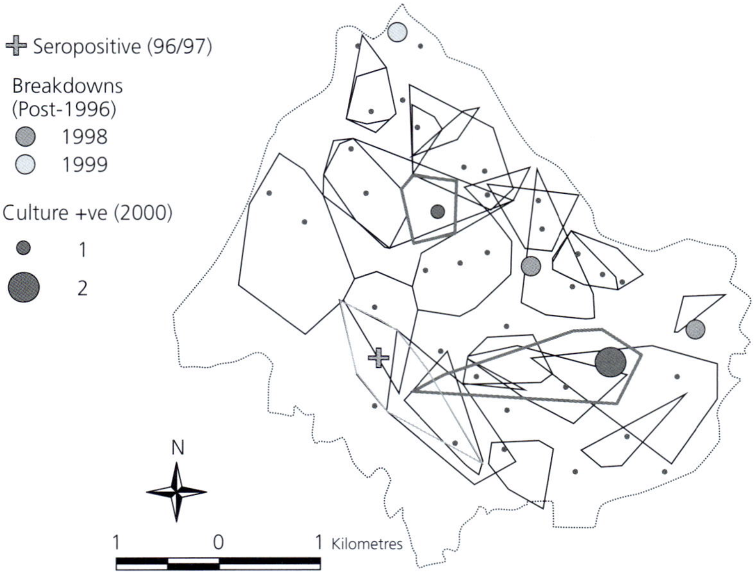

Figure 5.8 North Nibley badger TB infection post-culling. 1999 bait-marking ranges of badger social groups at North Nibley (95% minimum convex polygons). Numbers of infected animals (1 or 2) within groups denoted by variable sized dot markers indicating main sett locations. Location of breakdown farms (post-1996) also shown. Adapted from Macdonald et al. (2006) reproduced with permission from Elsevier.

none of the 258 badgers tested positive by culture[20], suggesting that culling had successfully eliminated bTB among North Nibley's badgers. However, despite indications that the badger population was bTB free, two farms in 1998, and one in 1999, suffered herd breakdowns (Macdonald et al. 2006; Fig. 5.8). By 2000, three badgers (of 64) were culture positive (culture prevalence 4.7%) (Fig. 5.8).

The TEE suggested at a local scale, as did the RBCT at a larger scale, that the prevalence of badgers excreting *M. bovis* increased post-culling; at the treatment site the culture prevalence increased from 5% (3 groups) to 15% (18 groups) (n = 53 groups although it also increased from 2.7% to 15.9% in the control sites; Riordan et al. 2011). Testing of culled badgers in the RBCT showed the same pattern, with infection prevalence in badgers rising after both reactive and proactive culling (Woodroffe et al. 2006a; Woodroffe et al. 2009b). In addition, in the TEE, culling was associated with a significant increase in detected *M. bovis* in cubs within the treatment population, from one excreting cub pre-culling (n = 101) to eight[21] cubs (n = 76) post-culling (Riordan et al. 2011).

[20] Culture test has a sensitivity of only *c.* 5%. One of the culture-negative badgers tested positive using ELISA, which is a more sensitive test.

[21] Four cubs tested positive for *M. bovis* from urine samples, three from sputum, and one from a faecal sample.

Furthermore, and also reinforcing the perturbation hypothesis, the effect of culling on the prevalence of detected *M. bovis* in badgers within reactive areas had the greatest increase within Neighbouring social groups, from 1% (sd = 4%) to 14% (sd = 12%), with the increase being greater in groups whose members had been involved in inter-group movements, and a decrease in Other and Removed groups (Riordan et al. 2011) (Fig. 5.9c). Indeed, movement between social groups had been identified as a potential factor in disease spread in badger populations: in Woodchester Park, the incidence of new TB cases increased in the years following high inter-group movement rates (Rogers et al. 1998). Likewise, Pope et al. (2007) showed that badgers with genetic profiles suggesting that they were immigrants were also more likely to test positive for *M. bovis*.

The distribution of bTB within the TEE treatment badger population was more scattered before culling (Fig. 5.9a) than after culling, when a cluster of infection arose in the west of the study area (Fig. 5.9b) (Macdonald et al. 2006). This pattern contrasts with that found more generally in the RBCT, where infection became less clustered in response to badger culling (Jenkins et al. 2007). The majority of infected badgers in this cluster showed signs of kidney or urinary tract infection, with urine samples testing positive for *M. bovis*, but also showing bTB-positive tracheal aspirate samples (sputum) in the north and east. Most of the infected cubs were found in this cluster.

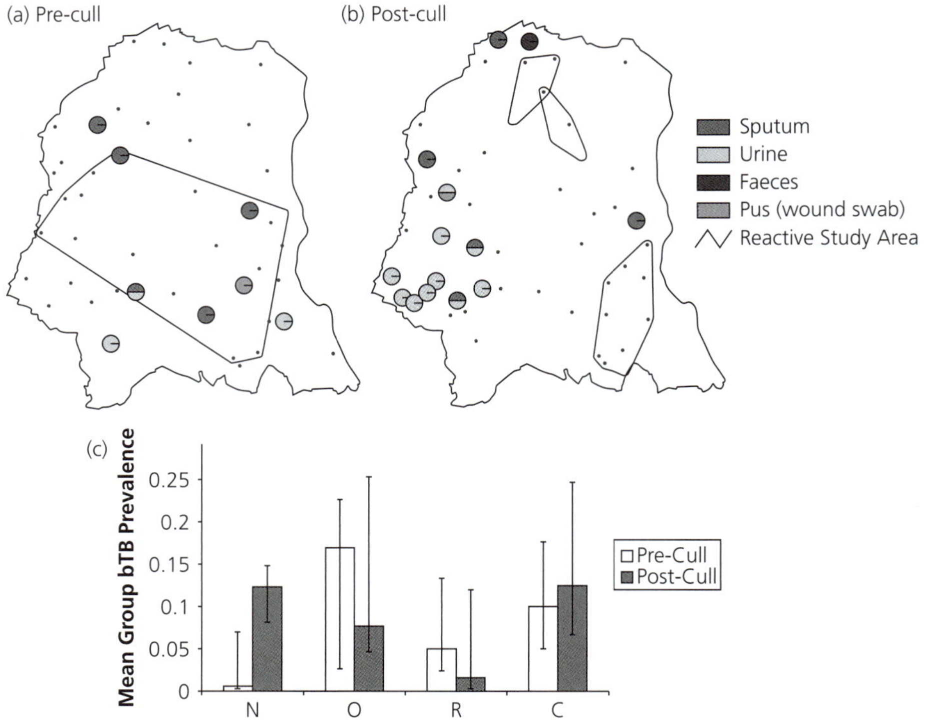

Figure 5.9 Distribution and prevalence of bTB among badgers pre- and post-cull at TEE. (a) Distribution of bTB (a) pre-cull and (b) post-cull. Infection status of badgers was determined by bacterial culture of samples of urine, faeces, sputum, and pus (n = 466 individuals). (c) Mean bTB prevalence in Removed (R), Neighbouring (N), Other (O), and Control (C) badger social groups before and after culling. Error bars show 95% binomial confidence intervals. From Riordan et al. (2011).

M. bovis was not detected among the 47 badger cubs tested in the control population; however, from 27 identified groups in the control in 2002, two contained one culture-positive individual, which increased to eight groups (nine individuals) by the time the cull was complete in 2003. This result may have been linked to circumstantial evidence of illegal badger control (Macdonald et al. 2006), to a nationwide increase in infection associated with the 2001 foot-and-mouth disease epidemic (Woodroffe et al 2006), or simply to chance. A conclusion of this work was that reactive culling at the spatial- and time-scales undertaken in the treatment areas was of no benefit for controlling bTB in badgers, and may have made it worse (Macdonald et al. 2006). This is consistent with the broader finding from the RBCT that reactive culling increased both the prevalence of *M. bovis* in badgers and the frequency and the duration of TB incidents in cattle herds (Karolemeas et al. 2011, 2012; Donnelly et al. 2003; Woodroffe et al. 2009b; Vial and Donnelly 2012).

Undisturbed comparators and the tenor of badger interactions

Work at Nibley and in the TEE was intended to describe the behaviour of individual badgers, and groups of badgers, to test the perturbation hypothesis and to elucidate the mechanisms that underlie it. The original comparator for undisturbed badgers on lowland British farmland was a model of group living, territorial exclusivity, developed from the pioneering work at Wytham by Kruuk (1978a,b; Chapter 4, this volume), and paralleled by the long-term work at Woodchester Park (Cheeseman et al. 1988; Delahay et al. 2000). In this archetype of badger society, mixed sex groups of up to 30 individuals share a communal territory and setts, where long-term dispersal is infrequent and territorial boundaries are marked by latrines, with minimal overlap, and group membership is stable over many years. This spatial exclusivity and stability was thought to restrict bTB transmission in comparison with the destabilized socio-spatial dynamics among badgers surviving culls. In accord with this expectation, bTB among badgers at Woodchester Park was highly localized and concentrated within particular territories (Cheeseman et al. 1988), and the same pattern of clustering was observed in RBCT areas (Woodroffe et al. 2005). Indeed, the expectation that culling would be associated with higher prevalence of bTB among the survivors also accorded with the results of the RBCT, including findings of the TEE.

However, exactly what element of perturbation causes this counter-productive effect bears further scrutiny. It was already clear to Tuyttens et al. (2000a,b) that the badgers at Nibley had been serially culled (12 BROs in previous 20 years) (Fig. 5.10), potentially explaining why, when the killing began there in 1995, the badger home ranges already tessellated less neatly than those at Wytham or Woodchester. The changes documented in badger behaviour were, therefore, not between undisturbed and disturbed, but between recovering from disturbance and worse disturbance; and the same applies to TEE, where there had also been a history of culling prior to the reactive culls of 2000. Indeed, the inefficiency of some culling in the TEE constituted mortality at a level (34–43%), perhaps twice that caused by road traffic accidents in some, otherwise undisturbed areas (Rogers et al. 1997; Macdonald and Newman 2002). So, while there is no doubt that the counter-productive effect predicted by the perturbation hypothesis occurs (even when the culling is undertaken at a very small scale, Bielby et al. 2014), and no doubt that the survivors and neighbours of culls behave differently, as predicted, exactly what aspects of the culling and of the badgers' responses lead to the greater prevalence of bTB is uncertain.

This point is clarified by the behavioural ecology of Wytham's badgers, summarized in Chapter 4, this volume. There, badger population density more than doubled from 83 badgers in 1987 to 272 in 1996, and since the occupied area had not changed, the population underwent a major spatial reorganization when, notably, in about 1996, they dug many more setts but, with more setts available, the occupants of each group range became more dispersed between them (Macdonald et al. 2004b), and other differences emerged.

This long-term monitoring has exposed that, although permanent dispersal (change in group affiliation) was infrequent in Wytham (*c.* 20% of individuals) (Fig. 5.11a), at any one moment, about one fifth of badgers of both sexes were found on visits to setts other than their principal affiliation (Fig. 5.11b) (Macdonald et al. 2008). Furthermore, analyses of genetic pedigree have revealed that these extra-territorial movements were associated with the fact that almost half the cubs born (48%: 315 of 655 cubs) were sired by males outside the mother's social group, and while 85% (268 of 315 cubs) of these extra-group fathers were resident in neighbouring territories, 15% were resident further away (Dugdale et al. 2007; Annavi et al. 2014). Retrospectively, our pedigree also indicated that extra-group paternity in Wytham had been commonplace since 1988 (Chapter 4, this volume), although some other relevant aspects of badger sociology had changed. For example, as the population had increased, the frequency of bite wounding had risen steadily (from on average 7% when the population was under 100 individuals in Wytham in 1990, to in excess of 20% by 1999 when the population rose above 200: Macdonald et al. 2004a), but thereafter it has fallen dramatically (unpublished),

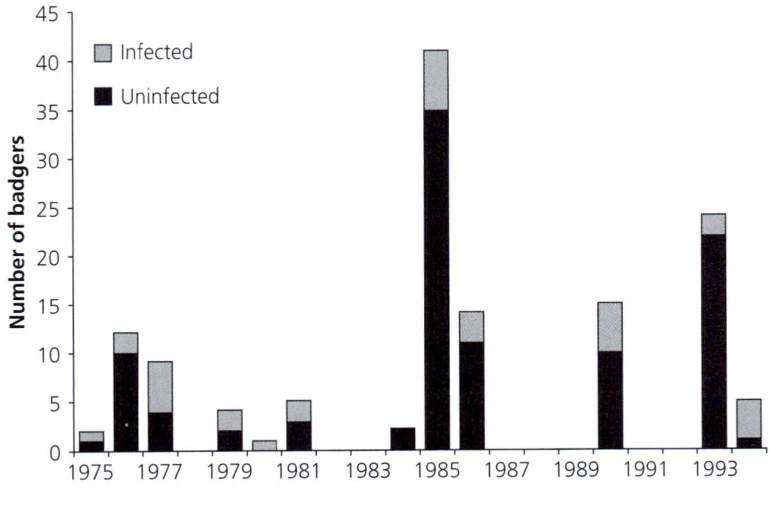

Figure 5.10 Badgers culled in the North Nibley area 1975–1994, showing numbers caught and examined for bTB by MAFF (excluding unrecovered gassed animals before end of gassing in 1982). Data provided by Defra. ©Crown Copyright.

perhaps as the badgers reduced 'crowding stress' by digging more setts (Macdonald et al. 2004b). Thus, if we are to take the badgers of Wytham Woods as a paradigm for unperturbed behaviour, it seems as if some features potentially relevant to epidemiology (e.g. bite wounding, dispersion among subsidiary setts) change with population density, but others (extra-group paternity and excursions) are characteristic of badger sociology in the lowland farmscape.

Nevertheless, Wytham's badgers, although relatively undisturbed, group living, and occupying ranges ringed by latrines, do engage in very substantial interaction between groups beyond the borders of these ranges (Fig. 5.11). There is little evidence that

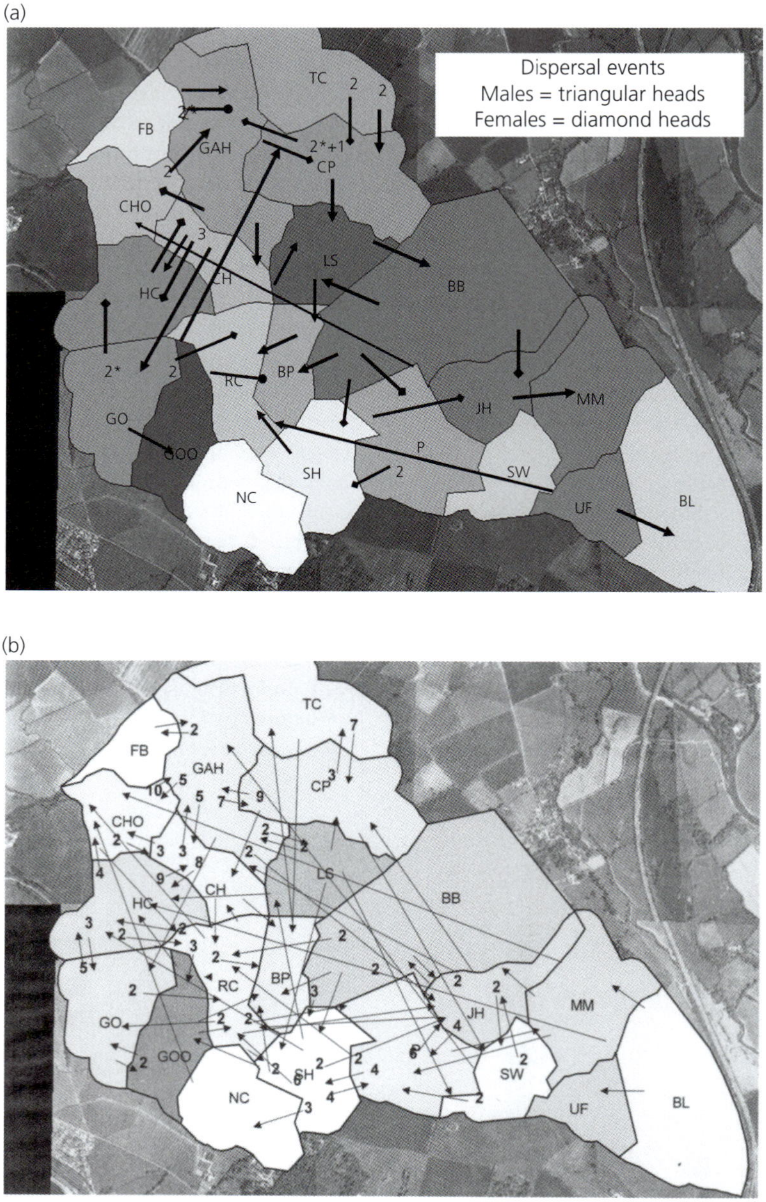

Figure 5.11 Badger dispersal and short movements at Wytham Woods. (a) Top: between 1987 and 2003, the dispersal patterns of 267 badgers (5255 trapping events) revealed that only 19.1% dispersed (28 males and 23 females); (b) Bottom: excursions were records of an individual at a social group different from its previous site of capture. From Macdonald et al. (2008).

this extra-group contact is aggressive (*vide* the lowered incidence of bite wounding—although when bite wounding was escalating we do not know the relative contributions of within- and extra-group biting). This raises the question of to what extent are the epizootiological consequences of culling a reflection, not just of increased contact between perturbed badgers, but rather of a change in the tenor of their contact.

It also raises questions about the nature of interactions, insofar as they affect bTB transmission, within badger communities recovering from culling yet below carrying capacity. These relationships may be sociologically, and thus epizootiologically, much more complex than a dichotomous distinction between culled and undisturbed communities (Fig. 5.12).

5.5.4 The aftermath of culling

Badger society, the emergent property of complicated interactions between its members, is dynamic and complex, and therefore one would expect the consequences of perturbation to reverberate for years: Carter et al. (2007) predict that some measures of sociality (e.g. range overlap, territoriality, immigration) can remain disrupted for up to eight years after culling. In the North Nibley study, the population took only three years to recover numerically to its previous density (Fig. 5.13) but, even after the numbers recovered, their distribution was markedly different from that before the culls (Macdonald et al. 2006; Fig. 5.13).

While RBCT proactive culling was underway, confirmed herd breakdowns decreased in proactive core areas (*relative* to no-cull areas), but increased on adjoining unculled lands (Donnelly et al. 2006). Although these elevated TB risks to cattle on adjoining land disappeared rapidly after culling ended, farmers in the 2 km perimeter still had a worsened breakdown rate over the entire period (Godfray et al. 2013; Table 5.1a), offsetting the relative benefits accrued by core farmers and inducing a cost on non-participating farmers. Extrapolations from RBCT findings suggest that five years of proactive culling of badgers within a 150 km^2 circular area would cause a 20–34% relative reduction in confirmed cattle herd breakdowns within the culled areas over nine years, but, crucially, if taking into consideration the core and the 2 km-wide perimeter surrounding the culling, the estimated benefit falls to 3–21% (Table 5.1b). It is a matter of judgement if this relative benefit is worthwhile to society, especially when failure to overcome the multiple logistical challenges of delivering badger culls of the required standard and scale is likely to increase cattle TB rather than reduce it.

5.5.5 The story unfolds

In 2007, the Independent Scientific Group on Cattle TB concluded, in relation to the RBCT, that: *These results, combined with evaluation of alternative culling methods, suggest that badger culling is unlikely to contribute effectively to the control of cattle TB in Britain.* (Bourne et al. 2007). On 7 July 2008, Hilary Benn, then Secretary of State for Environment, Food and Rural Affairs, made a statement to Parliament that the Government's policy would be neither to cull badgers nor to issue any licences to farmers to do so, while remaining open to the

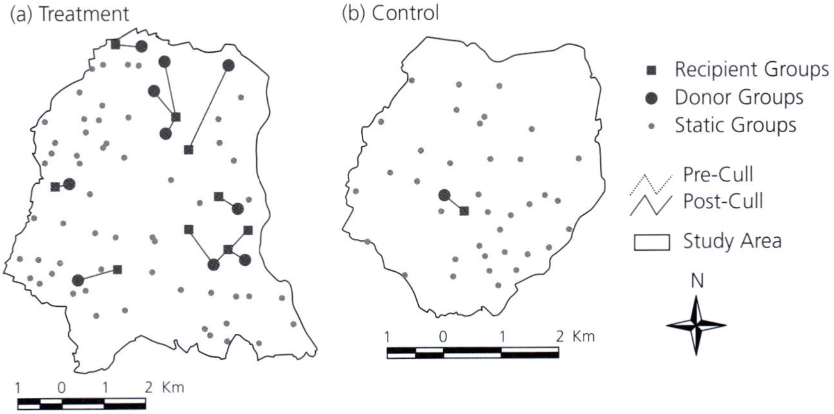

Figure 5.12 Inter-group movements (IGMs) of individual badgers from trapping records within the TEE. IGMs in (a) treatment and (b) control areas before and after Nov02/Jan03 culls. From Riordan et al. (2011).

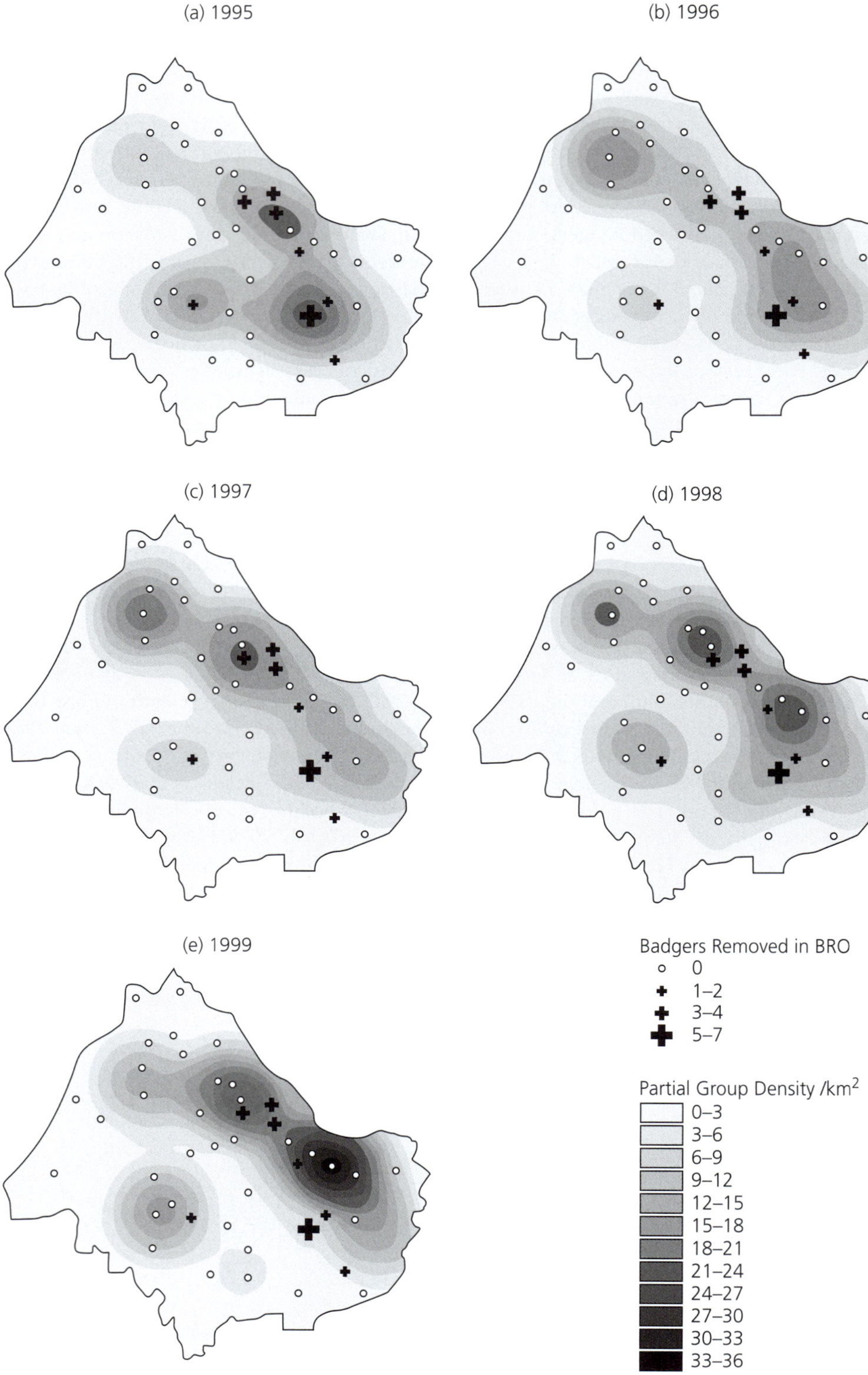

Figure 5.13 Interpolated visualization of badger density across the North Nibley study area between 1995 and 1999. Social groups removed in 1995 are shown as black crosses, with the symbol size indicating the number of animals culled. Social groups not culled in 1995 are shown as white dots. From Macdonald et al. (2006), reproduced with permission from Elsevier.

Table 5.1a Percentage change in herd breakdowns within the RBCT culling area and in a 2 km-wide perimeter surrounding it (in Godfray et al. 2013).

Time period	Average % change			
	Proactive culling area		Areas surrounding cull	
	Central estimate (%)	95% CI	Central estimate (%)	95% CI
During trial	−23	−12 to −33	+ 25	−1 to + 56
After trial*	−28	−15 to −39	− 4	−26 to + 24
Entire period	−26	−19 to −32	+ 8	−14 to + 35

*After trial period includes averages over five years of data

Table 5.1b Percentage change in herd breakdowns of core and perimeter over nine years (five annual culls and then four more years) extrapolated from estimates in Table 5.1a (estimates by C. Donnelly in Macdonald and Burnham 2011). (i) Five years from the initial proactive cull (assuming annual culling, with five such culls), (ii) The following four years: from 12 to 60 months after the final proactive cull, (iii) After nine years: from the completion of the initial proactive cull until 60 months after the fifth annual cull. (If the removal area had 'hard edges' completely impermeable to badgers and thus no perimeter perturbation, the reduction in bTB incidence might be 20–34%).

Extrapolated consequences of badger culling in minimum licenceable cull area for predicted herd breakdowns		
Time period	Average % change	
	150 km^2 circular culling area plus 2 km-wide perimeter surrounding it	
	Central estimate (%)	95% CI
(i) Cull years	−4	+ 9 to −17
(ii) Post-cull years	−21	−8 to −33
(iii) Entire period	−12	−3 to −21

possibility of revisiting this policy under exceptional circumstances, or if new scientific evidence were to become available. By 2010, the government had changed, and the new government launched a consultation regarding badger culls in England.

The context of the reconsideration was an imperative to tackle a crisis whereby *c.* 25,000 cattle had been slaughtered in England in 2010 due to TB (costing £90 million) and 23% of farms in the west and south west had been unable to move stock off their premises at some point (Defra 2011a). Making assumptions extrapolated in part from the RBCT, the proposed new badger culls would each cover at least 70% of land within a minimum of 150 km^2 (larger area reduces the relative size of the perturbed perimeter) and might each involve killing about 1500 badgers over four years. The culling would be done by licensed groups of farmers or their agents who would shoot free-ranging badgers.

With respect to killing badgers the salient facts were (1) that the RBCT had revealed that culling at that scale in England did not come even close to eliminating bTB in badgers (actually it increased during the RBCT) or cattle, indeed, any overall gain from it is estimated to be about a net 12–16% relative reduction in herd breakdowns over nine years, (2) in the early years culling was associated with a worsening of the situation in cattle (which was never turned to benefit at the perimeter where farmers remain at a net disadvantage), probably due in part to perturbation, (3) there was no precedent at all for judging the likely outcome of a policy based on shooting badgers by private individuals rather than cage-trapping by government teams, and (4) if the culls failed to deliver the required reductions in badger numbers, rapidly enough, over a large enough continuous area, then cattle TB risks would likely increase.

The government's estimate of the rewards from a 350 km^2 zone of five years' culling with four years' post-culling was a saving of £3.68 million in reduced cattle TB, at a cost of £4.56 million to deliver the cull (Defra 2011b). In practice, these calculations were found to have under-estimated the costs of delivering culling (because free shooting had to be supplemented by more expensive cage-trapping, and because widespread public protest more than doubled the policing costs).

Whether such culling is worthwhile not only depends on whether society judges these gains to merit the costs (financially and in badgers), and over what time scale. The huge practical task and financial commitment facing the farmers, the likelihood of protest and use of an untested method (shooting), and the international legality (of locally extirpating a scheduled species under the Bern Convention on the Conservation of European Wildlife and Natural Habitats)[22] of scaling up this type of control to a sufficient proportion of the 39,000 km^2 of Britain (*c.* 30% of England) in which bTB is endemic, to make significant inroads into the national problem (such scaling up might involve killing about 100,000 badgers over four years—there are an estimated 360,000 badgers in the UK—within an estimated number of 71,600 social groups in England and Wales; Judge et al. 2014), all needed to be considered.

Having considered these factors, the government decided to launch two pilot culls in Somerset and Gloucestershire, based, as said in 2012 by Defra's chief scientific adviser, Ian Boyd, and Chief Veterinary Officer, Nigel Gibbens, on 'the best available scientific evidence after more than 15 years of intensive research.' Although as the government's Chief Scientist, Sir John Beddington, wryly remarked, it was 'interesting' that exactly the same scientific evidence was used to decide for a cull in England and against a cull in Wales.

The culls, having been postponed in 2012 due to problems associated with setting cull targets, began in the two pilot areas in November 2013, with the intention of conducting a series of four annual six-week culls using free shooting. The use of free shooting was based on a desk study by the Game Conservancy Trust (now the Game & Wildlife Conservation Trust) (GCT 2006, 'Shooting as a potential tool in badger population control'), which suggested that shooting might be an effective way of reducing badger densities, albeit more slowly than cage-trapping. The licences required that neither too few nor too many badgers be killed: the minimum numbers to be killed were set to give 80% confidence that at least 70% of the estimated total would be removed, while maxima were also imposed to reduce the risks of local extinction. However, uncertainty in the estimates of population size meant that the minimum number to be killed exceeded the lower confidence limit on estimated population size in both areas: in principle, marksmen might contravene the Bern Convention by eliminating badgers from the area, while simultaneously contravening their licence conditions by failing to kill enough badgers (Donnelly and Woodroffe 2012). The proportion of land required to be covered by the shooting was 70%, raising questions as to whether enough landowners (who must pay for the night shooting, hiring of the trained marksman, and for the equipment to dispose of the carcasses) could be recruited and sustained for four years (Moody 2013). This criterion was based on the RBCT, in which officials were able to access 70% of the land for culling badgers (Bourne et al. 2007). Another key criterion was that culls had to be delivered rapidly, being completed within a maximum of six consecutive weeks each year. This condition was again based on the RBCT which revealed that, although all culls were associated with elevated infection in badgers, these increases were greater (by a factor of 1.7) if culls were conducted in a patchy and discontinuous manner over a period of >12 nights (Woodroffe et al. 2006b). So, even if this culling approach could deliver the anticipated reduction in badger densities, at the required scale and to the required timetable, it would be expected to provide only marginal gains in terms of TB control, and these would come at a very significant cost to two groups: (1) the farmers whose expense, effort, and hopes will be cruelly disappointed, and (2) the very large number of badgers that will have been killed. With so much at stake, what has science to offer in terms of confidence that even the forecasted gains will materialize?

At least three factors worsened the already unpromising odds of worthwhile reduction of bTB in cattle. First, public opposition to badger culling manifested as disruptive intervention. This increased the costs of policing the culls from a predicted £1 million for the first two pilot culls to £2.5 million in practice, altering both the chances and the costs of fulfilling assumptions on which benefits had been calculated.

A second concern was that imprecision, and possible error, arose from the requirement to estimate the population density of the badgers, in order to create a baseline against which the desired tally of 70% can be measured (along with the risk of infracting the Bern Convention). Counting badgers is difficult (e.g. Macdonald et al. 1996; Tuyttens et al. 2001)—so much so that in 2012 the farming consortium was unable to survey their setts satisfactorally, so this task was devolved to the Food and Environment Research Agency (FERA). Counting active setts (a definition which itself contains complexities) is much easier than counting the badgers that occupy them (yet the numbers and characteristics of setts themselves provide only very imprecise indicators of the numbers of badgers within; Tuyttens et al. 2001; Wilson et al. 2003). The numbers of badgers per active sett (itself a biologically dubious measure, since individual badgers occupy multiple setts) in the cull areas were estimated using DNA identification of individuals based on hair-traps. Donnelly and Woodroffe (2012) concluded that the uncertainty of the population estimates meant that killing an estimated 70% of badgers could in fact remove as few as 51% of

[22] Badgers are protected under Appendix III of the Bern Convention: it prohibits the use of means that could cause local disappearance or serious disturbance to a population of badgers, and exceptions can only be applied if there is a guarantee that it is not detrimental to the survival of the population of badgers (NE 2011b).

the resident population (risking an increase in BT in cattle), or as many as 100% (which would breach the Bern Convention). So, the addition of any further variability in the estimates could in principle be the straw that breaks the camel's back (in this metaphorical case, the fact that the camel is already laden with a veritable bale of straw does not detract from the truth that one more may break its back, if it is not already broken). Weather affects the badgers' propensity to attempt foraging (Noonan et al. 2014), and thus to emerge and be available for trapping (Noonan et al. 2015). Projecting best/worst-case scenarios between weather and body-condition can result in 8.6% ± 4.9 SD difference in seasonal trapping efficiency. This translates into a potential 55.0% population abundance under-estimation under the worst-case scenario; 38.6% over-estimation under the best-case (Noonan et al. 2015). Attempting to trap during sub-optimal weather may be amongst the reasons why, in 2014, only 341 of the targeted 316-435 badgers were culled in Somerset and only 174 of the targeted 615-1091 in Gloucestershire (Defra 2014b).

A third concern was that the proposed culling methods might not be able to deliver large enough reductions in badger numbers, rapidly enough, across the very large areas required by licencing conditions. Free shooting was an untested method, and deploying cage traps over the >200 km^2 of each area would be a logistically challenging operation likely to require over a thousand traps. Failing to meet the targets in terms of either numbers of badgers killed, or the time taken to achieve this, would undermine any likely benefits for the control of cattle TB, and might even increase TB risks for cattle.

At the time of writing, the predicted failure of the culls has been confirmed. The independent expert panel established to oversee the pilot culls concluded in 2014 that the culls had failed to meet their targets for both effectiveness and humaneness. Although culls will not be extended to new areas in 2014, the current Environment Secretary remains committed to expanding them in the future[23], and culling will continue in the two pilot areas, this time without independent monitoring. Indeed, as forecasted by scientific experts, the independent expert panel report on the pilot areas (Munro 2014) confirmed the shooting only managed to remove less than half of the 70% of badgers to be killed, and welfare concerns about free shooting were corroborated by up to 18% of badgers taking longer than five minutes to die after being shot (Munro 2014). Gassing of badgers has now been suggested, even when the use of hydrogen cyanide (HCN) had already been banned because of its inhumane consequences. C. Cheeseman, former Head of Wildlife Diseases at the Central Science Laboratory (now AHVLA), highlights that gassing would need investment to find out if new gases have any welfare concerns, how to gas setts in inaccessible places, how to avoid badgers under sublethal doses opening new exits, and how to avoid killing non-target species often utilizing these setts.

[23] Case, P. (2014).

Is this culling plan the least bad option, in comparison with, for example, vaccinating cattle (which requires developing a molecular tag to differentiate infected from vaccinated cattle), or vaccinating badgers (Corner et al. 2010)? It is difficult to make a judgement on this issue, considering the livelihoods and human anguish at stake and the conviction within much of the farming community that badger culling is required. Nevertheless, the uncertainties, the costs and seemingly poor return on capital, the impacts on badgers, and intense societal disquiet combined to make any past or future proposed badger culls unpromising.

This led Macdonald and Willis (2013), in an essay on conservation 'elephants in the room' to nominate as one such elephant: *Are there some problems which are simply intractable for the time-being – however sad it may be? For example, it appears that there is no tidy solution now to controlling bTB in badgers, at least not one that society is likely to judge acceptable. The understandable desire to do something should not be at the cost of doing the wrong thing*. This caveat, however, refers to controlling bTB in badgers, for which the motivation is to reduce the incidence of bTB in cattle. This prompts two wider questions: what basket of approaches might be combined to solve the bTB problem in cattle, and if killing badgers seems unpromising as a solution in itself, might it usefully be part of a wider integrated multiplicity of interventions? A cull-vaccination intervention, the test–vaccinate/remove (TVR) pilot study in which only test-positive badgers are culled but test-negative badgers are vaccinated, is to start in 2014 across 100 km^2 in Northern Ireland, but recent findings by Bielby et al. (2014) show that even culling at such small-scale causes social perturbation, and therefore risks increasing TB transmission. The British government has now proposed vaccination of badgers to create a buffer zone around high-risk areas,

5.6 Beyond perturbation: the bigger picture

Bovine tuberculosis is an international problem: *c.* 50 million cattle are infected worldwide, costing losses of $3 billion (Waters et al. 2012). In Ireland, Europe, New Zealand, and parts of the USA, where cattle TB has not yet been eradicated by test and slaughter, there is also a major wildlife reservoir. With respect to the success of badger culls, in Ireland, killing badgers in the four-area study (badger culling for five years in four counties with bTB incidence in cattle higher than the national average) reduced the prevalence of infection in badgers and was associated, after the fifth year, with a reduced rate of herd breakdowns (Griffin et al. 2005). Macdonald et al. (2006) warn against extrapolating these findings to culling strategies in England because, among many differences, badger densities in Ireland began at much lower levels, capture rates were influenced by the use of snares, the removal areas were atypical in that they were deliberately geographically isolated (for example, in some cases by coastline), had high levels of farmer and public compliance, were highly rural, with few substantial human settlements, and thus different risk factors for a by-catch of domestic pets and different attitudes to public access. By comparison, the proactive treatment in the Randomised Badger Culling Trial, with random allocation of treatments and no-culling controls, used cage-traps (not snares), had lower landowner participation, did not trap during the badger breeding season, and during the first year, removed between approximately 1.7 and 6.0 badgers/km^2 and reported cattle incidence from the entire culled area rather than only > 6 km inside the culling area boundaries.

In Britain, where the environment can support high densities of badgers, any attempt to reduce local densities below carrying capacity is likely to prompt immigration from neighbouring areas. Hence, perturbation is likely to be a feature of any attempt to reduce badger density, whether that attempt entails culling or contraception. Thus, perturbation is likely to undermine the disease control benefits of any reduction in badger density, short of virtual elimination of badgers across vast areas of Britain.

That perturbation acts as an impediment to disease control highlights the fact that badgers' normal social and territorial structure acts as an impediment to disease spread. Where badger populations are left undisturbed, infection is often highly localized in space and remarkably stable over time (Cheeseman et al. 1981, 1988; Delahay et al. 2000; Woodroffe et al. 2005). Badgers' infrequent movement between social groups at high densities appears to check the spread of TB across the landscape.

Interestingly, it is possible that the social structure which undermines the utility of badger culling might possibly improve the prospects for successful badger vaccination. Infections can only spread when infected animals have the opportunity to interact (directly or indirectly) with susceptible animals. The currently available vaccine, the Bacille Calmette-Guérin, or BCG, reduces the risk that individual vaccinated badgers will subsequently test positive for *M. bovis* infection (Chambers et al. 2011), suggesting that it might deplete the pool of susceptible animals in each social group and hence reduce disease transmission. Encouraging evidence of this process comes from the finding that vaccination of adults reduces the risk of infection in unvaccinated cubs within the same social group (Carter et al. 2012). By reducing the number of susceptible animals with which each infected animal interacts, badgers' social system might reduce the level of vaccine coverage required to control infection within the badger population. Tests are urgently needed to determine whether vaccination can reduce the prevalence of infection at the population scale, and to explore (and ultimately quantify) any impacts on TB in cattle.

Although vaccination has been dismissed in some quarters because of its cost, two factors mean that it is likely to be cheaper than culling. First, vaccination is unlikely to require policing. Second, a large number of volunteers are willing to contribute time and effort to badger vaccination, reducing manpower costs. These factors would need to be considered as part of a trial programme of badger vaccination.

5.6.1 Trade, testing, and monitoring: the UK and beyond

5.6.1.1 Cattle testing

Since 1973, British cattle producers have had to comply with EU rules regarding certification of TB testing on exported animals and TB freedom of herds. The single intra-dermal comparative cervical tuberculin (SICCT) test ('skin test') was introduced in 1947 and is the primary method of cattle testing, due to its specificity[24] of > 99.9% at an individual level (Strain et al. 2011). However, its sensitivity[25] gives cause for concern, being estimated at 49% at herd level (in Godfray et al. 2013). Increasing the frequency of the tests and the number of animals tested could increase sensitivity at the herd level, and hence reduce the number of herds wrongly classified as uninfected (false negatives), but to the

[24] The specificity of a test refers to the proportion of uninfected animals correctly identified. In this case, the test has 99.9% specificity at individual level, so the test identifies correctly 99.9% of individuals without the disease (true negatives) and 0.01% are incorrectly identified as testing positive (false positives).

[25] Sensitivity refers to the proportion of infected animals identified correctly (Krebs 1997). In this case, a mean sensitivity of 49% at herd level means that the test has identified 49% of infected herds (true positives) but 51% of infected herds have gone undetected (false negatives).

detriment of specificity, with the consequent economic loss for farmers of having herds wrongly classified as infected (false positives).

The sensitivity and specificity of the test can vary considerably at both the individual and herd levels (Godfray et al. 2013), and this may contribute to the 38% of herds cleared from movement restrictions which experience a recurrent incident within 24 months (Karolemeas et al. 2011). Conlan et al. (2012) have estimated that 50% of such recurrent breakdowns are due to infections missed during prior tuberculin testing. Recent research reveals multiple reasons for test failures. Claridge et al. (2012) estimated an under-detection rate of *M. bovis* of one third with the SICCT test when the animal was co-infected with the common pathogen *Fasciola hepatica*, which is likely to decrease the immune response to the test, producing a negative result. Such co-infection might partly explain why bTB continues to spread in spite of control measures. Second, a recent study has suggested that certain genetic lines of cattle may react less strongly to the test, and hence avoid slaughter even when infected (Amos et al. 2013). In addition, the interpretation in the field of the skin response to the SICCT is subjective, as it is determined by interpreting the size of two lumps generated as immunological responses to the *M. bovis* and *M. avium* components of the tuberculin test (in Godfray et al. 2013).

The interferon-gamma test (also certified; IFN-γ) is an *in vitro* test which measures the increased release of interferon gamma from T cells exposed to *M. bovis* over other *Mycobacteria* (Strain et al. 2011). The IFN-γ test is currently used on a discretionary basis as a follow-up to the tuberculin test; while it can detect infected cattle not identified by the skin test due to its higher individual-level sensitivity (estimated median of 67%; in Godfray et al. 2013), its lower specificity (estimated median of 98%; in Godfray et al. 2013) means that the high number of false positives make it currently unacceptable as a primary diagnostic text (Strain et al. 2011). The IFN-γ test only requires a single farm visit, and is evaluated in the laboratory, which may give more consistent results, although it does require specialist laboratory capacity and rapid sample processing. Slaughterhouse testing also provides important surveillance information, which in 2012 triggered the confirmation of a quarter of new TB incidents in Great Britain (European Commission 2012).

To encourage prompt testing, since 2012 lowered compensation is paid if herds are overdue for testing (e.g. 60–90 days overdue: 25% reduction in compensation, > 180 days: 95%) (Defra 2012b). bTB is a notifiable disease in cattle and farmers are required to notify the AHVLA of any outbreak.

5.6.1.2 Testing badgers

A recommendation of the 1986 Dunnet Report was that, in future, culling should be targeted at infected badgers. This approach was eventually attempted in the live test trial of 1994–1996 (which included one of the culls at North Nibley), using the Brock TB ELISA test. This test cannot distinguish animals actively excreting bacilli from those that are not, and, if acquired immunity to bTB occurs, such animals would also return positive results (Newell et al. 1997). The test is now available in a readily usable statpak formulation, which can immediately be used at the capture site, but although the specificity of the test is 93%, its sensitivity is poor (49% overall; Chambers et al. 2008). Test sensitivity is particularly low for animals with subclinical infections, whereas those with evident lesions at post-mortem were correctly identified in 66–78% of cases (Chambers et al. 2008). This low sensitivity has restricted the utility of this approach; in the live test trial it was recognized that sensitivity might be improved by classifying groups of badgers, rather than individual animals, as test-positive and targeting culling accordingly. Unfortunately, the sensitivity of the test was still too low to contribute usefully to disease control, and the decision to group badgers by sett, rather than social group (bearing in mind that each social group is likely to occupy multiple setts) meant that badgers which tested negative and were released at one sett were regularly culled at another sett (Woodroffe et al. 1999).

The IFN-γ is more sensitive but requires analysis shortly after sample collection in a laboratory and takes longer to perform (Drewe et al. 2010).

5.6.2 Improving biosecurity

Understanding the impact of biosecurity measures in reducing herd breakdown risk is still a knowledge gap. Currently, farmers are exhorted to attend to biosecurity, but there is very poor evidence to indicate which measures they should adopt.

Farmers have always been offered advice on minimizing transmission risks through better 'biosecurity', but this issue climbed the agenda when it became evident that 80% of foot-and-mouth disease incidents in the 2001 outbreak were associated with a lack of biosecurity implementation by farmers (Anderson 2002). As TB levels continued to rise, the Environment, Food and Rural Affairs Select Committee pushed Defra to commission an independent review of biosecurity. The resulting expert panel reviewed the role of good biosecurity in, for example, limiting direct contact between cattle and badgers (e.g. exclusion from setts and latrines by fencing), limiting badger access to cattle feed and water, and altering opportunities for interaction by changing grazing patterns (Phillips et al. 2003). In 2007, Defra revised existing bTB biosecurity guidance for farmers, but few of the recommendations are based upon evidence, and the National Farmers Union argues that some of these guidelines are impractical (Enticott 2008).

Neither the effectiveness nor the cost of these measures has been evaluated due to the large number of potential measures and the comparative rarity of TB incidents in herds. This lack of evidence is unlikely to persuade farmers to implement such measures (Wilson et al. 2011). However, advice and raising of awareness provided by veterinarians (in Wales) appear to have contributed to changing farmers' attitudes into positive responses to biosecurity (Enticott 2008).

5.6.2.1 Cattle movements and herd size

In Britain, cattle movements are recorded at an individual level: to comply with European Union legislation, cattle must be tagged with an approved eartag (in both ears within 20 days of birth, and in dairy animals, at least one tag must be fitted within 36 hours of birth; BCMS 2011).

Because cattle movements from high-incidence areas are associated with an increased risk of onward transmission (Gilbert et al. 2005), pre-movement testing is expected to reduce this risk. In 2006, a new regulation on cattle movements was introduced, requiring a compulsory pre-movement tuberculin test, conducted within 60 days of the planned movement, for all cattle over 42 days old which were to be moved out of herds which are on either an annual or biennial regime of regular testing. This tuberculin test may be supplemented by an interferon-gamma (IFN-γ) test in certain circumstances. In 2009 in Wales, and 2010 in England, it became compulsory to re-test any so-called inconclusive reactors (IRs). From 1 January 2013, the time allowed to move cattle from TB restricted herds was reduced from 60 to 30 days from the day of their last negative tuberculin skin test (Defra 2013b), and from October 2014, Defra will cease partial de-restrictions on holdings with TB-breakdowns (which had allowed some cattle that had tested negative to infection within infected farms, and were under restriction, to be tested earlier than the rest of the herd in order to be traded Defra 2014c). Since Scotland has made post-movement testing compulsory, farmers have been incentivized to purchase cattle from low-disease areas. However, interactions between cattle on contiguous premises, and even short-distance cattle movements, can be an unrecorded source of cattle-to-cattle transmission (Godfray et al. 2013).

Herd size has been recognized repeatedly as a risk factor for TB infection: the per-animal risk is higher on larger herds. For example, Brooks-Pollock and Keeling (2009) found that disease persistence was greater in larger herds: after a breakdown, 90% of herds of < 20 cattle tested clear of bTB after the two required tests, but for herds of > 400 cattle, only 55% passed both tests. There are many plausible explanations for this pattern (within a larger herd, there is a greater risk of one animal coming into contact with a source of infection, and also a greater opportunity for transmission within the herd). The trend for increased herd size (from 30 in 1973 to 123 in 2011) over time may well have contributed to the trend for increased TB rates in cattle (during the same period, the number of registered milk producers fell from 100,741 to 14,793: source: www. defra.gov.uk).

5.6.2.2 Management of badgers

Biologists are generally impressed by the intra-specific variation in carnivore behaviour under different ecological circumstances, and so it appeared likely that there would be links between agri-environmental factors on cattle farms and badger behaviour, and thus bTB risk. In a case-control study comparing farms with and without a recent history of bTB infection, Matthews et al. (2006b) found that some 'nature friendly' practices, such as the presence of ungrazed wildlife strips and the availability, width, and continuity of hedgerows, were associated with reduced bTB, perhaps because these management options reduce badger–cattle contact. However, as with other biosecurity measures the benefits to farmers are not well quantified.

Badgers sometimes visit farm buildings (Garnett et al. 2002; Ward et al. 2008). Judge et al. (2011) installed cameras, which confirmed that badger visits occurred in 59% of farms (n = 32), sometimes on > 60% of nights (over a year). Simple measures to exclude badgers from feed stores and cattle housing (e.g. galvanized aluminium sheeted metal panels) were 100% successful if properly used and maintained. However, such measures cannot prevent contact between badgers and cattle outdoors. The option of keeping cattle indoors, and thus out of contact with badgers, requires a much more costly approach to farming and is complicated by the ethical and welfare considerations of confining them.

5.6.3 Vaccination

5.6.3.1 Vaccination in cattle

A vaccine for cattle (*Mycobacterium bovis* Bacille Calmette-Guérin, or BCG) exists, but its use is currently not permitted under EU rules. This restriction is partly due to the possibility that vaccinated animals are not fully protected against bTB infection, but largely because BCG and the tuberculin test share the same antigens, and hence a vaccinated individual could possibly not be distinguished from an infected one (in Waters et al. 2012): six months after having administered BCG, 80% of experimentally vaccinated cattle tested positive to the tuberculin test[26] (Whelan et al. 2011).

Although BCG is unable to prevent mycobacterial infection, it nonetheless elicits a protective response which reduces the pathophysiology of infection and can lead to a reduction in pathogen burden, the progression, severity, and excretion of TB in cattle under experimental conditions (e.g. Wedlock et al. 2007; Hope et al. 2011; Tompkins et al. 2013). Similar levels of protection against bTB have been detected in calves administered with oral rather than parenteral vaccinations (Buddle et al. 2005). Field studies[27] have shown a reduction in transmission of disease to vaccinated animals: for example, a trial in Mexico showed a BCG efficacy of 59.4%, with non-vaccinated calves, 2.4 times as likely to become infected, and BCG preventing nasal excretion of *M. bovis* bacilli (Lopez-Valencia et al. 2009; Ameni et al. 2010). Godfray et al. (2013) highlight that it has been shown that, although vaccination has no effect on existing infections in cattle, if vaccinated cattle become infected, infectiousness will be limited and hence onward transmission to cattle and wildlife will be reduced. Used with existing control measures, cattle BCG would be expected to help to reduce the frequency and severity of bTB breakdowns in herds.

The current cost of BCG has been estimated at £8.25 per cattle dose (Defra 2011a), which needs to be administered annually (neonatal vaccination is the best protection but wanes after one year; in Godfray et al. 2013). The AHVLA is developing a diagnostic test to differentiate infected from vaccinated animals (DIVA test, costing £25.00; House of Commons 2013). The DIVA test is intended to accompany vaccinated animals to confirm whether a tuberculin skin test positive result is caused by vaccination or TB infection. Results of the DIVA test under research conditions show 95% relative sensitivity and 96% specificity, but these results have not yet been assessed in the field (in Godfray et al. 2013). Once this DIVA test is available, any future use in combination with the BCG cattle vaccine would require changes in EU legislation; the European Commission has suggested that trade of vaccinated animals is unlikely before 2023 (European Commission 2013).

5.6.3.2 Vaccination of badgers

Vaccination of badgers is attractive because, *inter alia*, it neither involves killing badgers nor risking a perturbation effect. Defra has invested over £16 million on research into badger vaccines since 1994. Since 2010 an injectable badger vaccine, BadgerBCG, has been authorized (Defra 2013a).

Although it does not prevent infection, intramuscular administration of (BCG) significantly reduces disease progression in captive individual badgers by 74% (decreasing its severity and the excretion of *M. bovis* pathogens which might sensitize or infect cattle) and in social groups of wild badgers (Chambers et al. 2011; Lesellier et al. 2011). In addition, Carter et al.'s (2012) field studies found that when more than a third of a social group had been vaccinated, the risk to unvaccinated cubs was reduced by 79% through an apparent herd immunity effect.

A vaccination scheme needs to be implemented yearly to protect new cubs, but it is unclear how long it would need to be applied for the disease to be eliminated. Models assessing the length and effectiveness of annual BCG vaccination campaigns have predicted that bTB would be eradicated in badgers by between 20–30 years with annual BCG vaccination rates of 0.7 (Wilkinson et al. 2004). However, projections are likely to be strongly influenced by the extent of transmission from cattle to badgers.

In 2012, the Welsh government abandoned a planned badger cull in favour of a pilot vaccination project, as part of a programme of 'intensive action' to drive down TB within a confined area. A total of 1424 badgers were vaccinated within an unreplicated 'intensive action area' (288 km^2) by the end of that year. This project may, alongside other vaccination projects, contribute field-based evidence on the use of vaccination to reduce TB incidence. The logistics of delivering badger vaccination—including training of lay vaccinators within the community—are being explored by the Defra-funded Badger Vaccine Deployment Project (BVDP), involving over 100 farms covering 100 km^2 of Gloucestershire (2000–2015). Vaccines can be administered after attending a four-day

[26] 90% revert to test negative within a year.
[27] In Ethiopia and Mexico.

course covering legal responsibilities, cage badger trapping, together with storage and handling of the BadgerBCG.

Although the labour costs of large-scale cage-trapping and vaccination by government employees and lay vaccinators would clearly be huge, vaccination could potentially be less costly than culling, because it would not involve the same level of policing, plus it might be an opportunity for very strictly accredited citizen scientists to be involved in the fieldwork (e.g. Chapter 15, this volume).

Clearly, manpower costs would be radically reduced by an oral vaccine added into bait that could simply be put into badger setts (Wilson and Carter 2013). However, there are factors impeding the oral badger vaccine, including ensuring that badgers take the required dose (enough bait consumed, although bait could be placed down badger sett entrances; Godfray et al. 2013) and the risks of BCG uptake by cattle, which would render them positive to the tuberculin skin test. The likelihood of contamination through badger faeces is extremely small since, although BCG can be shed in faeces for up to three weeks (Corner et al. 2010), the quantities (1 cfu/g faeces at 17 days post-oral vaccination; Corner et al. 2010) are far lower than the threshold of 10^7 cfu required to sensitize cattle (Robinson et al. 2012). In Ireland, a live oral vaccine is being trialled on cage-trapped, sedated badgers, based on formulations developed for possums in New Zealand (Case 2013).

5.7 The future

Successful control of bTB in cattle is likely to require multiple rather than single interventions, including vaccination, movement restrictions, and improved biosecurity of cattle, alongside vaccination of badgers (where there are also uncertainties, but at least more grounds for optimism). All of these approaches, separately or in combination, could be implemented in a programme of adaptive management which facilitates learning about the effectiveness of the measures implemented. Vaccination is likely to reduce the prevalence of infection in badgers, whereas the evidence indicates that, in England, culling increases this prevalence. Fearful that the focus on the badger factor has become disproportionate, we would also scrutinize possible policies in the light of a full life-cycle analysis of the wider economic scene of trading in cattle, to set the costs and benefits of various interventions in the context of risk-based trading nationally and internationally.

Each group of stakeholders will view the ocean of possibilities from the dynamic perspectives of different rising and falling crests of waves of opinion. Our point is less about the final outcome—there is no unarguably 'right' answer, and all options involve some losers—but more to illustrate the perplexing judgements that characterize almost all conservation issues. The badger case requires serious attention to issues as diverse as ecology, immunology, agriculture, economics, national and international law, civil rights, ethics, and politics.

Acknowledgements

We thank Christl Donnelly, Charles Godfray, and Chris Newman for comments on an earlier draft, Eva Raebel for assiduously collating information, and Ruth Feber for editorial advice. MAFF and Defra, successively, for conducting the RBCT and for funding work at Nibley and TEE, and PTES for funding the long-term study at Wytham.

References

Ameni, G., Vordermeier, M., Aseffa, A., Young, D.B., and Hewinson, R.G. (2010). Field evaluation of the efficacy of Mycobacterium bovis Bacillus Calmette-Guérin against bovine tuberculosis in neonatal calves in Ethiopia. *Clinical and Vaccine Immunology*, **17**, 1533–1538.

Amos, W., Brooks-Pollock, E., Blackwell, R., Driscoll, E., Nelson-Flower, M., and Conlan, A.J.K. (2013). Geneti predisposition to pass the standard SICCT test for bovine tuberculosis in British cattle. *PLOS ONE*, **8**, e58245.

Anderson, I. (2002). *Foot and Mouth Disease 2001: Lessons to be Learned Inquiry Report*, The Stationery Office, London.

Annavi, G., Newman, C., Dugdale, H.L., Buesching, C.D., Sin, Y.W., Burke, T., and Macdonald, D.W. (2014). Neighbouring-group composition and within-group relatedness drive extra-group paternity rate in the European badger (Meles meles). *Journal of Evolutionary Biology*, **27**(10), 2191–2203.

Bacon, P.J. and Macdonald, D.W. (1980). To control rabies: vaccinate foxes. *New Scientist*, **87**, 640–645.

BCMS (2011). *Cattle Keeper's Handbook*. British Cattle Movement Service, Rural Payments Agency, Defra.

Bielby, J., Donnelly, C.A., Pope, L.C., Burke, T., and Woodroffe, R. (2014). Badger responses to small-scale culling may compromise targeted control of bovine tuberculosis. *PNAS*, **111**, 9193–9198.

Bourne, J., Donnelly, C.A., Cox, D.R., Gettinby, G., McInerney, J.P., Morrison, I., and Woodroffe, R. (2007). *Bovine TB: The scientific evidence. A Science Base for a Sustainable Policy to Control TB in Cattle. An Epidemiological Investigation into Bovine Tuberculosis*. Final Report of the Independent Scientific Group on Cattle TB.

British Veterinary Association (2014). *Bovine TB situation in Scotland*. <http://www.bva.co.uk/>, accessed March 2014.

Brooks-Pollock, E. and Keeling, M. (2009). Herd size and bovine tuberculosis persistence in cattle farms in Great Britain. *Preventive Veterinary Medicine*, **92**, 360–365.

Buddle, B.M., Aldwell, F.E., Skinner, M.A., et al. (2005). Effects of oral vaccination of cattle with lipid-formulated BCG on immune responses and protection against bovine tuberculosis. *Vaccine*, **23**, 3581–3589.

Carter, S.P., Chambers, M.A., Rushton, S.P., et al. (2012). BCG vaccination reduces risk of tuberculosis infection in vaccinated badgers and unvaccinated badger cubs. *PLOS ONE*, **7**, 1–18.

Carter, S.P., Delahay, R.J., Smith, G.C., et al. (2007). Culling-induced social perturbation in badgers (*Meles meles*) and the management of TB in cattle: an analysis of a critical problem in applied ecology. *Proceedings of the Royal Society B*, **274**, 2769–2777.

Case, P. (2013). Oral badger vaccine field trial under way. *Farmers Weekly*, 7 March 2013, <http://www.fwi.co.uk/>, accessed March 2013.

Case, P. (2014). Badger cull to be rolled out in 2015, says Owen Paterson. *Farmers Weekly*, available at: <http://www.fwi.co.uk/>, accessed 22 April 2014.

Chambers, M.A., Crawshaw, T., Waterhouse, S., Delahay, R., Hewinson, R.G., and Lyaschenko, K.P. (2008) Validation of the BrockTB Stat-Pak assay for detection of tuberculosis in Eurasian badgers (Meles meles) and influence of disease severity on diagnostic accuracy. *Journal of Clinical Microbiology*, **46**, 1498–1500.

Chambers, M.A., Rogers, F., Delahay, R.J., et al. (2011). Bacillus Calmette-Guerin vaccination reduces the severity and progression of tuberculosis in badgers. *Proceedings of the Royal Society B-Biological Sciences*, **278**, 1913–1920.

Cheeseman, C.L., Jones, G.W., Gallagher, J., and Mallinson, P.J. (1981). The population-structure, density and prevalence of tuberculosis (*Mycobacterium bovis*) in badgers (*Meles meles*) from four areas in South-West England. *Journal of Applied Ecology*, **18**, 795–804.

Cheeseman, C., Mallinson, P.J., Ryan, J., and Wilesmith, J.W. (1993). Recolonisation by badgers in Gloucestershire. In T.J. Hayden, ed. *The Badger*, pp. 78–93. Royal Irish Academy, Dublin.

Cheeseman, C.L., Wilesmith, J.W., Stuart, F.A., and Mallinson, P.J. (1988). Dynamics of tuberculosis in a naturally infected badger population. *Mammal Review*, **18**, 61–72.

Choisy, M. and Rohani, P. (2006). Harvesting can increase severity of wildlife disease epidemics. *Proceedings of the Royal Society B-Biological Sciences*, **273**, 2025–2034.

Claridge, J., Diggle, P., McCann, C.M., et al. (2012). *Fasciola hepatica* is associated with the failure to detect bovine tuberculosis in dairy cattle. *Nature Communications*, **3**, 853.

Conlan, A.J.K, McKinley, C.J., Karolemeas, K., et al. (2012). Estimating the hidden burden of bovine tuberculosis in Great Britain. *PLOS ONE*, **8**, e1002730.

Corner, L.A.L., Costello, E., O'Meara, D., et al. (2010). Oral vaccination of badgers (*Meles meles*) with BCG and protective immunity against endobronchial challenge with *Mycobacterium bovis*. *Vaccine*, **28**, 6265–6272.

Courtenay, O., Reilly, L.A., Sweeney, F.P., et al. (2006). Is Mycobacterium bovis in the environment important for the persistence of bovine tuberculosis? *Biology Letters*, **2**, 460–462.

Crawshaw, T.R., Griffiths, I.B., and Clifton-Hadley, R.S. (2008). Comparison of a standard and a detailed badger necropsy protocol to determine the sensitivity of the standard protocol for detecting *Mycobacterium bovis*. *Veterinary Record*, **163**, 473–477.

Defra (2011a). *Bovine TB Eradication Programme for England*. Available from <http://www.defra.gov.uk/>, accessed April 2013.

Defra (2011b). *Measures to address bovine TB in badgers. Impact Assessment*. 30 November 2011. Available from <http://www.defra.gov.uk/>, accessed April 2013.

Defra (2012a). *Changes to Bovine TB Surveillance. Bovine TB Information Note 04/12*. <http://www.defra.gov.uk/>, accessed April 2013.

Defra (2012b). *Changes to cattle compensation payments. Bovine TB Information Note 01/12*. <http://www.defra.gov.uk/>, accessed March 2013.

Defra (2013a). *Bovine TB*. <http://www.defra.gov.uk/>, accessed April 2013.

Defra (2013b). *Changes to TB Cattle Movement Controls. Bovine TB Information Note 05/12. October 2012 (Updated March 2013)*. Available at: <http://www.defra.gov.uk>, *accessed March 2014.*

Defra (2014a). *Draft Strategy for Achieving Officially Bovine Tuberculosis-Free Status for England*. April 2014.

Defra (2014b). *Summary of badger control monitoring during 2014*. Defra.

Defra (2014c). *Changes to TB Cattle Movement Controls—ending the practice of de-restricting parts of TB-restricted holdings*. June 2014.

Defra and AHVLA (2014). *Latest statistics on the incidence of tuberculosis (TB) in cattle in Great Britain. Updated 12 March 2014*. Available at: <https://www.gov.uk>/, accessed March 2014.

Delahay, R.J., Langton, S., Smith, G., Clifton-Hadley, R.S., and Cheeseman, C.L. (2000). The spatio-temporal distribution of Mycobacterium bovis (bovine tuberculosis) infection in a high-density badger population. *Journal of Animal Ecology*, **69**, 428–441.

Delahay, R.J., Smith, G.C., Barlow, A.M., et al. (2007). Bovine tuberculosis infection in wild mammals in the South-West region of England: a survey of prevalence and a semi-quantitative assessment of the relative risks to cattle. *The Veterinary Journal*, **173**, 287–301.

Donnelly, C.A. and Hone, J. (2010). Is there an association between levels of bovine tuberculosis in cattle herds and badgers? *Statistical Communications in Infectious Diseases*, **2**.

Donnelly, C.A. and Nouvellet, P. (2013). The contribution of badgers to confirmed tuberculosis in cattle in high incidence areas in England. *PLOS Currents Outbreaks*. Edition 1. doi: 10.1371/currents.outbreaks.097a904d3f3619db2fe78d24bc776098.

Donnelly, C.A. and Woodroffe, R. (2012). Reduce uncertainty in UK badger culling. *Nature*, **485**, 582.

Donnelly, C.A., Woodroffe, R., Cox, D.R., et al. (2003). Impact of localized badger culling on TB incidence in British cattle. *Nature*, **426**, 834–837.

Donnelly, CA, Woodroffe, R, Cox, DR, Bourne, FJ et al. (2006). Positive and negative effects of widespread badger culling on tuberculosis in cattle. *Nature*, **439**, 843–846 .

Drewe, J.A., Tomlinson, A.J., Walker, N.J., and Delahay, R.J. (2010). Diagnostic accuracy and optimal use of three tests for tuberculosis in live badgers. *PLOS ONE*, **5**, e11196.

Dugdale, H.L., Macdonald, D.W., Pope, L.C., and Burke, T. (2007). Polygynandry, extra-group paternity and multiple-paternity litters in European badger (*Meles meles*) social groups. *Molecular Ecology*, **16**, 5294–5306.

Economic Advisory Council (1934). *Report of Cattle Diseases Committee*. H.M.S.O. London. Cmd. 4591.

Enticott, G. (2008). The spaces of biosecurity: prescribing and negotiating solutions to bovine tuberculosis. *Environment and Planning*, **40**, 1568–1582.

European Commission (2012). *Report of the 'Bovine tuberculosis' sub-group*. Unit G5-Veterinary Programme. <http://www.bovinetb.info/>, accessed May 2013.

European Commission (2013). *Bovine tuberculosis eradication programme 2013 Letter*. <http://www.defra.gov.uk/>, accessed March 2013.

Gallagher, J. and Clifton-Hadley, R.S. (2000). Tuberculosis in badgers; a review of the disease and its significance for other animals. *Research in Veterinary Science*, **69**, 203–217.

Game Conservancy Trust, GCT (2006). *Shooting as a potential tool in badger population control. Report to Defra.*

Garnett, B.T., Delahay, R.J., and Roper, T.J. (2002). Use of cattle farm resources by badgers (*Meles meles*) and risk of bovine tuberculosis (*Mycobacterium bovis*) transmission to cattle. *Proceedings of the Royal Society*, **269**, 1487–1491.

Gilbert, M., Mitchell, A., Bourn, D., Mawdsley, J., Clifton-Hadley, R., and Wint, W. (2005). Cattle movements and bovine tuberculosis in Great Britain. *Nature*, **435**, 491–596.

Godfray, H.C.J., Donnelly, C.A., Kao, R.R., et al. (2013). A restatement of the natural science evidence base relevant to the control of bovine tuberculosis in Great Britain. *Proceedings of the Royal Society B*, **280**, 20131634.

Good, M. and Duignan, A. (2011). Perspectives on the history of bovine TB and the role of tuberculin in bovine TB eradication. *Veterinary Medicine International*, **2011**, 410–470.

Goodchild, A.V., Watkins, G.H., Sayers, A.R., Jones, J.R., and Clifton-Hadley, R.S. (2012). Geographical association between the genotype of bovine tuberculosis in found dead badgers and in cattle herds. *Veterinary Record*, **170**, 259.

Gradmann, C. (2001). Robert Koch and the pressures of scientific research: tuberculosis and tuberculin. *Medical History*, **45**, 1–32.

Griffin, J.M., Hahesy, T., Lynch, K., Salman, M.D., McCarthy, J., and Hurley, T. (1993). The association of cattle husbandry practices, environmental factors and farmer characteristics with the occurrence of chronic bovine tuberculosis in dairy herds in the Republic of Ireland. *Preventive Veterinary Medicine*, **17**, 145–160.

Griffin, J.M., More, S.J., Clegg, T.A., et al. (2005). Tuberculosis in cattle: the results of the four-area project. *Irish Veterinary Journal*, **58**, 629–636.

Hope, J.C., Thom, M.L., McAulay, M., et al. (2011). Identification of surrogates and correlates of protection in protective immunity against *Mycobacterium bovis* infection induced in neonatal calves by vaccination with *M. bovis* BCG Pasteur and *M. bovis* BCG Danish. Clinical and Vaccine Immunology, **18**, 373–379.

House of Commons (2013). *Vaccination against bovine TB. Second Report of Session 2013–14. Volume I: Report, together with formal minutes, oral and written evidence*. House of Commons, Environment, Food and Rural Affairs Committee. Available at: <http://www.parliament.uk/>, accessed April 2014.

Jenkins, H.E., Cox, D.R., and Delahay, R.J. (2012). Direction of association between bite wounds and *Mycobacterium bovis* infection in badgers: implications for transmission. *PLOS ONE*, **7**, e45584.

Jenkins, H.E., Woodroffe, R., and Donnelly, C.A. (2010). The duration of the effects of repeated badger culling on cattle tuberculosis following the cessation of culling. *PLOS ONE*, **5**, e9090.

Jenkins, H.E., Woodroffe, R. Donnelly, C.A., Cox, D. R. et al. (2007). Effects of culling on spatial associations of Mycobacterium bovis infections in badgers and cattle. *Journal of Applied Ecology*, 44, 897–908.

Johnston, W.T., Gettinby, G., Cox, D.R., et al. (2005). Herd-level risk factors associated with tuberculosis breakdowns among cattle herds in England before the 2001 foot-and-mouth disease epidemic. *Biology Letters*, **1**, 53–56.

Johnston, W.T., Vial, F., Gettinby, G., et al. (2011). Herd-level risk factors of bovine tuberculosis in England and Wales after the 2001 foot-and-mouth disease epidemic. *International Journal of Infectious Diseases*, **15**, 833–840.

Judge, J., McDonald, R.A., Walker, N., and Delahay, R.J. (2011). Effectiveness of biosecurity measures in preventing badger visits to far buildings. *PLOS ONE*, **6**, e28941.

Judge, J., Wilson, G.J., Macarthur, R., Delahay, R.J., and McDonald, R.A. (2014). Density and abundance of badger social groups in England and Wales in 2011–2013. *Scientific Reports*, **4**.

Kamerbeek, J., Schouls, L., Kolk, A., et al. (1997). Simultaneous detection and strain differentiation of *Mycobacterium tuberculosis* for diagnosis and epidemiology. *Journal of Clinical Microbiology*, **35**, 907–914.

Karolemeas, K., Donnelly, C.A., Conlan, A.J.K., et al. (2012). The effect of badger culling on breakdown prolongation and recurrence of bovine tuberculosis in cattle herds in Great Britain. *PLOS ONE*, **7**, e51342.

Karolemeas, K., McKinley, T.J., Clifton-Hadley, R.S., et al. (2011). Recurrence of bovine tuberculosis breakdowns in Great Britain: Risk factors and prediction. *Preventive Veterinary Medicine*, **102**, 22–29.

Krebs, J. (1997). *Bovine tuberculosis in cattle and badgers*. MAFF, UK.

Kruuk, H. (1978a). Spatial organisation and territorial behaviour of the European badger (*Meles meles*). *Journal of Zoology*, **184**, 1–19.

Kruuk, H. (1978b). Foraging and spatial organisation of the European badgers, *Meles meles* L. *Behavioural Ecology and Sociobiology*, **4**, 75–89.

Lesellier, S., Palmer, S., Gowtage-Sequiera, S., et al. (2011). Protection of Eurasian badgers (Meles meles) from tuberculosis after intramuscular vaccination with different doses of BCG. *Vaccine*, **29**, 3782–3790.

Lopez-Valencia, G., Renteria-Evangelista, T., de Jesús Williams, J., Licea-Navarro, A., de la Mora-Valle, L., and Medina-Basulto, G. (2009). Field evaluation of the protective efficacy of *Mycobacterium bovis* BCG vaccine against bovine tuberculosis. *Research in Veterinary Science*, **88**, 44–49.

Loveridge, A.J., Searle, A.W., Murindagomo, F., and Macdonald, D.W. (2007). The impact of sport-hunting on the population dynamics of an African lion population in a protected area. *Biological Conservation*, **134**, 548–558.

Macdonald, D.W. (1980). *Rabies and Wildlife: A Biologist's Perspective*. Oxford University Press, UK.

Macdonald, D.W. (1984). Badgers and bovine tuberculosis—case not proven. *New Scientist*, **104**, 17–20.

Macdonald, D.W. (1995). Wildlife rabies: the implications for Britain. Unresolved questions for the control of wildlife rabies: social perturbation and interspecific interactions. In P.H. Beynon and T.B. Edney, eds., *Rabies in a Changing World. Proceedings of the British Small Animal Veterinary Association*. Cheltenham, pp. 33–48.

Macdonald, D.W. and Bacon, P.J. (1982). Fox society, contact rate and rabies epizootiology. *Comparative Immunology Microbiology and Infectious Diseases*, **5**, 247–256.

Macdonald, D.W. and Burnham, D. (2011). *The State of Britain's Mammals 2011*. PTES & WildCRU.

Macdonald, D.W. and Feber, R.E., eds. (2015). *Wildlife Conservation on Farmland. Managing for Nature on Lowland Farms*. Oxford University Press, Oxford.

Macdonald, D.W., Harmsen, B.J., Johnson, P.J., and Newman, C. (2004a). Increasing frequency of bite wounds with increasing population density in Eurasian badgers, *Meles meles*. *Animal Behaviour*, **67**, 745–751.

Macdonald, D.W., Mitchelmore, F., and Bacon, P.J. (1996). Predicting badger sett numbers: evaluating methods in East Sussex. *Journal of Biogeography*, **23**, 649–655.

Macdonald, D.W. and Newman, C. (2002). Population dynamics of badgers (Meles meles) in Oxfordshire, UK: numbers, density and cohort life histories, and a possible role of climate change in population growth. *Journal of Zoology*, **256**, 121–138.

Macdonald, D.W., Newman, C., Buesching, C.D., and Johnson, P.J. (2008). Male-biased movement in a high-density population of the Eurasian badger (*Meles meles*). *Journal of Mammalogy*, **89**, 1077–1086.

Macdonald, D.W., Newman, C., Dean, J., Buesching, C.D., and Johnson, P.J. (2004b). The distribution of badger Meles meles setts in a high-density area: field observations contradict the sett dispersion hypothesis. *Oikos*, **106**, 295–307.

Macdonald, D.W., Riordan, P., and Mathews, F. (2006). Biological hurdles to the control of TB in cattle: a test of two hypotheses concerning wildlife to explain the failure of control. *Biological Conservation*, **131**, 268–286.

Macdonald, D.W. and Willis, K.J. (2013). Elephants in the room: tough choices for a maturing discipline. In D.W. Macdonald and K.J. Willis, eds., *Key Topics in Conservation Biology 2*, pp. 469–494. Wiley-Blackwell, John Wiley & Sons Ltd., Oxford.

Mathews, F., Lovett, L., Rushton, S., and Macdonald, D.W. (2006b). Bovine tuberculosis in cattle: reduced risk on wildlife-friendly farms. *Biology Letters*, **2**, 271–274.

Mathews, F., Macdonald, D.W., Taylor, G., et al. (2006a). Bovine tuberculosis (Mycobacterium bovis) in British farmland wildlife: the importance to agriculture. *Proceedings of the Royal Society of London Series B, Biological Sciences*, **273**, 357–365.

Montes, I., Newman, C., Mian, R., and Macdonald, D.W. (2011). Radical health: ecological corollaries of body condition, transport stress and season on plasma antioxidant capacity in the European badger. *Journal of Zoology*, **284**, 114–123.

Moody, O. (2013). *Pressure on farmers threatens to halt badger-culling plan*. The Times: Wildlife, <http://www.thetimes.co.uk/>, accessed March 2013.

Munro, R. (2014). *Pilot Badger Culls in Somerset and Gloucestershire. Report by the Independent Expert Panel*. Presented to the Secretary of State for Environment, Food and Rural Affairs The Rt Hon Owen Paterson MP, March 2014.

NE (2011a). *Consultation on guidance to Natural England on the implementation and enforcement of a badger control policy*. <http://www.naturalengland.org.uk/>.

NE (2011b). *The impact of culling on badger (Meles meles) populations in England and measures to prevent their 'local disappearance' from culled areas. Supplementary advice provided under the Protection of Badgers Act 1992 and Wildlife & Countryside Act 1981 (as amended)*. Natural England, UK.

Newell, D.G., Clifton-Hadley, R.S., and Cheeseman, C.L. (1997). The kinetics of serum antibody responses to natural infections with *Mycobacterium bovis* in one badger social group. *Epidemiology and Infection*, **118**, 173–180.

Noonan, M.J., Markham, A., Newman, C., Trigoni, N., Buesching, C.D., Ellwood, S.A., and Macdonald, D.W. (2014). Climate and the individual: inter-annual variation in the autumnal activity of the European Badger (*Meles meles*). *PLOS ONE*, **9**, e83156.

Noonan, M.J., Rahman, M.A., Newman, C., Buesching, C.D., & Macdonald, D.W. (2015). Avoiding verisimilitude when modelling ecological responses to climate change: The influence of weather conditions on trapping efficiency in European badgers (*Meles meles*). *Global Change Biology*. (pre-press; on-line).

Overend, E.D. (1980). Badgers and TB—does gassing spread the disease? *Oryx*, **15**, 338–340.

Phillips, C.J.C., Foster, C.R.W., Morris, P.A., and Teverson, R. (2003). The transmission of *Mycobacterium bovis* infection to cattle. *Research in Veterinary Science*, **74**, 1–15.

Pope, L.C., Butlin, R.K., Wilson, G.J., Erven, K., Conyers, C.M., Franklin, T., Delahay, R.J., Cheeseman, C.L., Woodroffe, R., and Burke, T.A. (2007). Genetic evidence that culling increases badger movement: implication for the spread of bovine tuberculosis. *Molecular Ecology*, **16**, 4919–4929.

Public Health England (2013). *Tuberculosis in the UK: Annual report on tuberculosis surveillance in the UK, 2013*. Public Health England, London, August 2013.

Riordan, P., Delahay, R.J., Cheeseman, C., Johnson, P.J., and Macdonald, D.W. (2011). Culling-induced changes in badger (*Meles meles*) behaviour, social organisation and the epidemiology of bovine tuberculosis. *PLOS ONE*, **6**, 357–365.

Robinson, P.A., Corner, L.A.L., Courcier, E.A., et al. (2012). BCG vaccination against tuberculosis in European badgers (*Meles meles*): A review. *Comparative Immunology and Infectious Diseases*, **35**, 277–287.

Rogers, L.M., Cheeseman, C.L., and Mallinson, P.J. (1997). The demography of a high-density badger (*Meles meles*) population in the west of England. *Journal of Zoology*, **242**, 705–728.

Rogers, L.M., Delahay, R., Cheeseman, C.L., Langton, S., Smith, G.C., and Clifton-Hadley, R.S. (1998). Movement of badgers (*Meles meles*) in a high-density population: individual, population and disease effects. *Proceedings of the Royal Society of London, B Biological Sciences*, **265**, 1269–1276.

Smith, G.C., Cheeseman, C.L., Clifton-Hadley, R.S., and Wilkinson, D. (2001). A model of bovine tuberculosis in the badger Meles meles: an evaluation of control strategies. *Journal of Applied Ecology*, **38**, pp. 509–519.

Smith, N.H., Dale, J., Inwald, J., et al. (2003). The population structure of Mycobacterium bovis in Great Britain: clonal expansion. *Proceedings of the National Academy of Sciences of the United States of America*, **100**, 15271–15275.

Strain, S.A.J., McNair, J., and McDowell, S.W.J. (2011). *Bovine tuberculosis: a review of diagnostic tests for M. bovis infection in cattle*. Bacteriology Branch Veterinary Sciences Division, Agri-Food and Biosciences Institute.

Swinton, J, Tuyttens, F., Macdonald, D., Nokes, D.J., Cheeseman, C.L., and Clifton-Hadley, R. (1997). Comparison of fertility control and lethal control of bovine tuberculosis in badgers: the impact of perturbation induced transmission. *Philosophical Transactions of the Royal Society of London, Series B-Biological Sciences*, **352**, 619–631.

Tompkins, D.M., Buddle, B.M., Whitford, J., et al. (2013). Sustained protection against tuberculosis conferred to a wildlife host by single dose oral vaccination. *Vaccine*, **31**, 893–899.

Tuyttens, F.A.M., Barron, L., Mallinson, P.J., Rogers, L.M., and Macdonald, D.W. (2000a). Wildlife management and scientific research: a retrospective evaluation of two badger removal operations for the control of bovine tuberculosis. In H.I. Griffiths, ed. *Mustelids in a Modern World. Management and Conservation Aspects of Small Carnivore: Human Interactions*, pp. 247–265. Backhuys Publishers, Leiden.

Tuyttens, F.A.M., Delahay, R.J., Macdonald, D.W, Cheeseman, C.L., Long, B., and Donnelly, C.A. (2000b). Spatial perturbation caused by a badger (*Meles meles*) culling operation: implications for the function of territoriality and the control of bovine tuberculosis (*Mycobacterium bovis*). *Journal of Animal Ecology*, **69**, 815–828.

Tuyttens, F.A.M., Long, B., Fawcett, T., et al. (2001). Estimating group size and population density of Eurasian badgers *Meles meles* by quantifying latrine use. *Journal of Applied Ecology*, **38**, 1114–1121.

Tuyttens, F.A.M. and Macdonald, D.W. (2000). Consequences of social perturbation for wildlife management and conservation. In L.M Gosling and W.J. Sutherland, eds., *Behaviour and Conservation*. Cambridge University Press.

Tuyttens, F.A.M., Macdonald, D.W., Rogers, L.M., et al. (1999). Differences in trappability of European badgers, *Meles meles*, in three populations in England. *Journal of Applied Ecology*, **36**, 1051–1062.

Tuyttens, F.A.M., Macdonald, D.W., Rogers, L.M., Cheeseman, C.L., and Roddam, A.W. (2000c). Comparative study on the consequences of culling badgers (*Meles meles*) on biometrics, population dynamics and movement. *Journal of Animal Ecology*, **69**, 567–580.

Vial, F. and Donnelly, C.L. (2012). Localized reactive badger culling increases risk of bovine tuberculosis in nearby cattle herds. *Biology Letters*, **8**, 50–53.

Ward, A.I., Tolhurst, B.A., Walker, N.J., Roper, T.J., and Delahay, R.J. (2008). Survey of badger access to farm buildings and facilities in relation to contact with cattle. *Veterinary Record*, **163**, 107–111.

Waters, W.R., Palmer, M.V., Buddle, B.M., and Vordermeier, H.M. (2012). Bovine tuberculosis vaccine research: historical perspectives and recent advances. *Vaccine*, **30**, 2611–2622.

Wedlock, D.N., Denis, M., Vordermeier, H.M., Hewinson, R.G., and Buddle, B.M. (2007). Vaccination of cattle with Danish and Pasteur strains of *Mycobacterium bovis* BCG induce different levels of IFN gamma post-vaccination, but induce similar levels of protection against bovine tuberculosis. *Veterinary Immunology and Immunopathology*, **118**, 50–58.

Whelan, A.O., Coad, M., Upadhyay, B.L., Clifford, D.J., Hewinson, R.G., and Vordermeier, H.M. (2011). Lack of correlation between BCG-induced tuberculin skin test sensitisation and protective immunity in cattle. *Vaccine*, **29**, 5453–5458.

Wilesmith, J.W. (1983). Epidemiological features of bovine tuberculosis in cattle herds in Great Britain. *The Journal of Hygiene*, **90**, 159–176.

Wilkinson, D., Smith, G.C., Delahay, R.J., and Cheeseman, C.L. (2004). A model of bovine tuberculosis in the badger

Meles meles: an evaluation of different vaccination strategies. *Journal of Applied Ecology*, **41**, 492–501.

Wilkinson, D., Smith, G.C., Delahay, R.J., Rogers, L.M., Cheeseman, C.L., and Clifton-Hadley, R.S. (2000). The effects of bovine tuberculosis (Mycobacterium bovis) on mortality in a badger (Meles meles) population in England. *Journal of Zoology*, **250**, 389–395.

Wilson, G.S. (1943). The pasteurization of milk. *British Medical Journal*, **1**, 261.

Wilson, G. and Carter, S. (2013). House of Commons Oral Evidence taken before the Environment, Food and Rural Affairs Committee on Bovine TB Vaccination, Tuesday 12 February 2013. Uncorrected Transcript of Oral Evidence. Parliament UK.

Wilson, G.J., Carter, S.P., and Delahay, R.J. (2011). Advances and prospects for management of TB transmission between badgers and cattle. *Veterinary Microbiology*, **151**, 43–50.

Wilson, G.J., Delahay, R.J., de Leeuw, A.N.S., Spyvee, P.D., and Handoll, D. (2003). Quantification of badger (*Meles meles*) sett activity as a method of predicting badger numbers. *Journal of Zoology*, **259**, 49–56.

Woodroffe, R., Donnelly, C.A., Cox, D.R., et al. (2006a). Effects of culling on badger (*Meles meles*) spatial organization: implications for the control of bovine tuberculosis. *Journal of Applied Ecology*, **43**, 1–10.

Woodroffe, R., Donnelly, C.A., Gilks, P., Jenkins, H.E., Johnston, W.T., Le Fevre, A.M., Cox, D.R., Bourne, F.J., Cheeseman, C.L., Clifton-Hadley, R.S., Gettinby, G., Hewinson, R.G., McInerney, J.P., Mitchell, A.P., Morrison, W.I., and Watkins, G.H. (2009b). Bovine tuberculosis in cattle and badgers in localised culling areas. *Journal of Wildlife Diseases*, **45**, 128–143.

Woodroffe, R., Donnelly, C.A., Jenkins, H.E., et al. (2006b). Culling and cattle controls influence tuberculosis risk for badgers. *Proceedings of the National Academy of Sciences of the USA*, **103**, 14713–14717.

Woodroffe, R., Donnelly, C.A., Johnston, W.T., et al. (2005). Spatial association of *Mycobacterium bovis* infection in cattle and badgers. *Journal of Applied Ecology*, **42**, 852–862.

Woodroffe, R., Donnelly, C.A., Wei, G., et al. (2009a). Social group size affects *Mycobacterium bovis* infection in European badgers (*Meles meles*). *Journal of Animal Ecology*, **78**, 818–827.

Woodroffe, R., Frost, S.D.W., and Clifton-Hadley, R.S. (1999). Attempts to control tuberculosis in cattle by removing infected badgers: constraints imposed by live test sensitivity. *Journal of Applied Ecology*, **36**, 494–501.

Young, J.S., Gormley, E., and Wellington, E.M.H. (2005). Molecular detection of *Mycobacterium bovis* BCG (Pasteur) in soil. *Applied and Environmental Microbiology*, **71**, 1946–1952.

CHAPTER 6

Biology, ecology, and reproduction of American mink *Neovison vison* on lowland farmland

David W. Macdonald, Lauren A. Harrington, Nobuyuki Yamaguchi, Michael D.F. Thom, and Joanna Bagniewska

In all works on Natural History, we constantly find details of the marvellous adaptation of animals to their food, their habits, and the localities in which they are found.

Alfred Russel Wallace *A Narrative of Travels on the Amazon and Rio Negro*. 1853. (1972 reprint).

6.1 Introduction

The American mink *Neovison vison* is a predatory, semi-aquatic, medium-sized member of the Mustelidae, similar in size to the European polecat *Mustela putorius* (*c.* 1 kg), and between that of the otter *Lutra lutra*, which is larger (at 6–10 kg), and the stoat *M. erminea*, which is smaller (at 200–300 g). Like many of the mustelid family, mink are elongate with short legs (Dunstone 1993) and are sexually dimorphic (males, on average, weigh almost twice as much as females, see Table 6.1). In contrast to the other more terrestrial mustelids, mink have partially webbed toes (although polecats can also swim, mink are more efficient swimmers; Lodé 1999, Williams 1983). American mink are solitary, they breed once yearly (both males and females mating with several partners), generally between late February and early April (Birks 1981; Ireland 1990; Dunstone 1993), and females give birth to five to eight kits between late April and mid May. Juvenile mink disperse from their natal range at the age of about three months, in July or August (Dunstone 1993).

As its name implies, the American mink is native to North America, so why is it relevant to a book on conservation on lowland British farmland? The answer is that it is a paradigmatic invasive species in the UK (and beyond), its presence posing technically challenging questions to the conservation of its prey and competitors under circumstances where it illustrates far-reaching scientific, ethical, and thus, policy issues for the biology and management of invasive species. The American mink's luxuriant and lustrous fur has led to their exploitation in fur farms and, through this industry, their transportation to the UK (indeed, across the world) in the early twentieth century (Macdonald and Harrington 2003). At the peak of the UK fur trade, in the 1950s, there were at least 400 mink farms in the country[1]. By the early twenty-first century, fewer than 20 fur farms were in operation, the last of which closed down in 2002, following a ban under the Fur Farming (Prohibition) Act 2000[2], which came into effect on the 1 January 2003. American mink were first recorded breeding in the wild in the UK in 1956 in Devon (Linn and Stevenson 1980). Thus, much of the early work on American mink in the UK was based in the south west (e.g. Chanin and Linn 1980; Wise et al. 1981; Birks and Linn 1982), although an interest in their biology in marine ecosystems also developed in coastal environments in Scotland, where numerous fur farm escapees established in areas of high conservation importance (e.g. Birks and Dunstone 1984; Dunstone and Birks 1985, 1987).

As realization dawned of the American mink's impact on native wildlife in lowland England, and their

[1] From: <http://www.gwct.org.uk/>. Accessed July 2013.
[2] Available at: <http://www.legislation.gov.uk>. Accessed July 2013.

Wildlife Conservation on Farmland. Conflict in the Countryside. Edited by David W. Macdonald and Ruth E. Feber.
© Oxford University Press 2015. Published 2015 by Oxford University Press.

potency as a model invasive species from Patagonia to Siberia, we began a riparian mammal research programme based in the Upper Thames Valley. In Macdonald and Feber (2015), two chapters focus on the programme's work to conserve the endangered water vole *Arvicola amphibious* in the UK: the first describing an ecological restoration experiment to restore water voles to the Upper Thames (Macdonald and Feber 2015, Chapter 14) and the second using water voles as a case study to illustrate the conservation journey from the discovery that a species is declining, through diagnostic research into the causes, then development and testing of solutions, to the production and implementation of a strategy for the species' recovery (Macdonald and Feber 2015, Chapter 15). An important element of all of these studies was the impact on water voles of the invasive American mink and it is to this species that we now turn our attention in this volume. Over two chapters we give an account of the mink's behaviour and ecology in the Upper Thames Valley, and describe what we know of the interactions among mink and their native competitors, discussing practical issues associated with the control of this invasive predator.

Here, in this first chapter, we describe the American mink (hereafter "mink") on the River Thames, and their initial spread in the Upper Thames Valley and then ask: What densities of mink are supported by English lowland farmland? How are they organized spatially? We then explore their habitat preferences, diet, and predatory behaviour, which, of course, shapes their impact on prey (and conflict with naturalists, gamekeepers, and water bailiffs), and their place in the native mustelid guild. Finally, we investigate the minks' mating system on lowland farmland, because, invasive or not, these animals are deeply fascinating and a monument to adaptability.

6.1.1 River Thames mink

In our study area (a 24 km stretch of the River Thames, to the west of Oxford, Fig. 6.1), both male and female mink (n = 97 adults, live-trapped over six years between 1995 and 2008) appeared to be slightly larger and heavier than mink in their native North America, and those in British coastal[3] or island populations (with the exception of the large male mink on the island of North Uist), but similar in size to those in several other mainland invasive populations (Table 6.1). The size and body weight of mink varies considerably across their native and invaded range (e.g. Chapman and Feldhamer 1982; Dunstone 1993; Zalewski and Bartoszewicz 2012) but, although there are a number of hypotheses for variation in body size, including latitude (Zalewski and Bartoszewicz 2012), diet (Ferreras and Macdonald 1999; Harrington et al. 2009), and time since colonization (e.g. Sidorovich et al. 1999; Bonesi et al. 2006; Melero et al. 2012; Zalewski and Bartoszewicz 2012), the main drivers of body size (and of large size on the River Thames) are unclear, and, in a population derived from feral animals, complicated by unknown origin (and domestication of founders) (Kruska and Sidorovich 2003).

Inter- and intra-sexual variation on the River Thames was such that, while the heaviest male was twice the weight of the heaviest female, and 2.6 times that of the average female, the largest female was slightly heavier than the smallest male. River Thames mink appear to be highly dimorphic compared with most other populations (with the exception of North Uist, Table 6.1). On Hiiumaa Island (Estonia), for example, the heaviest male was only 1.5 times the weight of the heaviest female, and less than twice that of the average female. However, while the extent of sexual dimorphism varies among populations, both locally and globally (Table 6.1), inter-sexual overlap is such that all populations (not only the River Thames) include a proportion of big females that approach (or even exceed) the size (and weight) of the smaller adult males in the population (Table 6.1). One might ask whether these male-sized females are morphologically equivalent to a male, but measurement of a number of skull dimensions of mink carcasses collected from the Outer Hebrides and Hiiumaa Island, revealed that the answer is that they are not. Even after controlling for underlying differences in skeletal size, male mink had larger carnassial teeth, larger canines, and the shape of their skulls differed (features associated with killing and/ or feeding were larger, Thom et al. 2004a). Male mink (controlling for skeletal size) were also heavier than female mink. Sexual dimorphism, which is a feature of mink populations everywhere (and indeed common among other mustelids, e.g. Ralls and Harvey 1985; Dayan et al. 1989), has pervasive influences throughout the ecology and natural history of the species, and raises the possibility of inter-sexual niche partitioning, which we explore further below.

[3] However, in contrast to earlier analyses suggesting that larger riverine mink were in poorer condition than smaller coastal mink (Clode et al. 1995), we were unable to detect a difference in body condition between Thames mink and either coastal mink in Scotland or island populations on the Outer Hebrides (L.A. Harrington and E. Fraser, unpublished data).

Table 6.1 Body metrics for adult (≥ 1 year old) mink. HB = Head-body.

Site	Body mass (g)(min–max) (mean, SD, n)		HB length (mm)(min–max) (mean, SD, n)		Index of sexual dimorphism[1]	Source
	Female	Male	Female	Male		
R. Thames	480–938 722, 97.0, 37	900–1900 1383, 238.9, 56	327–393 360, 16.0, 39	370–500 429, 26.6, 56	1.91	Our data
Is. South Harris (UK)	420–753 566, 95.8, 18	673–1472 1032, 207.5, 22	316–360 334, 13.0, 18	340–422 384, 26.0, 22	1.82	Our data
Is. North Uist (UK)	490–670 568, 83.7, 6	990–1830 1386, 258.9, 10	330–365 354, 12.2, 7	390–451 428, 20.7, 10	2.44	Our data
Scottish mainland (river)	379–853 651, -, 12	813–1685 1221, -, 12	300–373 347, -, 12	370–433 407, -, 12	1.88	E. Fraser, unpub. data
Scottish mainland (coast)	518–582 550, -, 2	715–1433 1047, -, 12	335–345 340, -, 2	340–435 400, -, 12	1.90	E. Fraser, unpub. data
Is. Hiiumaa (Estonia)	540–930 739, 116.7, 14	940–1400 1200, 159.0, 8	355–406 381.6, 15.9, 14	403–486 451, 23.6, 9	1.62	Our data
Ross Peninsula (UK) (coastal)	560–800 656, -, 12	860–1530 1148, -, 20	-	-	1.75	Ireland 1990
Southern UK (river)	450–810 619, -, 32	850–1805 1153, -, 48	320–360 338, 10.1, 29	330–450 397, 23.8, 39	1.86	Chanin 1983
Belarus	440–930 715, -, 51	900–2000 1276, -, 50	- 370, 12.2, 23	- 430, 25.9, 42	1.78	Mass: Kruska and Sidorovich 2003 Length: Sidorovich et al. 1999
Poland	650–960 778, 83.8, 29	1000–1700 1182, 402.0, 53	360–440 398, 22, 29	410–510 467.7, 29.49, 45	1.52	A. Zalewski, pers. comm.
Iceland	- 629, 79, 188	- 1179, 224, 203	- 352, 16.1, 188	- 467.7, 29.5, 45	1.87	Skirnisson 1993
USA (Michigan) [native]	- 481, -, 8	- 910, -, 18			1.88	Reference population in Zwiernik et al. 2008
USA (Montana) [native]	- 600, -, 50	- 1150, -, 18			1.92	Mitchell 1961
USA (Maryland) [native]	- 563, -, 61	- 989, -, 108			1.76	Askins and Chapman 1984
USA (California) [native]	496–790 596, -, 5	1000–1451 1221, -, 15	298–368 342, -, 10	305–413 377, -, 10	2.05	Grinnell et al. 1937
Canada [native]	430–858 624, -, 47	595–1410 1126, -, 41			1.80	Kruska and Sidorovich 2003

[1] Mean male mass/mean female mass (Powell 1979).

6.1.2 Spread of mink in the Upper Thames Valley

Mink appeared in the Upper Thames Valley relatively late in their colonization of Britain—the first field signs were recorded in the early 1960s, but signs only became numerous during the 1980s (Strachan and Jefferies 1993). Across 161 survey sites throughout the region (each a 500 m section of river bank), mink occupancy increased, from 7% in 1975, to 25% in 1990 (by which time there were no longer any mink farms in the Thames catchment, all having closed by the late 1980s), and reaching 46% by 1995 (Fig. 6.1). Mink were found predominantly in sites with high

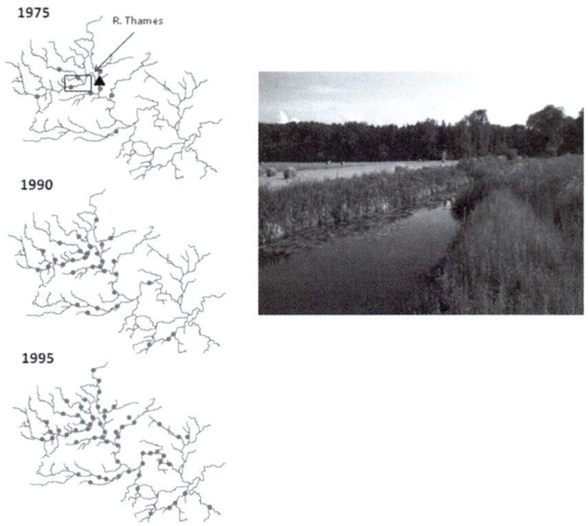

Figure 6.1 The Thames catchment showing the spread of American mink signs between 1975 and 1995. The River Thames (pictured), within our study area (delineated by the square in the top map), is 15–45 m wide, up to 3 m deep, with emergent reeds (bur-reed *Sparganium erectum*, common reed *Phragmites australis*, and reedsweet grass *Glyceria maxima*), fringed with trees (such as willow *Salix fragilis*, alder *Alnus glutinosa*, and ash *Fraxinus excelsior*), and vegetation (such as nettles *Urtica dioica*, rank grassland, willowherb *Epilobium* spp., bramble *Rubus fruticosus*, blackthorn *Prunus spinosa*, and hawthorn *Crataegus monogyna*), and bordered largely by agricultural pasture that is interspersed with woodland and occasional cereal fields. Along the River Thames, rabbits are abundant (but patchily distributed), small mammals are presumably abundant in the hedgerows traversing the surrounding farmland (see Macdonald and Feber 2015: Chapter 4); in the river, there are high densities of American signal crayfish *Pacifastacus leniusculus* (hereafter crayfish, see Chapter 8, this volume), a rich fish community, waterfowl, including coots *Fulica atra*, moorhen *Gallinula chloropus*, and mallards *Anas platyrhynchos*, are dispersed throughout the river's length. The black triangle shows the location of Oxford city. Adapted from Macdonald and Strachan 1999.

banks, unvegetated bars, exposed bedrock, and trees, but not in urban areas (Strachan and Jefferies 1993; Barreto et al. 1998; Macdonald and Strachan 1999). In the mid 1990s, population parameters of our study population on the River Thames (which was largely unmanaged, i.e. not subject to culling) suggest that it was a saturated, stable population, and mink continued to spread through the wider catchment, such that the upper reaches that were uncolonized at the time (providing enclaves for water voles, Barreto et al. 1998), were occupied by mink by the early 2000s (Harrington et al. 2009). By this time the water voles had largely disappeared (see Macdonald and Feber 2015: Chapter 14).

6.2 Mink populations on lowland farmland

6.2.1 What densities of mink are supported by English lowland farmland?

On the River Thames, adult mink densities in summer (when only residents are present) were approximately 0.4–0.9 mink/km (Harrington et al. 2009), increasing seasonally due to an influx of transient males during the mating season (January–March) and juveniles during the dispersal season (August–September) (Yamaguchi and Macdonald 2003). Elsewhere, densities range between approximately 0.2 (e.g. in Scotland, and inland areas of the Outer Hebridean islands) and > 1 (e.g. in Spain, coastal areas of Scotland, and other offshore islands) mink/km (Scottish Mink Initiative[4], Roy 2011). Adult sex ratios did not differ significantly from 1:1 (Harrington et al. 2009), litter size (at about two months after birth) averaged 2.5 (± 0.29, n = 4), and overall juvenile:adult ratio was 0.4:1 (Yamaguchi and Macdonald 2003)—an adult-bias typical of an unmanaged population (Bonesi et al. 2006).

6.2.2 How are mink organized spatially on lowland farmland?

Taking the River Thames as a model population, we asked whether invasive mink on lowland farmland follow spacing patterns typical of the long, thin mustelids elsewhere (i.e. intra-sexual territoriality, whereby individuals maintain territories only with respect to members of the same sex, and males overlap several females; Powell 1979). To answer this question, we fitted 36 mink (17 males, 19 females), with 10 g radio-collars[5], and radio-tracked them for between 4 and 48 days each (recording 8207 total radio-locations, average per mink per season[6] = 164 radio-locations) year-round between 1995 and 2000 (n = 25), and later (specifically to explore how sympatric mustelids fit into this spatial arrangement; see Chapter 7, this volume) in the autumns of 2004 and 2005 (n = 11). Across all study years, including all seasons, resident adult males occupied

[4] Scottish Mink Initiative Newsletter No. 4 December 2011. <http://www.scottishmink.org.uk/>. Accessed September 2013.

[5] Less than 2% of the body mass of the smallest individual in the study.

[6] Seasons, for this analysis, were defined as 'biological' seasons: Jan–Mar = mating, Apr–Jun = births, Jul–Sep = dispersal, Oct–Dec = winter.

linear seasonal home ranges[7] along the riverbank that averaged 3.44 km (median, range: 0.78–9.04 km, IQR: 2.88–5.21, n = 21)[8] and statistically significantly longer than the median 2.10 km (range: 0.21–4.45 km, IQR: 1.10–2.91, n = 29) used by radio-tracked resident adult females[9]. Initial analyses in the first three years of the study revealed that resident males (n = 5) did indeed overlap 1–3 females each, and thus mink conform to the typical mustelid pattern, but this led us to ask: why are male ranges so large?

6.2.3 Why are male home ranges so large?

Yamaguchi and Macdonald (2003) proposed that the answer lay in home range overlaps. In contrast with predicted intra-sexual territoriality, neither male nor female home ranges on the River Thames were completely intra-sexually exclusive (four females in two pairs of female–female neighbours had home ranges that were overlapped by between 47 and 100% of their area by their respective neighbours, and two neighbouring males had home ranges that were overlapped by 81–96% of their area by their neighbour: Yamaguchi and Macdonald 2003). Harrington and Macdonald (2008), examining home range overlaps among 11 mink in autumn (see Chapter 7, this volume), found that mink neighbours (whether opposite or same sex) tended to avoid each other within areas of overlap, either spatially (using different areas within their shared home range) or temporally (using the same areas at different times), and overlaps of core areas were rare. Nevertheless, averaging across both studies revealed that, per month, males (n = 11) shared their home range with 0.82 other males and 1.52 females; females (n = 13) with 1.18 males and 0.32 other females. Yamaguchi and Macdonald (2003) calculated that home range overlaps (perhaps better described as incursions) of this order would result in a loss, for males, of approximately 60% of the potential energy available due to resource loss to incursive neighbours, and approximately 40% of potential energy available for females. They proposed that male mink may be forced to maintain large ranges to counter the energetic burden of overlapping neighbours, and estimated that, to account for their larger body size as well as the energy lost to neighbours (due to resource depletion), a male would require a home range 1.73 times that of a female (and her dependent kits)—a value that fits remarkably well with relative home range sizes presented here[10].

While this summary is satisfying insofar as it fits with our initial explanation of large male home ranges, it is necessarily based on averages. Averages are useful for seeking generalities but can distract from a different reality, revealed by variation among individuals. While it is true that male ranges on average were larger than those of females (and several male ranges were very large), *some* male home ranges were very small (Fig. 6.2). Individual variation in home range size is likely influenced by local prey availability, and perhaps individual specialization and dispersion of the chosen resource type. However, the fact that only males held home ranges greater in size than 4.5 km, but that both males and females held home ranges smaller than 2 km, suggest that it is not that males *require* larger ranges, rather that they *can* occupy larger ranges (perhaps allowing them to select more widely dispersed resources). The question then is not, *why do males have larger home ranges than females?* but rather, *how are males able to occupy larger home ranges than females?* At least part of the answer may lie in sexual differences in mobility, and there is some evidence that males travel both further and faster than females. Males (n = 12, radio-tracked between 1995 and 2000) travelled a median distance of 35.5 m (IQR: 24.6–102.0) each hour at a speed of 0.48 km/ h (0.32–0.79), compared with 29.5 m (15.9–43.3, n = 13) at 0.36 km/ h (0.32–0.42) for females. However, variation in the data meant that the differences between sexes in either distance travelled or speed of movement were not statistically significant.

6.3 Habitat preferences, diet, and diving behaviour

Considering the typical lowland farmscape, and to assess what impact mink might have on native prey (Macdonald and Feber 2015: Chapter 14), and how they might fit into the native mustelid guild (see Chapter 7, this volume), we asked the following questions: Where are mink found? What do mink on lowland farmland

[7] Calculated as the length of watercourse occupied, between the farthest upstream and downstream locations (or the cumulative length of all watercourses, if an animal's range encompassed more than one stream or river).

[8] The number of home range estimates differs from the number of mink radio-tracked because some mink were radio-tracked in more than one season.

[9] Mann–Whitney test: $W = 529$, $p < 0.001$. Average home range estimates were the same even when mating season home ranges were excluded, and non-mating season (April to December) male home ranges were also statistically significantly larger than those of females ($W = 425$, $p < 0.001$, $n = 26$ females, 17 males).

[10] Median female home range of 2.10 X 1.73 = 3.63; actual median male home range = 3.44.

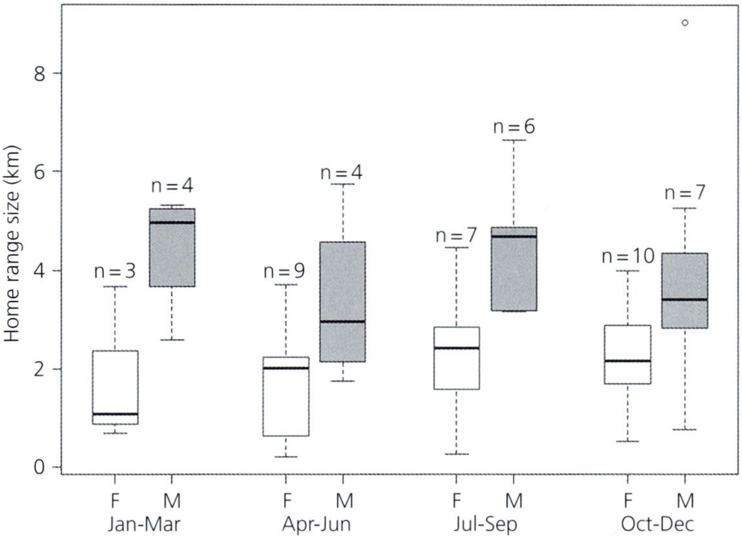

Figure 6.2 Home range size (calculated as the length of watercourse occupied in km) for female and male mink by season. Jan–Mar = mating season, Apr–Jun = births, Jul–Sep = dispersal, Oct–Dec = winter. Data are median, IQR, and range; sample sizes are shown above the bars.

eat? And what habitats do they use? Throughout, we consider the implications of the minks' sexually dimorphic nature, and explore possible indicators of inter-sexual niche partitioning.

6.3.1 Where are mink found?

Mink are versatile opportunists and can adapt to almost any body of water. We know this from our mink studies, which range from the rough rocky coastlines at the southern tip of Patagonia (Fasola et al. 2011; see also Valenzuela et al. 2013), the offshore islands of the Outer Hebrides (Clode and Macdonald 1995), and the fjords of western Iceland (Magnusdottir et al. 2012), to the extensive remote marshlands of Belarus (e.g. Sidorovich et al. 1999; Sidorovich and Macdonald 2001), and the tiny, seasonally dry, streams and ditches of Hiiumaa Island in Estonia (Macdonald et al. 2002). Others have studied mink (outside their native range) in rivers and lakes in Sweden (e.g. Gerell 1970), Mediterranean rivers (e.g. Melero et al. 2011), German fishponds (Zschille et al. 2010), the Baltic Sea archipelagos of southern Finland (e.g. Niemimaa 1995; Nordström et al. 2002), and reedbeds in south west England (e.g. MacPherson and Bright 2010).

In lowland England, mink predominantly occupy slow-flowing rivers and are found in well-vegetated areas (Dunstone 1993). In the Upper Thames Valley, we found that mink (on a 10 km stretch of river) were more numerous where there were more potential den sites—willow pollards, tree hollows, holes among tree roots, root plates, stick piles, washed up debris, and rabbit *Oryctolagus cuniculus* burrows (Halliwell and Macdonald 1996). That mink abundance on these rivers was not correlated with either fish density or availability, or with adult, juvenile, or total moorhen abundance (both fish and moorhen are taken locally for food—see below), raises the possibility that mink are limited by the availability and dispersion of dens, rather than by the availability of their prey.

6.3.2 What do mink on lowland farmland eat?

Generally, across their native and non-native range, mink feed on fish, small mammals, some larger mammals such as rabbits and muskrats *Ondatra zibethicus*, crustaceans, birds, amphibians, reptiles, and insects (reviewed in Macdonald and Strachan 1999), with diet reflecting local and seasonal prey availability. On the River Thames, identification of prey remains in 123 (Ferreras and Macdonald 1999; P. Ferreras unpub. data) and 78 (Harrington et al. 2009) mink scats collected through the year revealed that mammals were the main prey (41–50% relative occurrence[11]) followed by birds (21–38%) and fish (10–23%), with low proportions of crayfish (< 4%) and other invertebrates (dragonfly larvae, diving beetles e.g. *Dytiscus marginalis* 2–7%). Mammals consumed on the River Thames, and more widely in the Upper Thames Valley, included mostly rabbits (although fewer rabbits in later years of

[11] Relative frequency of occurrence = total number of occurrences of a particular prey item divided by the total number of items found (Conroy et al. 2005).

the study), woodmice *Apodemus sylvaticus*, field voles *Microtus agrestis*, and bank voles *Myodes glareolus*, as well as common rats *Rattus rattus*, house mice *Mus musculus*, and shrews *Sorex* spp. Water voles were not present in the area during our study, but do form a large proportion[12] of the diet of mink where they are locally available (see Macdonald and Feber 2015: Chapters 14 and 15), and other mammals, such as grey squirrels *Sciurus carolinensis*, and dormice *Muscardinus avellanarius*, may be taken in riparian habitats within woodland (e.g. Bonesi et al. 2004).

In coastal areas, mink predate heavily on ground-nesting birds during the nesting season (e.g. Clode and Macdonald 1995; Ibarra et al. 2009). Birds taken by mink in the Upper Thames Valley largely comprised coots and moorhens (adults, chicks, and their eggs; see Ferreras and Macdonald 1999), but also charadiformes (waders and gulls), anseriforms (ducks and geese), and even passerines (songbirds and corvids). In accordance with other studies elsewhere in the UK (e.g. Strachan et al. 1998), bird consumption by mink was higher in summer (46% relative occurrence, n = 44 scats)—when nesting birds and their chicks are most vulnerable—than in winter (27%, n = 54). A similar pattern has been observed for otters (Grant and Harrington in press).

Fish consumed by mink in the Upper Thames valley were predominantly percids (perch *Perca fluviatilis* and other similar species) and cyprinids (species such as roach *Rutilis rutilis* or dace *Leuciscus leuciscus*), also sticklebacks *Gasterosteus aculaetus*, and salmonids (e.g. brown trout *Salmo trutta*)—similar to those species taken in freshwater habitats elsewhere (Macdonald and Strachan 1999; Bonesi et al. 2004), and including many of the same species taken by otters in the same area (Grant and Harrington in press). We did not attempt to estimate the size of fish taken by mink, but generally they are relatively small. In Scotland, 97% of the salmonids taken by mink were < 25 cm, and most (72%) were < 12 cm (Cuthbert 1979). Mink are capable, however, of taking larger fish; for example, Wise et al. (1981) reported 30–60 cm pike being taken. Other freshwater fish consumed by mink include eels *Anguilla anguilla*, and bullheads *Cottus gobio* (e.g. Bonesi et al. 2004). The proportion of fish taken in the upper Thames area was somewhat lower than in other freshwater habitats (e.g. McDonald 2002), and considerably lower than the proportion consumed by coastal mink (e.g. in the Outer Hebrides) where mink consume mostly fish (51.8–112% occurrence) and a large proportion of crustaceans (36–54.2%, Clode and Macdonald 1995), or in Iceland, where 50.2% of the biomass of 636 stomach remains was fish (Magnusdottir et al. 2012; see also Brzeziński 2008). On the continent, and in Patagonia, crustacea are also a particularly important dietary component for riverine mink (Erlinge 1969; Gerell 1967, 1968; Brzeziński 2008; Fasola et al. 2011), so the apparently low consumption of crayfish by mink in the Thames Valley, where their availability is apparently very high, is surprising (see Chapter 8, this volume).

Elsewhere, others have reported females taking a greater proportion of fish and crustacea than males, and males taking more large prey (muskrats in the USA, Sealander 1943; lagomorphs in the UK, Birks and Dunstone 1985; or eider *Somateria mollissima* in Iceland, Magnusdottir et al. 2012). The suggestion is that smaller female mink are limited in their ability to take larger prey (either directly, due to their relatively small size, or indirectly, due to competition with the larger males), and have a wider niche breadth to compensate (Birks and Dunstone 1985; Magnusdottir et al. 2012). Indeed, sexual dimorphism in mink skulls is consistent with males specializing on larger prey (see Thom et al. 2004a). Similar inter-sexual differences in diet have been found in other mustelids (McDonald 2002), although one study, intriguingly, reported female polecats taking more rabbits—large prey—than male polecats (Lodé 2003). However, in general, few such studies have been carried out, largely because of the difficulty in attributing dietary evidence to individual mink, which is usually only possible by assessing the stomach contents of carcasses (e.g. Magnusdottir et al. 2012).

6.3.3 What habitats do mink use?

On the River Thames, and at a finer scale than the above analysis, the number of mink present (within a 200 m section of the river) was significantly positively correlated with tree cover, rank grasses, and the size of rabbit warrens, and negatively correlated with open areas characterized by farming activities (Yamaguchi et al. 2003). On average, within their home ranges, mink (n = 13, 1995–1997) used areas < 10 m from the water, and with more scrub/ grass coverage, more rank grass, banks with more tree cover, and bigger rabbit warrens (Yamaguchi et al. 2003), and, although nearly 60% of the mink's home ranges were open areas, not one radio-fix out of a total of 5152 fixes occurred

[12] Undigested prey remains of colonizing mink on 11 rivers in Derbyshire, Leicestershire, Staffordshire, and Nottinghamshire (Strachan et al. 1998), consisted of 32.2% (by volume) water vole remains.

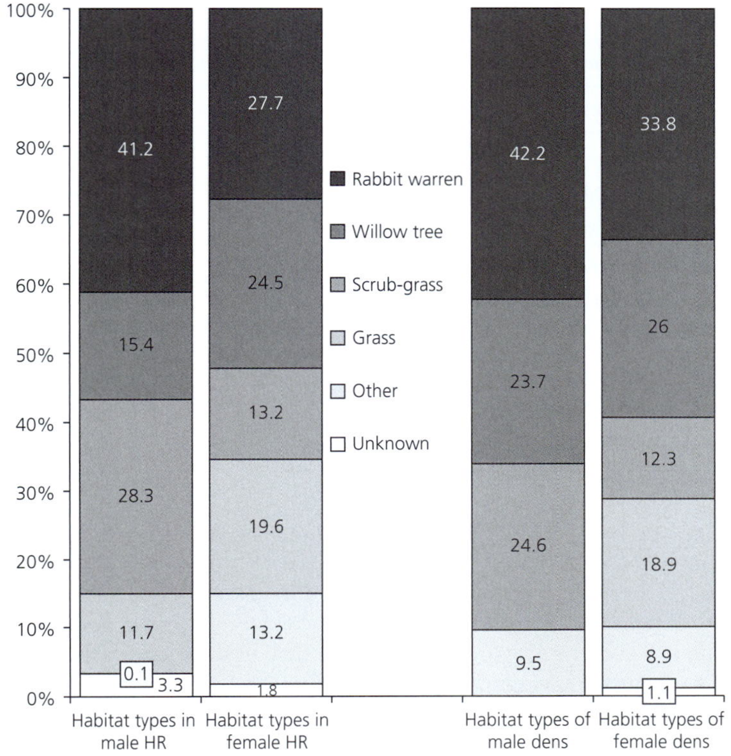

Figure 6.3 Habitat types within the home ranges (HR) of male and female mink, and associated with male and female (non-breeding) dens. Redrawn using data in Yamaguchi et al. (2003).

there. Rabbit warrens and hollow willow trees were favourite day-time den sites (38% and 25%, respectively) (Halliwell and Macdonald 1996; Yamaguchi et al. 2003), all of which were < 50 m from water, and most (91%) < 10 m from water (n = 119 dens belonging to 13 mink). During the breeding season, we confirmed five breeding dens, and located a further three dens which were probably used for breeding; all eight were in rabbit burrows, and in larger warrens than those used for non-breeding dens (n = 21) (mean number of entrances: breeding dens = 7.25 ± 3.4, non-breeding dens = 2.76 ± 2.49, Macdonald and Strachan 1999). Female mink with kits forage close to their dens (to reduce energetic costs and to protect kits), and abundant rabbits in a large warren close to the den is likely to be an important requirement for a mother mink whose energy requirements during lactation might rise five-fold (Dunstone 1993). Indeed, the number of kits captured was significantly influenced only by the size of rabbit warrens (r = 0.237) (and not by any other habitat variables). Outside the breeding season, males (and their dens) appeared to be more closely associated with rabbits than were females (Fig. 6.3), and males and their dens were more likely to be found > 10 m from water than were females and their dens, suggesting that males are influenced more by the presence of rabbits and their burrows than by the proximity of the river.

6.3.4 Do males and females prefer different habitats?

To explore inter-sexual differences in habitat use further, we analysed the habitat preferences (taking account of habitat availability)[13] of eight male and ten female mink between 1995 and 2000. Males selected home ranges with more riverbank woodland, and used riverbank woodland (where rabbits are concentrated) within their home range more than any other habitat, but the actual length of riverbank in their home range (and their usage of it) was ranked as less important (Table 6.2). Females selected similar home ranges, but appeared to use the riverbank more than males; riverbank was the most used habitat for females (followed by riverbank scrub) (Table 6.2). Intersexual niche overlap was high for the total habitat in

[13] Habitat quantified in terms of the length (in m) of each of nine habitat types (see Table 6.2) within each 100 x 100 m cell (which is more appropriate for mink than aerial measures of habitat types).

Table 6.2 Habitat selection, and usage, by male (n = 9) and female (n = 10) mink, ranked from most (1) to least (9) preferred. RB = riverbank.

Rank	Abundance of habitats within the home range		Habitat usage within the home range	
	Males	Females	Males	Females
1	RB woodland	Ditch	RB woodland	River bank
2	Ditch	RB woodland	RB ruderal	RB scrub
3	River bank	River bank	RB scrub	RB ruderal
4	Hedge	RB ruderal	River bank	RB reed
5	RB ruderal	RB reed	RB reed	RB woodland
6	RB open	RB open	Hedge	RB open
7	RB reed	RB scrub	RB open	Ditch
8	Road/fence	Hedge	Ditch	Hedge
9	RB scrub	Road/fence	Road/fence	Road/fence

home ranges (Renkonen percentage overlap (Krebs, 1989) = 92.1%), but lower for habitat use within home ranges (weighted by the number of radio-locations) (83.0%), suggesting that while home ranges contained similar habitats, habitat use within them differed, and up to a fifth of habitat use could be sexually distinct. We deduce from these results that some level of inter-sexual niche partitioning might exist, and that females (with an apparent preference for the riverbank) are likely to be more aquatic in their behaviour, and males (with a relative preference for woodland) more terrestrial (as suggested by Birks and Dunstone 1985). Our radio-tracking data were not of sufficiently fine resolution to distinguish between, for instance, a mink in the water and one on the riverbank, and we were thus unable to determine whether females on the riverbank were hunting for small mammals or birds in the vegetation bordering the river, or fishing *in* the river. More recently we have been able to describe the activity of mink in water in fine detail with the use of very small data loggers and novel analytical techniques, but generating complete activity budgets remains challenging.

6.3.5 Do male and female mink adopt different predatory strategies?

Given the possibility of inter-sexual differences in diet, and the differences in habitat selection (especially with respect to their proximity to the river) observed on the River Thames, we asked whether male and female mink adopt different predatory strategies. We explore that question here in terms of their diving behaviour (Fig. 6.4).

Figure 6.4 American mink with radio-collar and time-depth recorder (TDR). The TDR sits under the mink's chin, attached to the radio-collar, and records depth at 1 s intervals, over a period of 5–6 days, providing very detailed, second-by-second dive profiles (see Hays et al. 2007; Harrington et al. 2012). Photograph © A.L. Harrington.

Diving ability generally increases with body size (e.g. Halsey et al. 2006), therefore smaller females are predicted, on the basis of allometry, to be poorer divers than larger males, but within the Thames river system, we found no evidence that this was true for mink. There was no difference in maximum dive depth, or maximum duration, of individual dives between the sexes (possibly because in this shallow river system—maximum depth of the river is 3 m—diving was constrained by the habitat rather than by the animal's physiology). The maximum dive depth we recorded was 2.96 m (mean maximum dive depth per individual = 1.82 m, n = 14) and the maximum dive duration was 57.9 s (mean maximum duration per individual = 37.2 s, Harrington et al. 2012). However, female mink made significantly more dives per day (mean median dives per day = 61.0, n = 8) and spent more time diving per day (mean median time spent diving per day = 12.3 minutes) than males (mean median dives per day = 12.4, mean median time spent diving per day = 2.6 minutes, n = 5 and 4, respectively) (Harrington et al. 2012), which is consistent with a female preference for aquatic habitats and suggestions that females have a more aquatic diet. Because we do not know which dives were successful, we cannot say whether female mink were *better* divers (and caught more fish) than males, but they were more *persistent*[14] divers than the males. Females dived up to 28 times consecutively in a temporal cluster of dives, and males up to 18 times (mean of maximum dive cluster length per individual = 17.6 dives for females, 9.5 for males) (Bagniewska et al. in press).

The long, thin body shape of mink makes them vulnerable to heat loss (Brown and Lasiewski 1972), particularly so in water (Williams 1986), thus we were surprised to find that some of the highest diving rates recorded occurred during the coldest winter months (Harrington et al. 2012). Even more surprisingly, persistence was significantly greater at lower temperatures, but only for females (Bagniewska et al. in prep). Mink (of both sexes) have been observed to be less active generally in winter (Birks and Linn 1982; Niemimaa 1995; Zschille et al. 2010), as have other long, thin mustelids, such as American marten *Martes americana* (Buskirk et al. 1988) and black-footed ferrets *M. nigripes* (Richardson et al. 1987). Certainly, the mink that we were radio-tracking through fields around Oxfordshire seemed to dislike the rain! However, examination of their diving behaviour revealed that, while males tended to dive more when it was warm, females appeared to dive whatever the weather (Fig. 6.5). Several authors have noted increased fish consumption by mink in winter and attributed this to the ease by which slow swimming ectothermic fish[15] in cold water

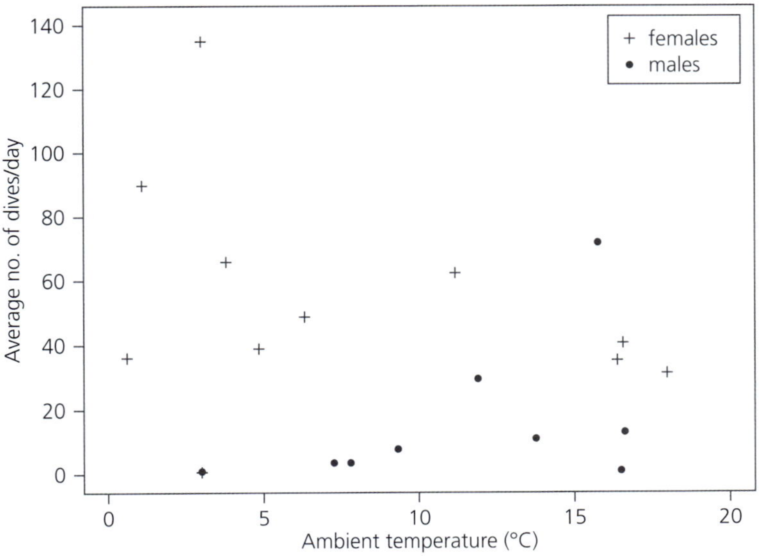

Figure 6.5 Mean number of dives per day (n = 5 or 6 days) against mean ambient temperature during the monitoring period, for male (circle) and female (cross) mink.

[14] Defined here as the number of dives within a cluster; dives classified into clusters using a hidden Markov model algorithm (method described in Bagniewska et al. 2013).

[15] Consuming fish at any time is beneficial because fish have a relatively high energy content (7 KJ g^{-1}) compared with other terrestrial prey such as rabbits or small mammals (6.21 and 6.34 KJ g^{-1}, respectively).

might be captured (Gerell 1968; Sidorovich 2000). It is tempting to conclude that female mink capitalize on this apparent winter paradox (while males do not), but small sample size meant that it was not possible to distinguish between an overall sex or season effect, or an interaction between the two. Only two males were tagged in winter and only one during a period when ambient temperatures fell below 7 °C. There were three particularly 'high-diving' individuals (two females in winter and one male in summer), which could be explained equally well by individualistic behaviour as differences between the sexes. Certainly, variation among individuals (at the level of individual dives) was extremely high: some individuals dived once or twice a day, others made over a hundred dives in a single day (and up to 70 dives per hour or seven dives per minute)[16]. However, variation was also high within individuals: one particular male dived only five times one day but 143 times another day, and the short monitoring periods (5–6 days) meant that it was not possible to distinguish between individual specialists and generalists temporarily exploiting a patchy resource (see Bolnick et al. 2002).

6.4 Reproduction and mating systems

Behavioural differences between the sexes are most apparent during the mating season, from late winter to early spring. That male mustelids within a solitary, intra-sexual territorial system often abandon their home ranges in the mating season (presumably in search of receptive females), is well known (Erlinge and Sandell 1986)[17]—and we observed that male mink on lowland farmland conform to this pattern. Indeed, divergent reproductive roles are hypothesized to be one of the main drivers of sexual dimorphism in mustelids (Erlinge 1979; Moors 1980; Powell and Leonard 1983) (although we suggest that niche separation may contribute to its maintenance, Thom et al. 2004a). While the impetus of our work on mink was biodiversity conservation, our field- and lab-based studies allowed us the opportunity to explore some of the details of the mating system of this fascinating mustelid. To provide context for the questions that we pose below, we first briefly review key aspects of the mating tactics of male and female mink.

6.4.1 Female reproductive physiology and mating tactics

Female mink have an unusual reproductive physiology, in that they exhibit both superfecundation (fertilization of two or more ova from a single ovulation by different males, which in mink must occur within 1–3 days following the first mating), and superfoetation (fertilization of ova from different ovulations within a single mating season, at *c.* 8-day intervals in mink; Shackleford 1952; Venge 1973). Both, theoretically, may lead to multiple paternity. Multiply sired litters are commonplace in other species, occurring in badgers *Meles meles* (Yamaguchi et al. 2006; Dugdale et al. 2007; Chapter 4, this volume), and in some rodents (Berteaux et al. 1999; Ratkiewicz and Borkowska 2000; Wolff and Dunlap 2002), insectivores (Stockley et al. 1993), ursids (Schenk and Kovacs 1995), and felids (Ishida et al. 2001). Natural superfoetation as a part of an animal's normal reproductive strategy (rather than a rare occurrence) has, in contrast, been reported in few other wild mammals (examples are the badger and brown hare *Lepus europaeus*, reviewed in Roellig et al. 2011; also the flying fox *Pteropus* spp., Fox et al. 2008)[18]. Mink are also unusual (among other mustelids) in that they, together with the striped skunk *Mephitis mephitis*, exhibit a short (up to 35 days) and variable delay in implantation after fertilization (Venge 1973; Wade-Smith et al. 1980). Delayed implantation (DI, or embryonic diapause, whereby the fertilized egg develops only partially and remains inactive for a period of time before being implanted in the uterus) is widespread among mammals (e.g. Renfree and Shaw 2000), but is particularly common among mustelids (34% of mustelids exhibit DI compared to < 0.05% of mammals overall[19]; Mead 1981, Sandell 1990). Long delays (up to 11 months) are

[16] Mostly dives occurred at a rate of 1–2 dives per minute (Harrington et al. 2012).

[17] Stoats in Sweden, for example, increased the area over which they ranged between the non-mating and the mating season by around 44 times (Erlinge and Sandell 1986); although martens appear to maintain their ranges year round (Zalewski and Jędrzejewski 2006).

[18] Superfoetation in the brown hare occurs without embryonic diapause, and the second oestrus occurs just before the birth of the young from the first set of ova (i.e. each set of ova develops separately), which is essentially two different pregnancies leading to two different births, and quite different from superfoetation occurring in association with embryonic diapauses (as in mink and badgers). In mink, fertilized ova from different ovulations remain in the reproductive tract as immature blastocysts until the last ova are fertilized and all implant (and thus are born) simultaneously. Roellig et al. (2011) suggest that these two phenomena should be reclassified as 'superconception' and 'superfertilization' respectively.

[19] All seals of the families (otariidae and phocidae) also exhibit DI, as do all the ursids (bears) (Lindenfors et al. 2003).

conventionally considered to provide a means of adjusting the period between mating and birth and thus to optimize the timing of both (reviewed in Thom et al. 2004b). The adaptive value of a short DI (which clearly does not have any seasonal benefit) may lie, instead, in its variability, which allows synchronization of parturition without necessitating synchronous mating (Stenson 1985)—in effect, an extended mating season (Thom et al. 2004b), allowing females access to several males, and potentially multiple paternity of their litters.

6.4.2 Male mating tactics

As in other studies (e.g. Dunstone 1993), resident females in our study site appeared to maintain their home ranges over the mating season; resident males left their home ranges by February, and there was a significant increase in the influx of previously unrecorded males from the beginning of January (Yamaguchi and Macdonald 2003; L. Harrington pers. obs). Males, generally, are thought to maximize their mating success by guarding females from rivals and by seeking extra-pair copulations (Brotherton and Manser 1997; Henzi et al. 1998; Komdeur et al. 1999) and it is assumed that resident males achieve most matings in carnivore populations (Sandell 1989, although this is not true for resident badgers within social groups: Dugdale et al. 2007; Chapter 4, this volume). Two (not necessarily mutually exclusive) hypotheses for males abandoning their territories during the mating season are that (1) males mate with local females before vacating (in search of additional matings), thereby retaining the reproductive benefit of maintaining a territory and maximizing additional opportunities, and/or (2) roaming males leave (perhaps to avoid mating with unrelated females)[20] and settle, *after* the mating season, in a territory overlapping females they have inseminated. The argument for the latter is that the male protects his investment from harassment, including infanticide by unrelated males (Wrangham and Rubenstein 1986), as has been suggested for the pine marten *Martes martes* (Schröpfer et al. 1997). The test for these hypotheses is, quite simply, that kits are fathered by males that overlap territories with their mothers either before or after the mating season, but few studies have provided molecular evidence to support either.

The spatial organization of mink (with respect to their mating system), their reproductive 'quirks', and the physiological links between them, give rise to four fundamental questions: (1) Does multiple paternity occur in the wild? (2) Why do males roam? (3) What is the functional advantage of a short DI? (4) Do female mink choose their mates (do they choose to mate multiply, and do they select the males with which they mate)? We provide evidence for (1) from our Thames population, explore (2) and (3), and adopt an experimental behavioural approach using a captive population of mink in Denmark to answer (4) (Box 6.1).

6.4.3 Does multiple paternity occur in wild mink populations?

Six litters were captured in the study area, only five of which consisted of two or more kits—but at least three of these were sired by two or more males, demonstrating that multiple paternity did occur in this mink population (Yamaguchi et al. 2004, Table 6.3). However, no tested offspring were sired by a male whose territory had overlapped that of their mother's before the mating season (n = 4 possible male–female combinations), and only in one case (n = 15 possible male–female combinations) did a male subsequently take up residence in a territory overlapping a female carrying his young. Thus, our paternity results did not support either of the proposed hypotheses for male roaming. In general, males sired litters beyond their territory boundaries, and their progeny were widely dispersed (Table 6.3).

6.4.4 Why do males roam?

So why do males roam? Yamaguchi et al. (2004) suggest that the explanation lies in the reproductive biology of female mink and multiple paternity. Because females are capable of superfoetation and are induced ovulators, any male that mates with an unguarded female during the mating season can potentially sire at least part of the resulting litter. However, because early matings result in fewer kits ('order effects', Shackleford 1952; Venge 1973), a male seeking to ensure he sires a reasonable proportion of a litter might have to guard the mated female for more than a month, which would be impossible for more than one female. Similarly, because a male's progeny may be widely and sparsely dispersed (indeed, we found this to be the case in the Upper Thames), and because the entire litter is unlikely

[20] Although the movement patterns of male mink suggest that males and females in overlapping ranges are unlikely to be closely related, most males move and establish new territories every year while females tend to stay in one area for longer periods (Hatler 1976; Ireland 1990), and in our study area, all juvenile males left the study area before their first mating season (Yamaguchi et al. 2004).

Table 6.3 Litters recorded in the Upper Thames Valley and their mothers and fathers. Unresolved indicates that the sire could not be determined, and Unsampled indicates that the sire was likely a male from which a sample was not obtained. Spatial overlap indicates whether the male was recorded within the mother's territory before, during, or after the breeding season. From Yamaguchi et al. 2004.

Year	Focal litter	Number of kits	Dam	Assigned sires (number of kits sired)	Spatial overlap
1995	Males tested: M1, M2, M3, M4, M5, M6				
	1	4	F1	M1 (1) Unresolved (3)	During
	2	3	F3	M1 (2) Unresolved (1)	During
	3	1	F6	Unresolved	
1996	Males tested: M3, M5, M6, M7, M8, M9, M10, M11, M12, M13, M14, M15				
	4	3	F3	M8 (1) M9 (1)	During During
	5	2	F4	M8 (1) M9 (1)	During During
	6	6	F10	M11 (2) Unsampled (3) Unresolved (1)	During and after

to be sired by a single male (and since not all kits survive, a male could not guarantee that the survivors were his), it is also neither practical nor worthwhile for a male to protect a female and her litter. In short, roam-and-mate tactics are likely to be more profitable for males than any type of guarding tactics of either the female or her litter (although superfecundation and order effects mean that mate guarding for the last 2–3 days of the mating season, when further fertilizations could still occur, might be expected in order to maintain the last—and likely most successful—mating position).

6.4.5 When should males leave?

Roam-and-mate tactics, however, prompt another question: *when* should a male leave his territory? Leaving earlier means longer to find and mate with more females, but transience risks aggressive encounters with other males. To explore the consequences of leaving early, we compared body condition and scars/wounds of resident males (defined as those that had been in the study area for three months or more) with those of transient males arriving in the study area at different times during the mating season. Transients were defined as T1s, new males that arrived in the study area at the beginning of the mating season in January, or T2s, if they arrived later in the mating season in March. Residents that remained in their territory at the onset of the mating season appeared to stay in their territory until March (i.e. they were late leavers), although their territories were invaded by up to four transient males from January (Yamaguchi et al. 2004).

Transients (T1s and T2s) bore significantly more scars and wounds than did residents (Fig. 6.6) and, although the body condition of both transient types appeared to be comparable to that of residents when they arrived, body condition of T1s had declined by March (the four T1s that were still present in March lost a median of 113 g in weight in the preceding two months, Fig. 6.6). Further, although we were unable to determine whether the previous residents had also lost weight by March (because they had now left), T1s appeared to be in poorer body condition than did T2s coming into the area, with whom they had to compete (Fig. 6.6). We assume that T2s were residents[21] from outside the study area that had left their home ranges late in the season (although we cannot exclude the possibility that they had been transients elsewhere since the beginning of the mating season), and we tentatively suggest that prolonged residency and late departure from the home

[21] There is no method for reliably determining the age of a live mink, but the length of the baculum (the penis bone) increases with age (Elder 1951) and thus can be used to assess relative age in males: baculum lengths of T2s were comparable to those of residents (63 mm ± 2.7 versus 62 ± 2.2, n = 4 and 7, respectively), whereas those of T1s (57 ± 2.3, n = 8, excluding one particularly large individual) were significantly shorter than those of both residents (W = 3, p = 0.026) and T2s (W = 10, p = 0.036), suggesting that T1s were younger animals, and that T2s were mature animals of similar age to resident males.

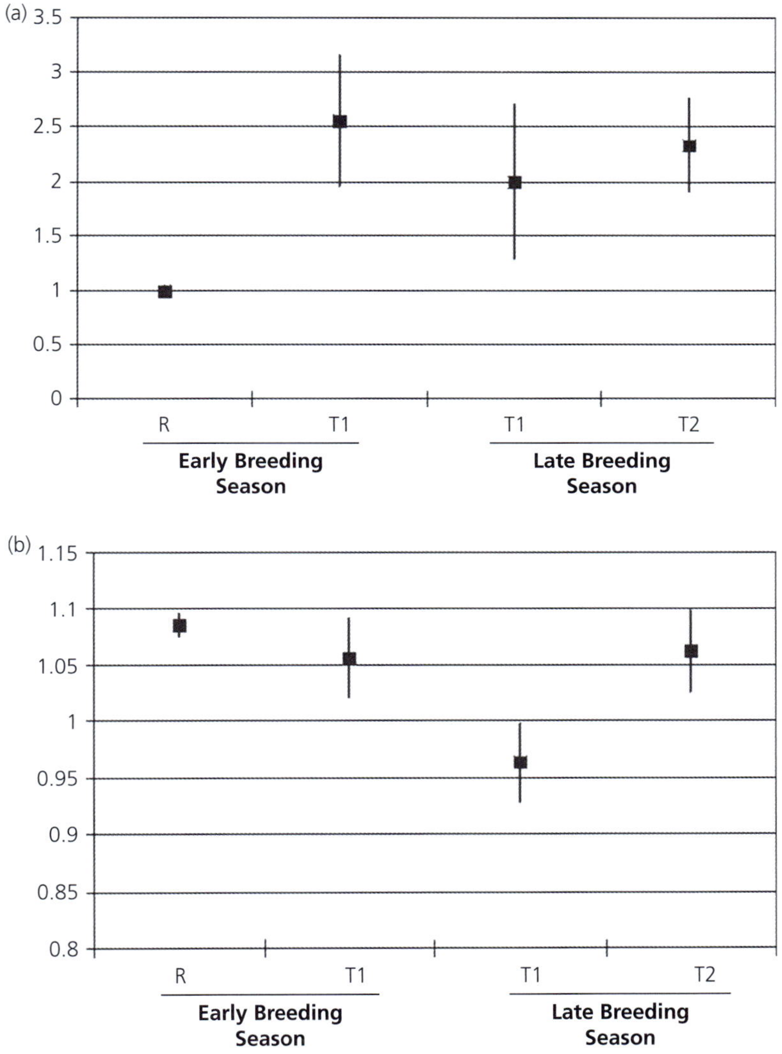

Figure 6.6 Scars and wounds, and body condition, of male mink during the mating season. (a) Scars and wounds assessed on a scale of 0–6 (6 = worst); (b) body condition = observed/expected body mass (where expected mass is based on a regression of body mass and head body length for all males). Data are mean ± SE. R = resident, T1 = transient that arrived in the study area early in the mating season, T2 = transient that arrived late in the mating season (see text). Residents n = 4, T1 n = 8, T2 n = 7.

range appears to be the best strategy for maintaining body condition (and presumably competitive ability). The potential benefit of the T1 strategy is unclear and it is possible that T1s did not choose to become transients early, but were simply young inexperienced individuals (see footnote 21) that did not yet hold a stable home range. Male shrews *Sorex araneus* also adopt different movement patterns during the mating season (Stockley et al. 1994) dependent on their maturity: later maturing males appeared to suffer reduced access to females but compensated with increased epididymal sperm counts (although early maturing males fathered more offspring, see Box 4.3 in Macdonald and Feber 2015: Chapter 4). Considering the area over which male mink can potentially range, we were unable to assess how many kits they had sired, but given the late departure of resident male mink from their home ranges, it is surprising that our paternity results (above) showed no evidence that they had sired kits with females from their home range before they left.

6.4.6 What is the functional advantage of a short delayed implantation (DI)?

For females, multiple mating may be advantageous to protect against infanticide[22] (through paternity confusion), avoid sexual harassment, assure fertilization,

[22] There is no evidence of male infanticide in mink, but it has been reported in other mustelids (Ebensperger 1998).

Box 6.1 Multiple mating: female choice or male coercion?

Although often assumed that females are relatively passive partners in polyandrous mating systems (Reynolds 1996), there is increasing evidence for female solicitation of multiple males (e.g. in rodents: Shapiro and Dewsbury 1986; Berteaux et al. 1999; Box 11.3 in Chapter 11, this volume], birds: Petrie et al. 1992; and fish: Evans and Magurran 2000). In a free-choice experiment, in which eight female mink had access to three males each (but not vice versa, Box Fig. 6.1.1), Thom et al. (2004c) showed that female mink mated multiply (seven of the eight, with all three males), that females were able to resist copulations, and inferred that polyandry is driven by female behaviour rather than male coercion.

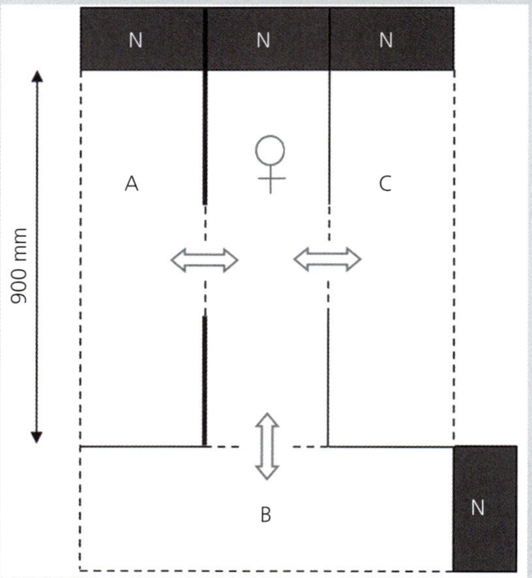

Box Figure 6.1.1 Plan view of a single experimental arena. The female (wild-type strain, farm-reared, yearling virgin) occupied the centre enclosure, and three males (A, B, and C, one familiar, two unknown) were placed in the contiguous enclosures. Females could access the male enclosures via openings (arrows) too small to allow males passage. Each animal had access to a separate nest box (N). Mesh walls are marked as dashed lines, solid walls unbroken lines. The female had an equal visible area of each male. Each of eight arenas was filmed continuously and the movements and mating behaviours of the female monitored for 19 days during the mating season. In the first experiment, eight females were exposed to a single set of three males each. The experimental set-up for the second experiment was the same except that only two of the three male enclosures were used, and each of the 16 females in this experiment were exposed to four different pairs of males, three times each.

A second set of similar experiments, in which 16 females were given access to one 'large' and one 'small' male (Box Fig. 6.1.1), provided little evidence that females were selecting matings on the basis of male morphological traits, except that females appeared to prefer males with large testes (however, there was no evidence that testis size was correlated with male size or body weight, as in some rodents; Breed and Taylor 2000). The male of the pair with larger testes (mean volume 6702 ± 429 mm^3) mated with 3.6 (± 0.4) females compared with 2.2 (± 0.4) for the male with smaller testes (mean volume 5951 ± 374 mm^3)—although the sample size was too small to show statistical significance, this trend held for seven of eight pairs. Selecting males with large testes could benefit females directly because larger testes produce more sperm (Blottner et al. 2006) (increasing the likelihood of fertilization or, possibly, larger litters) or indirectly, if testis size is heritable, by producing sons with larger testes (and a potential advantage in sperm competition).

Although male body size did not influence number of copulations, it did influence copulation duration: smaller males mated for longer (in total and per female, Box Fig. 6.1.2) than larger males. In mink, more sperm is transferred in longer copulations (Venge 1973; Jørgensen 1985), therefore long durations may be related to sperm competition (if duration is under male control, which we assume it is). If smaller males have less chance of acquiring matings through inter-male competition, they might invest more in each copulation that they do achieve (in a comparable manner to late maturing common shrews having higher sperm counts than do early maturing males, see text). If this were the case, we would expect large males to achieve more copulations than smaller males (based on their size advantage in contest competition), which in our experiments (designed to remove inter-male competition), they did not, but in the wild they may. There was some (weak) evidence that males that mated for longer sired more kits (Box Fig. 6.1.2), but eight males failed to sire any kits, despite copulating for up to 234 minutes per female. Neither number of females mated nor number of copulations per female, nor any male phenotypic variable, predicted the number of kits sired among successful males, nor success or failure at siring a litter. Similarly, litter size was uncorrelated with female mass, copulation duration, copulation number, or number of males mated.

How realistic are these findings? In the wild, females have nowhere to escape and thus forced copulations may be more frequent, although it is in the females' interest to seek

continued

Box 6.1 *Continued*

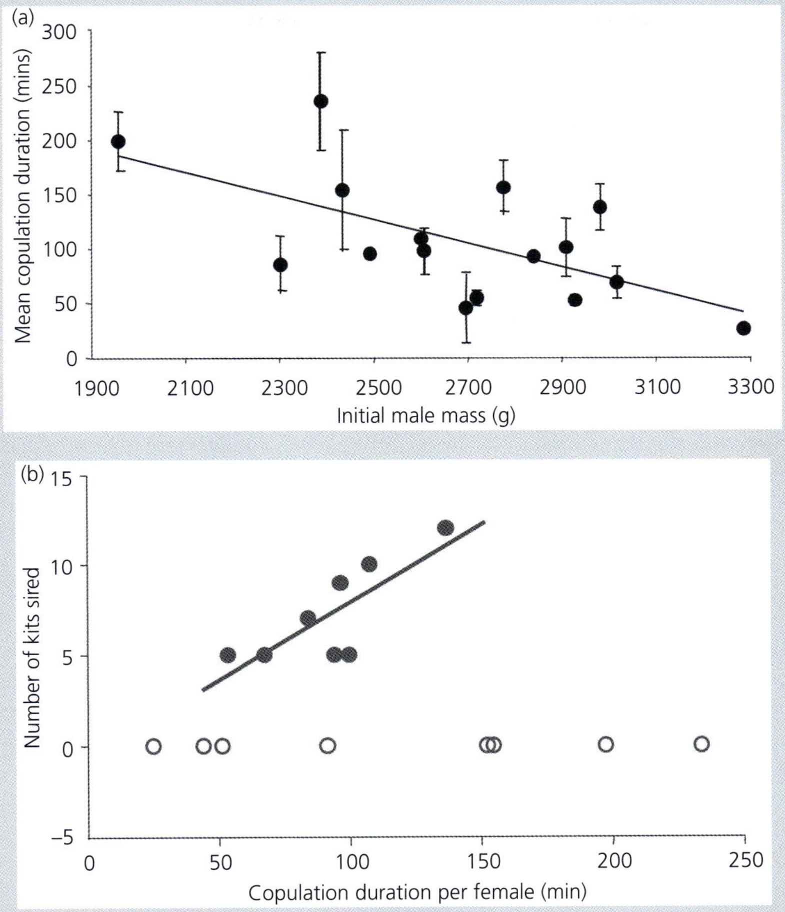

Box Figure 6.1.2 Several measures of male size were *negatively* correlated with copulation duration, which is assumed to be primarily under male control. (a) shows the mean duration of copulations (per female mated ± SE) decreasing with increasing male weight (points with no error bar represent males who mated with only one female). The line of best fit is shown for the mean values ($r = -0.620$, $p = 0.01$). (b) shows the positive correlation between the number of kits sired by successful males and mean copulation duration per female ($r = 0.787$, $p = 0.02$), but the relationship is weak because four of these eight males sired the same number of kits, and eight (of the total 16) males failed to sire any kits. Median number of kits sired per successful male = 7. Note that as little as 3 minutes may be sufficient to ensure fertilization (Enders 1952), and 10–20 minutes (Jørgensen 1985) is considered sufficient on commercial farms, which is considerably shorter than the copulations we observed, which were up to 280 minutes.

multiple matings. If it is beneficial for females to mate with males with large testes, it is less clear how such a selection would take place in a natural setting; in our experimental set-up, preferences were assessed simultaneously but, for a solitary species in the wild, it is more likely that sequential choices would be required (although Ireland (1990) reported that females often had multiple males in attendance). Questions remain regarding the role of sperm competition in this system, and it would be insightful to assess the relative mating success of wild males of different age, size, and/or status.

increase the genetic diversity (and compatibility) of the litter, and (potentially) allow cryptic female choice through sperm competition[23] (reviewed in Wolff and Macdonald 2004). Superfecundation allows a female to avoid devoting her entire reproductive effort to one male, but the window of opportunity for securing multiple fathers in one ovulation event is short (1–3 days), and within this time period one male may be able to monopolize the female. A female's opportunity for mating with multiple males is thus greatly enhanced by superfoetation (which, via multiple ovulations, extends a female's receptivity for about a month). Physiologically, superfoetation in mink (resulting in a single litter) is possible only under the hormonal conditions associated with DI[24] (Shackleford 1952) and DI seems to be the key to multiple mating. Taking the wider view of DI across all mustelids, Yamaguchi et al. (2006) propose this idea for badgers, and suggest that DI generally, in association with superfoetation, may have originated in response to sexual conflict, with longer delays to alter mating and parturition times evolving later in mustelids.

6.5 Conclusions

Thom et al. (2004a) argue that some degree of niche separation between male and female mink is likely to arise as a result of sexual dimorphism, and our ecological data—on habitat use, diet, and behaviour—provide tentative support for the occurrence of inter-sexual niche partitioning (and there is similar evidence supporting inter-sexual niche partitioning in other mustelids, e.g. polecats, Marcelli et al. 2003). However, variability in behaviour (shown by the considerable range in tendency to dive) also suggests individualism, which might in turn explain variation in other factors, such as home range size, facilitated by the adaptability and versatility of this species. There is increasing interest in the existence of individual specialists within generalist species (reviewed in Bolnick et al. 2003), and some support for individual specialists in mink is provided by Sidorovich et al.'s (2001) descriptions of individual diets.

From a practical point of view, the adaptability of American mink suggests that they are unlikely to be negatively influenced by farmland habitat management. Indeed, Lundy and Montgomery (2010) suggest that farmland conservation schemes will benefit mink. We do not completely understand to what extent mink might be limited by the availability of rabbits or availability of their dens. Certainly, there is evidence that, in some areas, their presence is supported by the presence of rabbits (Oliver et al. 2009)—but it seems likely that there are alternative den sites along rivers in lowland farmland, except in barren open areas. However, the use of linear home ranges along river banks means that mink are relatively easily targeted by trappers and thus that, potentially (dependent on manpower, resources, and commitment), they could be removed. Ethical considerations require us to consider their impacts on biodiversity (as justification for control efforts) and the effectiveness with which they can be removed, and we address these issues in Chapter 7, this volume, and in Macdonald and Feber (2015): Chapters 14 and 15.

Beyond our interests in biodiversity conservation, understanding the behavioural strategies used by mink, the conditions under which each might occur, and associated trade-offs, brings methodological challenges. Inferring foraging strategies from diving behaviour, for example, is complicated by the fact that mink probably use the water for other purposes (such as travel) and that dives are probably not always an attempt to catch a fish. Harrington et al. (2012) identified long, shallow dives (that did not fit the otherwise positive correlation between dive depth and duration) and proposed that these dives had a function other than foraging—perhaps 'searching' for prey prior to diving, or energetically efficient swimming[25] (that may or may not be part of a foraging session). Development of data loggers suitable for use on small, and shallow diving animals (see Hays et al. 2007) allows precise measurements of dives, but they cannot tell us the actual function of the dive. Nor, of course, can they tell us what the animal is doing when it is not diving. Further development (and miniaturization) of GPS, accelerometers, and integration of these with techniques such as stable isotope analysis and genetic paternity analysis will most likely be required for further insights. Key questions include: How do different foraging strategies differ in terms of energetic efficiency? How do different mating tactics differ in terms

[23] Competition (within the female's reproductive tract) between sperm from two or more males to fertilize the female's eggs.

[24] During diapause in mink, corpora lutea produce only low levels of progesterone, which allows a second oestrus and the ovulation of a second set of eggs capable of normal development (see Renfree and Shaw 2000).

[25] On the basis that swimming beneath the surface of the water minimizes drag, Williams (1983) calculated that mink would achieve a 7- to 10-fold reduction in energy costs by swimming underwater.

of mating/reproductive success? And are there links between foraging efficiency and reproductive success?

Acknowledgements

Through the many years of our work on mink in the UK, the Environment Agency (particularly Alastair Driver and Graham Scholey) have provided support and collaboration for which we are very grateful—we thank them, and the many other funders that have supported various stages of this work, including, perhaps most notably, The People's Trust for Endangered Species, for their long-term support. Other members of the team that contributed to our knowledge of this fascinating species included Liz Halliwell, Pablo Ferreras, Danielle Clode, as well as a whole host of volunteers, field assistants, and students. Paul Johnson assisted with statistical analysis throughout. Lastly, we would like to dedicate this chapter to Rob Strachan, who influenced our work on mink in many ways and with whom we all spent many happy hours chasing mink, or searching for their scats, along the River Thames. He will be sorely missed and remembered always.

References

Askins, G.R. and Chapman, J.A. (1984). Age determination and morphological characteristics of wild mink from Maryland, USA. *Zeitschrift für Säugetierkunde*, **49**, 182–189.

Bagniewska, J., Harrington, L.A., Hart, T., Harrington, A.L., and Macdonald, D.W. (in press). Persistence in diving American mink. Animal Biotelemetry.

Bagniewska, J., Hart, T., Harrington, L.A., and Macdonald, D.W. (2013). Hidden Markov analysis describes dive patterns in semi-aquatic animals. *Behavioral Ecology*, **24**, 659–667.

Barreto, G.R., Macdonald, D.W., and Strachan, C. (1998). The tightrope hypothesis: an explanation for plummeting water vole numbers in the Thames catchment. In *United Kingdom Floodplains*, R.G. Bailey, P.V. Jose, and B.R. Sherwood eds., pp. 311–326. Westbury Academic and Scientific Publishing, Otley.

Berteaux, D., Bety, J., Rengifo, E., and Bergeron, J.M. (1999). Multiple paternity in meadow voles (*Microtus pennsylvanicus*): investigating the role of the female. *Behavioural Ecology and Sociobiology*, **45**, 283–291.

Birks, J.D.S. (1981). Home range and territorial behaviour of the feral mink (*Mustela vison* Schreber) in Devon. PhD dissertation, Exeter University, UK.

Birks, J.D.S. and Dunstone, N. (1984). A note on prey remains collected from the dens of feral mink (*Mustela vison*) in a coastal habitat. *Journal of Zoology*, **203**, 279–281.

Birks, J.D.S. and Dunstone, N. (1985). Sex-related differences in the diet of the mink *Mustela vison*. *Holarctic Ecology*, **8**, 245–252.

Birks, J.D.S. and Linn, I.J. (1982). Studies of home range of the feral mink, *Mustela vison*. *Symposia of the Zoological Society of London*, **49**, 231–257.

Blottner, S., Schön, J., and Jewgenow, K. (2006). Seasonally activated spermatogenesis is correlated with increased testicular production of testosterone and epidermal growth factor in mink (Mustela vison). *Theriogenology*, **66**, 1593–1598.

Bolnick, D.I., Svanback, R., Fordyce, J.A., Yang, L.H., Davis, J.M., Hulsey, C.D., and Forister, M. (2003). The ecology of individuals: incidence and implications of individual specialization. *American Naturalist*, **161**, 1–28.

Bolnick, D.I., Yang, L.H., Fordyce, J.A., Davis, J.M., and Svanbäck, R. (2002). Measuring individual-level resource specialization. *Ecology*, **83**, 2936–2941.

Bonesi, L., Chanin, P., and Macdonald, D.W. (2004). Competition between Eurasian otter *Lutra lutra* and American mink *Mustela vison* probed by niche shift. *Oikos*, **106**, 19–26.

Bonesi, L., Harrington, L.A., Maran, T., Sidorovich, V.E., and Macdonald, D.W. (2006). Demography of three populations of American mink *Mustela vison* in Europe. *Mammal Review*, **36**, 98–106.

Breed, W.G. and Taylor, J. (2000). Body mass, testes size, and sperm size in murine rodents. *Journal of Mammalogy*, **81**, 758–768.

Brotherton, P.N.M. and Manser, M.B. (1997). Female dispersion and the evolution of monogamy in the dik-dik. *Animal Behaviour*, **54**, 1413–1424.

Brown, J.H. and Lasiewski, R.C. (1972). Metabolism of weasels: the cost of being long and thin. *Ecology*, **53**, 939–943.

Brzeziński, M. (2008). Food habits of the American mink *Mustela vison* in the Mazurian Lakeland, Northeastern Poland. *Mammalian Biology*, **73**, 177–188.

Buskirk, S.W., Harlow, H.J., and Forrest, S.C. (1988). Temperature regulation in American marten (*Martes americana*) in winter. *National Geographic Research*, **4**, 208–218.

Chanin, P.R.F. (1983). Observations on two populations of feral mink in Devon, UK. *Mammalia*, **47**, 463–466.

Chanin, P.R.F. and Linn, I. (1980). The diet of the feral mink (*Mustela vison*) in southwest Britain. *Journal of Zoology*, **192**, 205–223.

Chapman, J. and Feldhamer, G. (1982). *Wild Mammals of North America: Biology, Management, and Economics*. The Johns Hopkins University Press, Baltimore.

Clode, D., Halliwell, E., and Macdonald, D.W. (1995). A comparison of body condition in riverine and coastal mink. *Journal of Zoology*, **237**, 686–689.

Clode, D. and Macdonald, D.W. (1995). Evidence for food competition between mink (*Mustela vison*) and otter (*Lutra lutra*) on Scottish islands. *Journal of Zoology*, **237**, 435–444.

Conroy, J.W.H., Watt, J., Webb, J., and Jones, A. (2005). *A Guide to the Identification of Prey Remains in Otter Spraint*. Mammal Society, London, UK.

Cuthbert, J.H. (1979). Food studies of feral mink (Mustela vison) in Scotland. *Fisheries Management*, **10**, 17–25.

Dayan, T., Simberloff, D., Tchernov, E., and Yom-Tov, Y. (1989). Inter- and intraspecific character displacement in mustelids. *Ecology*, **71**, 1526–1539.

Dugdale, H.L., Macdonald, D.W., Pope, L.C., and Burke, T. (2007). Polygynandry, extra-group paternity and multiple-paternity litters in European badger (Meles meles) social groups. *Molecular Ecology*, **16**, 5294–5306.

Dunstone, N. (1993). *The Mink*, T & AD Poyser Ltd, London.

Dunstone, N. and Birks, J.D.S. (1985). The comparative ecology coastal, riverine and lacustine mink *Mustela vison* in Britain. *Zeitschrift für angewartde Zoologie*, **72**, 52–70.

Dunstone, N. and Birks, J.D.S. (1987). The feeding ecology of mink (*Mustela vison*) in a coastal habitat. *Journal of Zoology*, **212**, 69–83.

Ebensperger, L.A. (1998). Strategies and counterstrategies to infanticide in mammals. *Biological Reviews of the Cambridge Philosophical Society*, **73**, 321–346.

Elder, W.H. (1951). The baculum as an age criterion in mink. *Journal of Mammalogy*, **32**, 43–50.

Enders, R.K. (1952). Reproduction in the mink *Mustela vison*. *Proceedings of the American Philosophical Society*, **96**, 691–755.

Erlinge, S. (1969). Food habits of the otter (*Lutra lutra* L.) and the mink (*Mustela vison* Schreber) in a trout water in southern Sweden. *Oikos*, **20**, 1–7.

Erlinge, S. (1979). Adaptive significance of sexual dimorphism in weasels. *Oikos*, **33**, 233–245.

Erlinge, S. and Sandell, M. (1986). Seasonal changes in the social organization of male stoats, Mustela erminea: an effect of shifts between two decisive resources. *Oikos*, **47**, 57–62.

Evans, J.P. and Magurran, A.E. (2000). Multiple benefits of multiple mating in guppies. *Proceedings of the National Academy of Sciences*, **97**, 10074–10076.

Fasola, L., Muzio, J., Chehébar, C., Cassini, M., and Macdonald, D.W. (2011). Range expansion and prey use of American mink in Argentinian Patagonia: dilemmas for conservation. *European Journal of Wildlife Research*, **57**, 283–294.

Fererras, P. and Macdonald, D.W. (1999). The impact of American mink *Mustela vison* on water birds in the upper Thames. *Journal of Applied Ecology*, **36**, 701–708.

Fox, S., Spencer, H., and O'Brien, G.M. (2008). Analysis of twinning in flying-foxes (Megachiroptera) reveals superfoetation and multiple-paternity. *Acta Chiropterologica*, **10**, 271–278.

Gerell, R. (1967). Food selection in relation to habitat in mink (*Mustela vison* Schreber) in Sweden. *Oikos*, **18**, 233–246.

Gerell, R. (1968). Food habits of the mink (*Mustela vison* Schreber) in Sweden. *Viltrevy*, **5**, 119–211.

Gerell, R. (1970). Home ranges and movements of the mink *Mustela vison* Schreber in southern Sweden. *Oikos*, **21**, 160–173.

Grant, K.R. and Harrington, L.A. (in press). Fish selection by riverine Eurasian otters in lowland England. Mammal Research.

Grinnell, J., Dixon, J.S., and Linsdale, J.M. (1937). Furbearing mammals of California. Their Natural History, Systematic Status and Relations to Man. University of California Press, Berkeley, California.

Halliwell, E.C. and Macdonald, D.W. (1996). American mink *Mustela vison* in the Upper Thames catchment: relationship with selected prey species and den availability. *Biological Conservation*, **76**, 51–56.

Halsey, L.G., Blackburn, T.M., and Butler, P.J. (2006). A comparative analysis of the diving behaviour of birds and mammals. *Functional Ecology*, **20**, 889–899.

Harrington, L.A., Harrington, A.L., Yamaguchi, N., Thom, M., Ferreras, P., Windham, T.R., and Macdonald, D.W. (2009). The impact of native competitors on an alien invasive: temporal niche shifts to avoid inter-specific aggression? *Ecology*, **90**, 1207–1216.

Harrington, L.A., Hays, G., Fasola, L., Harrington, A.L., Righton, D., and Macdonald, D.W. (2012). Diving ability and behaviour of a small-bodied, semi-aquatic mammal in the wild. *Journal of Mammalogy*, **93**, 198–210.

Harrington, L.A. and Macdonald, D.W. (2008). Spatial and temporal relationships between invasive American mink and native European polecats in the southern United Kingdom. *Journal of Mammalogy*, **89**, 991–1000.

Hatler, D.F. (1976). The coastal mink on Vancouver Island, British Columbia. PhD dissertation, University of British Columbia, Vancouver, Canada.

Hays, G.C., Forman, D.W., Harrington, L.A., Harrington, A.L., Macdonald, D.W., and Righton, D. (2007). Recording the free-living behaviour of small-bodied, shallow-diving animals with data loggers. *Journal of Animal Ecology*, **76**, 183–190.

Henzi, S.P., Lycett, J.E., and Weingrill, T. (1998). Mate guarding and risk assessment by male mountain baboons during inter-troop encounters. *Animal Behaviour*, **55**, 1421–1428.

Ibarra, J.T., Fasola, L., Macdonald, D.W., Rozzi, R., and Bonacic, C. (2009). Invasive American mink *Mustela vison* in wetlands of the Cape Horn Biosphere Reserve, southern Chile: what are they eating? *Oryx*, **43**, 87–90.

Ireland, M.C. (1990). The behaviour and ecology of the American mink (*Mustela vison* Schreber) in a coastal habitat. PhD dissertation, Durham University, UK.

Ishida, Y., Yahara, T., Kasuya, E., and Yamane, A. (2001). Female control of paternity during copulation: inbreeding avoidance in feral cats. *Behaviour*, **138**, 235–250.

Jørgensen, G.E. (1985). *Mink Production*. Scientifur, Hilleroed, Denmark.

Komdeur, J., Kraaijeveld-Smith, F., Kraaijeveld, K., and Edelaar, P. (1999). Explicit experimental evidence for the role of mate guarding in minimising loss of paternity in the Seychelles warbler. *Proceedings of the Royal Society of London*, **266**, 2075–2081.

Krebs, J.R. (1989). *Ecological Methodology*. Harper and Row, New York.

Kruska, D.C.T. and Sidorovich, V. (2003). Comparative allometric skull morphometrics in mink (Mustela vison Schreber, 1777) of Canadian and Belarus origin; taxonomic status. *Mammalian Biology*, **68**, 257–276.

Lindenfors, P., Dalèn, L., and Angerbjörn, A. (2003). The monophyletic origin of delayed implantation in carnivores and its implications. *Evolution*, **57**, 1952–1956.

Linn, I.J. and Stevenson, J.H.F. (1980). Feral mink in Devon. *Nature in Devon*, **1**, 7–27.

Lodé, T. (1999). Comparative measurements of terrestrial and aquatic locomotion in *Mustela lutreola* and *M. putorius*. *Mammalian Biology*, **64**, 110–115.

Lodé, T. (2003). Sexual dimorphism and trophic constraints: prey selection in the European polecat (*Mustela putorius*). *Ecoscience*, **10**, 17–23.

Lundy, M.G. and Montgomery, W.I. (2010). A multi-scale analysis of the habitat associations of European otter and American mink and the implications for farm scale conservation schemes. *Biodiversity and Conservation*, **19**, 3849–3859.

Macdonald, D.W. and Feber, R.E., eds. (2015). *Wildlife Conservation on Farmland. Managing for Nature on Lowland Farms*. Oxford University Press, Oxford.

Macdonald, D.W. and Harrington, L.A. (2003). The American mink: the triumph and tragedy of adaptation out of context. *New Zealand Journal of Zoology*, **30**, 421–441.

Macdonald, D.W., Sidorovich, V.E., Maran, T., and Kruuk, H. (2002). European Mink, *Mustela lutreola*: Analyses for Conservation. Wildlife Conservation Research Unit, Oxford.

Macdonald, D.W. and Strachan, R. (1999). *The Mink and the Water Vole: Analyses for Conservation*. Wildlife Conservation Research Unit and the Environment Agency, Oxford.

Macpherson, J.L. and Bright, P.W. (2010). Movements of radio-tracked American mink (*Neovison vison*) in extensive wetland in the UK, and the implications for threatened prey species such as the water vole (*Arvicola amphibious*). *European Journal of Wildlife Research*, **56**, 855–859.

Magnusdottir, R., Stafansson, R.A., von Schmalensee, M., Macdonald, D.W., and Hersteinsson, P. (2012). Habitat- and sex-related differences in a small carnivore's diet in a competitor-free environment. *European Journal of Wildlife Research*, **58**, 669–676.

Marcelli, M., Fusillo, R., and Boitani, L. (2003). Sexual segregation in the activity patterns of European polecats (*Mustela putorius*). *Journal of Zoology*, **261**, 249–255.

McDonald, R.A. (2002). Resource partitioning among British and Irish mustelids. *Journal of Animal Ecology*, **71**, 185–200.

Mead, R.A. (1981). Delayed implantation in mustelids, with special emphasis on the spotted skunk. *Journal of Reproduction and Fertility Supplement*, **29**, 11–24.

Melero, Y., Palazón, S., Revilla, E., and Gosàlbez, J. (2011). Winter activity patterns in an invading Mediterranean population of American mink (*Neovison vison*). *Folia Zoologica*, **60**, 47–53.

Melero, Y., Santulli, G., Gómez, A., Gosàlbez, J., Rodriguez-Refojos, C., and Palazón, S. (2012). Morphological variation of introduced species: the case of American mink (*Neovison vison*) in Spain. *Mammalian Biology*, **77**, 345–3502.

Mitchell, J.L. (1961). Mink movements and populations on a Montana river. *Journal of Wildlife Management*, **25**, 48–54.

Moors, P.J. (1980). Sexual dimorphism in the body size of mustelids (carnivore): the roles of food habits and breeding systems. *Oikos*, **34**, 147–158.

Niemimaa J. (1995). Activity patterns and home ranges of the American mink *Mustela vison* in the Finnish outer archipelago. *Annales Zoologici Fennici*, **32**, 117–121.

Nordström, M., Högmander, J., Laine, J., Nummelin, J., Laanetu, N., and Korpimäki, E. (2002). Variable responses of waterfowl breeding populations to long-term removal of introduced American mink. *Ecography*, **25**, 385–394.

Oliver, M., Luque-Larena, J.J., and Xavier, L. (2009). Do rabbits eat voles? Apparent competition, habitat heterogeneity and large-scale coexistence under mink predation. *Ecology Letters*, **12**, 1201–1209.

Petrie, M., Hall, M., Halliday, T., Budgey, H., and Pierpoint, C. (1992). Multiple mating in a lekking bird: why do peahens mate with more than one male and with the same male more than once? *Behavioural Ecology and Sociobiology*, **31**, 349–358.

Powell, R.A. (1979). Mustelid spacing patterns: variations on a theme by *Mustela*. *Z. Tierpsychol.*, **50**, 153–165.

Powell, R.A. and Leonard, R.D. (1983). Sexual dimorphism and energy expenditure for reproduction in female fisher *Martes pennant*. *Oikos*, **40**, 166–174.

Ralls, K. and Harvey, P.H. (1985). Geographic variation in size and sexual dimorphism of North American weasels. *Biological Journal of the Linnean Society*, **25**, 119–167.

Ratkiewicz, M. and Borkowska, A. (2000). Multiple paternity in the bank vole (*Clethrionomys glareolus*): field and experimental data. *Zeitschrift für Säugetierkunde*, **65**, 6–14.

Renfree, M.B. and Shaw, G. (2000). Diapause. *Annual Review of Physiology*, **62**, 353–375.

Reynolds, J.D. (1996). Animal breeding systems. *Trends in Ecology and Evolution*, **11**, 68–72.

Richardson, L., Clark, T.W., Forrest, S.C., and Campbell, T.M. (1987). Winter ecology of black-footed ferrets (*Mustela nigripes*) at Meeteetse, Wyoming. *American Midland Naturalist*, **117**, 225–239.

Roellig, K., Menzies, B.R., Hildebrandt, T.B., and Goeritz, F. (2011). The concept of superfetation: a critical review on a 'myth' in mammalian reproduction. *Biological Reviews*, **86**, 77–95.

Roy, S. (2011). Strategies to improve landscape scale management of mink populations in the west coast of Scotland: lessons learned from the Uists 2001–2006. In Veitch, C.R., Clout, M.N., and Towns, D.R. eds., pp. 114–117. *Island Invasives: Eradication and Management*. IUCN, Gland, Switzerland.

Sandell, M. (1989). The mating tactics and spacing patterns of solitary carnivores. In *Carnivore Behaviour, Ecology and Evolution*, J.L. Gittleman, ed., pp. 164–182. Chapman and Hall, London, UK.

Sandell, M. (1990). The evolution of seasonal delayed implantation. *The Quarterly Review of Biology*, **65**, 23–42.

Schenk, A. and Kovacs, K.M. (1995). Multiple mating between black bears revealed by DNA fingerprinting. *Animal Behaviour*, **50**, 1483–1490.

Schröpfer, R., Weigand, P., and Hogrefe, H.-H. (1997). The implications of territoriality for the social system of the European pine marten, *Martes martes* (L., 1758). *Zeitschrift für Säugetierkunde*, **62**, 209–218.

Sealander, J.A. (1943). Winter food habits of mink in southern Michigan. *Journal of Wildlife Management*, **7**, 411–417.

Shackleford, R.M. (1952). Superfetation in the ranch mink. *American Naturalist*, **86**, 311–319.

Shapiro, L.E. and Dewsbury, D.A. (1986). Male dominance, female choice and male copulatory behaviour in two species of voles (*Microtus ochrogaster* and *Microtus montanus*). *Behavioural Ecology and Sociobiology*, **18**, 267–274.

Sidorovich, V.E. (2000). Seasonal variation in the feeding habits of riparian mustelids in river valleys of NE Belarus. *Acta Theriologica*, **45**, 233–242.

Sidorovich, V., Kruuk, H., and Macdonald, D.W. (1999). Body size, and interactions between European and American mink (*Mustela lutreola* and *M. vison*) in Eastern Europe. *Journal of Zoology*, **248**, 521–527.

Sidorovich, V. and Macdonald, D.W. (2001). Density dynamics and changes in habitat use by the European mink and other native mustelids in connection with the American mink expansion in Belarus. *Netherlands Journal of Zoology*, **51**, 107–126.

Sidorovich, V.E., Macdonald, D.W., Pikulik, M.M., and Kruuk, H. (2001). Individual feeding specialization in the European mink, *Mustela lutreola*, and the American mink, *M. vison*, in north-eastern Belarus. *Folia Zoologica*, **50**, 27–42.

Skirnisson, K. (1993). Minkur (in icelandic). *Villt íslensk spendýr* (eds. P. Hersteinsson and G. Sigbjarnarson), pp. 79–102. The Icelandic Natural History Society and Icelandic Environment Association, Reykjavík.

Stenson, G.B. (1985). *The Reproductive Cycle of River Otters*. University of British Columbia, Vancouver.

Stockley, P., Searle, J.B., Macdonald, D.W., and Jones, C.S. (1993). Female multiple mating behaviour in the common shrew as a strategy to reduce inbreeding. *Proceedings of the Royal Society of London*, **254**, 173–179.

Stockley, P., Searle, J.B., Macdonald, D.W., and Jones, C.S. (1994). Alternative reproductive tactics in male common shrews: relationships between mate-searching behavior, sperm production, and reproductive success as revealed by DNA fingerprinting. *Behavioural Ecology and Sociobiology*, **34**, 71–78.

Strachan, R. and Jefferies, D.J. (1993). The water vole *Arvicola terrestris* in Britain 1989–1990: its distribution and changing status. The Vincent Wildlife Trust, London.

Strachan, C., Jefferies, D.J., Barreto, G.R., Macdonald, D.W., and Strachan, R. (1998). The rapid impact of resident American mink on water voles, case study in lowland England. *Symposia of the Zoological Society of London*, **71**, 339–357.

Thom, M.D., Harrington, L.A., and Macdonald, D.W. (2004a). Why are American mink sexually dimorphic? A role for niche separation. *Oikos*, **105**, 525–535.

Thom, M.D., Johnson, D.D.P., and Macdonald, D.W. (2004b). The evolution and maintenance of delayed implantation in the Mustelidae (*Carnivora: Mammalia*). *Evolution*, **58**, 175–183.

Thom, M.D., Macdonald, D.W., Mason, G.J., Pedersen, V., and Johnson, P. J. (2004c). Female American mink (*Mustela vison*) mate multiply in a free-choice environment. *Animal Behaviour*, **67**, 975–984.

Valenzuela, A.E.J., Rey, A.R., Fasola, L., and Schiavini, A. (2013). Understanding the inter-specific dynamics of two co-existing predators in the Tierra del Fuego Archipelago: the native southern river otter and the exotic American mink. *Biological Invasions*, **15**, 645–656.

Venge, O. (1973). Reproduction in the mink. Royal Veterinary and Agricultural University, Copenhagen, Denmark, yearbook 1973: 95–146.

Wade-Smith, J., Richmond, M.E., Mead, R.A., and Taylor, H. (1980). Hormonal and gestational evidence for delayed implantation in the striped skunk *Mephitis mephitis*. *General and Comparative Endocrinology*, **42**, 509–515.

Williams, T.M. (1983). Locomotion in the North American mink, a semi-aquatic mammal. *Journal of Experimental Biology*, **105**, 283–295.

Williams, T.M. (1986). Thermoregulation of the North American mink during rest and activity in the aquatic environment. *Physiological Zoology*, **59**, 292–305.

Wise, M.H., Linn, I.J., and Kennedy, C.R. (1981). A comparison of the feeding biology of mink (*Mustela vison*) and otter (*Lutra lutra*). *Journal of Zoology*, **195**, 181–213.

Wolff, J.O. and Dunlap, A.S. (2002). Multi-male mating, probability of conception, and litter size in the prairie vole (*Microtus ochrogaster*). *Behavioural Processes*, **58**, 105–110.

Wolff, J.O. and Macdonald, D.W. (2004). Promiscuous females protect their offspring. *Trends in Ecology and Evolution*, **19**, 127–134.

Wrangham, R.W. and Rubenstein, D.I. (1986). Social evolution in birds and mammals. In *Ecological Aspects of Social Evolution*. D.I. Rubenstein and R.W. Wrangham, eds., pp. 452–470. Princeton University Press, Princeton University, New Jersey.

Yamaguchi, N., Dugdale, H.L., and Macdonald, D.W. (2006). Female receptivity, embryonic diapause, and superfetation in the European badger (Meles meles): implications for the reproductive tactics of males and females. *Quarterly Review of Biology*, **81**, 33–48.

Yamaguchi, N. and Macdonald, D.W. (2003). The burden of co-occupancy: intraspecific resource competition and spacing patterns in American mink, *Mustela vison*. *Journal of Mammalogy*, **84**, 1341–1355.

Yamaguchi, N., Rushton, S.P., and Macdonald, D.W. (2003). Habitat preferences of feral American mink in the Upper Thames. *Journal of Mammalogy*, **84**, 1356–1373.

Yamaguchi, N., Sarno, R.J., Johnson, W.E., O'Brien, S.J., and Macdonald, D.W. (2004). Multiple paternity and reproductive tactics of free-ranging American minks, *Mustela vison*. *Journal of Mammalogy*, **85**, 432–439.

Zalewski, A. and Bartoszewicz, M. (2012). Phenotypic variation of an alien species in a new environment: the body size and diet of American mink over time and at local and continental scales. *Biological Journal of the Linnean Society*, **105**, 681–693.

Zalewski, A. and Jedrzejewski, W. (2006). Spatial organisation and dynamics of the pine marten *Martes martes* population in Bialowieźa Forest (E Poland) compared with other European woodlands. *Ecography*, **29**, 31–43.

Zschille, J., Stier, N., and Roth, M. (2010). Gender differences in activity patterns of American mink *Neovison vison* in Germany. *European Journal of Wildlife Research*, **56**, 187–194.

Zwiernik, M.J., Kay, D.P., Moore, J., Beckett, K.J., Khim, J.S., Newsted, J.L., Roark, S.A., and Giesy, J.P. (2008). Exposure and effects assessment of resident mink (*Mustela vison*) exposed to polychlorinated dibenzofurans and other dioxin-like compounds in the Tittabawassee River basin, Midland, Michigan, USA. *Environmental Toxicology and Chemistry*, **27**, 2076–2087.

CHAPTER 7

Riparian mustelids in the Upper Thames: a case of alien invasives versus native competitors

Lauren A. Harrington and David W. Macdonald

". . . You see, Wonakake-dat's otter-got mad 'cause mink ketch um so many fish, so he chase mink for kill it, an' mink pooty scare. He all white then jus' same weasel in winter, so otter can see it great way off; an' mink can't hide. So he run in where fire burn tree an' rub hese'f on burnt tree so he all black. Den he turn roun' an' walk back, an' by'm by meet otter run hard. Otter ain't know dat black feller, an' ask it, 'You see mink go dis way?' Mink say no, he an't see it. Otter t'ink funny he can smell mink but can't see it, an' run on fast, but never ketch um mink. Mink like um color so well he always keep it, an' ketch 'em more fish as ever, 'cause fish can't see um so easy, an' so he be black now."

An Abenaki folktale, told in Rowland E Robinson's book, *Sam Lovel's Boy* (1901).

7.1 Introduction

Alien species[1], of many taxonomic groups, have been introduced across the globe in staggering numbers (e.g. Mooney 2005; Macdonald et al. 2007). Not all alien species become invasive[2], but understanding how those that do, do, is fundamentally connected with understanding how species co-exist under natural conditions (Melbourne et al. 2007). Stable co-existence of any set of species (native or non-native, mustelid or not) requires them to partition their use of available 'resources', whether they be food, space, or time (Schoener 1986). The success of an invader in establishing itself within a community thus depends, at least in part, both on the availability of 'niche opportunities' (defined as the potential within a given community for an alien species to increase from low density, which might occur due to the availability of resources, lack of enemies, or some favourable combination of resources, enemies, and environmental conditions) and on the (often location-specific) competitive superiority of the invaders relative to residents (Shea and Chesson 2002).

Biologists sometimes study invasions as 'natural experiments' with a view to testing ideas on competition theory (Macdonald et al. 2001). The current situation among riparian mustelids in the UK is particularly revealing because the roles of alien invader and native resident are reversed. Currently, it is the invader (American mink *Neovison vison*) that is established, and the native competitors (Eurasian otter *Lutra lutra* and European polecat *Mustela putorius*) (Fig. 7.1) that are recovering ('invading') within their natural, but now occupied, range (Harrington et al. 2009a).

Invasive species do have an intrinsic theoretical interest, but they also present a serious threat to native communities worldwide (e.g. Mooney 2005), including in the UK (Macdonald and Burnham 2010), and management must sometimes come first. There have now been a number of successful eradications of invasive species from islands of increasing size (Veitch and Clout 2002; Bellingham et al. 2010), but control of open populations (those on continents or larger islands where areas of control are surrounded by a sea of potential immigrants) is considerably more difficult (Simberloff 1997). The control of invasive small carnivores (that are wide ranging, fast moving, intelligent, and

[1] The term 'alien' refers to the presence of a species in an area outside its natural geographic range (usually transported by people).

[2] The invasiveness of an alien species relates to its rate of spread and its ability to increase in number (i.e. the extent to which it is successful in colonizing a new area).

Wildlife Conservation on Farmland. Conflict in the Countryside. Edited by David W. Macdonald and Ruth E. Feber.
© Oxford University Press 2015. Published 2015 by Oxford University Press.

Figure 7.1 (a) American mink *Neovison vison*, (b) European polecat *Mustela putorius*, (c) Eurasian otter *Lutra lutra*. Photographs © A.L. Harrington.

Figure 7.1 *Continued*

wary) appears to be particularly challenging (Simberloff 2001; King et al. 2009) and, in general, Europe has been lagging behind the rest of the world in tackling the problem of invasive species (Genovesi 2005).

In the UK, national-scale mink removal has hitherto been considered impossible or prohibitively expensive (Macdonald and Strachan 1999; Bonesi and Palazon 2007), so management has been restricted to small-scale, local removal projects. Yet there has been no scientific assessment of the success or efficiency of local operations. In this chapter, we describe our own attempt to do just this, using modelling techniques and a large-scale field experiment. We also discuss what we have learnt of the relationships between the non-native mink and native mustelids, the otter and the polecat, from our studies in the Upper Thames area, drawing on relevant snippets from our research further afield in Belarus and Argentina. We begin with a brief synopsis of the biology of the American mink and of the damage it does to native prey species.

7.2 The American mink

American mink are semi-aquatic and highly adaptable generalist predators (Dunstone 1993) that were transported around the world for fur farming in the early nineteenth century and now live wild in countries from Russia to Argentina (Macdonald and Harrington 2003). Brought to Britain in 1929, they were first recorded breeding in south Devon in 1956 (Linn and Stevenson 1980), and now occupy all counties throughout England and Wales, and all but the far northern highlands of Scotland (Harrington et al. 2010; Fraser et al. 2015).

Formerly classified as a species of *Mustela* (congeneric with several other species commonly recognized as 'weasels', 'ferrets', or 'polecats'), the taxon *M. vison* has recently been elevated to a distinct monospecific genus within the Mustelidae—*Neovison*—on the basis of biochemical, cytogenetic, molecular, and morphological differences (Kurose et al. 2008). For example, the diploid chromosome number of the American mink is $2n = 30$, whereas those of the other *Mustela* species range from 38 to 44. There are also substantial differences in the chemical composition of the anal gland secretion (Brinck et al. 1983) and cranial morphology (Youngman 1982; Abramov 2000) of American mink compared with that of all other *Mustela*. In fact, a new phylogeny based on genetic analyses suggests that the American mink diverged from the other *Mustela* species approximately 8–11 million years ago (Kurose et al. 2008). By contrast, the European mink *Mustela lutreola* (native to the European mainland but never known to occur in the UK), is thought to have diverged from the 'polecat group'—European polecat, Siberian polecat *M. eversmannii*—only very recently, and so is quite distantly related to the American mink (Davison et al. 2000), despite the obvious similarities in their ecology, behaviour, and appearance.

American mink average about 1 kg in body weight and about 40 cm in length and are highly sexually dimorphic (males may weigh twice as much as females, Thom et al. 2004a; Chapter 6, this volume). They occupy generally linear home ranges along river banks of 2–3 km (but up to 6 km in males) (Yamaguchi and Macdonald 2003; Harrington and Macdonald 2008). Mink have an intra-sexual territorial social system, in which several, non-overlapping, female home ranges are all overlapped by one male home range (Dunstone 1993), as in many other mustelid species (Powell 1979), although, intriguingly, male residency does not appear to be associated with the probability of fatherhood (Yamaguchi et al. 2004; Chapter 6, this volume). There is some overlap even between same-sex neighbours (Yamaguchi and Macdonald 2003; Harrington and Macdonald 2008), and the system breaks down during the single annual mating season between January and March (Yamaguchi and Macdonald 2003) when males roam over larger areas, presumably in search of receptive females (e.g. Erlinge and Sandell 1986; Chapter 6, this volume). Females, after a 35 day delay in implantation (Thom et al. 2004b), give birth to a litter of 4–6 (*c.* 5 g) kits between late April and mid May, that disperse from their natal range in August (Dunstone 1993).

7.3 Consequences of mink invasions on native prey species across the world

Mink eat almost anything that they can catch, whether covered in scales, carapace, feathers, or fur, in water or on land. Food items range from relatively small invertebrates (e.g. water beetles, *Dytiscus* spp.), which may or may not be taken incidentally while searching for a more substantial meal, to a 1–2 kg rabbit that may even exceed its captor in weight (Dunstone 1993). Consequently, non-native mink may potentially damage populations of a range of native species. Most notable, in the UK, is the loss of an estimated 88% of the national water vole *Arvicola amphibious* (formerly *A. terrestris*) population over just seven years during the early 1990s, that was largely correlated with an unprecedented increase in the distribution of the American mink population (Jefferies 2003; Macdonald and Feber 2015: Chapter 15). On the west coast of Scotland, mink predation is responsible for colony breeding failures and the loss of local colonies of ground-nesting gulls and terns (Craik 1997). On islands in the Baltic Sea, experimental mink removals have revealed negative consequences of mink predation on population size and/or density of several seabird species—e.g. arctic skua *Stercorarius parasiticus* and arctic tern *Sterna paradisaea*, waterfowl (shelduck *Tadorna tadorna*, tufted duck *Aythya fuligula*, and velvet scoter *Melanitta fusca*)—as well as bank voles *Myodes glareolus*, field voles *Microtus agrestris*, and common frogs *Rana temporaria* (Banks et al. 2008). There are fears that mink predation may similarly affect populations of ground-nesting birds in the Cape Horn Biosphere Reserve in southern Chile, where 58.5% of mink scats in summer (during the nesting season) contained bird remains (Ibarra et al. 2009). More desirable perhaps, at least from the viewpoint of invasive species management, is the decline of the introduced muskrat *Ondatra zibethicus* in Poland, that appears to have coincided with the increase in American mink there (Brzeziński et al. 2010).

7.4 Niche differentiation and co-existence with native competitors

In continental Europe, and in the former USSR, American mink not only prey on native species; they also out-compete the smaller (even the males usually weigh < 1 kg)[3] and highly endangered European mink (reviewed in Macdonald et al. 2002, and Macdonald and Harrington 2003). Female polecats in Belarus, where they are slightly smaller (507 ± 9 g, $n = 3$) than the local female American mink (781 ± 21 g, $n = 23$; Sidorovich et al. 1999), also appear to be excluded by American mink from riparian habitats (Sidorovich and Macdonald 2001). Otters, however, at 6–10 kg, are approximately seven times larger than American mink, and have not been affected by the arrival of American mink in Belarus (Sidorovich and Macdonald 2001). In their native North America, the mink is sympatric with the river otter *Lontra canadensis* (a different genus but ecologically similar to the Eurasian otter, Kruuk 2006).

During the initial spread of the American mink in the UK, the native Eurasian otter and European polecat were both absent from much of the country, having previously suffered substantial population declines due to, in the case of the Eurasian otter, pesticide poisoning in the late 1950s (predominantly the organochlorine insecticide—dieldrin) (probably accentuated by a long history of over hunting) (Strachan and Jefferies 1996) and, in the case of the European polecat,

[3] Mean body weights of European mink in Belarus before the arrival of American mink: male = 854 ± 25 g, $n = 14$, female = 537 ± 13 g, $n = 8$ (Sidorovich et al. 1999).

persecution by gamekeepers in the late nineteenth century (Birks and Kitchener 1999). Both otter and polecat are now recovering nationally (Birks 2008; Crawford 2010) as a result of increased water quality (following reductions in agricultural uses of dieldrin between 1962 and 1983[4]) and reduced gamekeeping pressure during World War I, respectively. All three species, the alien mink, native otter, and native polecat, have potentially overlapping diets (otters are more of a fish specialist but also take a wide variety of prey); all occupy similar habitats (polecats are more terrestrial than American mink, but are often found near rivers); and all are generally nocturnal (except otters in marine habitats) (Dunstone 1993; Birks and Kitchener 1999; Kruuk 2006). Mink and polecats in the UK rely particularly heavily on rabbits, and their burrows, both for food and dens (Halliwell and Macdonald 1996; Birks and Kitchener 1999; Harrington and Macdonald 2008).

Several studies have found inverse correlations between field signs of otter and mink (Strachan and Jefferies 1996; Jefferies 2003; Bonesi and Macdonald 2004a; Bonesi et al. 2006; McDonald et al. 2007), which led to the popular suggestion that encouraging the recovery of one native species might also solve the problem of the non-native mink. In particular, Bonesi and Macdonald (2004a) demonstrated that an experimental release of otters in the Upper Thames Valley, in 1999, was followed by a significant and rapid reduction in the signs of mink (from 77% to 23% of 44 survey sites occupied) and in the number of mink trapped (from a total of 11 mink trapped before the release of otters—nine per 1000 trap-nights, to three mink trapped afterwards—four per 1000 trap-nights). These authors suggested that otters caused the decline in mink by direct aggressive interference competition and that, in the presence of otters, mink densities would be reduced by about half. This decline in mink occurred in the first year after the release of otters, and at a time when otters were locally at high density (0.1 otters per km). On the River Thames, we were able to compare two time periods, one before the recovery of the otter during which it was still absent (in the 1990s), and the other approximately five years after its reappearance (in the mid 2000s) (at which time the otter occurred at an unknown but presumably more natural low density; Harrington et al. 2009a). By comparing trapping, radio-tracking, and dietary data from both periods, we aimed, first, to assess whether mink had declined between the two periods, and second, to explore potential behavioural changes in mink that might facilitate their co-existence with otters. Contrary to expectations, we found that mink were still abundant in the 2000s even in the presence of otters (*c.* 0.45–0.9 mink/km), and that there was little evidence of a decline in numbers between the two periods (Jolly–Seber estimate of the summer adult population = 9.3 ± 3.2 mink, mean ± SE, in 1995; 12.6 ± 2.3 in 2005). There was no statistical difference between either the home range sizes of mink in the 1990s and the 2000s (median home range, both sexes combined but excluding male home ranges during the mating season averaged 3.1 km in the 1990s, $n = 10$, 2.9 km in the 2000s, $n = 11$), or between the extent to which mink used the river habitat (> 60% radio-locations were on the main river in the 1990s and 2000s, n = 10 and n = 11 individuals[5], respectively).

Consistent with previous studies (Clode and Macdonald 1995; Bonesi et al. 2004), we found that mink ate fewer fish in the presence of otters (22.7% ($n = 123$ scats) relative occurrence of fish in mink scats in the 1990s, 10.3% ($n = 78$ scats) in the 2000s), although fish consumption by mink on the River Thames, both before and after the return of otters, was substantially lower than on some other rivers in the UK. On the River Thames, mink diet appeared to be predominantly mammalian (50.3% and 41.2% occurrence in mink scats, in the 1990s and the 2000s respectively, mostly rabbits, voles, and mice—there were no water voles on the River Thames during the study). This is similar to some parts of the United States, where muskrats *Ondatra zibethicus* form the majority of the diet (Errington 1946). Mink also consumed a high proportion of birds throughout the year, and an even higher proportion in the 2000s than in the 1990s (21.1% occurrence in the 1990s, 38.1% in the 2000s). In contrast, otter diet on the River Thames was predominantly fish, with birds only being consumed substantially in summer, meaning that dietary overlap between the two species was actually rather low (see Chapter 8, this volume).

More intriguingly, we found that, prior to the recovery of otters, mink on the River Thames were nocturnal (as expected, Dunstone 1993), but, in the presence of otters, were active mostly during the day (Fig. 7.2). General activity levels (as measured by the proportion of radio-tracking locations in which the mink was physically moving around) were similar between the two periods, averaging ≤ 5% to 20–25% active radio-locations[6] within a four-hour monitoring period. The timing of activity peaks, however, differed, occurring

[4] Dieldrin use in the UK was banned in 1989.

[5] $n = 20–297$ radio-locations (mean = 115) per individual.

[6] $n = 1–51$ radio-locations (mean = 13) per four-hour period, per individual.

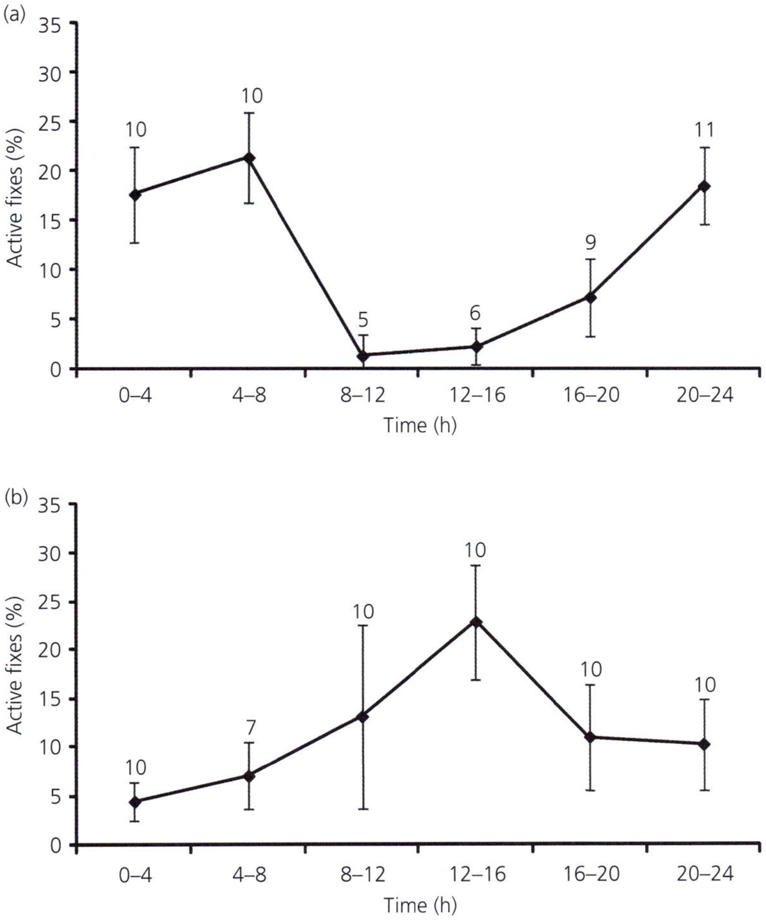

Figure 7.2 Daily activity patterns for mink on the River Thames (Oxfordshire, UK) (a) in the 1990s (in the absence of native competitors) and (b) in the 2000s (in the presence of otters and polecats). Data are mean percentage active radio-locations within a four-hour time period, error bars = SE, *n* = 12 (5 males, 7 females) and *n* =10 (4 males, 6 females) individuals in the 1990s and the 2000s respectively; number of individuals recorded in each four-hour time period shown above the error bars (not all individuals were recorded in all time periods). From Harrington et al. (2009a).

over night (between 8 pm and 8 am) in the 1990s, but during the middle of the day (between 12 am and 4 pm) in the 2000s; mink in the 1990s were very rarely recorded active around mid-day or early afternoon and likewise, mink in the 2000s were rarely active at night (Fig. 7.2). In short, mink may be able to avoid direct encounters with otters (when they are at more normal densities), at least in the short term[7].

An apparent abundance of mink that consume predominantly terrestrial prey in the presence of otters was, in fact, predicted by Bonesi and Macdonald (2004b), who suggested that mink and otters will co-exist where terrestrial resources are abundant. In Argentina, American mink co-exist with the native Patagonian otter *Lontra provocax*—co-existence, in this case, is probably permitted by a super-abundance of resources (macro-crustaceans in lakes, Fasola et al. 2009, and marine resources on the coast, Gomez et al. 2010). Indeed, it has been suggested that high densities of the invasive American signal crayfish *Pacifastacus leniusculuson* on UK rivers, might facilitate co-existence of mink and otter, but the generally low consumption of crayfish by mink in the Upper Thames Valley renders this suggestion unlikely (see Chapter 8, this volume). The changes in activity patterns of mink on the River Thames are unusual and were unexpected, and it is unknown whether mink adopt similar strategies in Argentina, although in southern Chile there is evidence that, where mink and otters use the same habitat

[7] Recent detailed analysis of the diving behaviour of mink in the Upper Thames Valley also showed that mink diving was restricted predominantly to daylight hours (Hays et al. 2007; Harrington et al. 2012a). It is not clear, however, whether daytime diving is an adaptation to the mink's poor underwater eyesight (Dunstone and Sinclair 1978) or to the presence of otters.

and consume the same prey, mink adopt diurnal activity (Medina-Vogel et al. 2013).

Between the 1990s and the 2000s, polecats also reappeared in the Upper Thames Valley, and established home ranges that overlap partially with those of the mink (Harrington and Macdonald 2008). Examining the extent of overlap, within and between species, revealed that mink shared their home ranges with 0.3–1.17 other mink (of both sexes) and 0.83–1.3 polecats; and also that polecats shared their home ranges with 0.6–1.0 polecats (of the opposite sex) and 1.6–2.0 mink. Polecats in our study, however, were nocturnally active (again, as expected), and so the diurnal activity of the mink, in the 2000s, meant that the two species were never active in the same area at the same time. We don't know if mink changed their activity patterns in response to otters, to polecats, or to some factor that we have not considered, but whatever the cause, it may now allow co-existence of the mink with both. Field demonstrations of temporal shifts in response to competitors are rare, although theoretically predictable (and there are similar cases among prey species in response to predators, e.g. Fenn and Macdonald 1995; Chapter 11, this volume). Among carnivores, temporal partitioning (where it is possible)[8] reduces the probability of inter-specific aggression, which is otherwise common within this guild (Macdonald and Sillero-Zubiri 2004; see e.g. Rasmussen and Macdonald 2012).

To elucidate a little further these inter-specific relationships, we experimentally tested the response of American mink to the odours of both otters and polecats (and an unfamiliar mink)[9]. Chemical communication is particularly important to solitary and nocturnal animals (Macdonald 1985), and residents of all three species leave prominently placed scats and anal gland secretions where others will find them (Hutchings and White 2000). The detection of sympatric competitors through chemical cues may allow individuals to assess competitive ability and avoid aggressive encounters (Gosling and Mckay 1990). For example, weasels *Mustela nivalis* are able to identify the odour of the larger stoat *M. erminea* and avoid it (Erlinge and Sandell 1988). Many animals avoid conspecifics on the basis of odour, but those that maintain territories may also be attracted to the odour of conspecifics, presumably to assess the level of threat present (Hurst et al. 1997). We predicted that mink would be attracted to the odour of unfamiliar mink, and that they would avoid the odour of otters (to reduce the chance of aggressive interactions with a larger, dominant competitor); we could not predict how mink might respond to the odour of polecats. In accordance with our first prediction, we found that mink were attracted to the odours of unfamiliar mink, as they were to the odours of polecats (Harrington et al. 2009b). But, contrary to our second prediction, mink showed little evidence (on average) of avoiding otter odour. Mink appeared to perceive a polecat as posing a threat similar to that of an unknown intruding mink (perhaps to be expected given their similarity in size). The response to otter scent was variable, which may suggest that a mink's perception of an otter depends on prior experience and/or local conditions.

Our study on the River Thames represents a brief snapshot of the ecological relationships between these three mustelid species. Powell and Zielinski (1983) suggest that co-existence between similar mustelids is never stable, but is temporary and fluctuates in response to fluctuations in their prey. We found that body weight, but not body length, of mink on the River Thames was significantly lower in the 2000s than in the 1990s (Table 7.1, Harrington et al. 2009a), suggesting that body condition (body mass in relation to body size) may have declined over this period. Reduced body condition may be related to the physiological costs of maintaining an activity pattern opposite to the species' natural rhythm (Kronfeld-Schor and Dayan 2003), or perhaps to stress associated with the presence of competitors (especially larger, dominant, and potentially lethal competitors). Whatever the cause of the decline in condition, further consequences are possible, for example, increased susceptibility to disease (Yamaguchi and Macdonald 2001; Harrington et al. 2012b) and/or reduced reproductive success; either or both of which could lead to population declines in the long term.

American mink and otters are both abundant in large rivers in Belarus (Sidorovich and Macdonald 2001); there, co-existence appears to be possible due to resource division and, probably critically, the use of the surrounding wetlands by mink. In Belarus, American

[8] Temporal partitioning is not always possible due to physiological, behavioural, or prey availability, constraints (Carothers and Jaksic 1984; Kronfeld-Schor and Dayan 2003), and may not be desirable if the costs (in terms of lost hunting opportunities) do not outweigh the benefits (of predator/competitor avoidance) (e.g. Cozzi et al. 2012).

[9] The response of wild mink to experimental odours was assessed by counting tracks (footprints) on rafts treated with anal gland secretions (of otter, polecat, and an unfamiliar mink), and compared with response to a 'control' raft, on two rivers in the Upper Thames Valley—the River Thames and the River Cherwell (n = 6 sites, four treatments presented at each site) (further experimental details in Harrington et al. 2009b).

Table 7.1 Body weight and length of American mink on the River Thames before (1990s), and after (2000s), the reappearance of native mustelids. Body length was measured from the head (the tip of the nose) to the base of the spine, along the back of the animal. Data are mean ± SD (n). Data from Harrington et al. (2009a).

Time period	Body weight	Body length
1990s	Male: 1368 ± 206.5 g (26) Female: 701 ± 98.5 g (15)	Male: 420 ± 19.9 mm (26) Female: 357 ± 19.0 mm (16)
2000s	Male: 1137 ± 214.5 g (12) Female: 629 ± 60.4 g (11)	Male: 413 ± 27.7 mm (13) Female: 364 ± 14.0 mm (11)

mink occupy a niche very similar to that of the (more terrestrial) polecat, which might explain their more damaging effects on polecats there. In the Upper Thames Valley, the surrounding landscape is predominantly open, grazed grassland, a habitat generally avoided by mink (Yamaguchi et al. 2003). Our radio-tracking data show that mink are generally restricted to the river itself and a narrow strip of riparian vegetation along the river bank, where they presumably hunt rabbits, small mammals, and birds. Until the recovery of the polecat, Dayan and Simberloff (1994) speculated that mink in the UK occupied the (at the time, empty) niche of the polecat. Now, in the presence of both otters and polecats (which, unlike in Belarus, are slightly larger on the River Thames than are mink), a more realistic view is probably that mink occupy a narrow niche squeezed between that of the otter (in the water) and the polecat (on the land). In this way, mink in the Upper Thames Valley could be considered to be restricted to a 'tightrope' by its competitors in much the same way that the water vole is restricted to a tightrope by its habitat (Barreto et al. 1998; Macdonald and Feber 2015: Chapter 15). The limited opportunity for spatial avoidance of competitors in this situation might well be the driving force behind the reversal of the mink's activity that, instead, allows temporal avoidance.

7.5 Hiding from competitors: have mink changed their marking behaviour?

Most of the evidence of a national decline in mink, coinciding with the recovery of the otter, is based on a decline in their scats rather than a decline in the mink themselves. Our observations, made during a larger study of the feasibility of mink control, suggest, however, that mink may reduce their marking behaviour in the presence of otters.

An extensive survey by the local Wildlife Trust had not detected otter spraint on the River Cherwell in the Upper Thames Valley in 2003 (Anon 2003), but in 2004–2005, an increase in both the abundance and the distribution of otter spraint suggested that otters were colonizing the river (Fig. 7.3)[10]. In parallel with the increase in otter spraints, we observed a decline in the distribution and abundance of mink scats, whereas the presence of mink tracks (detected on mink rafts)[11] increased after the first survey (use of rafts is often low initially, pers. obs.) and thereafter remained high (Fig. 7.3). The abundance of otter spraint varied throughout 2005–2006 (but was mostly high), while the abundance of mink scats remained low (Fig. 7.3).

Ideally, here, we would present a comparison of the numbers of mink with numbers of their scats in 2004–2005 and in 2005–2006. Unfortunately, because we had not designed the surveys to test explicitly the idea that scat deposition patterns in mink might change in the presence of potential competitors, we did not live-trap mink in 2004–2005. Nevertheless, in 2005–2006, we captured between nine and fifteen mink on *c*. 20 km of the River Cherwell (male:female ratio: 0.2:1.1), equating to about 0.5–0.8 mink/km, similar to the numbers of mink recorded on other rivers in lowland England in the absence of otters (0.86 mink/km, Dunstone and Birks 1985; 0.24, Halliwell and Macdonald 1996; 0.3, Yamaguchi and Macdonald 2003).

Prominently placed scats advertise the presence of a species, not only to their intended conspecific recipients, but also to potential predators (Wolff 2004; Hughes et al. 2010) and inter-specific competitors. Laboratory mice reduce their marking activity in the presence of predator odour (Roberts et al. 2001), and lone dispersing wolves do not scent mark (Rothman and Mech 1979). It is plausible that mink might also use this strategy to 'hide' from predators and/or

[10] Otters were not detected on the river during surveys in 2003, thus the increase in the occurrence of spraints observed between July and November 2004 probably represents the colonization of this river by otters. The low occurrence of otter spraints recorded in November 2005 was most likely an underestimate caused by loss of spraints due to severe floods immediately preceding the survey (rafts remain above the level of the water during flooding and at this time, 35% of rafts had otter spraint on top suggesting that otters were still present but that their spraints could not be found on the river bank).

[11] Mink rafts were designed as an efficient tool for the control of mink (see Section 7.6.1), but are also used for surveying (Harrington et al. 2008). Mink presence is detected by the presence of mink tracks on a tracking plate on the raft, and the relative abundance of mink is quantified as the percentage of rafts on which mink footprints are detected (% occupancy).

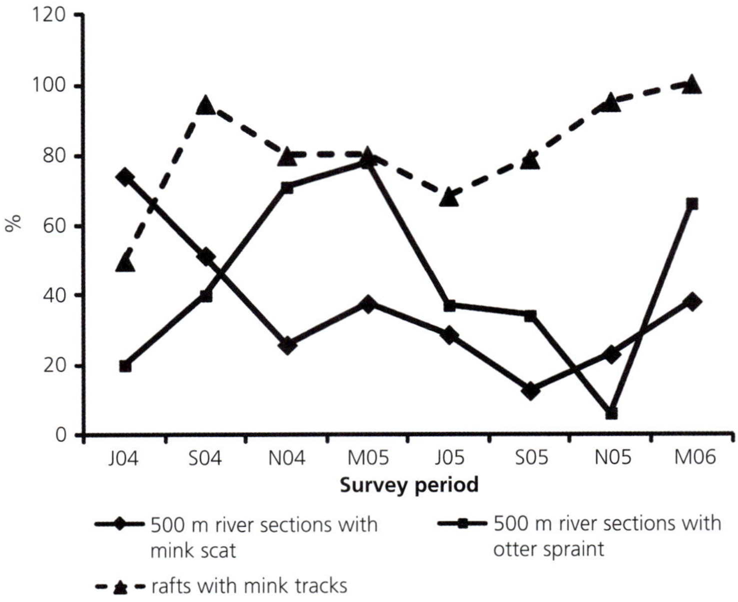

Figure 7.3 Mink scats and tracks on the R. Cherwell (Oxfordshire, UK) in relation to the occurrence of otter spraints. Data are % positive river sections (i.e. a 500 m river section where ≥ 1 mink scat or otter spraint was detected) or % positive rafts (a raft with mink tracks), $n = 36$ contiguous river sections or 20 rafts, in total. Surveys were carried out in Jul (J), September (S), November (N) and March (M).

competitors that are otherwise avoided (e.g. by adopting different temporal activity patterns). Although our observations were non-experimental, and we cannot be certain that numbers of mink on the River Cherwell did not decline, the consistently high incidence of mink tracks on rafts on the river, and the high numbers of mink trapped in 2005–2006, suggest that a decline in the level of marking by mink is a more parsimonious explanation than a decline in their numbers. Furthermore, the abundance of mink scat present on the River Cherwell after the arrival of the otter was similar to that found on other rivers within the catchment (all of which had otters), whereas the abundance of mink scats before the arrival of otters on the River Cherwell was almost twice as high (74% of 500 m river sections contained mink scats, $n = 36$ river sections) as those recorded anywhere else in the Upper Thames Valley in the 2000s (maximum 51%, median 19%). Likewise, although mink numbers apparently decreased following an increase in the abundance of otters on the River Teign (Bonesi et al. 2004; L. Bonesi, pers. comm.), the decrease in the number of mink scats observed was greater than the reduction in their numbers.

7.6 Is it possible to control American mink on farmland?

Even if mink are declining naturally, there is little evidence that they are doing so fast enough to permit the recovery of the water vole (see Macdonald and Feber 2015: Chapters 14 and 15). While mink appear to coexist with other mustelids, the problem regarding impacts on native prey species remains. The most crucial of these is the water vole. Water vole restoration has demonstrated that, without mink control, water voles will not survive. The harsh reality is, then, that if society wants water voles, it will have to accept and implement the killing of American mink to an extent which reduces their numbers. So the questions become: Is it possible to reduce mink numbers in lowland England? More specifically: Can it be done humanely? Can it be done efficiently? And at what level can it be effective?

7.6.1 Humane mink control using mink rafts

Mink are easily trapped in the UK, where they are generally restricted to linear riparian habitats. The traditional live-trapping method using cage traps set on riverbanks is time consuming and requires considerable field skill (Yamaguchi et al. 2002), but the use of traps on floating rafts designed for mink by the Game and Wildlife Conservation Trust (formerly the Game Conservancy Trust) offers a simple and highly effective approach (Fig. 7.4; Reynolds et al. 2004; Porteus et al. 2012). Mink rafts incorporate a 'tracking plate' that allows users first to detect the presence of mink (from footprints), and then to set live traps on those particular rafts, thus increasing efficiency (for the trappers) and

Figure 7.4 A mink raft deployed in the river. Mink, being curious creatures, climb onto the rafts from the water and walk through the tunnel, underneath which is a clay surface (see Reynolds et al. 2004 for construction details) that allows the tracks of any mink walking through to be recorded (see inset). Mink rafts were designed as an efficient trapping method (cage traps are only placed within the tunnel when mink tracks are detected), but they can also be used for monitoring the relative abundance of mink by recording the proportion of rafts (set at regular intervals—in our case, one every km—over a specified stretch of river) that have mink tracks (Harrington et al. 2008). Photograph © L.A. Harrington.

decreasing the risk of capture (for other species). Lethal traps, commonly used by gamekeepers for stoats and weasels, are not recommended for use on rafts because they would be inhumane for mink, would risk capturing non-target animals, and arguably, may be illegal in areas where polecats are known to be present (Macdonald and Baker 2006; J. Birks, pers. comm.). Polecats or other non-target captures in live traps should be released; mink should be quickly dispatched using an appropriate and humane method (guidelines for landowners are produced by the Game and Wildlife Conservation Trust). Indeed, Jonathon Reynolds of the GWCT, was awarded the first UFAW Wild Animal Welfare Award in 2004 for the development of the mink raft, and rafts are now deployed across the country from the northern highlands of Scotland to Norfolk, Somerset, and Dorset (Macdonald and Burnham 2008).

7.6.2 Modelling to identify an efficient mink control strategy

Control programmes are usually costly (e.g. Moore et al. 2003), success may be difficult to predict, and the optimal strategy to take is not always obvious. Modelling can help managers to assess the chances of success before making a financial commitment, and to experiment with different approaches in ways that would not be possible in the field. Bonesi et al. (2007) used a spatially explicit, stochastic, individual-based model to simulate mink living, dispersing, and encountering traps in the landscape. Assuming that adult male mink are more likely to encounter traps during the mating season (January–March), that adult females have a low probability of being trapped during lactation and weaning (April–June), and that juveniles are more likely to be trapped during dispersal (August–October), Bonesi et al. (2007) ran model simulations imitating conditions within a part of the Upper Thames Valley (containing 127 km of river) over a hypothetical 20 years. To achieve an arbitrary reduction in mink numbers of 80% (assuming that mink, at the start, were at carrying capacity), the model predicted that it would be necessary to trap every year, for at least three months each year. The model also predicted that annual trapping over a combination of three separate periods representing the mating season, the end of the dispersal season, and the beginning of the stable winter season (modelled as January, October, and November), is best for keeping mink at low densities.

Contrary to common suggestions that trapping during the dispersal season is a waste of effort because most of the juveniles it removes will die anyway, this modelling exercise suggested that targeting immigrating juvenile mink, in addition to resident adults, is important. However, the periods of trapping that worked best when trappability (the likelihood of a mink being trapped, which depends on the behaviour of both the mink and of the trapper) was low were not necessarily those that worked best when trappability was high (Fig. 7.5). At low levels of trappability, trapping during

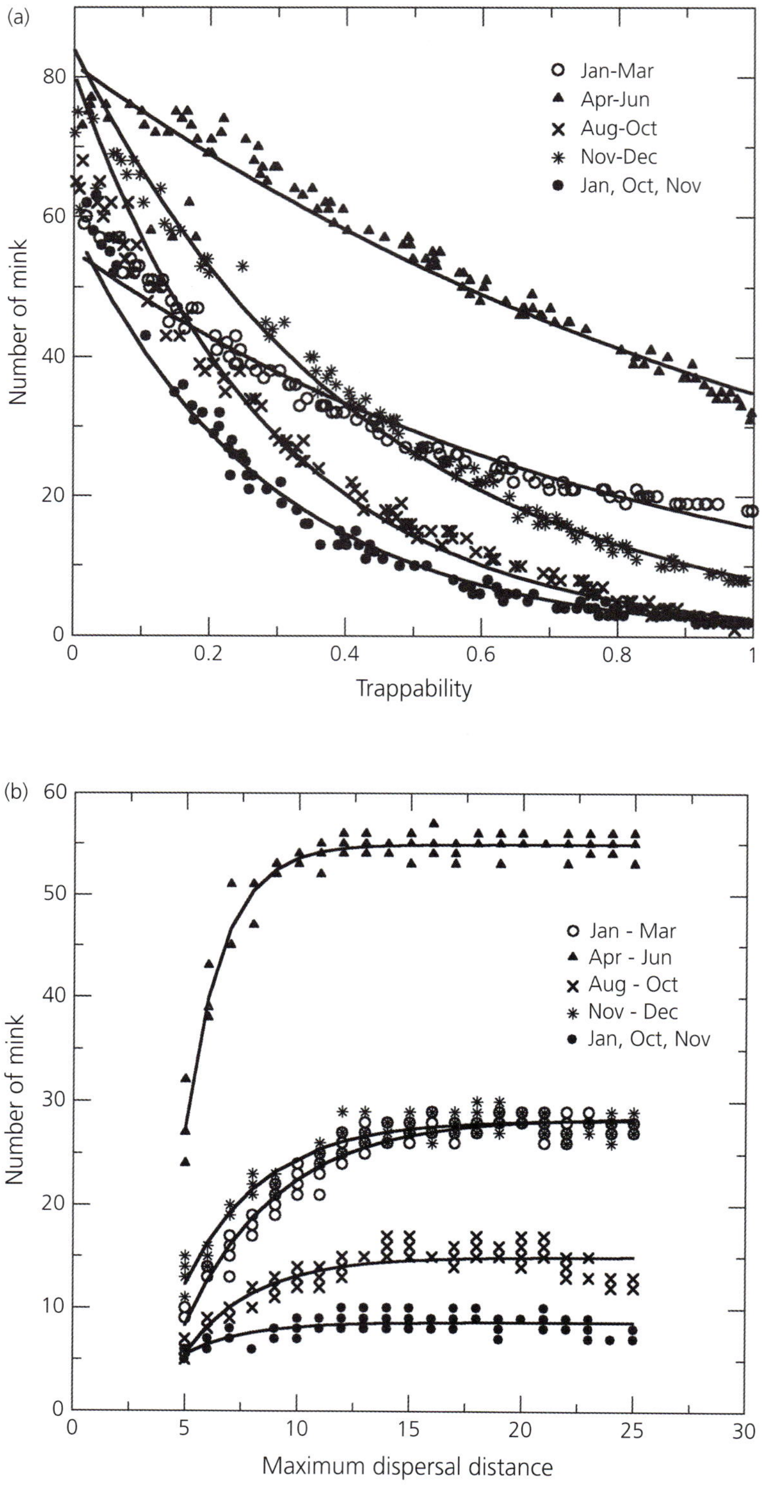

Figure 7.5 Predicted number of mink in the control area as a function of (a) their trappability and (b) immigrant pressure, in relation to the trapping period adopted. Exponential curves were fitted to each of the different periods of trapping in order to better visualize the relationships between the different data sets. Immigrant pressure is expressed as the dispersal distance divided by two to take into account the sinuosity of rivers. The carrying capacity of the whole area was assumed to be 0.7 mink/km, therefore, under the aim of achieving an 80% reduction in mink densities, the minimum monthly acceptable density was 0.13 mink/km or 17 mink over the whole area (shown by the dotted line in (a)). From Bonesi et al. (2007), reproduced with permission from Elsevier.

the mating or dispersal season produced similar results, whereas at high levels of trappability, trapping during the dispersal season became more effective. Similarly, when immigration pressure is low, a number of different trapping schedules are predicted to be equally effective, but as immigration pressure increases, differences between trapping schedules emerge, and some do not reduce numbers of mink at all (Fig. 7.5). Trapping in January, October, and November was, in all cases, predicted to be optimal, keeping mink numbers well below the desired 20% of carrying capacity, even under high immigration pressure (Fig. 7.5).

7.6.3 Testing the effectiveness of mink control in the field

Of course, a model is only a model, and its predictions depend not only on the values of life-history parameters it uses but also on other factors, such as habitat, that were not included in Bonesi et al.'s (2007) model. Harrington et al. (2009c) used a large-scale field experiment to test the effectiveness of Bonesi's predicted mink control strategy in reducing the relative abundance of mink in reality. We monitored the relative abundance of mink at three pairs of randomly selected sites (each a 20 km stretch of river within the Upper Thames Valley)[12] over four years, using mink rafts (*c.* 1 raft/km) for both monitoring and trapping. The control strategy (trapping for a minimum of three months, over the mating season, the end of the dispersal season, and the beginning of winter) was implemented at one of each pair (the removal site) after the first year of monitoring.

There were considerable differences in relative abundance of mink among the pairs of rivers (between 11 and 95 mean % occupancy, measured as the percentage of rafts with mink tracks, $n = 20$ rafts per site, 3–4 surveys per site), as well as variation within some sites (coefficient of variation = 0.15–0.61, $n = 6$ sites), yet we were able to show that at two of the three paired sites (pairs 1 and 3; Fig. 7.6), the relative abundance of mink declined significantly at the removal site (relative to the non-removal site), following the initiation of trapping. The response observed at pair 2 was not statistically significant, largely due to the high level of variation in occupancy levels before trapping began, but occupancy levels remained low once trapping had begun (Fig. 7.6b). At all removal sites, the relative abundance of mink was low (< 20% occupancy) but variable during the removal period, including high values of 40–50% occupancy experienced at two of the sites for short periods (Fig. 7.6). We estimated that during the removal period, at removal sites, there were between < 1 and 2–5 mink present per site, which relates to about 0.03–0.2 mink/km. Assuming perfect detectability[13] (i.e. that we would always detect mink if they were there), we estimated the removal sites to be 'mink-free' for approximately one third of the year.

Unfortunately, for a small water vole population, the presence of even one mink, for even a short period of time, could be devastating (see e.g. Barreto and Macdonald 2000). Our field experiment was run in parallel with a project investigating water vole reintroductions (Moorhouse et al. 2009; see Macdonald and Feber 2015: Chapter 14) and, at four sub-sites within our three mink removal sites, captive-bred water voles were reintroduced during the second year of mink control. Because these were very small, newly released propagules of a highly threatened species, we decided to trap mink as soon as they appeared at a water vole release site, rather than to follow the proposed 3-month trapping regime rigidly. Such rapid intervention trapping was necessary at one site during one summer, and we believe that it was crucial to the successful establishment of water voles at this site. Key considerations at this scale are, not only how many mink come into the site, but also how quickly can they be detected and removed. Continual monitoring, and the ability to react very quickly to the detection of mink signs, is absolutely crucial.

The feasibility of mink removal on a small scale, demonstrated in the Upper Thames, suggests not only that the protection of small sites for water vole restoration is possible (given a dedicated field crew and adequate finance), but also that many such small-scale projects, each creating a local 'sink' effect, could potentially add up to regional control (Reynolds and Tapper 1996). Theoretically, given sufficient resources and willingness, and well coordinated national-scale planning (Reynolds (2009) discusses the benefits of a national plan for mink control), it might even be possible to link adjacent 'control regions', with the eventual aim of achieving removal of mink from the UK. A step in this direction is indeed being taken in northern

[12] Pair 1: R. Thames (NR) and Lower R. Windrush (R); Pair 2: Upper R. Evenlode (NR) and Upper R. Windrush (R); Pair 3: R. Cherwell (NR) and Lower R. Evenlode (R). NR = non-removal site, R = removal site.

[13] Using occupancy models, Reynolds et al. (2010) provide support for this assumption by showing that the risk of failing to detect a mink that was present would be very low (< 5%) given a similar mink raft spacing.

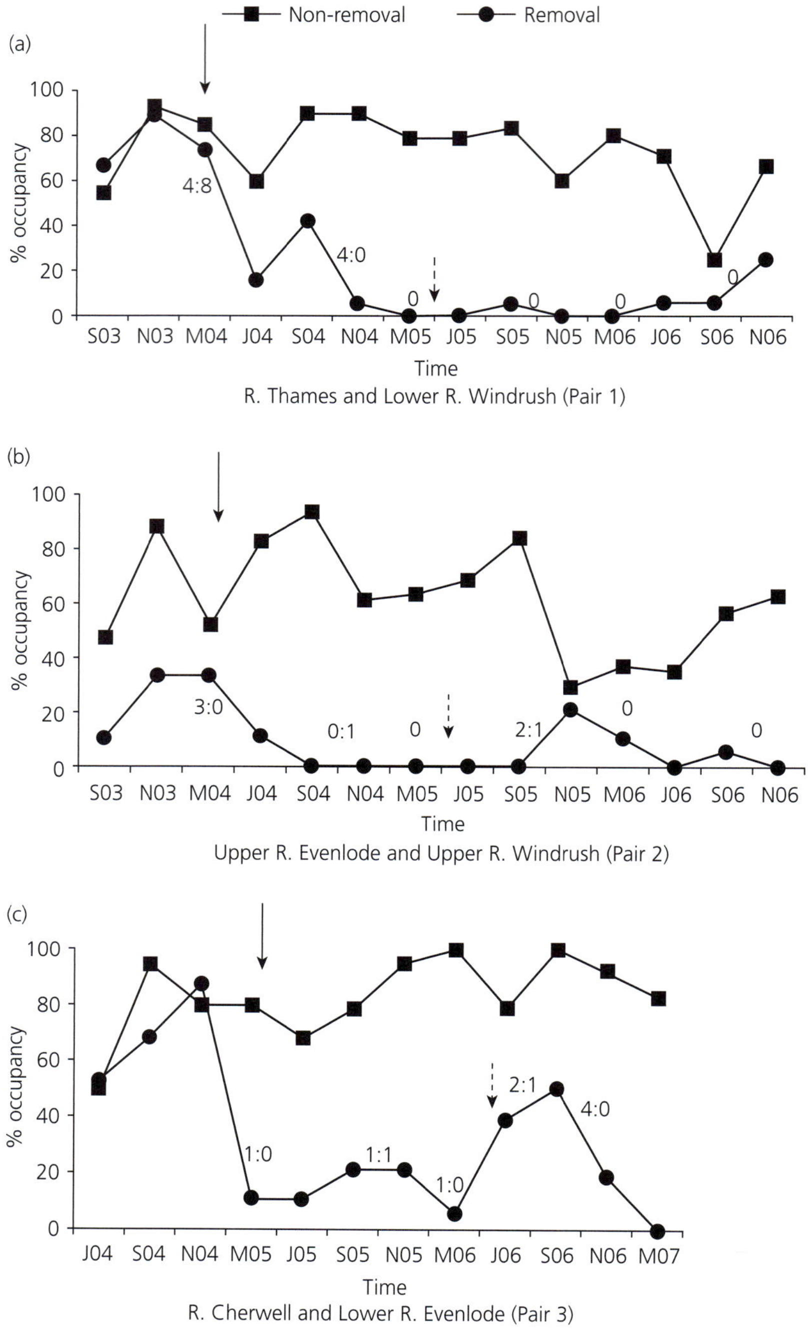

Figure 7.6 The effect of removal trapping on the relative abundance of mink. Graphs show the relative abundance of mink (% occupancy, derived from raft surveys, see Fig. 7.3) over time (S03 = September 2003, N = November, M = March, J = June) for each of three pairs of river sites. Removal trapping was carried out over one month in March and over two months in October–November, and repeated every year, in accordance with the optimal mink trapping strategy as defined by Bonesi et al. (2007) (numbers of individuals removed per trap session are shown on the graph as m:f). The start of trapping is marked by the solid arrows, water vole releases by the dashed arrows. Note that the decline observed on the Lower River Evenlode occurred just before trapping was started. The response of mink to trapping was statistically significant for pairs 1 ($p = 0.001$) and 3 ($p = 0.022$) but not for pair 2 ($p = 0.075$). From Harrington et al. (2009c), reproduced with permission from Elsevier.

Scotland where mink were successfully removed from a core area of 10,000 km^2, centred on the Cairngorms National Park, within a large community-based program (Bryce et al. 2011)—and, under the auspices of the Scottish Mink Initiative, this and neighbouring mink control efforts are being linked, with the aim of creating and maintaining a continuous mink-free area over much of northern Scotland. These initiatives follow the success of the Hebridean Mink Project, in the Outer Hebrides (off the west coast of Scotland), where mink are now scarce, following an intensive 11-year trapping programme[14].

7.7 Concluding remarks

The relationships between alien mink and native mustelids are far from simple, and are likely to depend crucially on local specificities, such as the type and availability of the habitat, its spatial configuration, and perhaps also local subtleties in size differences between those of similar size. In the UK, it is not certain whether or not mink really are declining at a national level or whether their marking behaviour has simply become less conspicuous. Further, although we can speculate about the possible implications of the apparent declining body condition of potentially stressed mink, we cannot rule out the possibility that the mink population really is declining in the long term.

In the meantime, local control projects are ever increasing in number and extent. The tools and the knowledge to tackle the 'mink problem' are now available at either local or regional scales, although long-term success will require extreme diligence, continuous commitment, and probably increasing community support. Paramount in these endeavours will be attention to both humaneness and (with difficult ethical choices in mind) effectiveness: if society wants water voles, then mink will have to be killed, but if we have to kill mink, we must ensure that it is done well.

Acknowledgements

We are extremely grateful to Andrew Harrington for his part in this work, and to the many volunteers who helped gather the data. We also thank Elaine Fraser, Laura Bonesi, Graham Scholey, Alastair Driver and, especially, Carolyn King for their insightful comments on an early draft.

[14] Efforts are currently underway to remove the last few individuals from the islands (<http://www.snh.gov.uk/>)

References

Abramov, A.V. (2000). A taxonomic review of the genus *Mustela* (Mammalia, Carnivora). *Zoosystematica Rossica*, **8**, 357–364.

Anon. (2003). Water vole survey of the River Cherwell and its tributaries 2002–2003. Ecoconsult Ltd., Oxfordshire.

Banks, P.B., Nordstrom, M., Ahola, M., Salo, P., Fey, K., and Korpimaki, E. (2008). Impacts of alien mink predation on island vertebrate communities of the Baltic Sea Archipelago: review of a long-term experimental study. *Boreal Environment Research*, **13**, 3–16.

Barreto, G.R. and Macdonald, D.W. (2000). The decline and local extinction of a population of water voles, *Arvicola terrestris*, in southern England. *Zeitschrift für Säugetierkunde*, **65**, 110–120.

Barreto, G.R., Macdonald, D.W., and Strachan, C. (1998). The tightrope hypothesis: an explanation for plummeting water vole numbers in the Thames catchment. In R.G. Bailey, P.V. Jose, and B.R. Sherwood, eds., *United Kingdom Floodplains*, pp. 311–327. Westbury Academic and Scientific Publishing, Otley.

Bellingham, P.J., Towns, D.R., Cameron, E.K., Davis, J.J., Wardle, D.A., Wilmshurst, J.M., and Mulder, C.P.H. (2010). New Zealand island restoration: seabirds, predators, and the importance of history. *New Zealand Journal of Ecology*, **34**, 115–136.

Birks, J.D.S. (2008). *The Polecat Survey of Britain 2004–2006. A report on the polecat's distribution, status and conservation*. The Vincent Wildlife Trust, Herefordshire, UK.

Birks, J.D.S. and Kitchener, A.C. (1999). *The Distribution and Status of the Polecat Mustela putorius in Britain in the 1990s*. The Vincent Wildlife Trust, London.

Bonesi, L., Chanin, P., and Macdonald, D.W. (2004). Competition between Eurasian otter Lutra lutra and American mink *Mustela vison* probed by niche shift. *Oikos*, **106**, 19–26.

Bonesi, L. and Macdonald, D.W. (2004a). Impact of released Eurasian otters on a population of American mink: a test using an experimental approach. *Oikos*, **106**, 9–18.

Bonesi, L. and Macdonald, D.W. (2004b). Differential habitat use promotes sustainable coexistence between the specialist otter and the generalist mink. *Oikos*, **106**, 509–519.

Bonesi, L. and Palazon, S. (2007). The American mink in Europe: status, impacts and control. *Biological Conservation*, **134**, 470–483.

Bonesi, L., Rushton, S.P., and Macdonald, D.W. (2007). Trapping for mink control and water vole survival: identifying key criteria using a spatially explicit individual based model. *Biological Conservation*, **136**, 636–650.

Bonesi, L., Strachan, R., and Macdonald, D.W. (2006). Why are there fewer signs of mink in England? Considering multiple hypotheses. *Biological Conservation*, **130**, 268–277.

Brinck, C., Erlinge, S., and Sandell, M. (1983). Anal sac secretion in mustelids. *Journal of Chemical Ecology*, **9**, 727–745.

Bryce, R., Oliver, M.K., Davies, L., Gray, H., Urquhart, J., and Lambin, X. (2011). Turning back the tide of American mink invasion at an unprecedented scale through community participation and adaptive management. *Biological Conservation*, **144**, 575–583.

Brzeziński, M., Romanowski, J., Zmihorski, M., and Karpowicz, K. (2010). Muskrat (*Ondatra zibethicus*) decline after the expansion of American mink (*Neovison vison*) in Poland. *European Journal of Wildlife Research*, **56**, 341–348.

Carothers, J.H. and Jaksic, F.M. (1984). Time as a niche difference: the role of interference competition. *Oikos*, **42**, 403–406.

Clode, D. and Macdonald, D.W. (1995). Evidence for food competition between mink (*Mustela vison*) and otter (*Lutra lutra*) on Scottish islands. *Journal of Zoology*, **237**, 435–444.

Cozzi, G., Broekhuis, F., McNutt, J.W., Turnbull, L.A., Macdonald, D.W., and Schmid, B. (2012). Fear of the dark or dinner by moonlight? Reduced temporal partitioning among Africa's large carnivores. *Ecology*, **93**, 2590–2599.

Craik, J.C.A. (1997). Long-term effects of North American mink Mustela vison on seabirds in western Scotland. *Bird Study*, **44**, 303–309.

Crawford, A. (2010). *Fifth Otter Survey of England 2009–2010*. Environment Agency, Bristol.

Davison, A., Griffiths, H.I., Brookes, R.C., Maran, T., Macdonald, D.W., Sidorovich, V.E., et al. (2000). Mitochondrial DNA and palaeontological evidence for the origins of endangered European mink, *Mustela lutreola*. *Animal Conservation*, **4**, 245–355.

Dayan, T. and Simberloff, D. (1994). Character displacement, sexual dimorphism, and morphological variation among British and Irish mustelids. *Ecology*, **75**, 1063–1073.

Dunstone, N. (1993). *The Mink*. T and AD Poyser Ltd., London.

Dunstone, N. and Birks, J.D.S. (1985). The comparative ecology of coastal, riverine and lacustrine mink *Mustela vison* in Britain. *Zeitschrift fur Angewandte Zoologie*, **72**, 52–70.

Dunstone, N. and Sinclair, W. (1978). Comparative aerial and underwater visual acuity of the mink (*Mustela vison* Schreber). *Animal Behaviour*, **26**, 14–21.

Erlinge, S. and Sandell, M. (1986). Seasonal changes in the social organization of male stoats, Mustela erminea: an effect of shifts between two decisive resources. *Oikos*, **47**, 57–62.

Erlinge, S. and Sandell, M. (1988). Coexistence of stoat, *Mustela erminea*, and weasel, *M. nivalis*: social dominance, scent communication, and reciprocal distribution. *Oikos*, **53**, 242–246.

Errington, P.L. (1946). Predation and vertebrate populations. *Quarterly Review of Biology*, **21**, 144–177.

Fasola, L., Chehebar, C., Macdonald, D.W., Porro, G., and Cassini, M.H. (2009). Do alien North American mink compete for resources with native South American river otter in Argentinian Patagonia? *Journal of Zoology*, **277**, 187–195.

Fenn, M.G.P. and Macdonald, D.W. (1995). Use of middens by red foxes: risk reverses rhythms of rats. *Journal of Mammalogy*, **76**, 130–136.

Fraser, E.J., Lambin, X., Travis, J.M.J., Harrington, L.A., Palmer, S.C.F., Bocedi, G. and Macdonald, D.W. (2015). Range expansion of an invasive species through a heterogeneous landscape – the case of American mink in Scotland. Diversity and Distributions doi:10.111/ddi.12303.

Genovesi, P. (2005). Eradications of invasive alien species in Europe: a review. *Biological Invasions*, **7**, 127–133.

Gomez, J.J., Gozzi, A.C., Macdonald, D.W., Gallo, E., Centron, D., and Cassini, M.H. (2010). Interactions of exotic and native carnivores in an ecotone, the coast of the Beagle Channel, Argentina. *Polar Biology*, **33**, 1371–1378.

Gosling, L.M. and Mckay, H.V. (1990). Competitor assessment by scent matching: an experimental test. *Behavioral Ecology and Sociobiology*, **26**, 415–420.

Halliwell, E.C. and Macdonald, D.W. (1996). American mink *Mustela vison* in the Upper Thames catchment: relationship with selected prey species and den availability. *Biological Conservation*, **76**, 51–56.

Harrington, L.A., Gelling, M., Simpson, V., Harrington, A., and Macdonald, D.W. (2012b), Notes on the health status of free-living, non-native, American mink, *Neovison vison*, in southern England. *European Journal of Wildlife Research*, **58**, 875–880.

Harrington, L.A., Harrington, A.L., Hughes, J., Stirling, D., and Macdonald, D.W. (2010). The accuracy of scat identification in distribution surveys: American mink, *Neovison vison*, in the northern highlands of Scotland. *European Journal of Wildlife Research*, **56**, 377–384.

Harrington, L.A., Harrington, A.L., and Macdonald, D.W. (2008). Estimating the relative abundance of American mink *Mustela vison* on lowland rivers: evaluation and comparison of two techniques. *European Journal of Wildlife Research*, **54**, 79–87.

Harrington, L.A., Harrington, A.L., and Macdonald, D.W. (2009b). The smell of new competitors: the response of American mink, *Mustela vison*, to the odours of otter, *Lutra lutra* and polecat, *M. putorius*. *Ethology*, **115**, 421–428.

Harrington, L.A., Harrington, A.L., Moorhouse, T., Gelling, M., Bonesi, L., and Macdonald, D.W. (2009c). American mink control on inland rivers in southern England: an experimental test of a model strategy. *Biological Conservation*, **142**, 839–849.

Harrington, L.A., Harrington, A.L., Yamaguchi, N., Thom, M., Ferreras, P., Windham, T.R., and Macdonald, D.W. (2009a). The impact of native competitors on an alien invasive: temporal niche shifts to avoid inter-specific aggression? *Ecology*, **90**, 1207–1216.

Harrington, L.A., Hays, G., Fasola, L., Harrington, A.L., Righton, D., and Macdonald, D.W. (2012a). Diving ability and behaviour of a small-bodied, semi-aquatic mammal in the wild. *Journal of Mammalogy*, **93**, 198–210.

Harrington, L.A. and Macdonald, D.W. (2008). Spatial and temporal relationships between invasive American mink and native European polecats in the southern United Kingdom. *Journal of Mammalogy*, **89**, 991–1000.

Hays, G.C., Forman, D.W., Harrington, L.A., Harrington, A.L., Macdonald, D.W., and Righton, D. (2007). Recording the free-living behaviour of small-bodied, shallow-diving animals with data loggers. *Journal of Animal Ecology*, **76**, 183–190.

Hughes, N.K., Korpimaki, E., and Banks, P.E. (2010). The predation risks of interspecific eavesdropping: weasel-vole interactions. *Oikos*, **119**, 1210–1216.

Hurst, J.L., Gray, S.J., Davey, P., Young, D., Corbishley, J., and Dawson, C. (1997). Social interaction alters attraction to competitor's odour in the mouse *Mus spretus* Lataste. *Animal Behaviour*, **54**, 941–953.

Hutchings, M.R. and White, P.C.L. (2000). Mustelid scent-marking in managed ecosystems: implications for population management. *Mammal Review*, **30**, 157–170.

Ibarra, J.T., Fasola, L., Macdonald, D.W., Rozzi, R., and Bonacic, C. (2009). Invasive American mink *Mustela vison* in wetlands of the Cape Horn Biosphere Reserve, southern Chile: what are they eating? *Oryx*, **43**, 87–90.

Jefferies, D.J. (2003). *The water vole and mink survey of Britain 1996–1998 with a history of the long-term changes in the status of both species and their causes*. The Vincent Wildlife Trust, London.

King, C.M., McDonald, R.M., Martin, R.D., and Dennis, T. (2009). Why is eradication of invasive mustelids so difficult? *Biological Conservation*, **142**, 806–816.

Kronfeld-Schor, N. and Dayan, T. (2003). Partitioning of time as an ecological resource. *Annual Review of Ecology, Evolution and Systematics*, **34**, 153–181.

Kruuk, H. (2006). *Otters. Ecology, Behaviour and Conservation*. Oxford University Press, Oxford.

Kurose, N., Abramov, A.V., and Masuda, R. (2008). Molecular phylogeny and taxonomy of the genus Mustela (Mustelidae, Carnivora), inferred from mitochondrial DNA sequences: New perspectives on phylogenetic status of the back-striped weasel and American mink. *Mammal Study*, **33**, 25–33.

Linn, I.J. and Stevenson, J.H.F. (1980). Feral mink in Devon. *Nature in Devon*, **1**, 7–27.

Macdonald, D.W. (1985). The carnivores: order Carnivora. In R.E. Brown and D.W. Macdonald, eds., *Social Odours in Mammals, Vol. 2*, pp. 619–689. Clarendon Press, Oxford.

Macdonald, D.W. and Baker, S. (2006). *The State of Britain's Mammals 2006*. People's Trust for Endangered Species, London.

Macdonald, D.W., Bryce, J.M., and Thom, M.D. (2001). Introduced mammals: Do carnivores and herbivores usurp native species by different mechanisms? In H.-J. Pelz, D.P. Cowan, and C.J. Feare, eds., *Advances in Vertebrate Pest Management II*, pp. 11–44. Filander Verlag, Fürth.

Macdonald, D.W. and Burnham, D. (2008). *The State of Britain's Mammals 2008*. People's Trust for Endangered Species, London.

Macdonald, D.W. and Burnham, D. (2010). *The State of Britain's Mammals—a Focus on Invasive Species*. People's Trust for Endangered Species, London.

Macdonald, D.W. and Feber, R.E., eds. (2015). *Wildlife Conservation on Farmland. Managing for Nature on Lowland Farms*. Oxford University Press, Oxford.

Macdonald, D.W. and Harrington, L.A. (2003). The American mink: the triumph and tragedy of adaptation out of context. *New Zealand Journal of Zoology*, **30**, 421–441.

Macdonald, D.W., King, C.M., and Strachan, R. (2007). Introduced species and the line between biodiversity conservation and naturalistic eugenics. In D. Macdonald and K. Service, eds., *Key Topics in Conservation Biology*, pp. 186–205. Blackwell Publishing, Oxford.

Macdonald, D.W., Sidorovich, V.E., Maran, T., and Kruuk, H. (2002). *European Mink, Mustela lutreola: Analyses for Conservation*. Wildlife Conservation Research Unit, Oxford.

Macdonald, D.W. and Sillero-Zubiri, C. (2004). Dramatis personae. In D.W. Macdonald, and C. Sillero-Zubiri, C., eds., *Biology and Conservation of Wild Canids*, pp. 3–36. Oxford University Press, Oxford.

Macdonald, D.W. and Strachan, R. (1999). *The Mink and the Water Vole: Analyses for Conservation*. Wildlife Conservation Research Unit and the Environment Agency, Oxford.

McDonald, R.A., O'Hara, K. and Morrish, D.J. (2007). Decline of invasive mink (*Mustela vison*) is concurrent with recovery of native otters (*Lutra lutra*). *Diversity and Distributions*, **13**, 92–98.

Medina-Vogel, G., Barros, M., Organ, J.F., and Bonesi, L. (2013). Coexistence between the southern river otter and the alien invasive North American mink in marine habitats of southern Chile. *Journal of Zoology*, **290**, 27–34.

Melbourne, B.A., Cornell, H.V., Davies, K.F., et al. (2007). Invasion in a heterogeneous world: resistance, coexistence or hostile takeover? *Ecology Letters*, **10**, 77–94.

Mooney, H.A. (2005). Invasive alien species: the nature of the problem. In H.A. Mooney, R.N. Mack, J.A. McNeely, L.E. Neville, P.J. Schei, and J.K. Waage, eds., *Invasive Alien Species. A New Synthesis*, pp. 1–15. Island Press, Washington.

Moore, N.P., Roy, S.S., and Halyar, A. (2003). Mink (*Mustela vison*) eradication to protect ground-nesting birds in the Western Isles, Scotland, United Kingdom. *New Zealand Journal of Zoology*, **30**, 443–452.

Moorhouse, T.P., Gelling, M., and Macdonald, D.W. (2009). Effects of habitat quality upon reintroduction success in water voles: evidence from a replicated experiment. *Biological Conservation*, **142**, 53–60.

Porteus, T., Short, M., Richardson, S., and Reynolds, J. (2012). Empirical development of strategy for the control of invasive American mink by trapping. *European Journal of Wildlife Research*, **58**, 403–413.

Powell, R.A. (1979). Mustelid spacing patterns: variations on a theme by Mustela. *Zeitschrift für Tierpsychologie*, **50**, 153–165.

Powell, R.A. and Zielinski, W.J. (1983). Competition and coexistence in mustelid communities. *Acta Zoologica Fennica*, **174**, 223–227.

Rasmussen, G.S.A. and Macdonald, D.W. (2012). Masking of the zeitgeber: African wild dogs mitigate persecution by balancing time. *Journal of Zoology*, **286**, 232–242.

Reynolds, J.C. (2009). American Mink: the art of the possible and national aspirations for biodiversity. *International Urban Ecology Review*, **4**, 74–82.

Reynolds, J.C., Porteus, T.A., Richardson, S.M., Leigh, R.J., and Short, M.J. (2010). Detectability of American mink using rafts to solicit field signs in a population control context. *Journal of Wildlife Management*, **74**, 1601–1606.

Reynolds, J.C., Short, M.J., and Leigh, R.J. (2004). Development of population control strategies for mink Mustela vison, using floating rafts as monitors and trap sites. *Biological Conservation*, **120**, 533–543.

Reynolds, J.C. and Tapper, S.C. (1996). Control of mammalian predators in game management and conservation. *Mammal Review*, **26**, 127–156.

Roberts, S.C., Gosling, L.M., Thornton, E.A., and McClung, J. (2001). Scent-marking by male mice under the risk of predation. *Behavioral Ecology*, **12**, 698–705.

Robinson, R.E. (1901). *Sam Lovel's Boy*. Houghton, Mifflin and Co., Boston, New York.

Rothman, R.J. and Mech, L.D. (1979). Scent-marking in lone wolves and newly formed pairs. *Animal Behaviour*, **27**, 750–760.

Schoener, T.W. (1986). Resource partitioning. In J. Kikkawa and D.J. Anderson, eds., *Community Ecology: Patterns and Processes*, pp. 91–126. Blackwell Scientific, California, USA.

Shea, K. and Chesson, P. (2002). Community ecology theory as a framework for biological invasions. *Trends in Ecology and Evolution*, **117**, 170–176.

Sidorovich, V., Kruuk, H., and Macdonald, D.W. (1999). Body size, and interactions between European and American mink (*Mustela lutreola* and *M. vison*) in Eastern Europe. *Journal of Zoology*, **248**, 521–527.

Sidorovich, V. and Macdonald, D.W. (2001). Density dynamics and changes in habitat use by the European mink and other native mustelids in connection with the American mink expansion in Belarus. *Netherlands Journal of Zoology*, **51**, 107–126.

Simberloff, D. (1997). Eradication. In D. Simberloff, D.C. Schmitz, and T.C. Brown, eds., *Strangers in Paradise. Impact and Management of Nonindigenous Species in Florida*, pp. 221–228. Island Press, Washington D.C.

Simberloff, D. (2001). Eradication of island invasives: practical actions and results achieved. *Trends in Ecology and Evolution*, **16**, 272–273.

Strachan, R. and Jefferies, D.J. (1996). *Otter Survey of England 1991–1994*. Vincent Wildlife Trust, London.

Thom, M.D., Harrington, L.A., and Macdonald, D.W. (2004a). Why are American mink sexually dimorphic? A role for niche separation. *Oikos*, **105**, 525–535.

Thom, M.D., Johnson, D.D.P., and Macdonald, D.W. (2004b). The evolution and maintenance of delayed implantation in the Mustelidae (Carnivora: Mammalia). *Evolution*, **58**, 175–183.

Veitch, C.R. and Clout, M.N., eds. (2002). *Turning the tide: the eradication of invasive species*. Proceedings of the International Conference on Eradication of Island Invasives, IUCN SSC Invasive Species Specialist Group. Gland, Switzerland.

Wolff, J.O. (2004). Scent marking by voles in response to predation risk: a field-laboratory validation. *Behavioral Ecology*, **15**, 286–289.

Yamaguchi, N. and Macdonald, D.W. (2001). Detection of Aleutian disease antibodies in feral American mink in southern England. *Veterinary Record*, **149**, 485–488.

Yamaguchi, N. and Macdonald, D.W. (2003). The burden of co-occupancy: intraspecific resource competition and spacing patterns in American mink, *Mustela vison*. *Journal of Mammalogy*, **84**, 1341–1355.

Yamaguchi, N., Rushton, S.P., and Macdonald, D.W. (2003). Habitat preferences of feral American mink in the Upper Thames. *Journal of Mammalogy*, **84**, 1356–1373.

Yamaguchi, N., Sarno, R.J., Johnson, W.E., O'Brien, S.J., and Macdonald, D.W. (2004). Multiple paternity and reproductive tactics of free-ranging American minks, *Mustela vison*. *Journal of Mammalogy*, **85**, 432–439.

Yamaguchi, N., Strachan, R., and Macdonald, D.W. (2002). Practical considerations for the field study of American mink *Mustela vison* in lowland England. *Mammal Study*, **27**, 127–133.

Youngman, P.M. (1982). Distribution and systematics of the European mink *Mustela lutreola* Linnaeus 1761. *Acta Zoologica Fennica*, **166**, 1–48.

CHAPTER 8

Crayfish management in the Upper Thames

Tom P. Moorhouse and David W. Macdonald

An ounce of prevention is worth a pound of cure.

Benjamin Franklin

8.1 Introduction

The introduction and spread of non-native species has become a global ecological and conservation crisis in both terrestrial and aquatic communities (Gurevich and Padilla 2004; Macdonald et al. 2006; Macdonald and Burnham 2011). Crayfish are among the most frequently introduced non-native aquatic organisms, and they often become *invasive*, defined as 'An alien species whose establishment and spread modifies ecosystems, habitats, or species' (Millennium Ecosystem Assessment 2005). Invasive crayfish have great potential to disrupt long established freshwater ecosystems formerly dominated by native species (Gherardi et al. 2001).

Britain currently has seven species of crayfish living in the wild, of which only one, the white-clawed crayfish *Austropotamobius pallipes*, is native (Holdich and Sibley 2009). This species is currently listed as Endangered in the IUCN Red List (Füreder et al. 2010), and, unless protected effectively, is predicted to become extinct by 2034 (Holdich et al. 2004). Outside of the UK, the western limit of the species' range is in north-western Spain, and Montenegro is the easterly limit, while Spain and Scotland are the southerly and northerly limits respectively. Its distribution is restricted in Austria, Corsica, Germany, Lichtenstein, and Montenegro (Füreder et al. 2010). Six non-indigenous crayfish species have established in Britain, but only two are widespread: the American signal crayfish *Pacifastacus leniusculus* and the narrow-clawed crayfish *Astacus leptodactylus*. Of these, the signal crayfish has a considerably larger distribution and far greater impacts upon the native white-clawed crayfish (Holdich and Sibley 2009).

Signal crayfish (Fig. 8.1) are the most widespread non-indigenous species of crayfish in Europe, currently found in 27 European countries (Holdich et al. 2009). They were deliberately introduced to Britain in 1976 to serve as a fishery and had colonised more than 250 British waters by 1988 (Lowery and Holdich 1988; see Box 8.1). They frequently carry crayfish plague *Aphanomyces astaci*, a fungal infection to which they are highly resistant (Alderman et al. 1990) but which is lethal to the white-clawed crayfish. Crayfish plague has devastated the British population of white-clawed crayfish (Holdich and Rogers 1997; Lozan 2000), but even in its absence, signal crayfish displace white-clawed crayfish over several years (Bubb et al. 2005). Aggressive inter-specific competition for in-stream refuges from predation, such as large cobbles and boulders, usually favours the larger signal crayfish (Bubb et al. 2006a). Remnant populations of the white-clawed crayfish are still widely distributed but increasingly fragmented in England, and are described as 'sparse' in Wales; Scotland has only two populations of this species, both introduced (Holdich and Sibley 2009).

Wild signal crayfish can be extremely abundant. In US and British habitats, estimates range from 0.9–20 individuals per m^2 (Abrahamsson and Goldman 1970; Goldman and Rundquist 1977; Bubb 2004). In 2005, commercial trappers caught over 100,000 signal crayfish from the River Lark in Suffolk over two months (data cited in Holdich and Sibley 2009), and our own trapping studies in the Upper Thames captured 15,793 individual signal crayfish from a 400 m total length of river over 64 sampling days. In this latter case, the estimated population density of signal crayfish with

Wildlife Conservation on Farmland. Conflict in the Countryside. Edited by David W. Macdonald and Ruth E. Feber.
© Oxford University Press 2015. Published 2015 by Oxford University Press.

Figure 8.1 An adult American signal crayfish *Pacifastacus leniusculus*. Photograph © T. Moorhouse.

a carapace length (CL) > 30 mm (approximately one year old or greater; the smallest size likely to be captured in a trap) was > 3000 individuals per 100 m length.

8.1.1 Signal crayfish in the Upper Thames

Prior to 1990, signal crayfish were recorded only at three locations in the Upper Thames, and large populations of white-clawed crayfish were present throughout the area (Environment Agency, unpublished data). By 2000, signal crayfish had spread to the majority of water courses, and white-clawed crayfish were increasingly limited to upstream locations, tributaries, or remnant populations enclosed by encroaching signal crayfish populations. At the inception of our crayfish project in 2008, signal crayfish dominated every major watercourse in the area and initial surveys found no evidence of the presence of white-clawed crayfish in any river or stream. While extremely small populations of white-clawed crayfish may still exist in the Upper Thames—particularly in isolated water bodies, disconnected from the main rivers—such populations remain exceedingly vulnerable either to overland colonization by signal crayfish (e.g. Holdich et al. 1999b), or through the accidental transference of crayfish plague by members of the public (e.g. by a boot, fishing line, or dog that has been in contact with infected waters) and wildlife.

Box 8.1 How signal crayfish arrived in the UK

In Britain (Northern Ireland remains free of alien crayfish), six non-native species of crayfish have been deliberately introduced since the 1960s, mostly for the restaurant and aquarium trades, for aquaculture, as angling bait, and for feeding fish stocks (Peay 2009). But it was the increase in Scandinavian demand for crayfish consumption that brought one of these, American signal crayfish *Pacifastacus leniusculus*, to the UK. In the 1950s, the high demand for the Scandinavian native noble crayfish *Astacus astacus* was partially met by the exploitation of the thin-clawed crayfish *A. leptodactylus* in Turkey. However, when crayfish plague hit Turkish stocks, the signal crayfish from North America was introduced (in 1969) to 70 Swedish lakes to replace it (Reynolds and Sauty-Grosset 2012).

The British Government saw a commercial opportunity to combat the 1970s recession, and strongly promoted the introduction of crayfish as a farm-business diversification strategy. In 1976, signal crayfish were introduced to England from a Swedish rearing station (Reynolds and Sauty-Grosset 2012). Since the aim was to serve the Scandinavian crayfish market[1] and high export prices were expected, controls were relaxed (Richards and Fuke 1977). Unfortunately several

continued

[1] In the 1970s, the narrow-clawed crayfish *Astacus leptodactylus*, was also imported from mainland Europe for culinary purposes and, after escaping, it also became established, but attempts to farm it were mostly unsuccessful (Rogers and Holdich 1995).

Box 8.1 *Continued*

authors (Goddard and Holdich 1979; Holdich et al. 1978, and Bowler 1979) were unsuccessful in recommending the exploitation of the native white-clawed crayfish *A. pallipes*, warning the government of the perils of introducing an alien crayfish.

It was publicised that crayfish production would reach 100 tonnes to meet demand (the market price at Billinsgate market, London, in 1979 was £3 per kg; Behrendt 1979), and crayfish farming peaked in 1980. In 1984, the British Crayfish Marketing Association (BCMA) was created, with 62 farmers being registered by 1989 (Holdich et al. 1995). At that time, trading fingerlings for stocking became profitable, and many crayfish escaped from farms, establishing in the wild (Reynolds and Sauty-Grosset 2012).

With demand expected to increase, the BCMA planned to set a price of £5.50 per kg throughout the season (early summer–end September), and created a scheme in which farmers paid 50p per half a kilo of crayfish to maintain a surplus stock, to be released in case demand and supply had to be evened out (Reynolds and Sauty-Grosset 2012). However, the table market in the UK never developed as predicted, and top grade (large) crayfish were sold to restaurants at a price of £12.10 per kg, which could not compete with imports[2] sold at £7.7 per kg (Peay 2009; Reynolds and Sauty-Grosset 2012). The BCMA collapsed in 1990 and, although the number of commercial operations registered as cultivating signals in Britain was as high as 99 in the 1990s (Rogers and Holdich 1995), by 2003 only five crayfish farms remained (producing 0.25 tonnes/year of signal crayfish), with 20–30 former farm sites with feral populations still being harvested (Holdich and Sibley 2009).

By 2009, all registered farms had ceased operating due to changes in legislation to protect white-clawed crayfish, and the ease with which signal and narrow-clawed crayfish could be harvested from the wild (Rogers and Holdich 1995; Everard et al. 2009). At present, wild populations of signal crayfish are being harvested by small-scale commercial trappers, but there is no record of the quantities produced. Since restaurants will only buy live crayfish and, despite the threat to the native species, the sale of live *P. leniusculus*, local and imported, is permitted in fish markets for human consumption across England and Wales, continued opportunities for escape into the wild still exist, even within white-clawed crayfish ark-site areas (Bean et al. 2006), which are meant to be threat-free isolated refuges, created to establish sustainable populations. A simple internet search reveals that live, frozen, and cooked crayfish can be easily bought on-line in England, with national stocks being sold at about £12 per kg.

8.2 Impacts of invasive signal crayfish

While the potential loss of a Red-Listed species is in itself justifiable cause for concern, the negative impacts of signal crayfish in the UK extend beyond this direct species replacement. Densities of signal crayfish are typically much larger than those of the white-clawed crayfish (up to four times higher in one study: Peay et al. 2009) and the negative consequences of their arrival in UK habitats are likely to be severe. Crayfish are benthic, omnivorous, and detritivorous feeders, feeding on a variety of items including amphibian eggs and larvae (Gamradt and Kats 1996; Axelsson et al. 1997), macroinvertebrates (Hanson et al. 1990; Guan and Wiles 1998; McCarthy et al. 2006), algae (Guan 1995 cited in Guan and Wiles 1998), macrophytes (Creed Jr. 1994; Lodge et al. 1994), fish, and their eggs and larvae (Guan and Wiles 1998; Guan and Wiles 1999), and dead leaves (Usio 2000). High densities of invasive crayfish may, therefore, severely damage a range of native biota which depend on these same resources (Holdich et al. 1999b).

In their native habitats, the activities and numbers of signal crayfish raise no concerns for managers of freshwater ecosystems, but elsewhere—potentially due to release from predators, pathogens, or due to the availability of naïve prey items—they behave quite differently. Invasive signal crayfish can affect fish species by direct predation, and also through competition for food and shelter, for example among benthic fish such as bullhead *Cottis gobio* and stone loach *Noemacheilus barbatulus* (Guan and Wiles 1997), brown trout *Salmo trutta* (Peay et al. 2009) and, under laboratory conditions, overwintering juvenile Atlantic salmon *Salmo salar* (Griffiths et al. 2004). Similarly, signal crayfish may destroy fish breeding sites through macrophyte reduction (Guan and Wiles 1998). Aquatic macrophytes are a food source for signal crayfish (Flint and Goldman 1975; Guan and Wiles 1998) and grazing by signal crayfish and other species has been shown to reduce substantially their standing crops (Flint and

[2] Imports of narrow-clawed crayfish *A. leptodactylus* (Turkish, Russian) and red swamp crayfish *P. clarkii* (Spain, Kenya).

Goldman 1975; Chambers et al. 1990). This may have indirect effects on other taxa because aquatic macrophytes provide important refuges for juvenile fish (those under one year old) (Copp et al. 1994), and can substantially increase the invertebrate diversity and abundance supported by rivers (O'Hare and Murphy 1999). High densities of invasive crayfish also typically cause a shift in macroinvertebrate composition towards relatively fast-moving and sediment-burrowing taxa (Hanson et al. 1990; Lodge et al. 1994; Crawford et al. 2006).

The presence of signal crayfish, as a hyper-abundant prey item, may alter the feeding behaviour of mammalian predators. Both otter *Lutra lutra* and mink *Neovision vision* are present in the upper Thames (Harrington et al. 2009a), and their ranges and diets overlap (see Chapter 7, this volume). Mink are an invasive species in the UK, and can have severe impacts upon native fauna, not least upon the native water vole *Arvicola amphibius* (see Macdonald and Feber 2015: Chapters 14 and 15). It was hoped that mink numbers might be at least partially controlled by a recent resurgence in numbers of otters, with which mink have range and dietary overlap, but seemingly this has not happened (Chapter 7, this volume). On the contrary, it is possible that the presence and hyper-abundance of signal crayfish may at least partially facilitate co-existence of these two carnivores, which is not a desirable conservation outcome from the point of view of mink control (but more desirable than the continued absence of otters).

Furthermore, signal crayfish may damage the physical riparian environment. Signal crayfish are not known as a burrowing species in their native habitat (Lewis 2002) but, at least in the UK, their extensive burrows in river banks can potentially cause bank collapse (Guan 1994). Moreover a growing body of evidence suggests that signal crayfish activities, such as walking, fighting, burrowing, and feeding, may severely disturb river sediments, possibly creating catchment-scale management issues including habitat degradation, mobilization of sediment-associated nutrients and contaminants, and sediment-related flood risks (Harvey et al. 2011).

8.3 Control measures for signal crayfish

Controlling the numbers of such an abundant and damaging invader poses significant problems for habitat managers. Numerous studies, mainly from the UK, have examined the feasibility of controlling nuisance populations of crayfish by (1) direct removal by hand, trapping, or electro-fishing, (2) biological control agents such as fish predators (e.g. eels *Anguilla anguilla*, or chub *Squalius cephalus*), (3) microbial insecticides and diseases, (4) physical means such as de-watering and habitat destruction, and (5) chemical means (see Holdich et al. 1999a; Peay 2001; Hyatt 2004); but no effective strategy for eradication has been developed (Peay 2001).

Manual removal has the advantage of being relatively cheap and, arguably, has fewer impacts on non-target species and their habitats. However, manual methods, in particular hand searching, seine netting, and trapping, have a well-reported bias towards removal of the largest individuals (e.g. Abrahamsson 1966; Guan and Wiles 1996; Holdich et al. 1999a; Westman et al. 1999), leaving the juvenile population almost entirely undisturbed (Bills and Marking 1988; Holdich et al. 1999a). They may remove only a small percentage of the population (Frutiger et al. 1999; Peay 2001) and are commonly considered to be adequate only to limit crayfish populations, but not to eliminate them (Holdich et al. 1999a; Peay 2001; Hyatt 2004).

Two studies, however, indicate that a combination of intensive trapping and predation by predatory fish may achieve a substantial decrease in crayfish population densities (Rogers et al. 1997; Hein et al. 2007). Rogers et al. (1997) reduced the estimated numbers of signal crayfish in carp ponds from 4000 to 1500 over 900 trap nights, and suggested that, if regular trapping had continued, fish predation on smaller individuals might have eliminated the population. Similarly, Hein et al. (2007) achieved an approximate 95% reduction in catch rate of rusty crayfish *Orconectes rusticus*, following intensive trapping in a large (64 ha) seepage lake, and speculated that the observed declines were partly due to an increase in large predatory fish (in this case sunfish *Lepomis spp.*) after the local fishery restricted bag sizes. These studies, however, both took place in unconnected water bodies. Control attempts of populations in connected habitats, such as rivers, are usually hampered by immigration from surrounding areas. Working on a population of signal crayfish in the River Thame, UK, Ibbotson et al. (1997) provided circumstantial evidence that intensive trapping of one section of a river may reduce numbers of large individuals in neighbouring sections due to immigration, but they were unable to provide evidence that the reduction in signal crayfish numbers affected any other species (plants, fish, other macroinvertebrates) either positively or negatively.

8.4 The responses of signal crayfish populations to manual removal operations

We designed a three-year project (2008–2011) with the aim of auditing the movements and growth rates of signal crayfish in response to a replicated manual control campaign conducted in the wild. The Upper Thames provided a perfect experimental situation to perform this study, given the availability of deep, slow flowing rivers (which make trapping possible) and the widespread, and extremely abundant, nature of signal crayfish in this area. We removed signal crayfish in two separate studies: in 2009, we studied the immigration rates of signal crayfish in response to removal trapping and, in 2010, we studied the impact of crayfish removals on both the growth rates and movements of the remaining signal crayfish, as well as testing for benefits for the other macroinvertebrate species. We also investigated the degree to which signal crayfish were present in the diet of otters and American mink, to assess the potential significance of signal crayfish for the co-existence of these two species. Below we describe the three studies designed to achieve these goals.

8.4.1 Immigration rates of signal crayfish in response to removal by trapping

In 2009, we selected four 1 km sections of the River Windrush (Oxfordshire, UK), and divided each into three sections: a 250 m upstream section, a 500 m middle section, and a 250 m downstream section. Of the four 500 m middle sections, two were designated as removal sections, from which all crayfish captured were removed and humanely destroyed by freezing (RSPCA 2009). The remaining two middle sections, and all the bordering upstream and downstream sections, were non-removal sites in which all crayfish captured were marked (by punching a hole in the uropods—calcareous tail-plates—with a small-gauge needle, a process which does not harm the crayfish; Guan 1997) and returned to the water at the point of capture. This design therefore enabled an assessment of the numbers of crayfish immigrating into the middle sections, and the distances they moved from the upstream and downstream borders, in response to the reduction in population density in the removal areas.

The entire length of each site was trapped once a month for six months between April and October 2009. During each capture session, traps were placed and set on Monday and checked each morning before removal on Friday. Traps[3] set for monitoring were spaced at 20 m intervals throughout, and traps in the removal sections were spaced at 15 m intervals to increase the number of crayfish removed (Moorhouse and Macdonald 2010).

We made a total of 16,737 captures of 14,510 individual crayfish over the six capture sessions. Of these captures, 7407 were from the upstream/downstream borders, and 9330 were from middle sections. Of those captured from the middle sections, 3760 were captured in non-removal sections, marked, and released, and 5570 individuals were captured, removed, and destroyed from the two removal sections. A total of 318 crayfish was recorded as immigrating into the middle sections (of both removal and non-removal sites), having been previously captured and marked in the upstream and downstream locations. We found no evidence that the number of individuals immigrating from the borders into the middle sections was larger in the removal treatments. Immigrant individuals comprised exactly 3.7% of the population trapped in the middle sections under both treatments, so data for both were pooled.

The probability of an individual being recaptured—an index of trappability—increased with its carapace length (CL) and decreased with the number of missing chelae (claws), confirming that the removals preferentially targeted relatively large, undamaged crayfish—the damage would be expected to limit crayfishes' ability to defend themselves and/or compete for food, meaning that they may be less likely to enter or remain in the baited traps. The mean size and weight of the crayfish that immigrated into the middle sections were both larger than the population means from the non-removal middle sections (CL of 61 mm as opposed to 56 mm, and weight, 75 g as opposed to 55 g, respectively; Moorhouse and Macdonald 2010). The crayfish classed as immigrants were necessarily recaptures, and so would be expected to be large relative to the general. However, the recapture rate of crayfish with a CL < 50 mm was 6% (41 of 686 marked individuals were recaptured), whereas only 0.1% (1 of 680 individuals) of individuals of that size were recorded as immigrating to the middle. We conclude, therefore, that although there will be a degree of recapture bias towards larger individuals, nearly all individuals which moved from a border into a middle section had a CL > 50 mm, and this is likely to be an accurate

[3] Commercially available cylindrical, plastic crayfish traps; Trappy™, Virserum, Sweden.

estimate of the minimum size of crayfish making such movements.

By marking the crayfish we were also able to see how far they moved within each river section. The mean distance that crayfish moved into the middle sections was significantly larger at removal sites than non-removal sites (239 m as opposed to 187 m; Fig. 8.2). Therefore, although the removal of crayfish from the middle sections appears not to have stimulated larger numbers to immigrate from the surrounding sections, it may have facilitated progress for those that did move. One explanation is that the higher capture probability, and therefore rates of removal, of large crayfish may have facilitated immigration, because it altered the dominance relationships among resident individuals. Social dominance hierarchies are soon established among various species of crayfish kept under laboratory conditions (Momot and Leering 1986; Ranta and Lindstrom 1992; Goessmann et al. 2000), and the outcome of aggressive interactions is principally determined by relative size (Ranta and Lindstrom 1992; Ranta and Lindstrom 1993; Figler et al. 1999), all other factors being equal (e.g. moult stage, habitat familiarity, etc.; Goessmann et al. 2000). The bias against larger individuals associated with manual removal could therefore be expected to minimize aggressive interactions, making large movements easier for immigrants. This interpretation is supported by our second study which demonstrated that larger crayfish move further. Alternatively, however, immigrants into the removal sites may simply have perceived that they were moving down a density gradient; for example, Bovbjerg (1959) demonstrated that crayfish are able to perceive, and move away from, areas containing a high density of conspecifics—in favour of settling at the point where densities were lowest, which we assume to have been in the centre of the middle section.

Irrespective of the underlying mechanism, these results support Ibbotson et al.'s (1997) speculation that the removal of crayfish may facilitate the immigration of relatively large individuals from elsewhere. We can add that the population from which relatively large individuals are preferentially removed will extend at least 200 m upstream and downstream of the limits of the trapped area, meaning that impacts will be diluted at the point of removal and spread beyond the immediately trapped area. The implications of this for crayfish control via manual removal in non-enclosed habitats are that any potential disbenefits following the removal of large individuals, such as release of smaller individuals and juveniles from competition, with a resulting increase in juvenile survival and growth (e.g. Momot 1993 and below), are therefore also likely to extend beyond the immediately trapped area. The distances and rates of immigration observed make it unlikely that removal strategies in river habitats could be as effective as those in isolated water bodies. This may imply that measures to control numbers of signal crayfish in running waters may be ineffective unless maintained over substantial lengths for prolonged time periods (e.g. intensive removals over tens of kilometres for numbers of years).

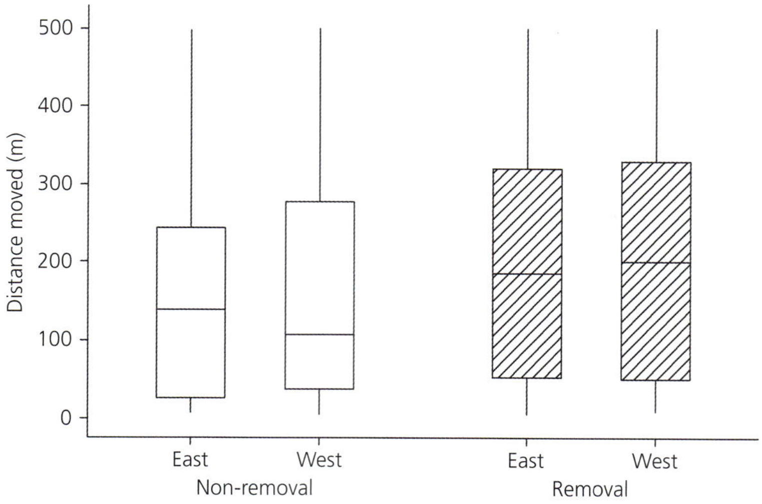

Figure 8.2 Boxplots showing the distance moved by crayfish (n = 313) that immigrated from the bordering sections into the middle sections at the non-removal (white boxes) and removal (hatched boxes) sites during the 2009 study. East and West refer to the arm of the River Windrush on which the site was located. Boxes represent the median and first-third quartiles, and whiskers represent the highest and lowest data values within their respective limits. From Moorhouse and Macdonald (2010), reproduced with permission from John Wiley & Sons.

8.4.2 Within-population movement distances and growth rates of signal crayfish in response to population reduction

Having demonstrated the effect of removing crayfish on immigration into the removal area, in the following year (2010) we studied the impact of intensive removal trapping on movements and body condition *within* a trapped population. We undertook this study on two different rivers, the Evenlode and the Thame. We chose four sections of river, each 100 m in length, representing two removal and two non-removal sections, divided between the two rivers such that each contained a removal and non-removal section, separated by a minimum distance of 800 m (to prevent individuals moving between treatments). In both cases the non-removal site was located upstream of the removal site to minimize the possibility of effects from the removals being propagated downstream and affecting the non-removal stretch. From the removal sections, half of the crayfish captured were marked with the position of the trap at which they were captured, and returned at the point of capture, and the other half removed and humanely destroyed (RSPCA 2009). Along the non-removal sections, all crayfish captured were marked and returned at the point of capture, and half were given additional marks describing the trap position (only half were so marked to match the numbers marked in the removal sections). Distances moved were calculated from recaptured crayfish using their previous trapped location. The design therefore enabled an assessment of movement distances and body condition of crayfish under two treatments: removal, in which the population density within the section was lowered by removal of half of the captured crayfish, and non-removal in which the density was maintained.

Both rivers were trapped every three weeks from May to August 2010. Removal and non-removal sections of the same river (Evenlode or Thame) were trapped simultaneously. Each capture session (with 84 traps) comprised eight consecutive days, making a total of 64 days of trapping over the course of the project. Trapping sessions at a given river were separated by a three-week period. Further trapping details are given in Moorhouse and Macdonald (2011a,b).

8.4.3 The impact of removal on crayfish numbers

In total, 27,354 captures were made of 15,793 individual crayfish (from only 400 m of river!). Of these, 7594 were captured in non-removal sections and 8199 from removal sections (comprising individuals immediately removed, and individuals captured and marked then subsequently recaptured and removed). A total of 6181 crayfish were removed from the two 100 m removal sections over the course of the study. On the first day of trapping in a given session, the mean numbers of crayfish captured per trap were 8.5 and 6.3 at non-removal and removal sites, respectively. By the last day of a trapping session, we captured on average 6.5 and 2.7 crayfish per trap at non-removal and removal sites respectively, a very significant reduction.

These findings have two implications. First, although the numbers of crayfish captured during a given session fell at both removal and non-removal sites (possibly because crayfish feeding on the bait became satiated and so less trappable), numbers fell further (by 57% as opposed to 24%) at the removal sites, confirming that the removals did reduce the number of crayfish present during the session. Second, in the three-week period between sessions the numbers at both removal and non-removal sites recovered, presumably due to the combined effects of immigration and an increase in trappability following the withdrawal of bait. This removal and recovery resulted in a pattern of density of crayfish in which, at the beginning of a trapping session, numbers of crayfish in the centre of the removal sites were substantially lower than at the edges (which were closer to a source of immigrants from up and downstream), whereas no such pattern existed in the non-removal sites.

The removals also affected the frequency distribution of body sizes within the population. Traps in the removal sections captured a smaller percentage of large individuals (with a CL > 50 mm) than those in the non-removal sections, but a larger percentage of relatively small individuals (CL < 45 mm) (Fig. 8.3; Moorhouse and Macdonald 2011a). Mean CLs were 54.5 mm and 52.8 mm for the non-removal and removal sites, respectively. There was, however, no evidence that CLs of captured crayfish became progressively smaller as trapping progressed in the removal sections, compared with those in the non-removal sections.

8.4.4 How did the removals affect movements in the remaining crayfish?

At sites where half the crayfish were removed, the mean distances travelled by the remaining crayfish during the capture session were smaller than at non-removal sites (10.8 m and 16.0 m respectively). Larger crayfish moved further than smaller crayfish: mean

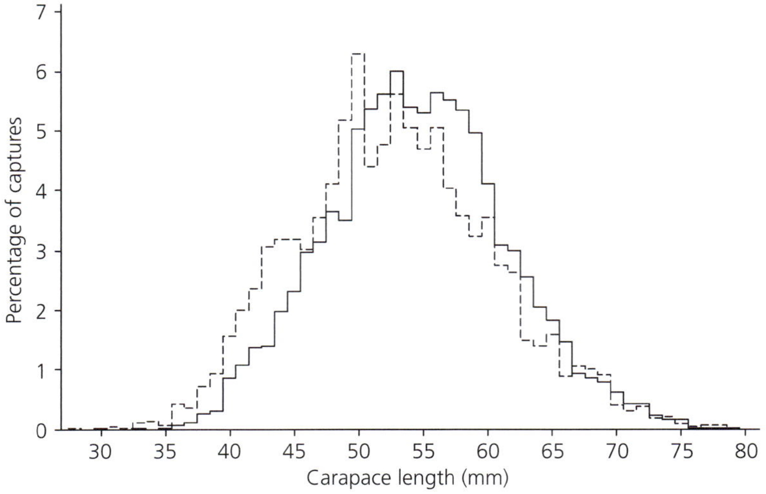

Figure 8.3 Frequency distribution of percentages of captures of a crayfish by carapace length (CL) from removal (dotted line) and non-removal sections (solid line) during the 2010 study. The two distributions are not significantly different. From Moorhouse and Macdonald (2011a), reproduced with permission from Elsevier.

movements of the smallest crayfish for which movements were recorded (CL = 30 mm) were 7.6 m, as opposed to 19.6 m for the largest (CL = 78 mm) (Moorhouse and Macdonald 2011b). Similarly, the distances that individuals moved over the three-week period *between* capture sessions were smaller at the removal sites than at the non-removal sites (14.5 m and 24.6 m, respectively) and, again, movement distances increased with increasing body size: mean movement distances for crayfish with CLs of 30 mm and 78 mm were 8.9 m and 32.6 m, respectively. A much greater proportion of individuals made movements of 10 m or less at the removal sites than at the non-removal sites (Fig. 8.4; Moorhouse and Macdonald 2011b).

The exact reason why movement distances were so much smaller in the removal sections is unclear, but is likely to include increased availability of food and shelter, and/or a direct response to decreasing population density. Since crayfish can perceive, and move away from, areas densely populated by conspecifics (Bovbjerg 1959), it may be that the removal sections were depopulated sufficiently to prevent this reaction. Food and shelter are both important resources for crayfish, and changes in the availability of both could alter movement distances (Capelli and Hamilton 1984). Fights are more frequent in the presence of a more valuable food resource (Bergman and Moore 2003), and are more intense (Hazlett et al. 1975) and

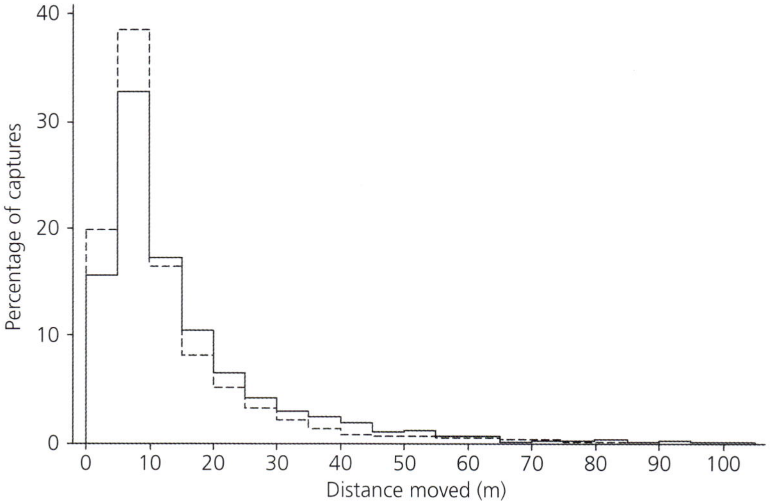

Figure 8.4 Frequency distribution of percentages of recaptured crayfish making movements of a given size in removal sites (dashed bars) and non-removal sites (solid bars) during the 2010 study. The two distributions are significantly different.

more likely to escalate (Stocker and Huber 2001) when crayfish are hungry. Hungry, virile crayfish *Orconectes virilis* move about more (Hazlett et al. 1975) and are less inhibited by alarm odours than are relatively satiated individuals (Hazlett 2003). Davis and Huber (2007) noted in rusty crayfish that time outside their shelter was mainly used for feeding, individuals frequently returned to the same shelter from which they had emerged, and that agonistic encounters generally occurred in the context of shelter acquisition or defence. Conflicts over shelters can be more intense than those over some food resources (Bergman and Moore 2003), and could plausibly have consequences for movement distances if evicted crayfish needed to locate an alternative shelter. Our removals of signal crayfish may have reduced competition for resources by making more food and shelter available to the remaining population, resulting in the smaller observed movements.

These results imply that trapping at signal crayfish invasion fronts could slow the rate of spread of signal crayfish populations. Bubb et al. (2006b) concluded from their study in a UK upland river that all signal crayfish of both sexes which are one year of age (typical CL = 28–34 mm; Guan and Wiles 1999) or over, might contribute to advancing an invasion. So trapping, which typically captures large adults, may have only limited success in reducing colonization because the removal of large adults would leave a large relatively untrapped cadre of potential dispersers. Our observations, however, imply that trapping at the edge of a river population may be sufficient to delay colonization of new sections, at least in the short term, by: (1) maintaining low densities and therefore reducing movement distances and (2) preferentially reducing the population of the large individuals which make the longest movements (Moorhouse and Macdonald 2011b). Over the long term it remains unlikely that any trapping programme can entirely prevent, rather than merely delay, colonization of a river by signal crayfish.

8.4.5 What is the effect of lowered density on crayfish weight?

In the first capture session, wet weights for a given CL were approximately identical between the removal and non-removal treatments (Fig. 8.5). By the final capture session, wet weights for a given CL were greater under both treatments than in the first capture session, but crayfish in the removal sections were significantly heavier for a given length than in the non-removal sections (Fig. 8.5). Mean weights for crayfish with the largest CL (78 mm) captured in session 4, were 2.1 g heavier in removal than non-removal sections (137.6 g as opposed to 135.5 g, respectively).

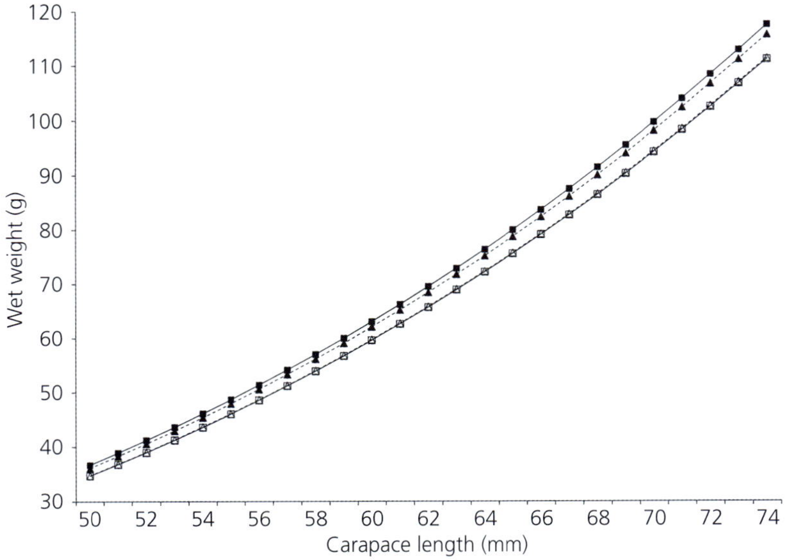

Figure 8.5 Weight–length curves for crayfish from removal (square markers and solid line) and non-removal (triangular markers and dotted line) sections in the first (open markers) and the fourth (filled markers) capture sessions. Data points were calculated from the coefficients of a regression analysis of log-weight versus log-length to illustrate the differing effects of removals over time between treatments. Error bars are omitted for clarity. From Moorhouse and Macdonald (2011a), reproduced with permission from Elsevier.

Finding that crayfish in the removal sections grew progressively heavier per unit length over the course of the study does not directly indicate increased growth rates among the crayfish in the removal sections—and we were unable directly to compare growth rates of crayfish between removal and non-removal sections, due to the lack of individually identifiable tagging. All the same, increased intermoult weights are correlated with larger moult increments (Jones et al. 1996) and therefore growth rates. Few studies have examined intermoult weights of crayfish as opposed to growth rates per se, but Jones et al. (1996) demonstrated that increased protein in the diet allows animals of a standard carapace length to develop significantly larger abdomens in two species of yabby (*Cherax albidus* and *C. destructor*), and larger claw sizes in *C. albidus*. Mean weight, percentage weight gain, and specific growth rate of both species were also positively correlated with quantities of dietary protein, and animals on the highest protein diet had a substantially shorter intermoult period and larger moult increment than those on lower protein diets (Jones et al. 1996).

The increased weights per unit length observed in this study are unlikely to be the result of immigration of large numbers of large, competitively dominant individuals to replace the removed crayfish, because immigrants comprised only 3.7% of the trapped population. Similarly there is no a priori reason to assume that immigrating individuals would be in better physical condition than the removed individuals. More plausibly, crayfish living in the removal sections may have had, before capture, access to a higher quality diet, because smaller numbers meant more food available per head, less interference competition, or both (e.g. Momot et al. 1978; Cobb et al. 1982; Herberholz et al. 2007).

The increase in weights may also stem from direct effects of a reduction in the numbers of large crayfish: growth rates of some crayfish species are reduced if surrounded by larger neighbours, even when there is no direct competition for food (Barki et al. 2006). The evidence suggests that the increased weights we observed are likely to have been permitted by the removal of large individuals releasing the remaining animals from some level of growth inhibition.

The implication of our findings for signal crayfish control by manual removal campaigns is that these have a serious and unintended consequence: they promote conditions that are favourable for individual growth of the remaining, untrapped population. This study was not long enough to determine whether the recruitment or survival of juveniles increased as a result of the reduced densities and improved body condition of the remaining stock, but the results do suggest a feedback mechanism whereby the effects of removals are at least partially counteracted by increased growth rates and improved body condition in the untrapped population (Moorhouse and Macdonald 2011a).

8.4.6 Did the removal of signal crayfish benefit the other macroinvertebrates?

We have described how the density of signal crayfish in the removal sites decreased towards the middle of the site, most likely due to immigration increasing the densities at the edges between capture sessions. We examined the effect of this density gradient on the numbers (counts) and taxon richness (number of taxa) of macroinvertebrates settling onto samplers (known as colonization samplers) in the removal and non-removal sites. The freshwater macroinvertebrates we studied were a diverse group of over 50 identified taxa, the most common of which were water beetles (e.g. *Helopherous brevipalpis*), midge larvae (members of the Chironomidae family), leeches (e.g. *Pisciola geometra*), and caddis flies (e.g. *Cyrnus trimaculatus*), but which also included molluscs (e.g. *Pisidum sp.*), alderflies (e.g. *Sialis lutaria*), and bugs (Gerridae family). All of these taxa—detritivore, herbivore, and carnivore—play crucial roles in the functioning of freshwater ecosystems (Covich et al. 1999).

Mean macroinvertebrate counts were inversely proportional to the number of crayfish trapped, ranging from 23–101 and therefore representing a fivefold increase in macroinvertebrate numbers at the lowest crayfish densities. Similarly taxon richness ranged from 5–14, meaning that the number of taxa represented was almost tripled at the lowest crayfish densities (Moorhouse et al., 2014). These findings are consistent with expectations if predation pressure from signal crayfish were a principal determinant of presence and abundance for a range of macroinvertebrate species, and if our crayfish removals diminished this predation pressure sufficiently to permit an increase in numbers and types of macroinvertebrates on the samplers.

8.4.7 Evidence for signal crayfish mediating the co-existence between otters and mink

We examined the dietary preferences of otters and mink in the Upper Thames by analysing prey remains

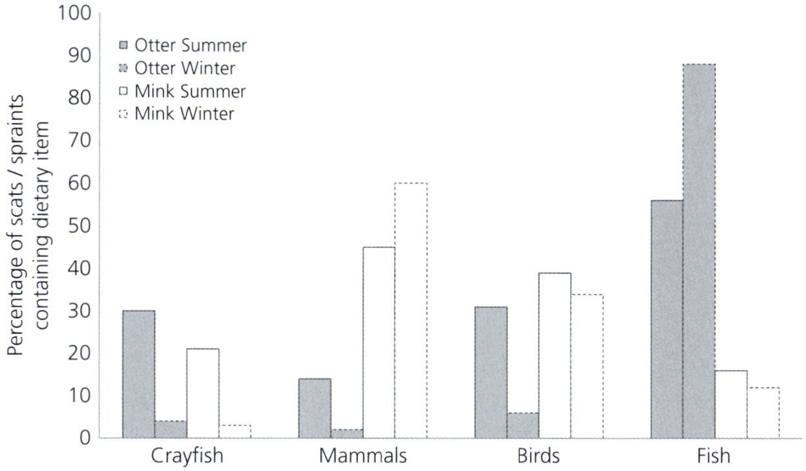

Figure 8.6 The percentage of otter spraints (grey bars) and mink scats (white bars) containing crayfish, mammals, birds, and fish during the summer (solid line) and winter (dashed line).

in 338 mink scats and 196 otter spraints, collected between 2004 and 2006, from five rivers in the Upper Thames. We found little direct evidence to suggest that the presence of the ubiquitous and hyper-abundant signal crayfish per se permits the two species to coexist (see Chapter 7, this volume). During summer months, crayfish remains were found in 21% (of 244) mink scats and 30% (of 140) otter spraints (Fig. 8.6), but only 3% (of 91 scats) and 4% (of 50 spraints), respectively, in winter (Fig. 8.6). Signal crayfish, therefore, while a regular food item for both species during the summer, are virtually ignored in winter, presumably because crayfish are largely inactive then, either in burrows or in the sediment. These findings confirm those of Beja (1996) who found that the introduction to Spain of the red swamp crayfish *Procambarus clarkii* increased food abundance for otters from spring to autumn but not over the winter.

The principal components of mink diet from the scats collected were mammals and birds (see Chapter 6, this volume), whereas otter diets mainly comprised fish and birds (the latter only during the summer months). During the winter only 15% of mink scats contained aquatic items (i.e. fish and crayfish), compared with 92% of otter spraints. These findings are consistent with those of a previous study showing that mink eat fewer fish and more birds in the presence of otters (Harrington et al. 2009b), and suggest that dietary competition with otters is primarily avoided through mink eating more birds and mammals, not signal crayfish, especially during the winter months.

8.5 Conclusions

Our findings comprise a mixture of good and bad news for those concerned by the impacts of signal crayfish. The bad news begins with the conclusion that all attempts to eradicate, by any conventional means, established signal crayfish populations from rivers in the Upper Thames, are likely to meet with failure. Even our intensive removals (of over 6000 individuals—or approximately 277 kg of crayfish—from 200 m of water course over four nine-day sessions) were sufficient only to reduce temporarily the catch per unit effort and the mean carapace length of captured crayfish. Worse, these effects were partially reversed by immigration between sessions and, in any case, affected only the trappable portion of the population (crayfish with a carapace length > 30 mm). Our results imply that in order to effect eradication from 1 km of river, extremely intensive trapping of a much larger length of river (perhaps 3 km in total) would have to be maintained for a prolonged period (multiple years) in the hope of reducing the population of breeding adults to the point where juvenile crayfish could be kept scarce by predatory fish, such as chub or eels. The chances of achieving such a programme over the long term are slim, implying that, for all practical purposes, eradication of signal crayfish from riparian habitat with current practices, and certainly by removal trapping, is impossible.

The bad news continues with the evidence of a feedback mechanism by which smaller individuals gain body weight in the removal sections, coupled with the evidence of long (> 200 m) immigration distances into removal areas. These findings have a potentially more worrying aspect in that they suggest that the removals have unintended consequences, firstly in indirectly benefiting the smaller individuals at the removal site (and thereby potentially increasing their survival rates and fecundity), and secondly in similarly benefiting smaller individuals in waters surrounding the removal site, due to the emigration of large individuals into the removal area.

Our studies do, however, provide some good news on several fronts. First, regardless of the numerous documented negative consequences of the presence of signal crayfish, it is reassuring that their presence in UK habitats does not appear to facilitate the co-existence of American mink with otters. Second, removal trapping may be effective in slowing the rate of colonization of unoccupied waters by signal crayfish, if it is conducted at a colonization front, because the removal of large individuals and the behavioural changes induced at low densities mean that the survivors make smaller movements. Lastly, the response of the non-crayfish macroinvertebrate community suggests that there remain tangible benefits for other aquatic species of any measure which may reduce the density of signal crayfish.

While there remain difficulties associated with crayfish trapping (for example the potential to injure other wildlife such as otters and water voles if conducted carelessly), the balance of the above evidence may suggest that encouraging the trapping of signal crayfish, particularly at commercial scales—with the concomitant benefits of providing employment in the harvesting of an abundant food resource—can provide conservation benefits (Box 8.2).

Box 8.2 Crayfish trade: could and should we eat British signal crayfish?

Global production of crayfish for human consumption has increased over the past 60 years (Lodge et al. 2012), with Europe producing 5–10,000 tonnes of crayfish per year (Reynolds and Sauty-Grosset 2012). Scandinavia, Germany, and France are the main consumers in Europe (Wickins and Lee 2002), where crayfish are sold as hard-shelled (whole live, whole frozen—cooked or uncooked, or fresh/frozen tail or claw meat), and soft-shelled (live or frozen) (Harlioğlu and Holdich 2001), at sizes over 9 cm total length, and are served whole with the claw/tail meat extracted at the table (Holdich 1993; Harlioğlu and Holdich 2001). Ninety-one per cent of global demand is provided for by non-indigenous species (with 87% being supplied by China; Lodge et al. 2012).

Since invasive wild signal crayfish are so abundant in British waters, they represent considerable potential for socio-economic gain. Currently, however, there is no strong tradition for eating native (and served whole) crayfish in Britain, and if commercially exploited, most crayfish would need to be exported. Since signal crayfish have largely replaced native crayfish in European markets, Swedish consumers are willing to pay double for their native crayfish compared to signal crayfish (Lodge et al. 2012), and so it is difficult to predict export prices (Kataria 2007). Some countries have succeeded in increasing native populations by encouraging the harvest of native crayfish as a means of conservation. Taugbøl (2004) maintains that exploitation and protection are closely linked (exploiters are concerned about sustaining the resources so they will protect them), but Holdich and Sibley (2009) clarify that this cannot work in Britain where the native species are protected from harvesting by law, and a consumer market is not established. Indeed, an experimental fishery of *A. pallipes* was set up in an Irish lake in 1987, and showed that, when catches were tested in a restaurant, crayfish was not only less popular than lobster *Nephorops norvegicus*, but that crayfish prices only increased when sea-fishing conditions were not suitable (Reynolds and Matthews 1993).

At the moment, restaurants will only buy live crayfish. As such costs of transport from point of capture are high, regulations for their storage and disposal are strict and costly, and escapes are still possible. A market does exist, however, for boiled-prepared meat. It has been noted[4] that one of the major sandwich companies in the UK, sells 1500 tonnes of red swamp crayfish *Procambarus clarkii* per year in their sandwiches, and this figure, if multiplied by all the companies that offer crayfish sandwiches, is truly colossal. Oddly, this immense consumption of crayfish meat is not met through exploitation of local populations, but, as with the majority of global demand, from exports from China (Lodge et al. 2012). Not only can China produce the required volume, but they can

continued

[4] Katy Salter, 'Why we should eat more crayfish'. *The Guardian*, 18 July 2012.

Box 8.2 *Continued*

process the crayfish (dis-assembling them by hand) on an industrial scale, which cannot be done cost-effectively in the UK.

The debate over whether we should exploit naturalized British signal crayfish is complicated by a number of factors. While a demand clearly exists for the meat, it is unclear whether domestic stocks can be exploited sufficiently cost-effectively to compete with overseas imports. Practically, trapping wild signal crayfish currently requires a licence from the Environment Agency that usually is limited to a particular location or set of locations. There is also potential for negative impacts on wildlife through by-catch, in particular otters and water voles which can drown in crayfish traps (which, in part, the licensing seeks to avoid). Negative impacts may occur also through trapping itself via the removal of large crayfish, resulting in populations of smaller individuals being released from competition, with a resulting increase in juvenile survival and growth, which is then likely to extend beyond the trapped area (Momot 1993; this chapter, see Section 8.4). Conversely, however, our most recent study shows that crayfish trapping achievable by commercial crayfisheries (not eradicating populations but locally reducing their density) can nonetheless be beneficial for a broad range of non-crayfish macroinvertebrates living in the benthos (i.e. the river bed, and which are therefore most susceptible to crayfish predation), leading to increases in their absolute numbers and numbers of taxa (Moorhouse et al. 2014). Similarly, although empirical evidence is currently lacking, the removal of signal crayfish could theoretically have the benefit of decreasing the turbidity of rivers in which their presence and activities may be increasing sediment loads (Harvey et al. 2011; Harvey et al. 2014).

In conclusion, while more research is definitely needed on the ecological and geomorphological impacts, positive or negative, of commercial crayfish trapping, given the large, under-utilized source of dietary protein that wild signal crayfish represents, coupled with their potential for generating revenue, and the emerging evidence for ecological benefits, perhaps it may be time to revisit the arguments in favour of a national wild crayfishery, carefully regulated and monitored to avoid undesired consequences.

Acknowledgements

Tom Moorhouse was funded by the Esmée Fairbairn Foundation. Many thanks to Alison Poole for help with study design and discussion of results and to Rosie Salazar, William Seamer, and William Linley-Adams for their help with data collection. Many thanks to the Freshwater Biological Association, Linear Fisheries, Smiths Bletchington, the Environment Agency, David Henman, and Richard Gurney for their support of the project.

References

Abrahamsson, S.A.A. (1966). Dynamics of an isolated population of the crayfish *Astacus astacus* Linné. *Oikos*, **17**, 96–107.

Abrahamsson, S.A.A. and Goldman, C.R. (1970). Distribution, density and production of the crayfish *Pacifastacus leniusculus* Dana in Lake Tahoe, California—Nevada. *Oikos*, **21**, 83–91.

Alderman, D.J., Holdich, D.M., and Reeve, I. (1990). Signal crayfish as vectors in crayfish plague in Britain. *Aquaculture*, **86**, 3–6.

Axelsson, E., Nystrom, P., Sidenmark, J., and Bronmark, C. (1997). Crayfish predation on amphibian eggs and larvae. *Amphibia-Reptilia*, **18**, 217–228.

Barki, A., Karplus, I., Manor, R., Parnes, S., Aflalo, E.D., and Sagi, A. (2006). Growth of redclaw crayfish (*Cherax quadricarinatus*) in a three-dimensional compartments system: Does a neighbor matter? *Aquaculture*, **252**, 348–355.

Bean, C.W., Maitland, P.S., and Collen, P. (2006). Crayfish in Scotland: a review of current status and legislative control. *Freshwater Crayfish*, **15**, 220–228.

Behrendt, A. (1979). Export push needed to open market for crayfish. *Fish Farmer 2*, January 1979, 44–46.

Beja, P.R. (1996). An analysis of otter *Lutra lutra* predation on introduced American crayfish Procambarus clarkii in Iberian streams. *Journal of Applied Ecology*, **33**, 1156–1170.

Bergman, D.A. and Moore, P.A. (2003). Field observations of intraspecific agonistic behavior of two crayfish species, *Orconectes rusticus* and *Orconectes virilis*, in different habitats. *Biological Bulletin*, **205**, 26–35.

Bills, T.D. and Marking, L.L. (1988). Control of nuisance populations of crayfish with traps and toxicants. *The Progressive Fish-Culturalist*, **50**, 103–106.

Bovbjerg, R.V. (1959). Density and dispersal in laboratory crayfish populations. *Ecology*, **40**, 504–506.

Bowler, K. (1979) Plague that has ravaged Europe. *Fish Farmer*, **2**, 34–35.

Bubb, D.H. (2004). Movement and dispersal of the invasive signal crayfish *Pacifastacus leniusculus* in upland rivers. *Freshwater Biology*, **49**, 357–368.

Bubb, D.H., Thom, T.J., and Lucas, M.C. (2005). The within catchment invasion of the non-indigenous signal crayfish

Pacifastacus leniusculus (Dana) in upland rivers. *Bulletin Francais de Peche et de Pisciculture*, 376–377, 665–673.

Bubb, D.H., Thom, T.J., and Lucas, M.C. (2006a). Movement, dispersal and refuge use of co-occurring, introduced and native crayfish. *Freshwater Biology*, **51**, 1359–1368.

Bubb, D.H., Thom, T.J., and Lucas, M.C. (2006b). Movement patterns of the invasive signal crayfish determined by PIT telemetry. *Canadian Journal of Zoology*, **84**, 1202–1209.

Capelli, G.M. and Hamilton, P.A. (1984). Effects of food and shelter on aggressive activity in the crayfish *Orconectes rusticus* (Girard). *Journal of Crustacean Biology*, **2**, 252–260.

Chambers, P.A., Hanson, J.M., Burke, J.M., and Prepas, E.E. (1990). The impact of the crayfish *Orconectes virilis* on aquatic macrophytes. *Freshwater Biology*, **24**, 81–91.

Cobb, J.S., Tamm, G.R., and Wang, D. (1982). Behavioural mechanisms influencing molt frequency in the American lobster, *Homarus americanus*. *Journal of Experimental Marine Biology and Ecology*, **62**, 185–200.

Copp, G.H., Guti, G., Rovný, B., and Černý, J. (1994). Hierarchical analysis of habitat use by 0 + juvenile fish in Hungarian/Slovak flood plain of the Danube River. *Environmental Biology of Fishes*, **40**, 329–348.

Covich, A.P., Palmer, M.A., and Crowl, T.A. (1999). The role of benthic invertebrate species in freshwater ecosystems—Zoobenthic species influence energy flows and nutrient cycling. *Bioscience*, **49**, 119–127.

Crawford, L., Yeomans, W.E., and Adams, C.E. (2006). The impact of introduced signal crayfish *Pacifastacus leniusculus* on stream invertebrate communities. *Aquatic Conservation: Marine and Freshwater Ecosystems*, **16**, 611–621.

Creed Jr., R.P. (1994). Direct and indirect effects of crayfish grazing in a stream community. *Ecology*, **75**, 2091–2103.

Davis, K.M. and Huber, R. (2007). Activity patterns, behavioural repertoires, and agonistic interactions of crayfish: A non-manipulative field study. *Behaviour*, **144**, 229–247.

Everard M., Gray J., Wilkins-Kindemba V., and Cowx I.G., (2009). Impacts of invasive species on ecosystem services: the case of the signal crayfish (Pacifastacus leniusculus). Environmental Law and Management, **21**, 250–259.

Figler, M.H., Cheverton, H.M., and Blank, G.S. (1999). Shelter competition in juvenile red swamp crayfish *Procambarus clarkii*: the influences of sex differences, relative size, and prior residence. *Aquaculture*, **178**, 63–75.

Flint, R.W. and Goldman, C.R. (1975). The effects of a benthic grazer on the primary productivity of the littoral zone of Lake Tahoe. *Limnology and Oceanography*, **20**, 935–944.

Frutiger, A., Borner, S., Busser, T., et al. (1999). How to control unwanted populations of *Procambarus clarkii* in Central Europe. *Freshwater Crayfish*, **12**, 714–726.

Füreder, L., Gherardi, F., Holdich, D., Reynolds, J., Sibley, P., and Souty-Grosset, C. (2010). *Austropotamobius pallipes*. In, *IUCN 2010. IUCN Red List of Threatened Species*. Version 2010.4, Vol. 2011. <http://www.iucnredlist.org>. Downloaded on 19 May 2011.

Gamradt, S.C. and Kats, L.B. (1996). Effect of introduced crayfish and mosquitofish on California newts. *Conservation Biology*, **10**, 1155–1162.

Gherardi, F., Renai, B., and Corti, C. (2001). Crayfish predation on tadpoles: a comparison between a native (*Austropotamobius pallipes*) and an alien species (*Procambarus clarkii*). *Bulletin Francais de la Peche et de la Pisciculture*, **361**, 659–668.

Goddard, S. and Holdich, D. (1979). Explore the native potential first. *Fish Farmer*, January 1979, p. 47.

Goessmann, C., Hemelrijk, C., and Huber, R. (2000). The formation and maintenance of crayfish hierarchies: behavioral and self-structuring properties. *Behavioral Ecology and Sociobiology*, **48**, 418–428.

Goldman, C.R. and Rundquist, J.C. (1977). A comparative ecological study of the California crayfish, *Pacifastacus leniusculus* (Dana), from two subalpine lakes (Lake Tahoe and Lake Donner). *Freshwater Crayfish*, **3**, 51–80.

Griffiths, S.W., Collen, P., and Armstrong, J.D. (2004). Competition for shelter among overwintering signal crayfish and juvenile Atlantic salmon. *Journal of Fish Biology*, **63**, 436–447.

Guan, R.Z. (1994). Burrowing behaviour of signal crayfish *Pacifastacus leniusculus* (Dana), in the river Great Ouse, England. *Freshwater Forum*, **4**, 155–168.

Guan, R.Z. (1995). *Ecological studies on the crayfish* Pacifastacus leniusculus (Dana). Ph.D. thesis, The University of Buckingham, UK.

Guan, R. Z. (1997). An improved method for marking crayfish. *Crustaceana*, 70, 641–652.

Guan, R.Z. and Wiles, P.R. (1996). Growth, density and biomass of crayfish, *Pacifastacus leniusculus*, in a British lowland river. *Aquatic Living Resources*, **9**, 265–272.

Guan, R.Z. and Wiles, P.R. (1997). Ecological impact of introduced crayfish on benthic fishes in a British lowland river. *Conservation Biology*, **11**, 641–647.

Guan, R.Z. and Wiles, P.R. (1998). Feeding ecology of the signal crayfish *Pacifastacus leniusculus* in a British lowland river. *Aquaculture*, **169**, 177–193.

Guan, R.Z. and Wiles, P.R. (1999). Growth and reproduction of the introduced crayfish *Pacifastacus leniusculus* in a British lowland river. *Fisheries Research*, **42**, 245–259.

Gurevich, J. and Padilla, D.K. (2004). Are invasive species a major cause of extinctions? *Trends in Ecology and Evolution*, **19**, 470–474.

Hanson, J.M., Chambers, P.A., and Prepas, E.E. (1990). Selective foraging by the crayfish *Orconectes virilis* and its impact on macroinvertebrates. *Freshwater Biology*, **24**, 69–80.

Harlioğlu, M.M. and Holdich, D.M. (2001). Meat yields in the introduced freshwater crayfish, *Pacifastacus leniusculus* (Dana) and *Astacus leptodactylus* Eschscholtz, from British waters. *Aquaculture Research*, **32**, 411–417.

Harrington, L.A., Harrington, A.L., Moorhouse, T.P., Gelling, M., Bonesi, L., and Macdonald, D.W. (2009a). American mink control on inland rivers in southern England: An experimental test of a model strategy. *Biological Conservation*, **142**, 839–849.

Harrington, L.A., Harrington, A.L., Yamaguchi, N., et al. (2009b). The impact of native competitors on an alien invasive: temporal niche shifts to avoid interspecific aggression? *Ecology*, **90**, 1207–1216.

Harvey, G.L., Henshaw, A. J., Moorhouse, T. P., Clifford, N. J., Holah, H., Grey, J. & Macdonald, D. W. (2014). Invasive crayfish as drivers of fine sediment dynamics in rivers: field and laboratory evidence. *Earth Surface Processes and Landforms*, **39**, 259–271.

Harvey, G.L., Moorhouse, T.P., Clifford, N.J., Henshaw, A.J., Johnson, M.F., Macdonald, D.W., Reid, I., and Rice, S.P. (2011). Evaluating the role of invasive aquatic species as drivers of fine sediment-related river management problems: the case of the signal crayfish (*Pacifastacus leniusculus*). *Progress in Physical Geography*, **35**, 517–533.

Hazlett, B.A. (2003). The effects of starvation on crayfish responses to alarm odor. *Ethology*, **109**, 587–592.

Hazlett, B., Rubenstein, D., and Rittschof, D. (1975). Starvation, energy reserves and aggression in the crayfish *Orconectes virilis* (Hagen, 1870) (Decapoda, Cambaridae). *Crustaceana*, **28**, 11–16.

Hein, C.L., Vander Zanden, M.J., and Magnuson, J.J. (2007). Intensive trapping and increased fish predation cause massive population decline of an invasive crayfish. *Freshwater Biology*, **52**, 1134–1146.

Herberholz, J., McCurdy, C., and Edwards, D.H. (2007). Direct benefits of social dominance in juvenile crayfish. *Biological Bulletin*, **213**, 21–27.

Holdich, D.M. (1993). A review of astaciculture: freshwater crayfish farming. *Aquatic Living Resources*, **6**, 307–317.

Holdich, D.M., Gydemo, R., and Rogers, W.D. (1999a). A review of possible methods for controlling nuisance populations of alien crayfish. In F. Gherardi, and D.M. Holdich, eds., *Crayfish in Europe as Alien Species: How to Make the Best of a Bad Situation?* pp. 245–270. A.A. Balkema, Rotterdam.

Holdich, D.M., Jay, D., and Goddard, J.S. (1978). Crayfish in the British Isles. *Aquaculture*, **15**, 91–97.

Holdich, D.M., Reynolds, J.D., Souty-Grosset, C., and Sibley, P.J. (2009). A review of the ever increasing threat to European crayfish from non-indigenous crayfish species. *Knowledge and Management of Aquatic Ecosystems*, **11**, 394–395.

Holdich, D.M. and Rogers, W.D. (1997). The white-clawed crayfish, *Austropotamobius pallipes*, in Great Britain and Ireland with particular reference to its conservation in Great Britain. *Bulletin Francais de la Peche et de la Pisciculture*, **347**, 597–616.

Holdich, D.M., Rogers, W.D., and Reader, J.P. (1995). Crayfish Conservation R&D note 466, National Rivers Authority, Bristol.

Holdich, D.M., Rogers, W.D., and Reynolds, J.D. (1999b). Native and alien crayfish in the British Isles. In F. Gherardi and D.M. Holdich, eds., *Crayfish in Europe as Alien Species: How to Make the Best of a Bad Situation*? pp. 221–236. A.A. Balkema, Rotterdam.

Holdich, D.M. and Sibley, P.J. (2009). ICS and NICS in Britain in the 2000s. In J. Brickland, D.M. Holdich, and E.M. Imhoffeds. Crayfish Conservation in the British Isles. In *Proceedings of a conference held on 25th March 2009 in Leeds, UK*, pp. 13–33.

Holdich, D.M., Sibley, P.J., and Peay, S. (2004). The white-clawed crayfish—a decade on. *British Wildlife*, **15**, 153–164.

Hyatt, M.W. (2004). Investigation of crayfish control technology. Final report to the Arizona Game and Fish Department, Wildlife Management Division.

Ibbotson, A.T., Tapia, G., Furse, M.T., et al. (1997). *Impact of the signal crayfish Pacifastacus leniusculus and its associated crayfishery on the River Thame*. Open Investigation Report to the Environment Agency.

Jones, P.L., De Silva, S.S., and Mitchell, B.D. (1996). Effect of dietary protein content on growth performance, feed utilization and carcass composition in the Australian freshwater crayfish, *Cherax albidus* Clark and *Cherax destructor* Clark (Decapoda, Parastacidae). *Aquaculture Nutrition*, **2**, 141–150.

Kataria, M. (2007). A cost-benefit analysis of introducing a non-native species: the case of the signal crayfish in Sweden. *Marine Resources Economics*, **22**.

Lewis, S.D. (2002). Pacifastacus. In D.M. Holdich, ed. *Biology of Freshwater Crayfish*, pp. 511–540. Blackwell, Oxford.

Lodge, D.M., Deines, A., Gherardi, F., Yeo, D.C.J., et al. (2012). Global introductions of crayfish: evaluating the impact of species invasions on ecosystem services. *Annual Review* of *Ecology and Systematics*, **43**, 449–472.

Lodge, D M., Kershner, M.W., Aloi, J.E., and Covich, A.P. (1994). Effects of an omnivorous crayfish (*Orconectes rusticus*) on a freshwater littoral food web. *Ecology*, **75**, 1265–1281.

Lowery, R.S. and Holdich, D.M. (1988). *Pacifatacus leniusculus* in North America and Europe, with details of the distribution of induced and native crayfish species in Europe. In D.M. Holdich and R.S. Lowery, eds., *Freshwater Crayfish: Biology, Management and Exploitation*. pp. 283–308. CMM Helm Ltd., London.

Lozan, J.L. (2000). On the threat to the European crayfish: A contribution with the study of the activity behaviour of four crayfish species (Decapoda: Astacidae). *Limnologica*, **30**, 156–161.

Macdonald, D.W. and Burnham, D. (2011). The state of Britain's mammals: A focus on invasive species. People's Trust for Endangered Species.

Macdonald, D.W. and Feber, R.E., eds. (2015). *Wildlife Conservation on Farmland. Managing for Nature on Lowland Farms*. Oxford University Press, Oxford.

Macdonald, D.W., King, C.M., and Strachan, R. (2006). Introduced species and the line between biodiversity conservation and naturalistic eugenics. In D.W. Macdonald and K. Service, eds., *Key Topics in Conservation Biology*. pp. 187–206. Blackwell Publishing, Oxford.

McCarthy, J.M., Hein, C.L., Olden, J.D., and Vander Zanden, M.J. (2006). Coupling long-term studies with meta-analysis to investigate impacts of non-native crayfish on zoobenthic communities. *Freshwater Biology*, **51**, 224–235.

Millennium Ecosystem Assessment (2005). *Appendices*. <http://www.millenniumassessment.org/>, accessed 19/05/2011.

Momot, W.T. (1993). The role of exploitation in altering the processes regulating crayfish populations. *Freshwater Crayfish*, **9**, 107–117.

Momot, W.T., Gowing, H., and Jones, P.D. (1978). The dynamics of crayfish and their role in ecosystems. *American Midland Naturalist*, **99**, 10–35.

Momot, W.T. and Leering, M. (1986). Aggressive interaction between *Pacifastacus leniusculus* and *Orconectes virilis* under laboratory conditions. *Freshwater Crayfish*, **6**, 87–93.

Moorhouse, T.P. and Macdonald, D.W. (2010). Immigration rates of signal crayfish (*Pacifastacus leniusculus*) in response to manual control measures. *Freshwater Biology*, **56**, 993–1001.

Moorhouse, T.P. and Macdonald, D.W. (2011a). The effect of removal by trapping on body condition in populations of signal crayfish. *Biological Conservation*, **144**, 1826–1831.

Moorhouse, T.P. and Macdonald, D.W. (2011b). The effect of manual removal on movement distances in populations of signal crayfish (*Pacifastacus leniusculus*). *Freshwater Biology*, **56**, 2370–2377.

Moorhouse, T.P., Poole, A.E., Evans, L.C., Bradley, D.C., and Macdonald, D.W. (2014). Intensive removal of signal crayfish (*Pacifastacus leniusculus*) from rivers increases numbers and taxon richness of macroinvertebrate species. *Ecology and Evolution*, **4**, 494–504.

O'Hare, M.T. and Murphy, K.J. (1999). Invertebrate hydraulic microhabitat and community structure in *Callitriche stagnalis scop*. patches. *Hydrobiologia*, **415**, 169–176.

Peay, S. (2001). *Eradication of alien crayfish populations*. Report to the Environment Agency. R&D Technical Report W1–037/TR1.

Peay, S. (2009). Invasive non-indigenous crayfish species in Europe: recommendations on managing them. *Knowledge and Management of Aquatic ecosystems*, **3**, 394–395.

Peay, S., Guthrie, N., Spees, J., Nilsson, E., and Bradley, P. (2009). The impact of signal crayfish (*Pacifastacus leniusculus*) on the recruitment of salmonid fish in a headwater stream in Yorkshire, England. *Knowledge and Management of Aquatic Ecosystems*, **394–395**, 1–15.

Ranta, E. and Lindstrom, K. (1992). Power to hold sheltering burrows by juveniles of the signal crayfish, *Pacifastacus leniusculus*. *Ethology*, **92**, 217–226.

Ranta, E. and Lindstrom, K. (1993). Body size and shelter possession in mature signal crayfish, *Pacifastacus leniusculus*. *Annales Zoologici Fennici*, **30**, 125–132.

Reynolds J.D. and Matthews M.A. (1993). Experimental fishing of Austropotamobius pallipes in an Irish midlands lake. *Freshwater Crayfish*, **9**, 147–153.

Reynolds, J. and Sauty-Grosset, C. (2012). *Management of Freshwater Biodiversity: Crayfish as Bioindicators*. Cambridge University Press.

Richards, K. and Fuke, P. (1977). Freshwater crayfish: the first centre in Britain. *Fish Farming International*, **4**, 12–15.

Rogers, W.D. and Holdich, D.M. (1995). Crayfish production in Britain. In *Freshwater Crayfish 10*, (eds. M.C. Geddes, D.R. Fielder and A.M.M. Richardson), pp. 583–595. Baton Rouge, Louisiana state university, LA, USA.

Rogers, W.D., Holdich, D.M., and Carter, E. (1997). *Crayfish Eradication*. Report for English Nature, Peterborough.

RSPCA (2009). *Humane electrical stun/killing of Crustacea*. RSPCA, <http://www.rspca.org.uk/>, accessed 10/2010.

Stocker, A.M. and Huber, R. (2001). Fighting strategies in crayfish *Orconectes rusticus* (Decapoda, Cambaridae) differ with hunger state and the presence of food cues. *Ethology*, **107**, 727–736.

Taugbøl T. (2004). Exploitation is a prerequisite for conservation of A. astacus. *Bulletin Francais de la Peche et de la Pisciculture*, **372–373**: 275–279.

Usio, N. (2000). Effects of crayfish on leaf processing and invertebrate colonisation of leaves in a headwater stream: decoupling of a trophic cascade. *Oecologia*, **124**, 608–617.

Westman, K., Savolainen, R., and Pursiainen, M. (1999). Development of the introduced North American signal crayfish, *Pacifastacus leniusculus* (Dana), population in a small Finnish forest lake in 1970–1997. *Boreal Environment Research*, **4**, 387–408.

Wickins, J.F. and Lee, D'O. (2002). *Crustacean Farming—Ranching and Culture*. Blackwell Science. 446 pp.

CHAPTER 9

Small farmland bird declines, gamebird releases, and changes in seed sources

Alan Larkman, Ian Newton, Ruth E. Feber, and David W. Macdonald

Let it also be borne in mind how infinitely complex and close-fitting are the mutual relations of all organic beings to each other and to their physical conditions of life.

Charles Darwin, *On the Origin of Species*

9.1 Introduction

The overall number of wild breeding birds in Britain, all species combined, has remained relatively stable over the last four decades, but many species associated with farmland have markedly declined or contracted in range (Eaton et al. 2011). Wide-ranging and profound changes in agricultural practices over this period are thought to have caused or accentuated these declines, but various causal mechanisms have been implicated, some of them complex (Fuller et al. 1995; Robinson and Sutherland 2002). Understanding these mechanisms has been an important research priority in recent years (Wilson et al. 2009). Despite this research focus, and a large financial input to agri-environment schemes, many farmland bird species continue to decline or have shown only slight increases after large declines (Davey et al. 2010; Baillie et al. 2013; Baker et al. 2012; Musgrove et al. 2013). Population declines among farmland birds appear to have been more severe in the UK than in any other European country (Donald et al. 2001, 2006).

The UK's annually published Farmland Bird Index (or Indicator; FBI), which describes trends in the breeding numbers of common farmland birds, is used to set conservation targets and to monitor progress (Gregory et al. 2004; Macdonald and Feber 2015: Chapter 5). The FBI is derived from data for 19 species of farmland bird, largely from two long-running, national-scale surveys, the Common Bird Census and the Breeding Bird Survey (Freeman et al. 2007; Baillie et al. 2013). Between 1970 and approximately 1977, the FBI increased modestly; however, since 1977, the FBI has decreased by approximately 50% (Defra 2013a). Declines among the different FBI species have not, though, been uniformly severe. The FBI comprises two subsets of species (Defra 2013a): a specialist subset (12 species restricted to, or highly dependent on, farmland habitats) and a generalist subset (7 species less dependent on farmland). The generalist subset has shown little long-term change while the specialist subset increased by around 10% between 1970 and 1977, after which it steeply declined to approximately 30% of its 1977 level (Fig. 9.1a).

It has proved difficult to link agricultural changes to the time course of changes in farmland bird populations. Progressive, and essentially linear, increasing trends in cereal yields were reported for Britain between 1945 and 2000 (Robinson and Sutherland 2002) and also for 15 European Union member states between 1961 and 1999 (Donald et al. 2001). Cereal yields are commonly taken as an index of overall agricultural productivity or 'intensity'. For seed-eating farmland birds, reductions in food supplies have been implicated in population declines and, in most such species, reduced annual survival seems to have been the most important demographic process involved (Robinson and Sutherland 2002; Gregory et al. 2004; Newton, 2004; Siriwardena et al. 2008). However, these findings offer no obvious explanation for the abrupt downturn in the trend of the FBI for specialist species that occurred around 1977.

Wildlife Conservation on Farmland. Conflict in the Countryside. Edited by David W. Macdonald and Ruth E. Feber.
© Oxford University Press 2015. Published 2015 by Oxford University Press.

In a detailed study, Chamberlain et al. (2000) attempted to link changes in farmland bird populations in England and Wales to changes in agricultural practices between 1962 and 1995. Thirty-one agricultural variables were used to derive a composite index of agricultural intensification (the 'HABDCA' score), which was compared with an index of bird population changes. They found that the HABDCA scores corresponded poorly with changes in their bird population index; indeed the two measures were largely out of phase, with periods of rapid decline in bird numbers (notably 1976–1982) occurring during times of relative stability in agricultural practice. We plotted the FBI for specialist species alone against Chamberlain et al.'s (2000) HABDCA score for the previous year and found a similar result. In particular, in the period 1970–77 the HABDA score increased rapidly but coincided with a small increase in the FBI, while in the period 1977–82 a slow increase in the HABDCA score coincided with a sharp fall in the FBI for specialist species.

The lack of temporal correspondence between changes in the HABDCA and the bird population indices of Chamberlain et al. (2000) and the FBI for specialist species raises the possibility that one or more additional factors operating at a national scale may have been important in driving bird population changes. In this chapter, through literature review and simple analyses of the available data on the population trends of farmland birds in Britain, we highlight two possible additional candidate factors. The first relates to the increase in the annual release of captive-reared, non-native gamebirds by the commercial shooting industry, while the second concerns changes in the sources and types of seed food available to farmland birds during winter and early spring. We focus on the period since 1966, during which breeding population trends are available for common bird species from the British Trust for Ornithology, largely based on the two long-running surveys mentioned earlier (Freeman et al. 2007; Baillie et al. 2013).

9.2 Intensification of gamebird shooting

Following the catastrophic decline in the native grey partridge *Perdix perdix* (Macdonald and Feber 2015: Chapter 13), the UK shooting industry increasingly moved towards the shooting of released, captive-reared, non-native species, particularly the pheasant *Phasianus colchicus* and, to a lesser extent, the red-legged partridge *Alectoris rufa*. Data from the Game & Wildlife Conservation Trust (GWCT) indicate that the density of captive-reared pheasants released each year increased from approximately 60 birds per 100 ha in 1966 to 290 per ha in 2005 (Fig. 9.1b), while the density of released red-legged partridges increased from virtually zero prior to 1970 to approximately 80 birds per ha in 2005. Increase in the release of pheasants could be divided into three phases, with a slow increase between 1966 and 1975, a rapid and approximately linear increase from 1976 to 1990, and a slower increase between 1991 and 2000. Over this period, there was also a widening divergence between the density of birds released and the proportion shot: between 1966 and 1975 there was little change between these figures, but since 1975, as the numbers of released birds increased, the proportion shot showed a progressive linear decline. In other words, the relationship between these variables appeared to be density dependent. The density of captive-reared red-legged partridges released per year increased progressively over the period involved without abrupt changes.

A rough estimate of the absolute numbers of pheasants and red-legged partridges released each year can be obtained from the annual release densities cited earlier, together with the estimates of 35 million pheasants and 6.5 million red-legged partridges for 2004 provided by the PACEC report (PACEC 2006). This procedure suggests that the annual pheasant releases increased from approximately 8 million in 1966 to 36.5 million in 2005, while red-legged partridge releases increased from approximately zero in 1966 to 6.7 million in 2005. Similarly, the difference between the number of pheasants released and those shot increased from approximately 0.6 million in 1966 to over 20 million in 2005. Recent estimates suggest a continuing increase in the numbers released; for example, 40 million pheasants and 6–8 million red-legged partridges were reportedly released in 2009 (Shooting Times and Country Magazine 2009), consistent with 50.5 million (in 2011) and 50.3 million (in 2012) pheasants and partridges combined, declared in the Great Britain Poultry Register (DEFRA 2013b).

9.3 Correlations between gamebird releases and farmland bird population trends

The period of greatest acceleration in gamebird releases appeared to coincide with the onset of steepest decline in the FBI for specialist farmland bird species. This led us to question whether the large-scale release of gamebirds was in some way connected with the population trends of specialist farmland birds.

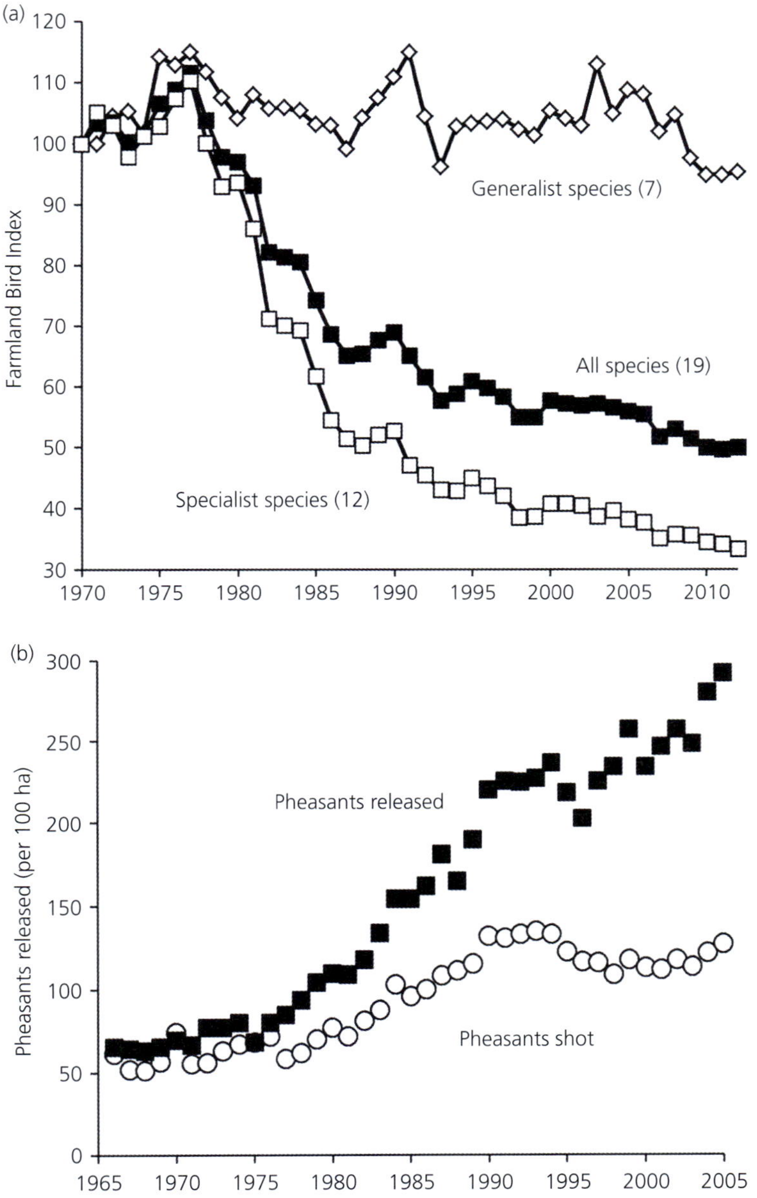

Figure 9.1 (a) The UK Farmland Bird Index for all 19 species (filled squares), the 7 generalist species (open diamonds) and the 12 specialist species (open squares). Data from Defra (2013a); (b) annual density of captive-reared pheasants released (filled squares) and shot (open circles) in the UK. Graph based on data from the Game & Wildlife Conservation Trust (<http://www.gwct.org.uk>, accessed November 2012).

9.3.1 Population trends among resident seed-eating farmland birds

Farmland birds vary considerably in their diets and seasonal movements. Both pheasants and red-legged partridges are primarily seed-eaters, so if these non-native species were at all involved in the declines of farmland birds, one might expect that wild seed-eaters would be the most vulnerable. We therefore focused on twelve wild species, each of which met three criteria: (1) seeds form an important part of their winter diet; (2) they are generally resident in Great Britain during winter; and (3) their numbers are more likely to have been influenced by agricultural activities than by large-scale disease outbreaks or garden bird feeding. Our selection of species was based on information from Robinson (2005), Newton (2010), Harrop et al. (2013), Baillie et al. (2013), and Lawson et al. (2012).

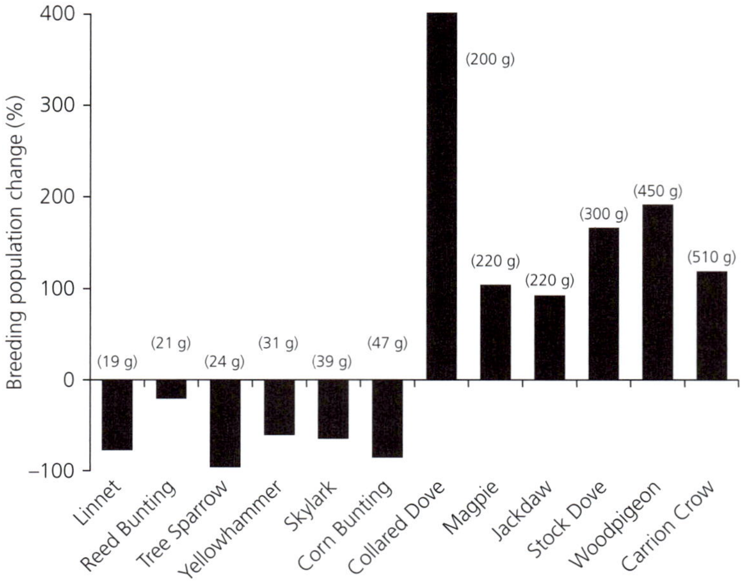

Figure 9.2 Relative changes in breeding population among 12 species of resident, seed-eating farmland birds in England 1966–2010 (data from Baillie et al. 2013), selected according to criteria given in the text. Numbers in brackets above columns are mean live body weight for each species (data from Robinson 2005), mean of male and female given where different. For collared dove, the relative change from 1972 is shown; the corresponding value for the change in the UK from the small population present in 1966 is 2074%.

Our twelve selected species varied considerably in their breeding population changes during the period 1966 and 2011, ranging from a decrease of 96% for tree sparrow *Passer montanus* to an increase of at least 400% for collared dove *Streptopelia decaocto* (Fig. 9.2). Interestingly, the direction of population change was clearly related to body weight, with all six of the small seed-eating species (< 50 g; linnet *Carduelis cannabina*, reed bunting *Emberiza schoeniclus*, tree sparrow, yellowhammer *Emberiza citrinella*, skylark *Alauda arvensis*, and corn bunting *Emberiza calandra*) showing declines, and the six large seed-eating species (≥ 200 g; collared dove, magpie *Pica pica*, jackdaw *Corvus monedula*, stock dove *Columba oenas*, woodpigeon *Columba palumbus*, and carrion crow *Corvus corone*) all showing increases. Any explanation of these population changes over the 46-year period would have to account for why the breeding populations of the large seed-eaters had increased, while those of the small seed-eaters had declined. This made it unlikely that a general reduction in the availability of seed food in the farmed environment was responsible.

Using analyses based on the smoothed breeding population trends produced by the BTO (Freeman et al. 2007; Baillie et al. 2013)[1], we gained further insights into the time course of population changes in our twelve species.

9.3.2 Tree sparrow

The species that showed the greatest relative decline since 1966 was the tree sparrow, its breeding population decreasing by approximately 96%, from over 5.5 million pairs in 1966 to about 250,000 pairs in 2011. The population showed a small decline between 1966 and 1970, then a period of relative stability until 1975, followed by a rapid and profound decline until approximately 1995 (Fig. 9.3a). Since the late 1990s the population has increased slightly, possibly due to local conservation efforts in remnant areas of suitable habitat, which often involved provision of food and/or nest boxes (Fig. 9.4). Breeding performance did not appear to be the key demographic factor driving the population decline in tree sparrows, since the numbers of fledglings per nesting attempt increased over this period, as was also recorded for other small seed-eating species (Siriwardena et al. 2000; Baillie et al. 2013).

Changes in the tree sparrow breeding population did not correspond closely to changes in the HAB-DCA index, and the relationship showed several abrupt changes in slope, (Fig. 9.3b). In contrast, however, the sparrow population trend did correspond closely with the density of captive-reared pheasants released per year, with no abrupt changes in the slope of the relationship (Fig. 9.5). Over the entire 40-year

[1] We converted the relative indices given by Baillie et al. (2013) to absolute numbers using the recent population estimates of Musgrove et al. (2013); for some species the number of breeding pairs was taken to be equal to the number of occupied territories.

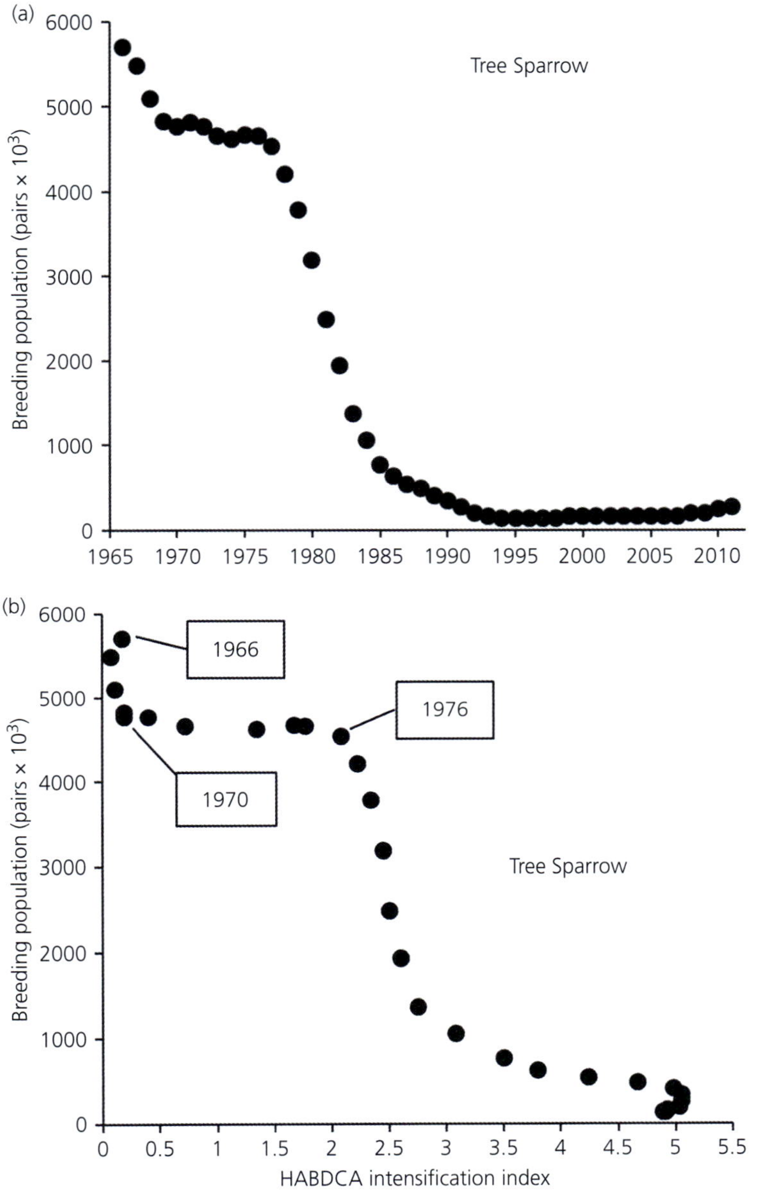

Figure 9.3 (a) smoothed annual estimate of the breeding population of the tree sparrow, calculated from Baillie et al. (2013); (b) annual tree sparrow breeding population plotted against the HABDCA index of agricultural intensification for the previous year (data obtained from Chamberlain et al. 2000).

period analysed, in which the tree sparrow breeding population declined by about 97%, the relationship could be well approximated by a single exponential decay function with only minimal systematic deviations. This relationship suggested a halving of the tree sparrow population for every increase of approximately 40 pheasants released per 100 ha.

The tree sparrow breeding population was also correlated with the density of pheasants shot, with the difference between the density of pheasants released and those shot, and with the density of red-legged partridges released. But in each case the correlations were less close than for the density of pheasants released.

Figure 9.4 Tree sparrow *Passer montanus* at a nest box provided by a local conservation scheme in Oxfordshire. Photograph ©A. Larkman.

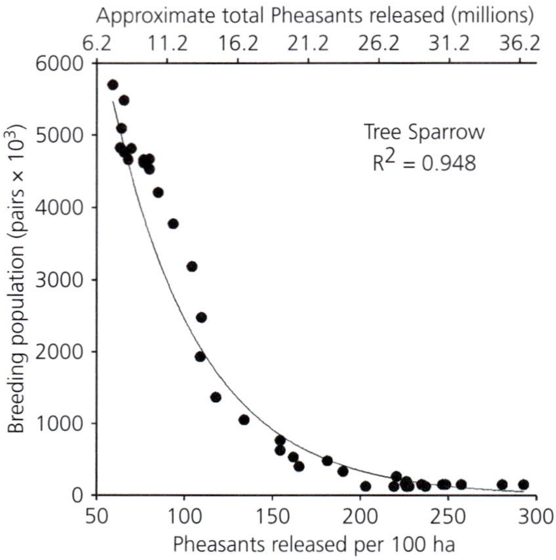

Figure 9.5 Annual tree sparrow breeding population plotted against the density of pheasants released during the previous year. Upper horizontal axis shows an approximate estimate of the total number of pheasants released. Solid line is a fitted single exponential function.

9.3.3 Combined small seed-eating species

We also calculated the combined breeding population for the other five species of small, resident, seed-eating farmland birds that met the criteria given above. This measure differed from the overall FBI not only in the fewer, but more homogeneous, set of species considered, but also in the method of calculation. In the FBI, all species are given equal weighting, but in our measure the breeding populations of the component species were summed for each year; in effect, the species were weighted according to their abundance.

Overall, the combined breeding population of these five species decreased from approximately 8.1 million pairs in 1966 to 2.9 million pairs in 2011, an overall decline of 74% over the 46 years. The time course of this overall decline was remarkably similar to that of the tree sparrow, but less severe. Again, this overall population trend for the five species did not correspond closely with the HABDCA index (Fig. 9.6a), but corresponded closely with the density of pheasants released during the previous year (Fig. 9.6b). The relationship between the combined small seed-eating bird population and the density of pheasants released could be approximated by a simple power function, and the pattern of minor systematic deviations from the smooth function was similar to that shown by the tree sparrow.

The explanation favoured by Chamberlain et al. (2000) for the lack of close correspondence between bird population trends and the HABDCA index was that a time lag of approximately six years may have operated before agricultural changes began to affect bird populations. Such a lag would have brought the main period of rapid decline (1976–1982) into line with a period of rapid agricultural intensification (1970–1976). However, a long lag period may not pertain for small seed-eating birds. Most agricultural changes would be expected to have a rapid effect on bird food supplies; for example, herbicide spraying or autumn stubble ploughing would result in an almost immediate reduction in surface weed seeds. Secondly, small seed-eaters are short-lived birds that breed in their first year and have typical lifespans of only 2 or 3 years (Robinson 2005). Furthermore, reduced winter survival has been identified as a key factor underlying small farmland bird declines (Robinson and Sutherland 2002; Gregory et al. 2004; Newton 2004), leading to the assumption that this would have an impact on the size of the breeding population in the following spring. Finally, the breeding populations of some small seed-eaters actually increased over the

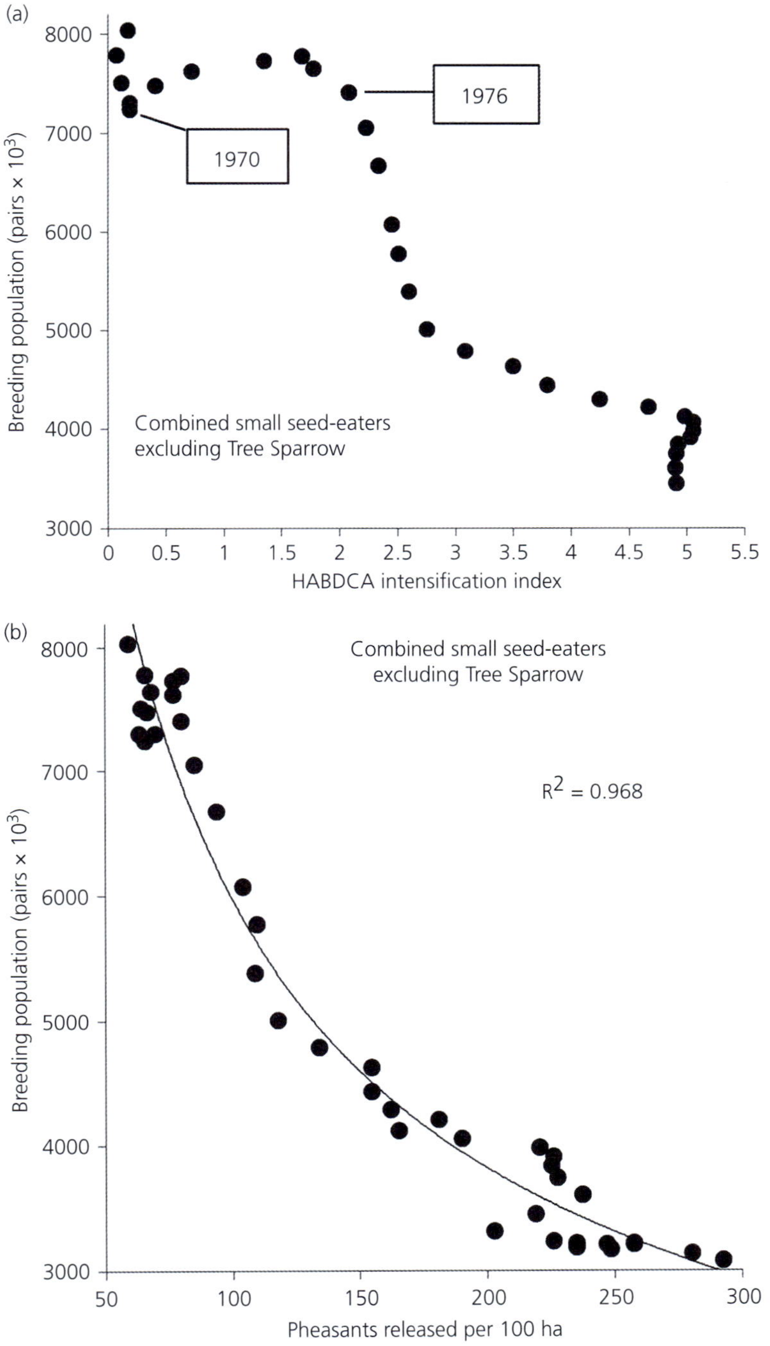

Figure 9.6 Combined breeding population for five resident small species of seed-eaters (linnet, reed bunting, yellowhammer, skylark, corn bunting; tree sparrow excluded) plotted against: (a) the HABDCA index of agricultural intensification; and (b) the density of pheasants released during the previous year. Solid line shows a fitted simple power function of the form $y = cx^r$ where c and r are constants.

period 1970–1975, and the timing of the rapid decline was similar in species that had increased (e.g. corn bunting), decreased (linnet), or remained relatively stable (skylark) in the preceding years. It seems to us more likely that the post-1975 declines of small seed-eaters were driven by some factor(s) other than agricultural change that came into operation or increased in importance close to 1975. Could the increase in the release of captive-reared gamebirds be one such factor?

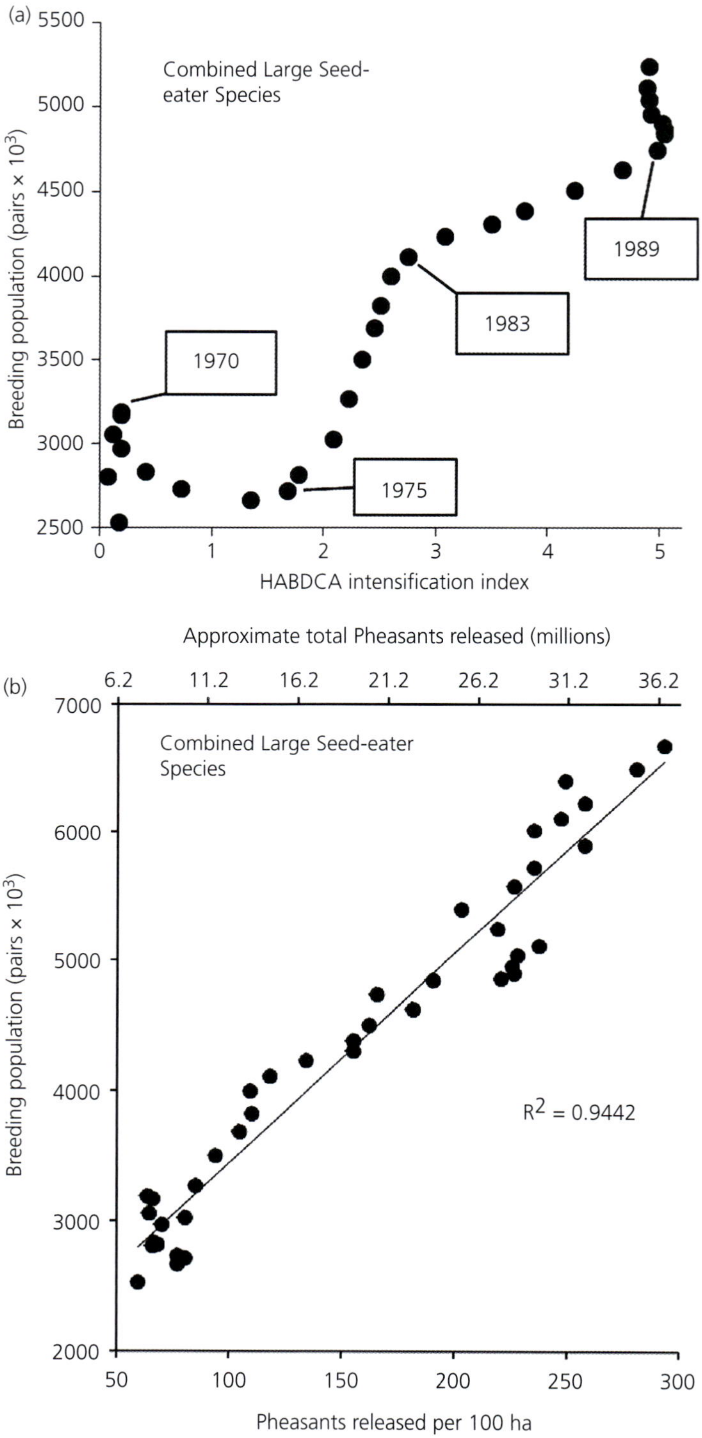

Figure 9.7 (a) Combined breeding population for three resident large seed-eating species included in the Farmland Bird Index (woodpigeon, jackdaw, stock dove) plotted against the HABDCA index of agricultural intensification in the previous year; (b) combined breeding population for three resident large seed-eating species plotted against the density of pheasants released during the previous year. The upper horizontal axis shows an approximate estimate of the total number of pheasants released each year. The solid line is a fitted linear function.

9.3.4 Large seed-eating species

In a similar way, we calculated the combined breeding population for the three large resident seed-eating species in the FBI which met our criteria for inclusion (woodpigeon, stock dove, and jackdaw). In stark contrast to the small seed-eaters, each of these large species showed substantial increases in their breeding populations between 1966 and 2011. Their combined total increased from approximately 2.5 million pairs in 1966 to > 7 million pairs in 2011. The time course of the increase was approximately linear, except for a somewhat steeper increase during 1966–1970 followed by a decrease during 1971–1975.

As in the small species, the combined population trend of these large species did not correspond closely with changes in the HABDCA index (Fig. 9.7a). Interestingly, the increase in the three large species corresponded closely with the density of pheasants released per year, and the relationship could be well approximated by a linear function with only minimal systematic deviations (Fig. 9.7b). For large species, then, their relationship with pheasant densities resembled an inverted form of the corresponding one for small seed-eating species, the large species increasing while the small ones declined.

The relationships between breeding population trends and both the HABDCA index and the density of pheasants released were explored for three additional groupings of large, resident seed-eating species: (1) an extended large seed-eater group, consisting of the three species from the FBI (woodpigeon, jackdaw, stock dove) plus the collared dove, carrion crow, and magpie; (2) a combined columbid group (woodpigeon, stock dove, collared dove); and (3) a combined corvid group (jackdaw, carrion crow, magpie). For none of these three groupings did the combined population trends correspond closely to the HABDCA index (data not shown). But the combined population trends for all three groupings showed close linear relationships with the density of pheasants released. Graphs for the three groupings were remarkably similar; even the minor deviations from linear behaviour were similar in the non-overlapping columbid and corvid species groups (which differ considerably in their ecology). Since released gamebirds are supplied with considerable quantities of food (e.g. cover crops and grain in hoppers, see below) which are exploited by corvids and columbids (Potts 2012), it may be that gamebird release density has served as a proxy for the quantity of food provided by the shooting industry, which may have had a direct effect on the population sizes of large seed-eating species. In a later section we discuss why this increased food supply may not have benefited small seed-eating species.

9.4 Autumn numbers of resident seed-eating birds

Is the scale of gamebird releases sufficiently large to have had a wide-ranging influence on the numbers of several native bird species? Numbers of gamebirds released in recent years seem large relative to the breeding numbers of native resident species. However, breeding populations, as at the start of the breeding season, generally represent the lowest point in the annual population cycle, before any young of the year have appeared (Newton 1998). In contrast, gamebird releases occur in late summer when the numbers of native resident birds would be expected to be substantially larger than their breeding populations. In order to view gamebird release numbers in the context of late-summer populations of native species, we derived an index of post-breeding numbers (Post-Breeding Index; PBI) for each species, calculated from the breeding population and published national-scale data on breeding productivity (Robinson 2005)[2]. Data on breeding success for wild pheasants are available from the BTO nest record scheme between 1950 and 1985, and from records collected by gamekeepers for 1978 and 1979 (Robertson 1991)[3]. The PBI represents the approximate upper limit of the number of birds likely to be present in early autumn. It does not take account of adult or fledgling mortality during or soon after the breeding season, which could be substantial. Overestimation of autumn numbers by the PBI is likely to be greatest for small bird species, which generally produce large numbers of young with high mortality. The PBI can be calculated in a consistent way for each species each year, and it takes account of available data

[2] The PBI was calculated for each species for a given year according to: $PBI = (BP \times 2) + (BP \times N_{broods} \times FS) + N_{released}$, where BP is the breeding population in pairs, N_{broods} is the average number of broods produced per year, FS is fledging success (the average number of fledglings produced per nesting attempt), and $N_{released}$ is the number of captive-reared birds released per year (if any). Estimates of fledging success and numbers of broods for native species were taken from BTO national surveys (Robinson 2005; Baillie et al. 2013).

[3] BTO records indicated a low success rate of 0.74 fledglings per female, while gamekeepers reported a higher figure of 1.69. In the present analysis, we used the lower BTO figure in order to be conservative, and the same low fledgling success rate was also used for red-legged partridge.

on changes in breeding population size and performance over time. The numbers of each species will decrease throughout autumn and winter, progressively tending towards the size of their subsequent breeding populations.

We calculated PBIs for the 12 species of native, resident seed-eating birds, and for pheasant and red-legged partridge, including the numbers released. Dramatic changes in the figures occurred between 1966 and 2011. In 1966, the three species with the highest PBI values were tree sparrow, skylark, and linnet. But by 2011, the species with the highest PBI by a considerable margin was the pheasant, followed by the woodpigeon. Remarkably, the combined PBI for all 14 species (including gamebirds) hardly changed over this period (decreasing slightly from around 115 million birds in 1966 to 112 million in 2011), despite the striking changes in the balance between small and large species. The combined PBI for the six small seed-eating species decreased from approximately 92 million in 1966 to 21 million in 2011, while that for the six large native species increased from 11 million to 38 million. The PBI for non-native gamebirds increased from 12 million in 1966 to at least 53 million in 2011.

Use of the PBI to estimate the relative numbers of small and large birds in early autumn does not take account of differences in body size, which may influence their impact on the environment. We therefore estimated the autumn biomass for different species, by multiplying their PBI by their mean live body weight (Fig. 9.8), and found that the total biomass of these various seed-eating birds more than trebled between 1966 and 2011. A striking change was in the balance between small and large seed-eating species. The combined autumn biomass of the six small species decreased from approximately 2500 tonnes in 1966 to 625 tonnes in 2011, while the combined biomass of the large species (including gamebirds) increased from approximately 18,000 tonnes to 72,000 tonnes. By 2011, the combined autumn biomass of non-native gamebirds exceeded that of all the native seed-eating species added together by a factor of > 4. Inclusion of other native seed-eating farmland birds that did not meet our inclusion criteria, such as goldfinch, greenfinch and rook, would not substantially alter this figure.

Our analyses suggest that the release of captive-reared gamebirds has been on a scale that has the potential to have had a marked impact on native seed-eating species, with overlapping dietary and habitat needs. The large increase in the combined biomass of native and non-native species implies that the quantity of food available for seed-eating birds as a whole, during at least the early part of the autumn/winter period, may have increased substantially since 1966.

On the basis of these results, we advance the hypothesis that changes in shooting practices, combined

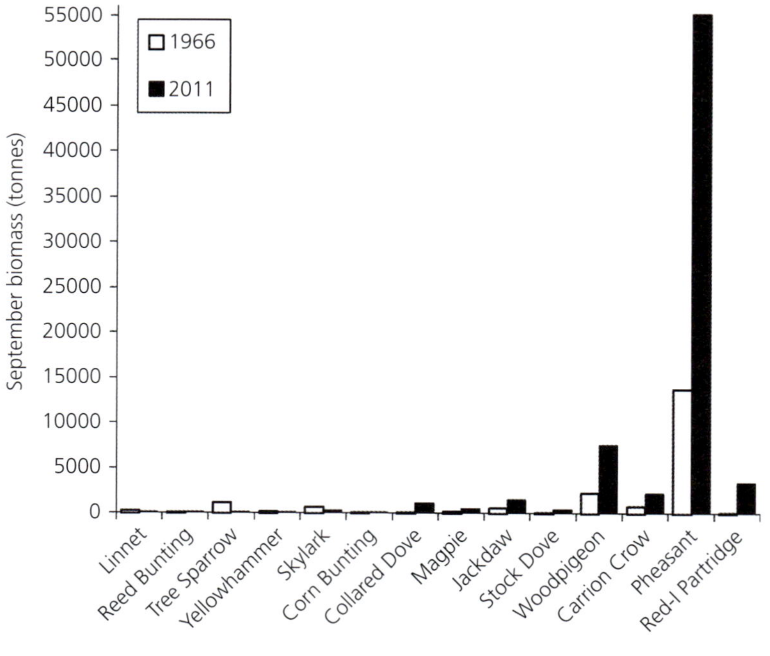

Figure 9.8 September biomass of seed-eating species, derived from the Post-Breeding Index (see text) and the mean live body weight for each species, in 1966 and 2011.

with on-going agricultural intensification, have been important in both the observed declines in small seed-eating species and the population increases in large species. How might changes in game shooting have affected small seed-eating farmland birds? One possible mechanism is that traditional sources of seed food such as overwintered stubbles have been replaced by novel sources such as game cover plots and the provision of grain for gamebirds in hoppers. We hypothesize that these novel sources have been highly suitable not only for gamebirds, but also for a range of large, native seed-eating birds. However, they have not been an adequate replacement for traditional seed sources for small species. In the second half of this chapter, we review published evidence and perform simple quantitative analyses relating to changes in the quantity, type, and density of seed food available for birds during winter over the last five decades, to investigate how these changes might have differentially affected large and small seed-eating species.

9.5 Energy and food requirements of large and small seed-eating species

9.5.1 Daily food demand

The food requirement of birds of different sizes can be estimated from their field metabolic rate (FMR), defined as the total energy cost a wild individual pays during the course of a day (e.g. through basal metabolism, thermoregulation, locomotion, feeding, digestion). FMR in birds is largely determined by body weight (e.g. Nagy 1987, 2005). It can be estimated (in kJ/day) from body weight (BW, in g) by equations such as that given by Stephens et al. (2003):

$$\log(\text{FMR}) = 1.037 + 0.640 \log(\text{BW})$$

[Equation 1].

Using this equation, we calculated the FMRs of our 12 native seed-eating species, and for pheasant and red-legged partridge. Values ranged from 72 kJ/day for linnet to just over 1000 kJ/day for pheasant. From the known energy content of different seed types (Green 1978; Robinson 2004), we estimated the weights of carbohydrate-based seed needed to maintain individuals of the different species in energy balance. We then calculated the numbers of seeds of different types that would have to be found and consumed each day by birds of different sizes. For example, if feeding exclusively on wheat seeds of mean weight 45 mg, a tree sparrow would need to eat about 120 seeds per day to maintain energy balance, and a pheasant about 1500 seeds/day, and while feeding on weed seeds (each of 1 mg) a linnet would need to find about 4800 seeds per day, and a pheasant 67,000 seeds/day. Under conditions of low seed density, much time could be spent searching for each seed, so that large species may have difficulty in getting enough food each day if the available seeds are small and at low density.

9.5.2 Total winter food demand

We went on to calculate the total food demand over the winter period using the PBI and the daily food demand for each species. The time profile of autumn and winter mortality is not known for most species, so we assumed that the population decreased linearly over this period from the hypothetical maximum of the autumn index to the low of the breeding population[4]. We calculated the total winter food demand (WFD) in terms of wheat equivalent from the beginning of September to the end of March (taken to be 210 days), based on the area of a right trapezoid using:

$$\text{WFD} = [(\text{PBI} + \text{BP}) \ / \ 2] \times 210 \times \text{Daily ration}$$

[Equation 2],

where PBI is the Post Breeding Index and BP the number of individuals in the breeding population.

Winter food demand was calculated for the 14 seed-eating species using population data for 1966 and 2011. The estimates showed profound changes over this 46-year period (Figs. 9.9a,b). The total winter food demand for all 14 seed-eating species combined increased from 234,000 tonnes in 1966 to 595,000 tonnes in 2011. The combined demand for the six small species decreased from approximately 75,000 tonnes in 1966 to 18,000 tonnes in 2011. By contrast, the combined demand of the six large native species increased from 58,000 tonnes to 188,000 tonnes, and the demand of the non-native gamebirds increased from 102,000 tonnes to 390,000 tonnes. The proportion of the total winter food demand attributed to non-native gamebirds increased from 43% in 1966 to 65% in 2011, while that attributed to small species decreased from 32% to 3%.

9.6 Changes in available seed sources during the last 50 years

While the total winter food demand for birds has changed (increased) over the last 50 years, the seed

[4] Research by the GWCT showed that autumn and winter mortality among 325 radio-tagged, captive-reared pheasants was close to linear over time (<http://www.gwct.org.uk/>).

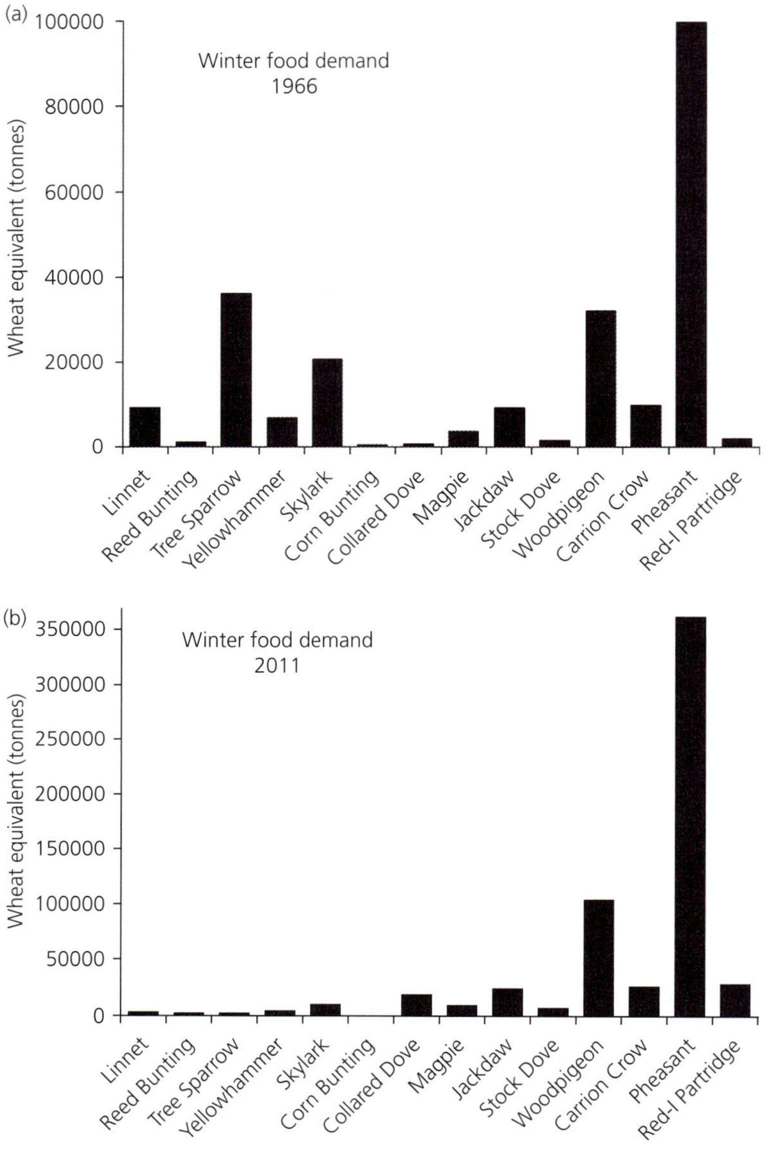

Figure 9.9 Total winter food demand, calculated using Equation 2 and expressed as tonnes of wheat equivalent, in (a) 1966 and (b) 2010, for individual species of resident, seed-eating farmland birds. The food demand of corn bunting in 2011 was too small to visualize.

sources available to birds on farmland have also changed with some traditional sources being reduced or lost, and some other sources being gained.

9.6.1 Losses of seed sources

Overwintered stubble fields are crucial sources of winter food for seed-eating birds (Wilson et al. 1996; Hancock and Wilson 2003; Gillings et al. 2005; Butler et al. 2010), but the switch from spring-sown to autumn-sown cereal growing has greatly reduced the area of stubbles available[5]. Herbicide use is also likely to have reduced weed seed production substantially, and hence the seed density in stubbles (Robinson and Sutherland 2002). For example, Butler et al. (2010) found markedly greater cumulative post-harvest seed in fields with low herbicide use (~40,000 seeds/m^2) and those not sprayed with herbicide soon before harvest

[5] Spring-sown wheat decreased by approximately 240,000 ha and spring-sown barley by approximately 180,000 ha between the mid 1960s and the mid 1990s in England and Wales: Chamberlain et al. (2000).

(~25,000 seeds/m^2), compared with fields with high herbicide use (~8000 seeds/m^2).

Taking the value of 25,000 seeds/m^2 and a mean weed seed weight of 1 mg, gives a cumulative seed mass of 25 g/m^2 or 250 kg/ha. If this were representative of the total seed available in weedy stubbles during the 1950s and 1960s, the reduction in overwintered stubble area of 420,000 ha represents a loss of ~105,000 tonnes of potential weed seed food since that time. Our estimate of the total winter food demand for all small, resident seed-eating species combined in 1966 was approximately 75,000 tonnes (see earlier). This suggests that the weedy stubble available in Britain in the mid 1960s had the potential to provide sufficient small weed seed to meet the entire winter food demand of resident small farmland birds (seed production and shedding continues for four and five months post-harvest; Butler et al. 2010).

Farming operations associated with the spring sowing of cereals also bring seeds within the seed bank to the soil surface and thus makes them available to small birds. The quantity and density of seed provided in this way may be relatively low, but it occurs at a crucial time during the late winter and early spring when other sources of seed are scarce. The use of smaller machinery may also have prolonged the period each year over which seeds were made available, thus spreading their beneficial effect. The decline in spring-sown cereal production will have reduced seeds available from this source, as will the use of herbicides in progressively reducing the seed-bank in the soil.

Further seed losses have occurred due to modern harvesting equipment reducing seed wastage. In 1970, around 250 cereal grains/m^2 were being left on fields (Robinson and Sutherland 2002). In spite of increasing crop yields, this figure had decreased to around 60 grains/m^2 by 1990, representing a loss of approximately 85 kg/ha for grains of 45 mg each. Taking account of a slight increase in the combined area of wheat and barley grown (as reported in Chamberlain et al. 2000), this represents a reduction in the amount of grain spilled during harvest of approximately 200,000 tonnes per year in England and Wales over the period 1965 to 1990.

However, grain spilled at harvest may not have been especially important in the winter survival of farmland birds, occurring briefly as it does during August–September (in the UK). Spilled seed has a short residence time, being particularly susceptible to avian and non-avian predators (Holmes and Froud-Williams 2005). We suggest that spilled seeds probably did not make a major contribution to the survival of small birds during the season of their greatest food shortage, in winter and early spring (Siriwardena et al. 2008).

Several other factors may have led to a reduction in sources of winter food, including more secure storage of grain and animal feed and reduced outdoor feeding of cereals and seed-rich hay to livestock in winter, but the magnitude of seed provision from these activities is difficult to quantify.

9.6.2 Seed source gains

At the time of the PACEC report (PACEC 2006), there were an estimated 61,000 providers of game shooting in the UK, 83% concerned mainly with released pheasants or partridges. In 2004, £33 million was spent on food put down for gamebirds, equivalent to 450,000 tonnes of wheat at the prevailing price of ~£70/tonne (Agristats; <http://www.ukagriculture.com>) (to this should be added an unknown quantity of wheat tailings and other low-value cereal products which are also fed to gamebirds). The GWCT's Lowland Shoot Survey (<https://www.gwct.org.uk/>) indicated that 89% of shoots provided supplementary food for gamebirds, the majority of which (93%) used feed hoppers and 72% also provided food scattered by hand; overall, an average of ~24 tonnes of food per shoot was provided, but it is not clear whether this survey was representative of all shoot providers, including small ones. The most commonly presented food was wheat (87%), but barley, triticale, maize, and proprietary pellets were also used.

More recent information from the GWCT suggests that shoots now spend £2.05 per bird released on feed, with a grain price now of £152 per tonne (<https://www.gwct.org.uk/>). If all the expenditure on feed were spent on grain, an annual release of 42 million pheasants would imply that some 566,000 tonnes of grain per year are currently provided.

Provision of seed for gamebirds is reduced after the end of January, when the shooting season ends (Siriwardena et al. 2008). We have found no information on how supplementary grain provision has changed over time, but it may well have increased broadly in line with the numbers of captive-reared gamebirds released.

In addition to feed dispensed in hoppers or by hand scattering, the PACEC survey (2006) reported that shoots maintained an overall 93,000 ha of game cover crops. The most commonly planted cover crop was maize or maize mix, with cereals, linseed, oilseed rape, millet, and quinoa as additional ingredients. We have not found published estimates of the seed yield from these plots, but yields of > 2 tonnes/ha from such crops can be reasonably assumed. Overall, > 180,000 tonnes of seed may be produced each year by game cover crops, giving an energy density probably > eight-fold higher than a weedy stubble. Game cover crops are typically ploughed at the end of the shooting season in early February, so provide little or nothing for seed-eaters after this date (Gillings et al. 2010).

It can thus be concluded that the shooting industry now dispenses very large quantities of seeds into the UK countryside, perhaps amounting to > 700,000 tonnes per year. This may be compared with our estimate of 595,000 tonnes of wheat or its equivalent as the total combined winter food need of the 14 species of resident farmland seed-eating birds that we considered (including non-native gamebirds). This food is provided in the form of large (very large in the case of maize) and medium-sized seeds (such as millet), presented at very high densities. Much of it may be taken by non-target species, including corvids and columbids (Potts 2012), as well as non-avian species such as brown rats *Rattus norvegicus* (Lambert et al. 2008; Chapter 11, this volume). Indeed, a recent study by the GWCT suggested that approximately 'half of the grain provided in pheasant feed hoppers is eaten by non-target species, notably rats, corvids, squirrels, and deer' (<https://www.gwct.org.uk/>). All these species are frequently seen at food hoppers, and it seems inevitable that such a huge amount of food will affect the wildlife of the wider UK countryside.

Agri-environment schemes (AES) offer various ways of increasing food supplies for seed-eating birds. They include wild bird seed plots, sown with a mixture of seed-bearing crops (such as barley, triticale, kale, quinoa, or millet) and left unharvested to provide winter food for birds. These are similar to the game cover plots used for many years by the shooting industry (Henderson et al. 2004; Stoate et al. 2004). In 2009, approximately 12,000 ha were planted with wild bird seed mixes in England (Natural England 2009). These plots are usually fertilized to raise the seed yield. Depending on the mix of plants grown, this yield could be > 1.5 tonnes/ha, or around 18,000 tonnes in 2009, in the form of large and medium-sized agricultural seeds presented at high densities.

9.7 Effects of seed provision at different seed densities

9.7.1 Required feeding rates

Whether birds can obtain their daily seed needs depends on their rate of intake. The estimated intake rates required to provide the daily needs of different species feeding on four different types of seed during an eight-hour winter foraging day are shown in Table 9.1. The required intake rates for large seeds (e.g. wheat) are modest for all the species concerned. On an eight-hour day, the largest of the 'small' species, the corn bunting, needs only to maintain a rate of approximately 0.4 seeds/minute, and even the largest species, the pheasant, needs to consume only approximately 3 seeds/minute. By contrast, for small weed seeds, the smallest species, like the linnet, must find approximately 10 seeds/minute, while woodpigeon must find 75 seeds/minute, and pheasant 140 seeds/minute. If seeds are eaten mainly one at a time, the high feeding rates required by large birds eating small seeds may limit the range of seed sources that such species can exploit effectively. A similar argument has been applied to terrestrial carnivorous mammals of different sizes (Carbone et al. 1999).

9.7.2 Insights from functional response modelling

Functional response models relate feeding rates to available food densities. One simple functional response model defines feeding rate (FR) in terms of three parameters, according to:

$$FR = aD / (1 + aDH)$$

[Equation 3]

where a is the searching efficiency (the area of substrate searched with complete efficiency per second, in m^2/s), D is the food density in seeds/m^2, and H is the handling time (seconds taken to select and swallow one seed) (Stephens et al. 2003). This model provides good fits to experimental data for seed-eating passerines (Smart et al. 2008; Baker et al. 2009). It predicts that at low seed densities feeding rate will be determined primarily by searching efficiency, and at high seed densities by handling time.

Several published studies have provided estimates of handling time and search efficiency in seed-eating birds (Murton et al. 1964; Green 1978; Robinson 1997; Atkinson 1998; Whittingham and Markland 2002; Smart et al. 2008; Baker et al. 2009, 2010; Butler et al. 2010). In these studies, handling time per item ranged from 0.85 s to 10.4 s, with those for small species generally being longer than those for large, probably because small birds usually remove the husks from seeds before swallowing (Newton 1967). The lowest handling times reported relate to the woodpigeon (1.3 s) and grey partridge (0.85 s), both feeding on cereal grains. If handling times of around 1 s pertain for large

Table 9.1 Mean food item intake rates (seeds/minute) required to provide an adequate daily ration for species of different sizes consuming different seeds, assuming eight hours of daylight foraging time.

Species	Required intake rate (seeds/minute)			
	Wheat[a]	Rapeseed[b]	Millet[c]	Weed seeds[d]
Linnet	0.22	1.49	1.99	9.96
Reed bunting	0.23	1.59	2.12	10.60
Tree sparrow	0.26	1.73	2.31	11.56
Yellowhammer	0.30	2.04	2.73	13.63
Skylark	0.35	2.35	3.13	15.65
Corn bunting	0.40	2.67	3.55	17.77
Collared dove	1.00	6.74	9.00	45.0
Magpie	1.06	7.16	9.54	47.7
Jackdaw	1.06	7.16	9.54	47.7
Stock dove	1.29	8.73	11.63	58.1
Woodpigeon	1.68	11.32	15.08	75.4
Red-legged partridge	1.77	11.95	15.96	79.8
Carrion crow	1.81	12.26	16.33	81.7
Pheasant	3.12	21.06	28.08	140.4

[a] Assuming available energy content of 15 kJ/g and mean seed weight of 45 mg.
[b] Assuming available energy content of 20 kJ/g and mean seed weight of 5 mg.
[c] Assuming available energy content of 15 kJ/g and mean seed weight of 5 mg.
[d] Assuming available energy content of 15 kJ/g and mean seed weight of 1 mg.

species eating small seeds, then species larger than stock dove would be unable to meet their daily ration on a diet of small weed seeds alone. They would require an average feeding rate of > 70 seeds/minute, which is practically impossible where seeds must be picked individually (Table 9.1).

Searching efficiency would be expected to vary between different seeds and substrates, and may be influenced by factors such as the size, colour, texture, and reflectance of the seed, and the complexity of the substrate (Whittingham and Markland 2002; Butler et al. 2005; Baker et al. 2009). Published estimates of searching efficiency in seed-eating birds vary widely, with a range spanning four orders of magnitude. Most studies have used large seeds such as cereal grains, sometimes on simple artificial substrates. We would predict that searching efficiencies for small weed seeds in natural weedy stubble would be close to the lowest published values.

Examples of relationships between feeding rate and seed density obtained using Equation [3] are shown in Fig 9.10a. In these calculations, we used a value for searching efficiency for small weed seeds in natural weedy stubble of 0.001 m^2/s, which is midway between the two lowest values from our literature survey. The lower curve is intended to be representative of small seed-eating species and uses a handling time of 2.0 s. The upper curve is intended to be representative of large species and uses a faster handling time of 0.85 s, the lowest published value we found. At high seed densities the feeding rates level off at around 30 seeds/minute for small species and 70 seeds/minute for large species.

9.7.3 Critical seed density

We next explored the minimum seed density that would enable birds of different sizes to find and ingest seeds at a sufficient rate to maintain their daily energy balance. At or below this seed density, birds must leave the feeding patch or risk starvation (Sutherland and Allport 1994). This density may be termed the critical (or threshold) seed density, and is the seed-eating equivalent of the leaving (or giving-up) food density widely used in models of feeding strategies in wading birds and wildfowl (Sutherland and Allport 1994;

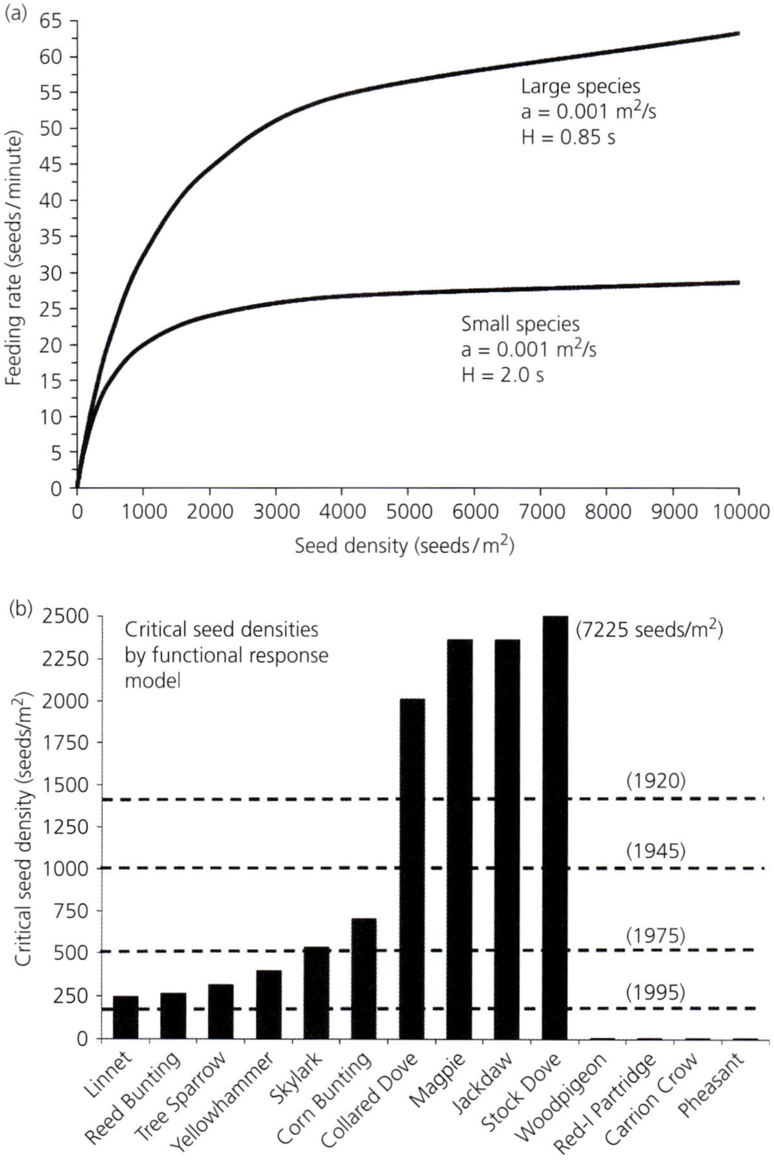

Figure 9.10 (a) Predicted relationship between seed density and feeding rate for simple two-parameter models based on Equation 3 (see text) using plausible values for searching efficiency, a and seed handling time, H for large and small native seed-eating species; (b) predicted critical weed seed densities for different bird species obtained using the functional models illustrated in (a). Horizontal dotted lines show average autumn weed seed densities in arable stubble fields at different dates, based on the regression analysis of Robinson and Sutherland (2002). Predicted critical densities for woodpigeon, carrion crow, pheasant, and red-legged partridge are effectively infinite.

Goss-Custard et al. 2003; Gyemesi et al. 2012). Some relevant figures have been published for small seed eaters. For example, linnets and reed buntings were rarely found in stubble fields with dietary seed densities < 250 seeds/m^2, while yellowhammers were rarely fed on wheat seeds at grain densities < 10 seeds/m^2 (Moorcroft et al. 2002; Stephens et al. 2003).

9.7.3.1 Critical density by FMR scaling

In this section, in the absence of empirical data, we assume that the critical density for a particular seed will vary between species according to their FMR. For example, if the critical seed density for linnets feeding on weed seeds is taken to be 250 seeds/m^2 (Moorcroft et al. 2002), estimates for larger species can be obtained by adjusting this value according to their relative FMR. This approach predicts that the critical weed seed density for skylark would be approximately 400 seeds/m^2, which is broadly in line with the figure of 500 seeds/m^2 derived from Stephens et al. (2003). The corresponding values for woodpigeon would be 1900 weed seeds/m^2 and for pheasant 3500 weed seeds/m^2.

Similar critical densities for other seed types can be estimated, under the same assumptions, by scaling the weed seed densities according to the relative energy contents of the seeds concerned. This approach suggests that the critical density for yellowhammer feeding on wheat seed would be 8 seeds/m^2, which is in good agreement with the figure of 10 seeds/m^2 derived from the empirical data of Stephens et al. (2003). The predicted critical density for tree sparrow feeding on rapeseed is 44 seeds/m^2, which is rather higher than the value of 25 seeds/m^2 for the similarly sized chaffinch, based on the observations of Baker et al. (2009). By way of context, a typical modern rapeseed crop, yielding 3 tonnes/ha, contains approximately 60,000 seeds/m^2, although only a small proportion of this total would be left on the ground after harvest.

9.7.3.2 Critical seed density from a functional response model

Another way of estimating the critical seed density is to take the calculated feeding rate required by each species to provide its daily needs, and then see what seed density is needed to support that feeding rate, using a functional response model. The model illustrated in Fig. 9.10a predicts critical weed seed densities of approximately 240 seeds/m^2 and 260 seeds/m^2 for linnet and reed bunting (Table 9.2), similar to the 250 seeds/m^2 observed for these species (Moorcroft et al. 2002). The prediction for yellowhammer feeding on wheat was lower than the value presented by Stephens et al. (2003), possibly because the parameter values we selected as relevant to foraging for small weed seeds are not appropriate for the much larger wheat seed (Pulliam 1985). Nevertheless, the model shows good agreement with published observations relating to critical seed densities, at least for relatively small seeds. It substantiates the great increase in critical density for weed seeds between the corn bunting (the largest of the small species, with a critical density of 700 seeds/m^2), and the collared dove (the smallest of the large species, with a critical density of 2000 seeds/m^2). With the larger agricultural seeds, relatively low seed densities are sufficient to exceed the critical density, even for large bird species. For example, on the basis of the model, rapeseed at 500 seeds/m^2 would exceed the critical density for all the species considered, including pheasant (Table 9.2). Consistent with this prediction, we have observed that medium-sized (~5 mg) seeds scattered

Table 9.2 Critical seed densities for different species predicted using Equation 2 in text, with parameter values for searching efficiency (a) of 0.001 m^2/s and handling time (H) of 2.0 s for small species and 0.001 m^2/s and 0.85 s, respectively, for large species. Other assumptions as for Table 9.1.

Species	Predicted critical seed density (seeds/m^2)		
	Weed seeds	Rapeseed	Wheat
Small species model: a = 0.001 m^2/s, H = 2.0 s			
Linnet	240	26	4
Reed bunting	260	28	4
Tree sparrow	310	31	4
Yellowhammer	390	37	5
Skylark	530	43	5
Corn bunting	700	48	6
Large species model: a = 0.001 m^2/s, H = 0.85 s			
Collared dove	2010	120	17
Magpie	2360	130	18
Jackdaw	2360	130	18
Stock dove	7225	160	21
Woodpigeon	∞	220	28
Red-legged partridge	∞	260	30
Carrion crow	∞	270	31
Pheasant	∞	490	53

Figure 9.11 (a) and (b) Large and small seed-eating birds feeding on a seed mixture consisting of equal volumes of millet, rapeseed, and linseed, scattered by hand on short grass at a density of approximately 5 g/m^2 (~1000 $seeds/m^2$). Photographs ©A. Larkman.

on the ground at a density of approximately 5 g/m^2, corresponding to ~1000 $seeds/m^2$, are highly attractive to large as well as small seed-eating species (Fig. 9.11).

9.7.3.3 Seed densities in arable fields

Recent published estimates of weed seed densities in arable fields vary widely, probably due to crop, management and/or methodological differences (Robinson and Sutherland 1999; Moorcroft et al. 2002; Heard et al. 2003; Robinson et al. 2004; Bohan et al. 2005; Butler et al. 2010). However, they generally lie in the range 300–1200 $seeds/m^2$. Perhaps one of the most useful analyses of time trends is that of Robinson and Sutherland (2002), who gathered together data on the density of seeds available to birds (soil surface and top 1 cm) in arable fields in autumn from studies over the last 90 years or so. The regression line they fitted to the data points indicated a reduction in seed density of approximately 17 $seeds/m^2$ per year, from some 1400 $seeds/m^2$ in 1920 to 1000 $seeds/m^2$ in 1945, 500 $seeds/m^2$ in 1975, and 170 $seeds/m^2$ in 1995. This represents an 88 % reduction in available weed seeds over this period.

Using the critical weed seed densities for different bird species calculated using the functional response model, together with the earlier seed data, it seems that, between 1920 and 1945, the autumn seed densities in arable fields exceeded the critical density for all of the small seed-eating bird species, but for none of the large ones (Fig. 9.10b). By 1975, seed density had fallen just below the critical density for the largest of the small species (skylark and corn bunting), and by 1995 the mean seed density was predicted to be inadequate for even the smallest seed-eating species considered. Analyses based on the FMR-scaling method gave similar results.

These analyses suggest that, prior to approximately 1975, weedy arable stubbles could provide small seed food at densities in autumn that were suitable for small species but, at least since 1945 (and possibly well before that), they have been inadequate for large species. Stubbles therefore represented a large-scale, low-density food resource that could be exploited by small species, relatively free from interference and resource competition from large seed-eating species. However, at some point prior to 1995, the autumn seed density, on average, fell below the level required to support even the smallest species. Thus, not only did the area of overwintered stubbles decrease dramatically with agricultural intensification, but most of the remaining areas became inadequate as food sources for even small birds. Within each winter, we can also expect that the progressive depletion of this seed resource would have made arable stubbles steadily less suitable, first for the largest species and then in turn for the smaller ones.

9.7.4 Conclusions on seed density

Between 1966 and the present, the breeding populations of several small, resident, seed-eating farmland bird species declined markedly in Britain. However, owing to increasing game bird management, both the total potential winter seed food demand and supply increased substantially over the same period. At the same time, the proportion of the total demand attributable to small species decreased from approximately 32% to 3%. So why did small birds fare so badly during a period of apparently increasing food supply? Our tentative suggestion is that, while the quantity of large seeds provided at high density has increased dramatically, the quantity of small seeds from low-density sources declined dramatically. These changes were driven primarily by the increased food provision for released gamebirds on the one hand, and the loss of overwinter weedy stubbles on the other.

Among birds, larger body size would be expected to confer a greater ability to survive periods of food shortage and adverse weather during winter in Britain. Advantages of large body size include the lower surface-to-volume ratio (reducing the relative energetic cost of maintaining body temperature), greater fasting endurance due to larger body reserves (Blem 1976), dominance of large species over smaller ones at high-density food sources and, in some seed-eaters, a greater ability to exploit non-seed food sources (such as invertebrates, carrion, or leaves), and a greater scratching and probing ability to obtain food covered by litter, soil, or snow.

These advantages lead us to ask how small seed-eating birds have been able to compete with and survive alongside larger birds despite considerable potential overlap in their diets. We speculate that a possible advantage of small body size is the lower absolute daily energy requirement, reducing the required rate of seed finding. Under conditions of widespread, low-density seed provision, this advantage could be crucial. In the past, weedy overwinter stubbles might have provided small seeds at a density where small species could obtain their daily food ration but large species could not. This could have given a degree of niche separation, allowing small birds to flock and feed in relative freedom from interference and resource depletion from larger species. The progressive switch from low-density to high-density seed sources may have reduced the effectiveness of this separation and reduced the ability of small species to co-exist alongside larger seed-eating birds.

9.8 Concluding remarks

The temporal correlative evidence set out above suggests that increased releases of non-native, captive-reared gamebirds in recent decades could have been involved in both the declines of small native seed-eating species and the increases of large ones observed since 1966. We have proposed possible mechanisms for both phenomena, involving the action of inter-specific competition and interference, together with changes in the densities of both small and large seeds. Our evidence on the role of released gamebirds does not invalidate the substantial published evidence on the role of agricultural change in influencing bird population trends; it simply adds another factor of probable importance. As yet, the evidence for the involvement of gamebird releases is largely correlative, and cannot on its own prove causality. There is clearly need for further work, particularly field experiments designed to test hypotheses.

The only small, resident seed-eating farmland bird whose national population has clearly benefited substantially from conservation measures on farmland in Britain is the cirl bunting *Emberiza cirlus* (mean weight 25 g, similar to tree sparrow). Between 1970 and 1980 the British population collapsed, attributed primarily to the loss of overwintered weedy stubbles, and its range was reduced to a small area in south west England (Evans and Smith 1994). A recovery project, based mainly on the provision of weedy stubbles created from cereals grown under a low-input herbicide regime (Bradbury et al. 2008), led to a > seven-fold increase in its breeding population by 2009 (Davies et al. 2011). Our analyses imply that similar measures could conserve other species of small, seed-eating farmland birds. On the other hand, measures that involve provision of medium-sized or large seeds at high density, as in game crops, could benefit large seed-eating birds and might further disadvantage small species over the longer term. Small seed-eating birds, as well as large ones, are attracted to game crops or wild bird cover crops, but the fact that their seeds tend to be exhausted or are ploughed under by February means that they are of no great value to small birds later in the winter and in early spring when other seed sources are also largely exhausted. Some of the larger seed-eaters can turn to other foods at this time of scarcity: for example, corvids to invertebrates and carrion, or pigeons and game birds to the green leaves of rape, clover, and other plants. But these alternative options are not open to smaller seed-eaters.

We accept that other changes associated with gamebird releases could also have affected small birds, for example the increased numbers of predators, such as rats and corvids, themselves benefiting from high density food provision, as described above. Our findings do not exclude the operation of other factors which changed over the years in parallel with the rise in gamebird releases.

In conclusion, we believe that our hypothesis that the contrasting population trends in large and small resident seed-eating birds have been driven, not only by the intensification of agriculture, but also by changes in the practices and intensity of the gamebird production is worthy of further, independent investigation. Better understanding of the drivers of population change in declining small farmland species could inform efforts to conserve them.

Acknowledgements

We are grateful to Peter Brotherton for useful discussions.

References

Atkinson, P.W. (1998). *The wintering ecology of the twite Carduelis flavirostris and the consequences of habitat loss*. PhD Thesis, University of East Anglia, Norwich, UK.

Baillie, S.R., Marchant, J.H., Leech, D.I., Massimino, D., Eglington, S.M., Johnston, A., Noble, D.G., Barimore, C., Kew, A.J., Downie, I.S., Risely, K., and Robinson, R.A. (2013). Bird Trends 2012: trends in numbers, breeding success and survival for UK breeding birds. BTO Research Report No.644. BTO, Thetford. <http://www.BTO.org/birdtrends>

Baker, D.J., Freeman, S.N., Grice, P.V., and Siriwardena G.M. (2012). Landscape-scale responses of birds to agri-environment management: a test of the English Environmental Stewardship scheme. *Journal of Applied Ecology*, **49**, 871–882.

Baker, D.J., Stillman, R.A., and Bullock, J.M. (2009). The effect of habitat complexity on the functional response of a seed-eating passerine. *Ibis*, **151**, 547–558.

Baker, D.J., Stillman, R.A., Smith, B.M., Bullock, J.M. and Norris, K.J. (2010). Vigilance and the functional response of granivorous foragers. *Functional Ecology*, **24**, 1281–1290.

Blem, C.R. (1976). Patterns of lipid storage and utilization in birds. *American Zoologist*, **16**, 671–684.

Bohan, D.A., Boffey, C.W.H., Brooks, D.R., Clark, S.J., Dewar, A.M., Firbank, L.G., Haughton, A.J., Hawes, C., Heard, M.S., May, M.J., Osborne, J.L., Perry, J.N., Rothery, P., Roy, D.B., Scott, R.J., Squire, G.R., Woiwood, I.P., and Champion, G.Y. (2005). Effects on weed and invertebrate abundance and diversity of herbicide management in genetically modified herbicide-tolerant winter-sown oilseed rape. *Proceedings of the Royal Society Series B*, **272**, 463–474.

Bradbury, R.B., Bailey, C.M., Wright, D., and Evans, A.D. (2008). Wintering Cirl Buntings *Emberiza cirlus* in south-west England select cereal stubbles that follow a low-input herbicide regime. *Bird Study*, **55**, 23–31.

Butler, S.J., Mattison, E.H.A., Glithero, N.J., Robinson, L.J., Atkinson, P.W., Gillings, S., Vickery, J.A., and Norris, K. (2010). Resource availability and the persistence of seed-eating bird populations in agricultural landscapes: a mechanistic modelling approach. *Journal of Applied Ecology*, **47**, 67–75.

Butler, S.J., Whittingham, M.J., Quinn, J.L., and Cresswell, W. (2005). Quantifying the interaction between food density and habitat structure in determining patch selection. *Animal Behaviour*, **69**, 337–343.

Carbone, C., Mace, G.M., Roberts, S.C., and Macdonald, D.W. (1999). Energetic constraints on the diets of terrestrial carnivores. *Nature*, **402**, 286–288.

Chamberlain, D.E., Fuller, R.J., Bunce, R.G.H., Duckworth, J.C., and Shrubb, M. (2000). Changes in the abundance of farmland birds in relation to the timing of agricultural intensification in England and Wales. *Journal of Applied Ecology*, **37**, 771–788.

Davey, C.M., Vickery, J.A., Boatman, N.D., Chamberlain D.E., Parry, H.R., and Siriwardena, G.M. (2010). Assessing the impact of Entry Level Stewardship on lowland farmland birds in England. *Ibis*, **152**, 459–474.

Davies, M., Grice, P., Evans, A., Jeffs, C., Lock, L., and Wotton, S. (2011). The Cirl Bunting recovery story: celebrating the success of a model species recovery project. *British Wildlife*, **22**, 263–271.

Defra (2013a). *Population of Wild Birds: 1970 to 2012*. Department for Environment, Food and Rural Affairs, London, UK.

Defra (2013b). *Great Britain Poultry Register (GBPR) Statistics: 2013*. Department for Environment, Food and Rural Affairs, London, UK.

Donald, P.F., Green, R.E., and Heath, M.F. (2001). Agricultural intensification and the collapse of Europe's farmland bird populations. *Proceedings of the Royal Society of London, Series B*, **268**, 25–29.

Donald, P.F., Sanderson, F.J., Burfield, I.J., and van Bommel, F.P.J. (2006). Further evidence of continent-wide impacts of agricultural intensification on European farmland birds, 1990–2000. *Agriculture, Ecosystems* and *Environment*, **116**, 198–196.

Eaton, M.A., Balmer, D.E., Cuthbert, R., Grice, P.V., Hall, J., Hearn, R.D., Holt, C.A., Musgrove, A.J., Noble, D.G., Parsons, M., Risely, K., Stroud, D.A., and Wotton, S. (2011). *The State of the UK's Birds 2011*. RSPB, BTO, WWT, CCW, JNCC, NE, NIEA and SNH, Sandy, Bedfordshire, UK.

Evans, A.D. and Smith, K.W. (1994). Habitat selection of Cirl Buntings *Emberiza cirlus* wintering in Britain. *Bird Study*, **41**, 81–87.

Freeman, S.N., Noble, D.G., Newson, S.E., and Baillie, S.R. (2007). Modelling population changes using data from different surveys: the Common Bird Census and the Breeding Bird Survey. *Bird Study*, **54**, 61–72.

Fuller, R.J., Gregory, R.D., Gibbons, D.W., Marchant, J.H., Wilson, J.D., Baillie, S.R., and Carter, N. (1995). Population declines and range contractions among lowland farmland birds in Britain. *Conservation Biology*, **9**, 1425–1441.

Gillings, S., Henderson, I.G., Morris, A.J., and Vickery, J.A. (2010). Assessing the implications of the loss of set-aside for farmland birds. *Ibis*, **152**, 713–723.

Gillings, S., Newson, S.E., Noble, D.G., and Vickery, J.A. (2005). Winter availability of cereal stubbles attracts declining farmland birds and positively influences breeding population trends. *Proceedings of the Royal Society Series B*, **272**, 733–739.

Goss-Custard, J.D., Stillman, R.A., Caldow, W.G., West, A.D., and Guillemain, M. (2003). Carrying capacity in overwintering birds: when are spatial models needed? *Journal of Applied Ecology*, **40**, 176–187.

Green, R. (1978). Factors affecting the diet of farmland skylarks, *Alauda arvensis*. *Journal of Animal Ecology*, 47, 913–928.

Gregory, R.D., Noble, D.G., and Custance, J. (2004). The state of play of farmland birds: population trends and conservation status of lowland farmland birds in the United Kingdom. *Ibis*, **146** (Suppl. 2), 1–13.

Gyemesi, A., Franken, M.S., Feige, N., and Nolet, B.A. (2012). Human disturbance of Bewick's Swans is reflected in giving-up net energy intake rate, but not in giving-up food density. *Ibis*, **154**, 781–790.

Hancock, M.H. and Wilson, J.D. (2003). Winter habitat associations of seed-eating passerines on Scottish farmland. *Bird Study*, **50**, 116–130.

Harrop, A.H.J., Collinson, J.M., Dudley, S.P., Kehoe, C., and the British Ornithologists' Union Records Committee (BOURC) (2013). The British list: a checklist of birds of Britain (8th edition). *Ibis*, **155**, 635–676.

Heard, M.S., Hawes, C., Champion, G.T., Clark, S.J., Firbank, L.G., Haughton, A.J., Parish, A.M., Perry, J.N., Rothery, P., Scott, R.J., Skellern, M.P., Squire, G.R., and Hill, M.O. (2003). Weeds in fields with contrasting conventional and genetically modified herbicide-tolerant crops. I. Effects on abundance and diversity. *Philosophical Transactions of the Royal Society of London B*, **358**, 1819–1832.

Henderson, I.G., Vickery, J.A., and Carter, N. (2004). The use of winter bird crops by farmland birds in lowland England. *Biological Conservation*, **118**, 21–32.

Holmes, R.J. and Froud-Williams, R.J. (2005). Post-dispersal weed seed predation by avian and non-avian predators. *Agriculture, Ecosystems and Environment*, **105**, 23–27.

Lambert, M.S., Quy, R.J., Smith, R.H., and Cowan, D.P. (2008). The effect of habitat management on home-range size and survival of rural Norway rat populations. *Journal of Applied Ecology*, **45**, 1753–1761.

Lawson, B., Robinson, R.A., Colville, K.M., Peck, K.M., Chantry, J., Pennycott, T.W., Simpson, V.R., Toms, M.P., and Cunningham, A.A. (2012). The emergence and spread of finch trichomonosis in the British Isles. *Philosophical Transactions of the Royal Society of London, Series B*, **367**, 2852–2863.

Macdonald, D.W. and Feber, R.E., eds. (2015). *Wildlife Conservation on Farmland. Managing for Nature on Lowland Farms*. Oxford University Press, Oxford.

Moorcroft, D., Whittingham, M.J., Bradbury, R.B., and Wilson, J.D. (2002). The selection of stubble fields by wintering granivorous birds reflects vegetation cover and food abundance. *Journal of Applied Ecology*, **39**, 535–547.

Murton, R.K., Westwood, N.J., and Isaacson, A.J. (1964). The feeding habits of the woodpigeon *Columba palumbus*, stock dove *C. oenas* and turtle dove *Streptopelia turtur*. *Ibis*, **106**, 174–188.

Musgrove, A., Aebischer, N., Eaton, N., Hearn, R., Newson, S., Noble, D., Parsons, M., Risely, K., and Stroud, D. (2013). Population estimates of birds in Great Britain and the United Kingdom. *British Birds*, **106**, 64–100.

Nagy, K.A. (1987). Field metabolic rate and food requirement scaling in mammals and birds. *Ecological Monographs*, **57**, 111–128.

Nagy, K.A. (2005). Field metabolic rate and body size. *Journal of Experimental Biology*, **208**, 1621–1625.

Natural England (2009). Agri-environment schemes in England 2009: a review of results and effectiveness. Available from: <http://www.naturalengland.org.uk/publication/46002>

Newton, I. (1967). The adaptive radiation and feeding ecology of some British finches. *Ibis*, **109**, 33–96.

Newton, I. (1998). *Population Limitation in Birds*. Academic Press, London, UK.

Newton, I. (2004). The recent declines of farmland bird populations in Britain: an appraisal of causal factors and conservation actions. *Ibis*, **146**, 579–600.

Newton, I. (2010). *Bird Migration*. HarperCollins, London, UK.

PACEC (2006). *The Economic and Environmental Impact of Sporting Shooting*. <http://www.pacec.co.uk>

Potts, G.R. (2012). *Partridges: Countryside Barometer*. Harper Collins, London, UK, p. 296.

Pulliam, H.R. (1985). Foraging efficiency, resource partitioning, and the coexistence of sparrow species. *Ecology*, **66**, 1829–1836.

Robertson, P.A. (1991). Estimating the nesting success and productivity of British pheasants *Phasianus colchicus* from nest-record schemes. *Bird Study*, **38**, 73–79.

Robinson, R.A. (1997). *Ecology and conservation of seed-eating birds on farmland*. PhD Thesis, University of East Anglia, Norwich, UK.

Robinson, R.A. (2004). The diet of seed-eating birds on lowland farmland. *British Birds*, **97**, 464–467.

Robinson, R.A. (2005). *BirdFacts: profiles of birds occurring in Britain & Ireland* (*BTO Research Report 407*). BTO, Thetford.

Robinson, R.A., Hart, J.D., Holland, J.M., and Parrott, D. (2004). Habitat use by seed-eating birds: a scale-dependent approach. *Ibis*, **146** (Suppl. 2), 87–98.

Robinson, R.A. and Sutherland, W.J. (1999). The winter distribution of seed-eating birds: habitat structure, seed density and seasonal depletion. *Ecography*, **22**, 447–454.

Robinson, R.A. and Sutherland, W.J. (2002). Post-war changes in arable farming and biodiversity in Great Britain. *Journal of Applied Ecology*, **39**, 157–176.

Shooting Times and Country Magazine (2009). Gamebird exporters under the microscope. *Shooting Times and Country Magazine*, 22 July 2009, 6–7.

Siriwardena, G.M., Baillie, S.R., Crick, H.Q.P., and Wilson, J.D. (2000). The importance of variation in the breeding performance of seed-eating birds in determining their population trends on farmland. *Journal of Applied Ecology*, **37**, 128–148.

Siriwardena, G.M., Calbrade, N.A., and Vickery, J.A. (2008). Farmland birds and late winter food: does seed supply fail to meet demand? *Ibis*, **150**, 585–595.

Smart, S.L., Stillman, R.A., and Norris, K.J. (2008). Measuring the functional responses of farmland birds: an example for a declining seed-feeding bunting. *Journal of Animal Ecology*, **77**, 687–695.

Stephens, P.A., Freckleton, R.P., Watkinson, A.R., and Sutherland, W.J. (2003). Predicting the response of farmland bird populations to changing food supplies. *Journal of Applied Ecology*, **40**, 970–983.

Stoate, C., Henderson, I.G., and Parish, D.M. (2004). Development of an agri-environment scheme option: seed-bearing crops for farmland birds. *Ibis*, **146** (Supplement 2), 203–209.

Sutherland, W.J. and Allport, G.A. (1994). A spatial depletion model of the interaction between bean geese and wigeon with the consequences for habitat management. *Journal of Animal Ecology*, **63**, 51–59.

Whittingham, M. and Markland, H. (2002). The influence of substrate on the functional response of an avian granivore and its implications of farmland bird conservation. *Oecologia*, **130**, 637–644.

Wilson, J.D., Evans, A.D., and Grice P.V. (2009). *Bird Conservation and Agriculture*. Cambridge University Press, Cambridge, UK.

Wilson, J.D., Taylor, R., and Muirhead, L.B. (1996). Field use by farmland birds in winter: an analysis of field type preferences using resampling methods. *Bird Study*, **43**, 320–332.

CHAPTER 10

In the wake of buzzards: from modelling to conservation and management in a changing landscape

Eduardo M. Arraut, David W. Macdonald, and Robert E. Kenward

Ignorance more frequently begets confidence than does knowledge.

Charles Darwin

10.1 Raptors in British countryside: 600 years of conflict

At the time of writing, one might almost think that for a vertebrate's name to begin with B is a blight—badgers, with their tuberculous perturbations, bats despoiling belfries, and beavers (and boar) with their controversial return to the wild; our team has engaged with each of these turbulent species (see Macdonald and Feber 2015: Chapters 9 and 16; Chapter 5, this volume) and now, to add to this benighted assemblage, we turn to buzzards. The background is that after a long history of persecution, buzzards *Buteo buteo* are now recovering strongly in most parts of the UK and this is, again, bringing them to the heart of controversy (Fig. 10.1). The perception of some wild raptors as the enemy, contrasting unhappily with falconers' delight in other species, is part of the UK's history. Reputedly, an Act by James II of Scotland in the fifteenth century already declared buzzards as vermin (Holloway 2010), and Henry VIII excluded them when he made egg-stealing of species used in falconry illegal in the 1500s. Shakespeare later wrote: on hearing of the slaughter of his wife and children, 'all my pretty chickens and their dam', MacDuff proclaims in grief 'O, hell-kite', in reference to Macbeth and his 'one fell swoop' murderous act. Buzzards too were particularly vilified. In the Taming of the Shrew, Petruchio teases Kate with 'O slow-winged turtle, shall a buzzard take thee?'. In the nineteenth century, the shooting of 'driven' grouse was so widespread and popular that one moor owner, the Marquess of Bute, required his keepers to take an oath to 'destroy all birds of prey with their nests', a commitment mostly aimed at the hen harrier *Circus cyaneus* (Watson 1977), for which illegal killing still remains controversial in the UK (RSPB 2013).

For twenty-first-century buzzards, the nub of the controversy is that, having benefited from protection laws and changed public attitudes to fight back from rarity and become relatively numerous, they are now again perceived as pests by some of those involved in the production of gamebirds and free-range poultry. The overarching questions we will untangle here are what factors limit the numbers and distribution of buzzards, the recolonization of their former range, and what, if any, management interventions are feasible and appropriate both at the landscape and point-location scales. Before tackling these questions, and the complex of biological, ethical, and policy issues that cascade from them, it is important first to set the context in terms of the recent history of (a) raptors and (b) game-shooting in the British countryside, both histories being interwoven with that of the buzzards themselves.

10.1.1 Raptors populations in the twentieth century

By the early 1900s, five[1] of the fifteen British breeding species of diurnal raptors had been extirpated (RSPB

[1] Goshawk *Accipiter gentilis*, marsh harrier *Circus aeruginosus*, honey buzzard *Pernis apivorus*, white-tailed eagle *Haliaeetus albicilla*, and osprey *Pandion haliaetus*.

Wildlife Conservation on Farmland. Conflict in the Countryside. Edited by David W. Macdonald and Ruth E. Feber.
© Oxford University Press 2015. Published 2015 by Oxford University Press.

Figure 10.1 Common buzzard *Buteo buteo*. Photograph ©Arend CC BY 2.0.

2013). Contributory factors were years of persecution by gamekeepers, unintentional poisoning during campaigns aimed at controlling other animals (e.g. foxes *Vulpes vulpes*), and fashions for collecting eggs, skins, and feathers (Newton 1979; RSPB 2008). These were exacerbated, for some species, by deforestation for fuel, timber, and conversion to farmland, and intensified cultivation methods in the twentieth century causing prey reductions (see Feber and Macdonald 2015: Chapter 1; Newton 1979). The vagaries of history also played their role: while the World Wars brought about an increase in sparrowhawk *Accipiter nisus* numbers when gamekeeping declined, it also blighted peregrine falcons *Falco peregrinus*, when the UK government ordered their killing to protect carrier-pigeons *Columba livia* (Ratcliffe 1990).

Dramatic declines of some raptors were precipitated by the use of organochlorine (e.g. DDT) and cyclodiene insecticides used on crops in the 1950s and 1960s, which persisted in the environment (e.g. the half-life of DDT in soil is estimated at 57 years; Cooke and Stringer 1982) and accumulated up the food chain (e.g. as reviewed for mammals by Harrington and Macdonald 2002). Famously, peregrine falcons that fed on pigeons that had consumed grain contaminated with DDT accumulated the toxins and laid eggs with fatally thin shells. By the end of the 1960s, peregrines and sparrowhawks had suffered a widespread and severe decline (Newton 1998). The same insecticides had indirect effects, such as reducing invertebrate food supplies (Bright et al. 2008). The Protection of Birds Acts 1954 protected all birds of prey (except sparrowhawk, which was added in 1967), but it was not until the furore over DDT (which was totally banned in the UK in 1984) and the increased attention to the killing of raptors through the Wildlife and Countryside Act 1981[2], which also protected against the destruction of their nests, that some species started to recover. However, for sparrowhawks it took 17 years for the residues of organochlorine (dieldrin and DDE) pesticides in their livers to fall significantly (Lee et al. 2008). Since then, natural recovery of some raptors has been possible, abetted by reintroductions, notably of the goshawk *Accipiter gentilis* (Kenward 2006), red kite (Welsh Kite Trust 2011), and the white tailed eagle *Haliaeetus albicilla* (RSPB 2013).

Nevertheless, illegal poisoning still looms large. There is a revealing contrast for red kites between those introduced to the Chilterns in England, where around 800 breeding pairs now flourish (Welsh Kite Trust 2011), and those simultaneously introduced to Scotland, of which only 50 pairs struggle to maintain a foothold; this difference may be explained by illegal poisoning in the vicinity of the Scottish release site, in areas managed for intensive driven grouse shooting (Smart et al. 2010). Furthermore, Hughes et al. (2013) found that 70% (n = 114) of red kite recovered carcasses showed evidence of secondary poisoning by anticoagulant rodenticides, and 10% may have died as a direct result.

10.1.2 Game rearing and birds of prey

To understand the particular controversy surrounding buzzards it is necessary to outline twentieth century trends in game shooting. Intensification of farming after World War II brought about changes in habitat management—application of chemicals, changes in cropping patterns, the removal of hedgerows, and the conversion of woodland and coppices into farmland—that had detrimental impacts on wild gamebirds (Martin 2011), including the main shooting quarry in the lowlands, the native grey partridge

[2] And the Wildlife Order 1985 in Northern Ireland, and the Nature Conservation Act 2004 in Scotland.

Perdix perdix (Potts 2012; Macdonald and Feber 2015: Chapter 13). However, following the Rearing of Pheasants (Revocation of Prohibition) Order in 1949 (Martin 2011), the total number of birds shot increased due to the artificial rearing and release of non-native common pheasant *Phasianus colchicus* and red-legged partridge *Alectoris rufa*. The value of the game shooting industry (birds and mammals) in Britain is now estimated at £1.6 billion per year (PACEC 2006) and, of 19 million gamebirds and wildfowl shot in 2004, 80% were pheasants[3] (PACEC 2006). This enormous economic revenue makes profitable the release of 35 million pheasants[4] each year (GWCT 2012). The non-native pheasant is now the most abundant bird in Britain, accounting for 30% of landbird biomass (Dolton and Brooke 1999; see also Chapter 9, this volume), with poults at release costing £2.50 each, with each pheasant shot having an economic value of up to £33 (Fera 2012).

Against this background, in June 1995, the then UK Environment Minister, Robert Atkins, established the UK Raptor Working Group. Its aim was to reconcile conflicts between the recovery of bird of prey populations and their impact on gamebirds (but also on moorland management, on racing pigeons, and on songbirds). The resulting survey of gamebird losses to raptor predation in 1996 attributed 36% to sparrowhawks, 20% to buzzards, and 17% to tawny owls *Strix aluco* (Harradine et al. 1997). The report highlighted the conviction among gamekeepers that buzzards were a threat to pheasants (and partridges), especially on lowland farmland release pens[5], although it concluded that sparrowhawks and tawny owl *Strix aluco* together were taking more pheasants than the buzzards at that time.

[3] 15 million pheasants and partridges ended in the food chain (99%) versus 750 million birds provided for the food chain by the UK poultry industry (PACEC 2006).

[4] *c.* same number as the size of the egg production flock in the UK (Martin 2011).

[5] Pheasant poults are introduced into the wild through an open-topped release pen so that birds can become acclimatized to their environment, and learn to roost in trees while protected from mammals. Pens are situated within woodland and are made of fine wire netting in a perimeter that also includes ground cover. When small, birds can go in and out at ground level through slits too narrow for foxes, and can also fly out as well as up to roost when large enough (which is why roofs are open). They are encouraged to leave the pen before 1 October, when the shooting season starts (JNCC 2000). The Game & Wildlife Conservation Trust recommends a pheasant release pen density of ≤ 1000 pheasants/ha (Fera 2012).

10.2 Raptor predation and pheasant mortality: actual numbers

So, how big a problem are raptors in general, and buzzards in particular, to the game industry now? In the mid 1970s, Lloyd (1976) estimated that, on 90% of shoots nationally, raptors caused losses of ≤ 1% (with only 1 in 30 shoots losing > 5%; but a few reporting > 10%). However, buzzard numbers have since increased. Kenward et al. (2001b) found that predation by a re-established buzzard population in Dorset accounted for around 4% of pheasant mortality in 28 release pens across two years, though 4–5 pens lost no pheasants to buzzards each year and a similar number lost 10–35%; 3% of deaths were attributed to foxes, with less than 1% each to sparrowhawks and tawny owls. The risk to pheasants diminishes with their age, since buzzards seldom kill poults older than 14 weeks (Lloyd 1976).

Away from release pens, other causes of death remain more important than raptors. In a survey by Turner and Sage (2003), of 486 reared pheasants killed away from pens, < 1% were killed by raptors in comparison to 36% killed or scavenged by foxes. Lloyd (1976) had also found that 18 of 22 mass kills in pheasant release pens were by mammalian predators while, in a study of predation on chickens in France, raptors were involved in more attacks than were foxes but, as they were less prone to surplus killing than foxes, raptors accounted for the death of fewer chickens overall (Stahl et al. 2002). Moving to Ireland, with few raptors, Robertson (1988) found that 67% of known losses of radio-tagged pheasants to predation were due to foxes. With a wider perspective, Lees et al. (2013) attributed 5–13% of pheasant deaths in the UK to road accidents, a figure supported by a Post Office (2008) estimate that 3 million (of the 35 million released) die annually on roads.

In a review of over thirty years of research on conflicts between birds of prey and pheasants at release pens, Fera (2012) concluded that 'UK losses of pheasant poults in and around release pens to raptor predation have been low', but that 'a small number do suffer higher losses'. Even for those, the authors could not endorse lethal control because non-lethal options are available and there was insufficient research to give comfort that killing raptors would not worsen the situation. They noted that licences can be issued to kill birds of prey under the Wildlife and Countryside Act (1981), but only to prevent 'serious' damage, and 'seriousness' is not defined.

The Fera (2012) report also concluded that, in general, significant impacts on poults are unlikely to be associated purely with population densities of buzzards.

Kenward et al. (2001b) showed that predation occurred most often in the absence of ground-cover, where overhanging trees provided the buzzards with perches, and where small numbers of pheasants were housed in large pens. Of 136 radio-tracked buzzards, only 8% of young buzzards spent much time near pheasant pens (Kenward et al. 2001b), a figure which resonated with the finding that about 10% of radio-tagged young buzzards were killed (illegally) near pens early in the (1990–1998) project (fewer were killed latterly, perhaps due to improvements in pen design in the meantime, or because of increasing tolerance among gamekeepers).

Valkama et al. (2005) reviewed 13 studies on the diet composition at nests of raptors across seven European countries and found the mean representation of gamebirds in buzzard diet to be only 3.3% (although, as in the Fera (2012) report, the review acknowledged that particular cases could reach 15%). In the UK, buzzards eat mainly rabbits *Oryctolagus cuniculus*, small rodents, birds, reptiles, amphibians, larger insects, and earthworms (Box 10.1); radio-tagged buzzards observed continuously in Dorset ate little else in winter other than worms and possibly other grassland invertebrates (Hodder 2001). In spite of the early view that food items over 500 g would be weak, diseased, or carrion (Cramp 1980), two Dorset buzzards were seen killing adult pheasants, while others brought to nests healthy rabbits of at least this weight and one regularly brought adult woodpigeons *Columba palumbus* (another favoured rooks *Corvus frugilegus*). These larger prey may not always be carried to nests and thus not recorded among food remains there. Dietary studies cannot clearly discern whether a pheasant was killed or scavenged (Allen and Feare 2003) unless there is an autopsy, although Sim (2003) suggested that gamebirds eaten by buzzards are usually scavenged. However, there is no reason why buzzards that can take rabbits, pigeons, and rooks could not kill pheasant poults. The association of heavier predation with vegetation factors that would favour predation but not necessarily scavenging (Kenward et al. 2001b), and the reduction in deaths once poults become large enough (Lloyd 1976), certainly makes plausible the contention of managers that at least some birds are killed by buzzards.

Box 10.1 Buzzards: biology, ecology, population trends

Buzzards are generalist predators whose various hunting techniques enable them to adapt to changes in local prey and prey densities. They typically scan for prey from branches, but also soar, glide, and walk across the ground (Tubbs 1974). They eat mainly mammals (rabbits and voles) during the breeding season but also birds, reptiles, amphibians, larger insects, and earthworms in seasons when these are readily available.

Female buzzards typically lay clutches of 3–4 eggs, which they incubate for up to 38 days; the eggs hatch asynchronously and young depend on their parents for 47–57 days, remaining in the nest before fledging. They typically start breeding in their second to fourth year, depending on population density (they breed later at high density). In Dorset, a study carried out at the end of the 1990s found that only one in four birds bred every breeding season, which meant that there were three times more non-breeders than breeders in the population (Kenward et al. 2000). Population modelling shows that the high survival of young to adulthood in relatively sedentary UK populations gives reasonable resilience to additional attrition, potentially of more than half the young in healthy habitats, but less resilience to loss of adults (Kenward et al. 2007).

Buzzards are recovering from historical declines. They bred throughout Britain in the 1800s, but by the end of the century started to decline and latterly survived only in the west, their lowest ebb being between 1900 and 1914. Thereafter they had begun to recover when, in 1955, the decimation of rabbits by myxomatosis, at a time when alternative vertebrate prey were in many cases contaminated with organochlorines, had a further detrimental effect on the buzzard population. Their recovery during the 1960s was slow, but helped by the re-establishment of rabbits and positive changes in attitudes of gamekeepers (game-rearing methods improved and reduced losses) and the general public towards buzzards (Balmer et al. 2013; Prytherch 2013). However, by the mid 1970s there was already some surprise that buzzards were not spreading more rapidly (Tubbs 1974), with suggestions that a relative absence of raptors in the east of England might be linked to gamekeeper density increasing eastward (Moore 1957; Newton 1979). Nevertheless, in the 1990s buzzards did begin to increase rapidly in numbers and expand their range eastward/southward in England and the east of Scotland: for example, between 1995 and 2011, buzzards increased by over 1000% in the East Midlands and the east and south of England, though they remained absent from Yorkshire and the North East (Risely et al. 2013).

The *Breeding Bird Survey* of 2012 (Risely et al. 2013) reports that buzzards have achieved considerable densities in western UK, and recolonized most of central Britain, but that in the east they remain quite rare (Box Fig. 10.1.1).

continued

Box 10.1 *Continued*

The buzzard population[6] has increased by 80% in the UK (167% in England; 31% in Scotland) from 1995–2011, with a 6% increase from 2011–2012 (12% in England, but –4% in Scotland): a recent (2011–2012) significant increase in England but not at national level. The British Trust for Ornithology (Musgrove et al. 2013) estimates that in 2009 there were 57,000–79,000 breeding pairs in Britain. They are confirmed breeders in 2033 of 100 km^2 hectads and probably breed in 2450, more than any other British raptor (Balmer et al. 2013). Buzzards are in fact currently the most abundant diurnal raptor in Britain (Balmer et al. 2013) and across Europe (Burfield and van Bommel 2004).

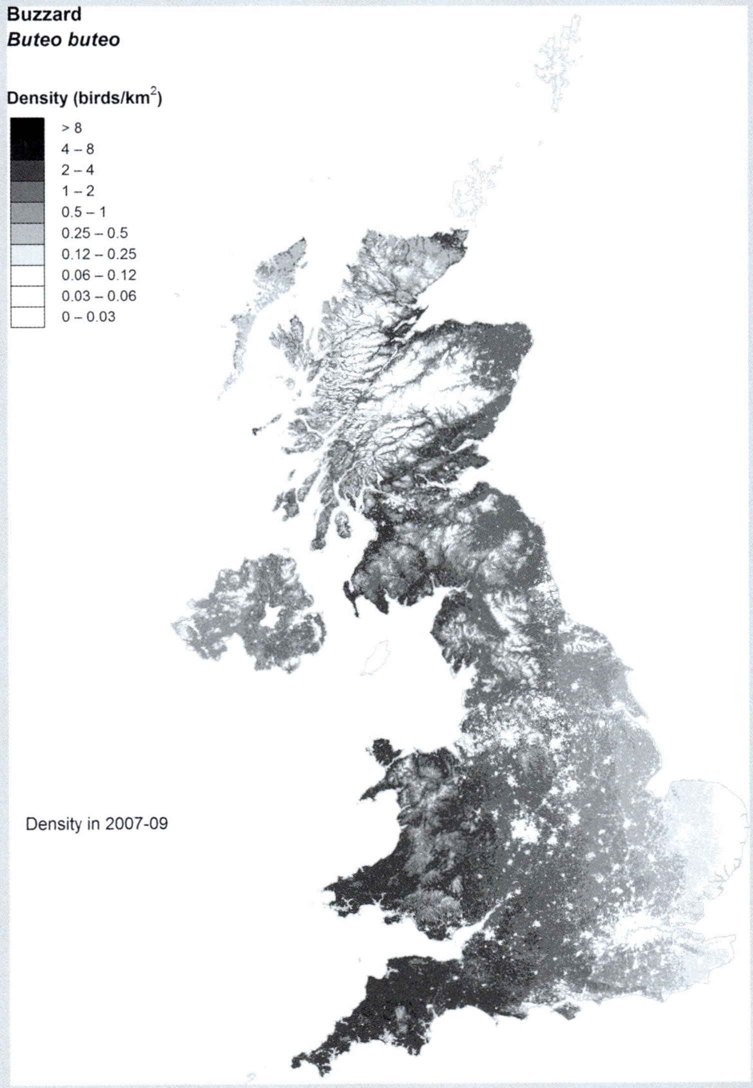

Box Figure 10.1.1 Distribution of buzzard population densities across UK. Note how buzzard density forms a gradient of variation > 260% from western to eastern Britain. Downloaded from <http://www.bto.org/>, with permission from BTO (2014).

[6] Trends are percentage changes, where the 95% confidence limits of the change do not overlap zero (indicating that there has been a significant change). The sample is the mean number of squares per year on which the species was recorded during 1994–2012: UK sample = 919 squares; England = 611; Scotland = 141 (Risely et al. 2013).

10.3 Actions taken so far

People's behaviour is determined by perceptions, which may become distorted in the context of human–wildlife conflict. Although 30 years of research suggests that buzzards take only relatively few pheasant poults, a survey undertaken by the National Gamekeepers Organisation in 2011 found that 76% of gamekeepers believed that buzzards had harmful effects on gamebirds. People have long memories, especially for episodes exceptional enough to affect individual livelihoods, and denial of the occasional problem tends to cause resentment and vigilantism. On that basis, one case, where a gamekeeper estimated a loss of 25%, was cited by Defra (2012) as a justification for acting against buzzards. The Fera (2012) report states that the number of applications to Natural England (NE) to kill buzzards had increased from three applications during the eight year period of 2003–2010, to six applications in 2011. None was granted until 2013.

Against a background of deepening controversy, in 2012, Defra outlined a government-funded £375,000 research proposal[7], including, inter alia, trials of buzzard nest destruction. However, public concern over lethal methods of control caused this proposal to be abandoned.

In 2013, Natural England issued a licence to trap two buzzards at a farm where a pair were deemed to be taking 9% of the free-range flock[8]. One young bird was trapped without effect, but predation ceased when an adult was trapped and given to a falconer, who later found this buzzard to be an effective hunter of rabbits and woodpigeons[9]. In the same year, Natural England judged that the conditions were met to licence[10] the destruction of four buzzard nests and eggs at the request of a pheasant shooting estate. The evidence appeared to be that a reduction in pheasant numbers at the pens coincided with a time when buzzard numbers increased[11].

[7] Available on-line as *Buzzard Control Experiment Overview: The development of management techniques to reduce predation of pheasant poults by buzzards*, Defra, June 2012.

[8] Natural England licence (2013): <http://www.rspb.org.uk/>, accessed March 2014.

[9] Irving, G. 2014. News and views: the problem buzzard issue. Newsletter of the British Falconers Club 47:9.

[10] *Myth buster: Full details of buzzard nest control licence*, Natural England, May 2013: <http://www.naturalengland.org.uk/>, accessed March 2015.

[11] The licensing was criticised by the RSPB because: (1) NE did not subject the decision to public scrutiny—it was exposed through an Environmental Information Regulation request, whereupon NE made the licences public through their website; (2) the 'serious damage test' lacks transparency and testability.

In this context, there is a hierarchy of questions to be asked about any proposal to attempt to solve a human–wildlife conflict by lethal means (Chapter 13, this volume). Under what conditions is killing any perceived 'pest' rational? We suggest that three questions should be asked, the answers to which would all need to be yes. First, does the farmer or gamekeeper suffer loss to the pest? Second, does action taken reduce that loss? Third, is it cost effective?

Applying the three tests to the proposal of destroying the nests of problematic buzzards, the answer to the first question might be yes, insofar as some gamekeepers may suffer loss to particular buzzards. The answer to the second question could be no, because removing the nest does not remove a buzzard that is in any case causing problems after the main brood-rearing period (and if a previous youngster caused problems there is no reason to assume that the next generation of young will be similarly problematic), and so, regarding question three, nest destruction would not be cost effective.

However, if applied to the removal of a problematic adult bird, the answer to all three questions could be yes. In that case, if there were no other satisfactory solution, a government could choose to register a derogation under Article 9 of the EU Birds Directive, to prevent serious damage to crops or livestock, to protect fauna or flora, for research, or for the capture, keeping, or other judicious use of certain birds in small numbers. However, before such action was taken, a modern government should feel obliged to obtain the evidence, not only that action was in each case justified, but also that populations of buzzards and other raptors would not be adversely affected, either by the individual cases or by *carte blanche* effects that might follow from recognition by government of a genuine predation problem. This requires study of these populations, and of human relationships with raptors, in a broader context.

10.4 Buzzard population dynamics and dispersal

Repeatedly in this book we reveal how the complexities of behaviour and ecology make the solution of human–wildlife conflict trickier than at first meets the eye (the perturbation effect and the control of bTB in badgers being a prime example: Chapter 5, this volume). The general lesson is that wildlife management can be better undertaken, with greater effectiveness and confidence in the outcome, if it is based on an understanding of the biology of the main actors (in Macdonald and Feber 2015: Chapter 15 we trace this journey from problem to solution for the recovery of the water vole, and in Chapter 6, this volume, we summarize relevant knowledge about the problematic mink, on the principle of 'know thy enemy'). So, to understand and mitigate any problems that are perceived from buzzards, what do we need to know, and having answered the key questions, what should be done?

A first question is what factors determine the distribution and abundance of buzzards in the UK? Although at least some and, importantly in a democracy, perhaps most, people rejoice in the now large numbers of buzzards in some parts of Britain (up to 79,000 breeding pairs in 2009, Musgrove et al. 2013), there is nonetheless an interesting question as to why there are not even more, particularly in certain areas where to date they remain rare.

In the early twentieth century, the common buzzard had become restricted to areas with little game rearing, in the west. In the 1990s they spread eastwardly, recolonizing central and northern Britain, but what has kept their population at such low numbers in eastern Britain? Furthermore, how might buzzards in the UK respond to new pressures imposed by control measures, land management, and climate change? The latest ecological theory indicates that the key to answering all these population-level questions lies in understanding the requirements of individual buzzards. One approach to tackling this problem has been through computer modelling, and this was stimulated by lessons learned in modelling the population processes of another raptor, the goshawk (see Box 10.2)[12]. The remainder of this chapter will focus on demonstrating how modelling can help us to answer these questions.

10.4.1 Population modelling of buzzards in the UK

Having found a population of buzzards on the Dorset coast, 140 km south west of Oxford, Sean Walls and Robert Kenward (Walls and Kenward 1995, 1998, 2001) set about studying them in an area where a 6 × 22 km north–south strip encompassed all the landscapes found in southern Britain (chalk downland, heathland, and conifer woods, water-meadows, and deciduous woods, with more intensively farmed land to the north). From 1990 to 1997, almost all nestling buzzards in this area were fitted with four-year radio-tags, and tracked from birth to death. These data populate the models we describe below.

10.4.1.1 Lethal control did not hinder expansion or increase in abundance

Remembering that buzzards have been puzzlingly slow to spread to, and then to increase in abundance, in the east of England, an obvious question was whether this was a consequence of illegal killing, as poor habitat was a possible alternative explanation. After all, buzzards in Dorset were spending much of the year seeking worms (their main winter food, Hodder 2001) on grasslands, and the emphasis in Eastern England is arable farming. Or could both keepers and habitat be involved?

The buzzard population model for Dorset revealed that only one in four of the buzzards alive in spring needed to breed for the population to be stable. Reassuringly, fieldwork involving transect counts confirmed that there were indeed three non-breeding buzzards for every paired bird at nests with eggs (Kenward et al. 2000). From information on the number of nests, the model predicted an overall density of 2 buzzards/km^2. At first this was puzzling, because home ranges averaged more than 1 km^2. However, peripheral areas of home ranges overlapped, with buzzards commonly foraging in company with their neighbours for worms on pastures; this tolerance ceased abruptly in core areas of about 0.5 km^2 (Walls and Kenward 2001). It was the mutually exclusive core ranges that dictated the predicted population density of 2 buzzards/km^2.

Elliot and Avery (1991) analysed reports of illegal killing of buzzards between 1975–1989 in Britain, and deduced that a greater proportion of buzzards died from this cause towards the periphery of their geographical range than did so in their strongholds. Using these risk factors to calibrate the matrix model, it was possible to investigate whether illegal attrition was impeding the recolonization of eastern England by buzzards. On the contrary, the model estimated that, in addition to the background natural annual mortality of 30% (including the low observed rate of illegal killing), the population could withstand a further 50%

[12] Ideally, population models are rigorously tested in the field (after all, they have value only if they make fairly accurate predictions). However, the northern goshawk had become extinct in Britain in the nineteenth century in association with persecution due to game rearing and almost complete loss of its woodland habitat, whereas in southern Britain, at the time Robert Kenward was developing the models, goshawks were too scarce for it to be feasible to study them effectively (and where they occurred landowners were reluctant to grant access for fear of incurring a heavy regulatory burden) (see Box 10.2).

Box 10.2 Early models: lessons from goshawks

Our approach to modelling buzzard ecology is built on lessons learnt from the loosely parallel problems of another raptor, the northern goshawk *Accipiter gentilis*. During the early 1970s, at Oxford's Edward Grey Institute of Field Ornithology, Robert Kenward radio-tracked goshawks as they recolonized southern Britain after an absence of 100 years. A practical outcome was to reveal the goshawks' impact on the woodpigeon *Columba palumbus*, a pest of brassica crops. This work started in and around Wytham Woods at the time when David Macdonald was tracking foxes nearby (Chapter 2, this volume); we often tell stories of unhappy occasions when foxes (in one case radio-tagged) ate radio-tagged goshawks (Kenward 2006), so the roots of our current collaboration on buzzards are deep! The sizes of the goshawks' home ranges were determined by the availability of woodland-edge habitat for hunting, and were smaller where woodland edge and prey were more abundant (Kenward 1982); clearly goshawks adjusted their hunting to cover enough ground to provide sufficient food.

However, a methodological complication was that home range size is much affected by the method used to calculate it (see Macdonald et al. 1980), so the analysis of goshawk range sizes and the factors affecting them could advance only hand in hand with the development of better techniques and software (Kenward 1987; Kenward et al. 2001a). After 20 years, this culminated in Resource-Area-Dependency Analysis (RADA), which was initially published without data in a manual for radio-tagging (Kenward 2001). RADA is based on the possibility of identifying important habitats *h* and estimating areas required *a* from a linear regression of log proportional habitat use against log range area for a sample of animals[13]. Possible complications arising from the model (e.g. several types of range area estimators, differences in habitat quality, more than one habitat being important, and habitat proportion and range area not being independent variables) can be taken into account and can even provide additional information on, for example, animal behaviour.

RADA goes beyond previous habitat analyses in three ways:

1. The use of habitat relationships over the continuum of range size in the utilization distribution is an extension beyond preference analysis, which compares use with availability at one or two discrete spatial scales.
2. As the analysis relies on proportions of habitats within home ranges estimated via radio-tracking data, it is based solely on areas of 'used' habitats, thus avoiding the need to make ad hoc definitions of the habitat that is 'available' to the animal. This latter issue has been shown to be an inherent problem of preference/selection analysis methods, which, all else being equal, may yield preference or avoidance for certain habitats depending solely on a change in the extent of the study area (often the measure of what is 'available' to the animal)—showing that conclusions of such analysis depend not only on the biology of the animals, but subjective definitions that researchers make when defining study areas (Beyer et al. 2010).
3. RADA seeks to identify how much habitat is required by each animal, for example, to use in landscape suitability studies or for providing empirical information for individual-based modelling.

Meanwhile, analysis of predation by goshawks on game birds had resulted in development of models for predicting predator and prey population dynamics from an individual raptor's rate of food intake and their subsequent breeding success (Kenward and Marcström 1988). Matrix-model concepts, developed in raptors to discover what level of breeding (especially when depressed by DDT) was needed to offset mortality (Newton 1979), were refined by estimations of mortality and its causes from large numbers of radio-tagged raptors (Kenward et al. 1991, 1999, 2001b, 2007). These models and the RADA could then be linked to predict the density of hawks, their breeding success, mortality, and even predation rates.

[13] Described by the equation $\ln(a/A) = b\ln(A) + c$, in which the slope $b = -1$ and the minimum area of required habitat is obtained from the area intercept as $a = \exp(c)$, and where a plot of the proportion of the habitat (a/A) against range area is hyperbolic, and a plot with both variables log-transformed is linear.

mortality among young buzzards without declining (Kenward et al. 2007). This made it unlikely that any slowing of recolonization in the 1990s was due to illegal killing.

10.4.1.2 Habitat alone did not hinder spread, but limits abundance

The next step was to ask whether habitat was in some way responsible for the slow eastward spread and increase in abundance. There were no obvious differences that seemed to be strong candidates in terms of the buzzard's needs, and this has led to our current research agenda, as we describe here.

First, we revisited and updated the RADA analysis. A decade ago, Robert Kenward's RADA model was based on the expectation from the goshawk study (Box 10.2, Kenward 1982), that an animal will adjust the size of its home range to encompass a sufficient area of those habitat types that provide it with the resources that it needs to survive (other habitats being encompassed within the territory as an inert matrix). This expectation, in the context of the evolution of group living in spatio-temporally heterogeneous resources, also underlies the resource dispersion hypothesis (Macdonald 1983) as applied in this book to foxes (Chapter 2, this volume) and badgers (Chapter 4, this volume).

Revisiting the original model, Kenward et al. (submitted) found that buzzard home ranges, despite wide variation in size, all contained a roughly similar amount of a particular habitat (or combination of habitats); this is a clue that these are the resource-critical habitats (exactly this argument applies to many mammalian examples of spatial groups, for example, the areas of beetle-rich wadis in Blanford fox *Vulpes cana* ranges (Geffen and Macdonald 1993), or detached gardens in red fox ranges (Chapter 2, this volume), or water holes in lion *Panthera leo* ranges (Davidson et al. 2012), or dry land in capybara *Hydrochoerus hydrochaeris* ranges (Herrera and Macdonald 1989)).

For buzzards, three types of habitat in lowland UK were identified as crucial proxies for resources—the first being a well-known buzzard requirement and the latter two obtained via RADA analysis (Kenward et al. submitted): (1) patch of deciduous or coniferous wood (even one tree is enough), which is necessary for roosting (and eventually building a nest), (2) rough ground, for hunting vertebrates, and (3) seasonally long grassland, typically cut for hay in summer, and good for hunting earthworms when grass is short in winter. Specifically, RADA predicted that a buzzard's minimum viable home range would encompass 0.56 ha of rough ground (incorporating also scrub orchard, marshland and grass heath) and 15 ha of seasonally long grass, and it would defend all the rough ground and the long grass - as long as the minimum requirement of long grass area is safeguarded, buzzards may share with other buzzards any extra patches of long grass that fall within the home range (the pattern of earthworm availability lends itself to sharing, as illustrated for foxes in Chapter 2, this volume, and badgers in Chapter 4, this volume). There are indeed areas of flat fenland in East Anglia with little seasonally long grass; these tend to be 'agrideserts' which are also relatively devoid of rough ground and trees for buzzards. As buzzards, when necessary, can eat a wide variety of foods, and grassland can be quite abundant in Suffolk, coastal Norfolk, not to mention Sussex and Kent to the south, and counties to the west of Cambridgeshire, it seemed unlikely that habitat alone was the problem, though it could provide part of the explanation as to why buzzards are rare in the east—these low-quality habitats, where resource availability fluctuates a lot seasonally, may only allow adults to survive the entire year at very low densities.

10.4.1.3 Dispersal, phylopatry, and the rate of buzzard population spread

The puzzles of the buzzard's slow eastward spread and increase in abundance deepened. Illegal killing was not a promising candidate as a constraint, and habitat alone did not seem to tell the whole story, so our attention turned to dispersal (South et al. 2002). Our earlier modelling work on buzzard dispersal had shone some light on this too, revealing that colonization models are very sensitive to a number of parameters (South and Kenward 2001), and predict relatively slow colonization, along a rolling front (if birds which disperse ahead of the front do not travel far) from the home ranges they first settled, when it is time to find mates. Young buzzards occasionally settle in home ranges after dispersing over 50–100 km in good weather (Walls et al. 2005), but more often they settle only around 15 km from where they fledged (Walls and Kenward 1998); this means that, considering their soaring flight, they can probably often see their birthplace from their new home range. Moreover, once they had settled, older buzzards seldom strayed far (Kenward et al. 2001c); the median distance of pre-nuptial dispersal from winter home range centres was less than 1 km. Indeed, far from exploring further for mates after finding a good, but distant, wintering area, buzzards showed quite strong natal philopatry and 14 out of 18 radio-tagged birds that dispersed long distances returned briefly in spring to their natal areas (Walls and Kenward 1995, 1998), though this was never associated with successful breeding there. In other words, any adventurous pioneers were nevertheless unadventurous Romeos that would rather mate close to where they were hatched (perhaps bringing the advantages of local knowledge) if there was no mate adjacent to their wintering areas. To test the consequences of this neighbourhood settlement policy, Walls et al. (2004) had released buzzards 50 km east of the area where the species was already nesting; these birds survived well after their assisted passages, found unpaired mates already settled there, and bred at an early age.

The overall picture is that the buzzard population seems to spread eastward much more like a front than like a set of colonizers that will strive to thrive alone in previously unoccupied areas. The mechanisms underlying the rate of population spread are linked to the philopatric behaviour of individuals. Moreover, if the spread is to areas where habitat quality is poor, population densities should remain low. Interestingly, as we have seen throughout this chapter, both this population spread and the increase in local density are patterns that actually emerge from the behaviour of individuals. Thus, a starting point would be to develop a model to discern how buzzards populate a landscape. Hence, we used the buzzard radio-tracking data, together with the Land Cover Map of Great Britain—LCMGB (Fuller et al. 1994), to develop a new type of model, one that takes information from relationships between real buzzards and their habitats and uses that to create virtual buzzards whose individual behaviours, and their consequences to population-level patterns, we can explore.

10.4.2 The development of Resource-Area-Dependence Analysis Individual-Based Model (RADA-IBM)

Against this background, we developed a new spatially explicit, individual-based model (IBM) to explore further the links between the behaviour and ecology of individual buzzards and the emergent properties of their populations. This family of models have proven potent tools for elucidating the conservation of vertebrates on UK farmland (e.g. Macdonald and Rushton 2003), and recent theoretical, conceptual, and computational developments have radically improved their sophistication.

IBMs have become the main tool in investigating complex adaptive systems (CAS), and while they involve complex computing, and indeed capture complex biology, they have the attraction that the outputs are often not complicated and are intuitively understandable. It should be noted that CAS should not be thought of as dealing with the same problems as those that are the metier of ecology. They are a means of seeing the world from a higher level of abstraction—just like mathematics, in which the number '2' may be used to represent a pair of whales, a pair of aeroplanes, or a pair of planets. CAS are concerned with how behaviour of individuals (or agents) govern or influence group structure and dynamics. They thus provide a theoretical framework within which ecosystems, financial systems (individuals = banks or companies), immune systems (individuals = white blood cells and other cells), and even the human mind (individuals = neurons and other cells), can be investigated (Levin 1998, 1999). They reveal how complex patterns can emerge from extremely simple rules. For example, the beautifully coordinated murmurations of a flock of starlings is not guided by a master command centre, but, rather, emerges from each individual moving independently while following three simple rules: (1) avoid collision with nearby neighbours, (2) attempt to match their velocity, and (3) attempt to stay close to your nearby neighbours (Reynolds 1987).

These models have been catapulted forward by astonishing advances in satellite imagery, geographical information systems, animal tracking technologies, and computer processing. An example from our work concerns the Amazonian manatees *Trichechus inunguis*, one of the most cryptic species alive and inhabitant of an almost impenetrably remote wilderness, but which we tracked through the dark waters and dense forests of the Western Amazon, using a blend of ecological and GIS analysis (Arraut et al. 2010). We were able to answer questions about their seasonal migrations and resource bases that broadly parallel those we now pose for buzzards in a much softer landscape.

IBMs, in contrast to models based on statistical relationships, are based on mechanistic rules and are anchored to the detail of animal behaviour (Railsback and Grimm 2012). These models thus create a virtual reality in which to test the likely responses of animals to change—in this case, the impact on buzzards of different land use, or even buzzard management policies (Evans et al. 2012).

10.4.2.1 A RADA-IBM to test habitat change impacts on buzzards

The immediate purpose of the model was to predict, at the individual level, territory, home range size, and shape, and at the population level, maximum buzzard

Box 10.3 Resource Area-Dependence Analysis Individual-Based Model (RADA-IBM) for buzzards

In the RADA-IBM, individual home ranges, and population densities, emerge from interactions between individuals and their habitats. In the model, entities are individual buzzards, which move through the landscape, and habitat patches, which remain in place. All buzzards are assumed to be juveniles at the time of dispersal and, because no significant difference in habitat use was observed among sexes (Kenward et al. 2001c), no sexes are represented. Buzzard state variables are the amount of rough-ground and long-grass habitats that each individual needs to acquire in order to establish a home range, and the maximum radii in which buzzards can find a patch from where to start to search for neighbour core and outer range habitat. Habitat patches were classed according to the Land Cover Map of Great Britain (LCMGB; CEH 1990, see Arraut et al. submitted for details). The spatial units (i.e. the grid cells) represented 25 × 25 m squares that were classified habitats from the LCMGB. The spatial extent of the study area was a 22 km × 6 km quadrat in Dorset, southern UK, where an example of most habitats present in the UK's landscape occurred. The temporal scale of the model was not explicitly represented; buzzards colonize the landscape randomly and the model stops when there is no more space for new buzzards to occupy.

The area of rough-ground and long-grass habitat that each buzzard needs to acquire in order to settle was estimated from the RADA analysis of the buzzards (see Section 10.4.1). The three habitats, and the area of each habitat that buzzards required to settle, became model parameters.

How does the model work? In the virtual world, the buzzards first search for a potential settling site, i.e. a patch of deciduous wood (one pixel equivalent to 25 × 25 m is enough, not already within the territory of another buzzard). Having found this gate-keeping resource, the buzzard then searches for a pixel of rough ground, closest to its settling site that is buzzard-free, within a maximum core search distance from the settling site. If it finds a rough ground pixel, it moves to this 'seed' pixel, which then becomes the central patch of the loop that is now described. In the loop, it searches for and, if it finds, incorporates, any additional neighbouring rough-ground pixels (the eight neighbours of a pixel in a square grid) that are also free. If it finds a free neighbouring pixel, it will move to it, incorporate it to its territory, and then move back to the central patch. It repeats this search-move-to-incorporate-and-move-back process until there are no more neighbouring free rough-ground pixels. It then assesses whether any rough-ground pixels that have already been incorporated have neighbouring rough-ground pixels that do not belong to another virtual buzzard. If it finds any such pixels, it randomly chooses one and moves to it. It then makes it the central patch for the neighbours search and restarts the loop. At each step the virtual buzzard checks whether it has been able to acquire the required amount of rough ground. If, for example, it finds no more rough-ground patches with free rough-ground neighbours, it searches for the closest free rough-ground pixel within the pre-established maximum search distance from its settling site and moves to it; if it finds one, it moves to it and starts the neighbour-acquisition loop again. If it doesn't, it emigrates from the area. A similar process is followed for acquisition of the required amount of long grass and so on.

This searching behaviour for rough-ground pixels by modelled buzzards closely mirrors real-life buzzard searching behaviour for rough-ground patches within a landscape. The size of the virtual buzzard's territory becomes an emergent property of the distribution in the landscape of the rough-ground patches that the buzzard has acquired—if, for example, it settled on a patch that was surrounded by free rough ground and it could thus acquire the required amount at a very close distance from the nest, the territory would be small, but if it had to travel far for that, the territory would be larger. Note too that the maximum rough-ground search radius and maximum meadow search radius set the maximum distance at which a 'seed' patch can be found, not the maximum distance at which, respectively, rough ground or long grass can be acquired; this characteristic of the model allows for individual variation in the size of buzzard home ranges, which emerges, as for the rough-ground search radius, from the landscape configuration.

Having been successful in acquiring the woodland pixel for settling, plus the required amounts of rough ground and long grass, the buzzard then settles and the model continues running by making another virtual buzzard settle in a random, but still free, woodland pixel within the study area. At a point when virtual buzzards discover that no more free patches of deciduous woods exist, the population has reached carrying capacity for the landscape.

population size/density and spatial distribution for a certain geographical area (the workings of the model are described in Box 10.3). However, in a longer term, such models have potential to address many more questions.

10.4.3 Buzzard model results

The RADA-IBM predicted, with precision, the mean core-percentage areas up to 85% (Fig. 10.2). There was a slight qualitative shift in the pattern of increase in area for the 90% and 95% cores of real buzzards, which became more pronounced for the 100% cores, which the model captured only partially. These outer cores correspond to the range in which real buzzards undergo excursions for reasons other than habitat (e.g. to visit potential mates); such processes were not within the scope of this model. The pattern for mean core% perimeters is similar to that of areas, although modelled buzzards consistently show core% perimeters that are, up to 85%, slightly larger than those of real buzzards, with an inversion of this pattern for cores above 90%. Overall, however, virtual buzzards had home range sizes that were very similar to those of real buzzards (Fig. 10.2).

We also conducted a sensitivity analysis by varying the values of each of the four parameters between 20% and 67% and analysing proportional changes to the outputs (Railsback and Grimm 2012; for full details see Arraut et al. submitted).

We found that increasing the radius within which virtual buzzards could find a 'rough-ground seed' had a negligible effect on mean core% area and perimeter, and maximum population density (Fig. 10.3). This is because buzzards can often obtain all the rough ground they need from only one or two patches, which are generally spaced by distances that are smaller than the 150 m rough-ground search radius within which buzzards search (Arraut et al. submitted).

A small proportional increase in meadow search radius resulted in a large proportional increase in mean core% area, a smaller increase in mean core% perimeter, and only a small proportional increase in maximum population density (Fig. 10.3). This is because a larger meadow search radius allowed modelled buzzards to roost in woodland areas that were further away from rough-ground and long-grass patches (hence core% areas and perimeter increase). However, because individuals settling further away (model with larger meadow search radius) would defend the same habitat patches that would be defended by individuals that settled closer to these patches (in model with smaller meadow search radius), the effect on population size is negligible. The reciprocal is true for decreasing meadow search radius.

Changes in the amount of rough ground required had negligible influence on home range area and perimeter, but did result in changes to maximum population density (Fig. 10.3). The area of rough ground corresponded to only 0.04% of the habitat required by buzzards, so change in this percentage had little effect on overall habitat requirement. However, because increasing the amount of rough ground that is defended by each buzzard implied a direct decrease

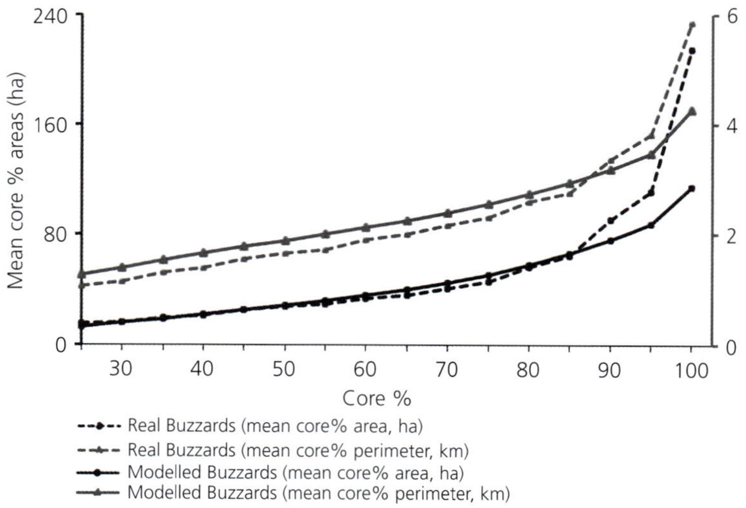

Figure 10.2 Comparison of mean core% areas from 25% to 100% cores of real and modelled buzzards as estimated by the minimum convex polygon method with arithmetic range centre.

in the amount of rough ground that was available to other buzzards, fewer buzzards were able to settle in the area, and hence maximum population density decreased (the reciprocal being true for decreases in requirement of rough ground area).

A small proportional increase in the area of long grass required by buzzards resulted in a comparatively large proportional increase in mean core% area, a smaller increase in perimeter, and a decrease in maximum population density, that was proportional to the increase in mean core% area (Fig. 10.3). This happened because long grass areas correspond to 95% of the areas of habitat that each buzzard needs, and the outer limits of the home range are dictated by the outermost long grass habitat patches, so increases in the area of required long grass tend to push the boundaries of the home ranges outwards, thus increasing their areas. When every individual increases the requirement of long grass in such a way, less long grass is left available for other buzzards, and therefore fewer buzzards can fit on the landscape (i.e. maximum population density is lower).

Overall, the analysis showed that the area of long grass required was the main factor controlling buzzard maximum population density. The model also showed that the relationship between the individual (core% area and perimeter) and the population scale was not straightforward. Marked changes to the home ranges of individual buzzards may result in negligible population effects or, conversely, maximum population density may change dramatically due to, or despite, changes in the structure of the home ranges (Fig. 10.3).

The home ranges of real and modelled buzzards are shown in Fig. 10.4. As we used the sole data set of radio-tracked buzzards in lowland UK, it is not known how, or if, the home ranges of individuals might change in response to changes in population density. Thus, for the sake of simplicity, our model allows for no individual variation in habitat requirements—the observed variation therefore emerges solely from the interplay between the simple deterministic rules that the individuals follow when establishing their home ranges and the structure of the landscape.

Table 10.1 shows the predictions for maximum population density and size in the study area. Note that, for the mechanistic model proposed in this study, the reported 'range' at which the density and population size predictions are yielded does not represent a

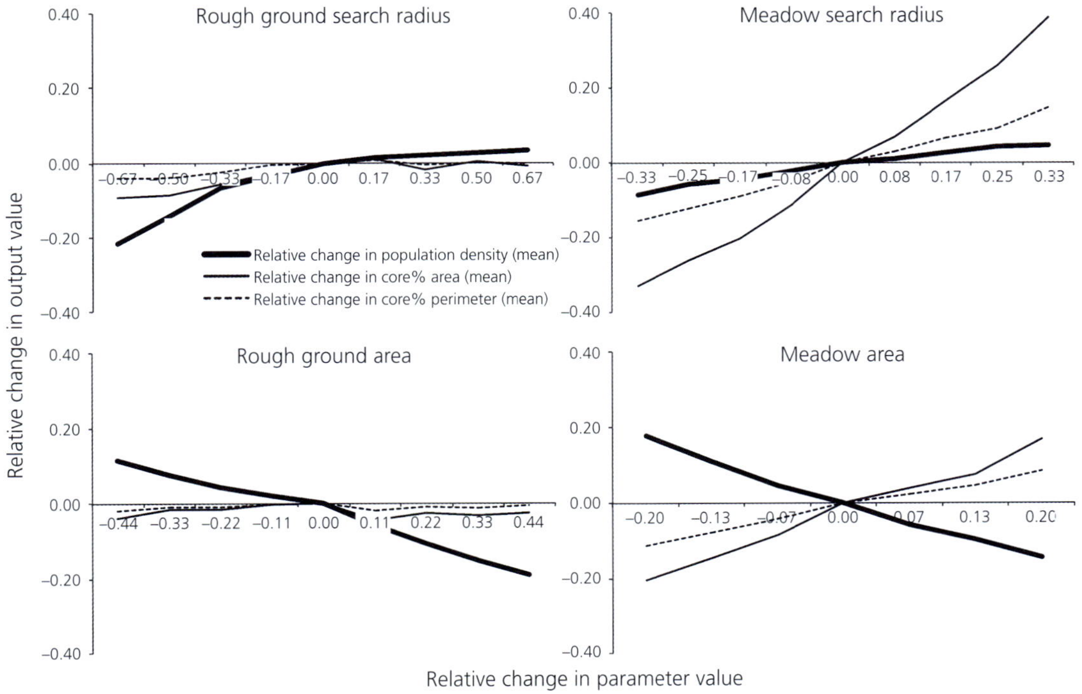

Figure 10.3 Sensitivity analysis, with variation in rough-ground search radius, meadow search radius, required rough ground area and required meadow area. Values in the *x*-axis represent proportional change to the parameter with respect to the chosen parameterization. Values in the *y*-axis represent proportional variation in mean core% area and maximum population density (Arraut et al. submitted)

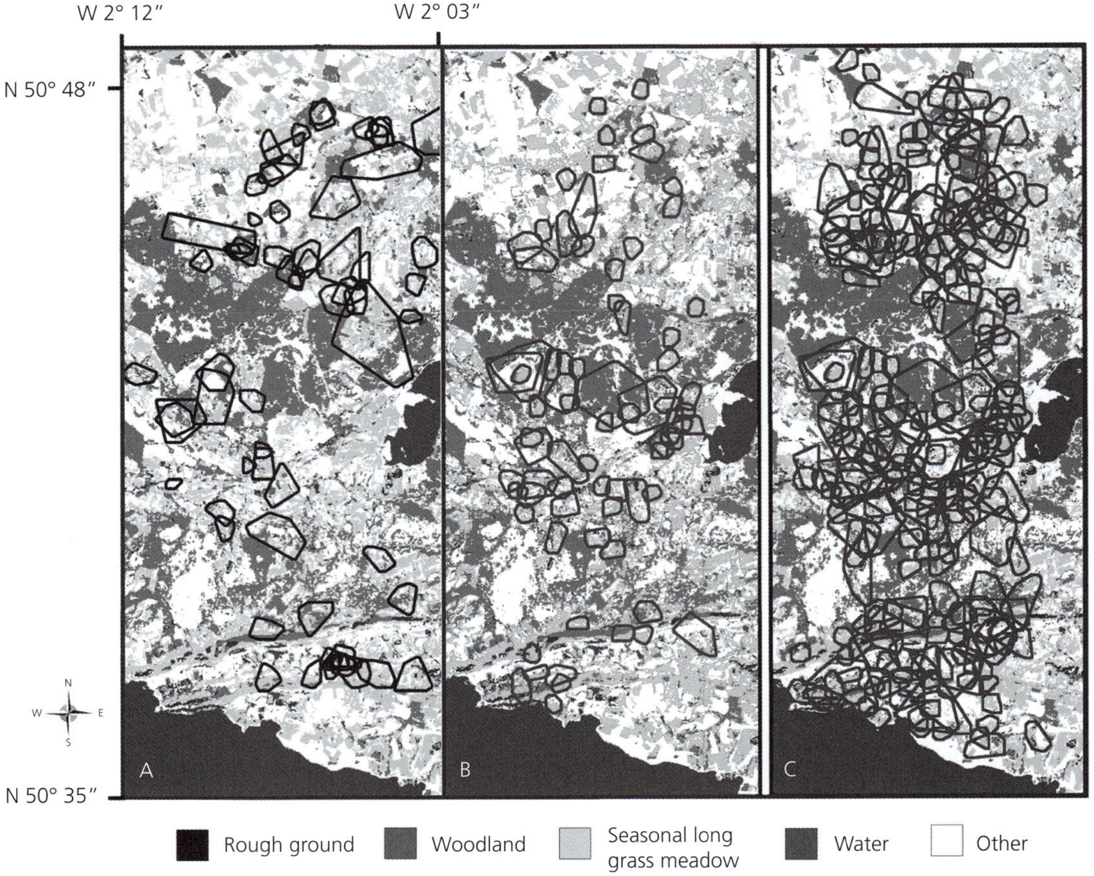

Figure 10.4 Comparison between home ranges of real and virtual buzzards (before reading the rest of this caption, one question for you: can you tell which are the real buzzards and which are the virtual buzzards?). (A) home ranges of 72 real radio-tracked buzzards, with (B) home ranges of 74 virtual buzzards, and (C) home ranges from same model run as shown in (B) but after population reached carrying capacity. Parameters were: rough-ground search radius = 150 m, meadow search radius = 1200 m, required rough ground = 0.56 ha and required long grass = 15 ha. Population carrying capacity was 220 individuals (density of 1.71 individuals/km^2).

Table 10.1 Comparison of estimates of population density and size of common buzzards in the study region (Arraut et al. submitted).

	Common buzzard IBM*	Field estimate (transect)**	Field estimate (mark-resighting)**
Density (range)	1.64–1.76	–	–
Density (mean)	1.70 (1.70–1.71)	1.87 (1.11–3.18)	1.82 (0.60–3.05)
Population size (range)	211–226	–	–
Population size (mean)	218.70 (218–219)	256 (152–435)	250 (82–417)

*This study, prediction of population mean and 95% confidence interval of carrying capacity, from 100 model runs.
**Kenward et al. (2000). Estimates are for mean and 95% confidence interval of population between 1990 and 1994, after recalculating field values considering same study area extent as that used in modelling study.

measure of uncertainty, but rather a prediction of what would be the population carrying capacity were it possible to run an experiment in which the landscape was cleared of buzzards 100 times and in each case the population was allowed to reach carrying capacity before a total count was carried out.

The modelling process has moved us forward in understanding how best to conserve this abundant raptor. The matrix modelling suggests that illegal killing is not now a problem for this species in southern Britain. The individual-based modelling suggests that buzzard density is controlled by landscape configuration, and that the landscape in eastern Britain, with few trees and much arable land, is what keeps buzzards in low numbers there. This intimate link between buzzard population abundance and landscape configuration connects buzzards with issues pertaining to landscape management, agriculture, and climate change, among others, in so far as these exert strong effects upon landscape configurations and the distribution of suitable buzzard habitats.

The evidence that slow spreading is a rolling-front phenomenon is relevant to management of other species, because it suggests that reintroductions are most important for species which do not search far from home for mates. Kites, for example, have spread very slowly from reintroduction sites. More work is needed on other species, such as hen harriers *Circus cyaneus*, to understand how much dispersal traits hinder their spread from remaining strongholds.

This RADA-IBM has been constructed for a small area in Dorset, in contrast with the dispersal modelling used to examine the observed relatively slow spread on a rolling front across southern Britain. To get maximum benefit from both types of models the next step could be to combine them, then test them for predicting buzzard colonization and density (at a small scale) elsewhere. They could also be applied to other species. The models could be used to examine what may happen for raptors at a fine scale (such as on large estates), or for smaller species at field or garden scale, while restoring desirable species or removing problematic individuals.

10.5 Conservation scenarios

So, to recap the salient points from which we started. First, we know that only a small proportion of the total number of game managers and poultry farmers suffer appreciable losses from buzzards and that only a small number of buzzards actually develop the habitat of eating pheasants regularly. Second, we know that a large percentage of buzzard individuals (50% of the annual production) can be removed without limiting existing buzzard populations (this is in addition to the background natural mortality—30%, Kenward et al. 2000). Third, a healthy lowland buzzard population had on average three non-breeding buzzards for every paired bird at nests (Kenward et al. 2000). At plausible levels of proposed interventions, such as egg removal and nest destruction, our modelling shows that in areas where buzzards have achieved higher densities it is likely that immigration from close-by, non-breeding, buzzards would soon repopulate empty areas. At a point location, even at rather large 'points', this phenomenon (loosely characterized as a vacuum effect) is well known in predator control—for example the sometimes many dozens of foxes that can be killed (arguably fruitlessly) on a single shooting estate, without any detectable diminution on the influx of replacements (see Chapter 3, this volume). For buzzards too, a combination of high density and extensive overlap of non-defended areas (e.g. Fig. 10.4) indicates that point removal may be slow to affect overall numbers present.

Fourth, and for the above three reasons, the debate on buzzard control is not about population reduction. It is about occasions where, despite application of other potential management solutions (e.g. in terms of pen design and poult release techniques), problematic individual buzzards may still associate with pheasant pens and kill poults there. In that case, even localized measures based on removal of buzzard nests may have limited efficacy, because if an adult is the problem it will still kill some pheasants to feed itself, albeit perhaps killing fewer pheasants than would be required if also rearing its young. Selective removal of problematic and atypical individuals killing full grown game or poultry may then be the solution that is most effective and least risky for raptor populations (Kenward 2002), as long as the removed bird is not replaced by an equally problematic successor, e.g. through persistence of poor pen management.

10.5.1 Management solutions for buzzards in the UK

Our scenarios uncover a swirl of cascading effects and unintended consequences that apply to buzzards much as they do to attempts to control foxes (Rushton et al. 2006; Chapter 3, this volume). For both predators these include ecological consequences, such as taking regulatory pressure off rabbit populations (Lees et al. 2013), but there can also be unintended socio-environmental consequences. Persecution to extirpate predators was formerly an acceptable management technique, but is no longer socially acceptable, thanks not only to legislation but also to more tolerant attitudes among those at risk of adverse impacts. However, if documented problems that affect livelihoods are ignored, there is risk of a backlash, with loss of tolerance that leads to a renewal of widespread illegal killing.

One can reasonably assert that there is no major problem with buzzard predation in lowland Britain, as it is, on average, a small cause of mortality among released pheasants. Occasionally, though, buzzard individuals take game or poultry and it seems likely that these individuals learn to kill prey that others find difficult. Such specialization is the same as for goshawks, of which some females regularly kill adult hares *Lepus* sp. while others kill none (Kenward 2006). Specialization on game by a minority of buzzards may partly explain why a survey of officials from governments and other organizations across Europe assessed buzzards as second only to goshawks in raptor–human conflicts (Kenward 2004). The same survey also found that where constraints on raptor management were least in relation to the seriousness of conflicts, the deliberate poisoning of raptors appeared least prevalent.

Poisoning is an unselective and inhumane management method but, ironically, is easier to hide for illegal use than live-capture traps, which are safe and selective. On one estate in Sweden, where goshawk predation of released pheasants was costing more than £2000 annually (Park et al. 2008), spring nets set on killed pheasants caught only the young hawks that were killing pheasants and not other radio-tagged hawks that were passing through the area. Moreover, few young hawks returned to Swedish estates after their release more than 30 km away (Marcström and Kenward 1981). Accordingly, live-trapping and translocation was permitted under a general Swedish 'hunting protection' licence. Game managers and other conservation interests were working together and there was no incentive for illegal killing.

What then, is the overall solution to conflict between game or farm managers and buzzards? Measures to further reduce buzzard kills at release pens discussed by Fera (2012) include ensuring sufficient cover beneath shrub and herb layers, brash piles, or 'wigwams'; minimizing opportunities for birds of prey to perch and monitor birds; stocking older poults at densities of 700–1000 per hectare; using reflective tape and other visual deterrents; removing right-angled corners to pens to avoid perching, and diversionary feeding. Refuges can also be created outside pens, for example, at exposed locations such as feeding stations and open crossing points.

However, what if, despite their best stock-management efforts, a few wildlife managers and poultry farmers still suffer serious problems from buzzards? If there are occasions where these remedies prove insufficient, buzzards would best be managed by live-trapping on kills, for example on fresh pheasant kills, which is likely to be selective of those individuals eating pheasant.

Once caught, and as a non-lethal control, translocation after live-capture has been mooted as a solution for problem buzzards. Indeed, Harradine et al. (1997) reported that of 996 gamekeepers surveyed by the British Association for Shooting and Conservation (BASC), 11% would have wanted supposedly trouble-making individual raptors to be relocated. However, translocations might have to be very far flung for buzzards, which naturally make long excursions and are natally philopatric (Walls and Kenward 1995). They may therefore be able to find their way back from release sites less than 100–200 km away.

The FERA (2012) review also raised the option of temporary housing for problematic caught raptors until pheasant poults have matured and dispersed (something they view as akin to housing wounded raptors into bird sanctuaries), or permanently removing problematic raptors to, for example, 'an aviculture collection'. This proposal which, with respect to wild-caught birds, has welfare considerations and could apply to only small numbers of birds, is similar to the (abandoned) 2012 Defra research proposal for housing delinquent buzzards in falconry centres. As buzzards capable of killing a problematic number of pheasants could be of interest for training by falconers, and are indeed likely to be few in number, transfer for active falconry would be another non-lethal approach, as was proposed also for delinquent goshawks (Kenward 2002, 2006).

If relocation after live trapping on kills is impractical, buzzards could then be euthanized. Other lethal solutions (e.g. shooting in flight) are undesirable, due not only to the possibility of killing rarer species (e.g. goshawks) by mistake, but also to the possible perception of a *carte blanche* for renewed general persecution. In contrast, live-capture of non-breeders or adults with nests after youngsters have fledged is selective, humane, and practical at points where genuine problems occur. We suggest it may be better to allow, under exceptional circumstances, a licensed removal option with no other risk to local raptor presence than to follow a coercive path which, at best, risks diverting public opinion from the more serious issues of changing land use and climate (also hindering co-operation across interests on these issues), and at worst, leads to increased illegal persecution of raptors.

10.5.2 Conservation conclusions

According to the 'World Conservation Strategy' published in 1980 and developed by the United Nations Environment Programme (UNEP) in conjunction with the International Union for Conservation of Nature (IUCN) and the World Wildlife Fund (WWF) 'Conservation is the management of human use of the biosphere so that it may yield the greatest sustainable benefit to present generations while maintaining its potential to meet the needs and aspirations of future generations. Thus conservation is positive, embracing preservation, maintenance, sustainable utilization, restoration, and enhancement of the natural environment'. Therefore conservation includes both protection and management of wildlife, in the interest of humans. How does this reconcile with our laws and research?

Our laws follow from International Conventions. The Habitats and Birds Directives of the EU flow from the 1979 Convention on the Conservation of European Wildlife and Natural Habitats (Bern Convention), which is primarily protective, and permits derogations for management. The 1992 Convention of Biological Diversity focuses more on sustainable use and restoration, with additions emphasizing adaptive management, but these aspects are not yet properly integrated into the laws of most European countries (notably excepting Norway).

In 1979, with many raptor populations newly decimated by organochlorine pesticides after decades of management by extirpation, raptors needed protection more than management. Now, however, with lowland raptors recovered, it is other bird taxa whose declines are dependent on changing farmland and woodland that are of more concern (Balmer et al. 2013; Macdonald and Feber 2015: Chapter 5), although there are still concerns for hen harrier in the uplands. In the future, it seems likely that habitat management will be key to the persistence of raptors and their prey. Such habitat management will require complex adaptive management, for which co-operation will be needed from all interests: governments and NGOs representing various citizen interests, landowners, and managers of the various products and recreational resources, including game species. Adaptive governance will be needed to support that co-operation. The sophisticated modelling that we have described in this chapter, and especially the RADA-IBMs that can handle habitat-based maintenance, restoration, and other management of species at scales from field level to national scale, puts us in a much better position for predicting effects of new governance, testing in practice, and then adapting as necessary.

Acknowledgements

We gratefully acknowledge comments on earlier drafts by Eva Raebel, Ruth Feber, Peter Brotherton of Natural England, Staffan Roos, Jeff Knott, and David Gibbons of the RSPB. EM acknowledges the support of the Brazilian National Research Council (CNPq), the Brazilian National Institute of Science and Technology for Climate Change Research (INCT-MC), and the Brazilian Network on Global Climate Change Research (Rede CLIMA) convênio FINEP/ Rede CLIMA 01.08.0405.01.

References

Allen, D.S. and Feare, C.J. (2003). Birds of prey and lowland gamebird management. In D.B.A. Thompson, S.M. Redpath, A.H. Fielding, M. Marquiss, and C.A. Galbraith, eds., *Birds of Prey in a Changing Environment*, pp. 427–442. The Stationery Office, Edinburgh.

Arraut, E.M., Marmontel, M., Mantovani, J.E., Novo, E.M., Macdonald, D.W., and Kenward, R.E. (2010). The lesser of two evils: seasonal migrations of Amazonian manatees in the Western Amazon. *Journal of Zoology*, **280**, 247–256.

Arraut, E.M., Walls, S.S, Macdonald, D.W., and Kenward, R.E. (submitted). Virtual animals in real landscapes: an empirical modelling approach for conservation. *Methods in Ecology and Evolution*.

Balmer, D.E, Gillings, S., Caffrey, B.J., Swann, R.L., Downie, I.S., and Fuller, R.J. (2013). *Bird Atlas 2007–11: the Breeding and Wintering Birds of Britain and Ireland*. BTO Books, Thetford.

Beyer, H.L., Haydon, D.T., Morales, J.M, Frair, J.L, Hebblewhite, M., Mitchell, M. and Matthiopoulos, J. (2010). The interpretation of habitat preference metrics under use-availability designs. *Philosophical Transactions of the Royal Society* B, **365**, 2245–2254.

Bright, J.A., Morris, A.J., and Winspear, R. (2008). *A review of Indirect Effects of Pesticides on Birds and Mitigating Land-management Practices*. RSPB, Sandy, UK.

BTO (2014). British Trust for Ornithology. <http://www.bto.org/volunteer-surveys/bbs/latest-results/maps-population-density-and-trends>, accessed 23 October 2014.

Burfield, I. and van Bommel, F., eds. (2004). *Birds in Europe—Population Estimates, Trends and Conservation Status*. Birdlife International, Cambridge, UK.

CEH, Centre for Ecology and Hydrology (1990). Land Cover Map of Great Britain 1990. http://www.ceh.ac.uk.

Cooke, B.K. and Stringer, A. (1982). Distribution and breakdown of DDT in orchard soil. *Pesticide Science*, **13**, 545–551.

Cramp, S. (1980). *Handbook of the Birds of Europe, the Middle East and North Africa. Volume II. Hawks to Bustards*. Oxford University Press.

Davidson, Z., Valeix, M., Loveridge, J.A., et al. (2012). Environmental determinants of habitat and kill site selection in a large carnivore: scale matters. *Journal of Mammalogy*, **93**, 677–685.

Defra (2012). *Buzzard control experiment overview. The development of management techniques to reduce predation of pheasant poults by buzzards*. https://raptorpersecutionscotland.files.wordpress.com/2012/05/buzzard-control-experiment-overview.pdf.

Dolton, C.S. and Brooke, M. de L. (1999). Changes in the biomass of birds breeding in Great Britain, 1968–1988. *Bird Study*, **46**, 274–278.

Elliott, G.D. and Avery, M.I. (1991). A review of reports of buzzard persecution 1975–1989. *Bird Study*, **38**, 52–56.

Evans, R.E., Norris, K.J., and Benton, T.G. (2012). Predictive ecology: systems approaches. *Philosophical Transactions of the Royal Society* B, **367**, 163–169.

Fera (2012). *Reviews of selected wildlife conflicts and their management Annex B: Approaches to mitigating bird of prey conflicts with pheasants at release pens, outdoor poultry and lambs.* A report to Defra by The Food and Environment Research Agency.

Fuller, R.M., Groom, G.B., and Jones, A.R. (1994). The Land Cover Map of Great Britain: an automated classification of Landsat Thematic Mapper data. *Photogrammetric Engineering and Remote Sensing*, **60**, 553–562.

Geffen, E. and Macdonald, D.W. (1993). Activity and movement patterns of Blanford's foxes. *Journal of Mammalogy*, **74**, 455–463.

GWCT (2012). *Common pheasant*. <https://www.gwct.org.uk/>, accessed February 2013.

Harradine, J., Reynolds, N., and Laws T. (1997). *Raptors and Gamebirds—a Survey of Game Managers Affected by Raptors*. BASC, Wrexham.

Harrington, L.A. and Macdonald, D.W. (2002). *A Review of the Effects of Pesticides on Wild Terrestrial Mammals in Britain*. WildCRU, Oxford.

Herrera, E.A. and Macdonald, D.W. (1989). Resource utilization and territoriality in group-living capybaras (Hydrochoerus hydrochaeris). *Journal of Animal Ecology*, **58**, 667–679.

Hodder, K.H. (2001). The Common Buzzard in Lowland UK: relationships between food availability, habitat use and demography. PhD thesis, University of Southampton.

Holloway, S. (2010). *The Historical Atlas of Breeding Birds in Britain and Ireland 1875–1900*. Bloomsbury.

Hughes, J., Sharp, E., Taylor, M.J., Melton, L., and Hartley, G. (2013). Monitoring agricultural rodenticide use and secondary exposure of raptors in Scotland. *Ecotoxicology*, **22**, 974–984.

JNCC (2000). *The Report of the UK Raptor Working Group*. Stationary Office, London.

Kenward, R.E. (1982). Goshawk hunting behaviour, and range size as a function of habitat availability. *Journal of Animal Ecology*, **51**, 69–80.

Kenward, R.E. (1987). *Wildlife Radio Tagging*. Academic Press, London, pp. 222.

Kenward, R.E. (2001). *A Manual for Wildlife Radio-tagging*. Academic Press, London.

Kenward R. (2002). *Management tools for reconciling bird hunting and biodiversity*. European Concerted Action within the 5th Framework Program: Reconciling Gamebird Hunting and Biodiversity (REGHAB).

Kenward, R.E. (2004). Management tools for raptors, pp. 329–339. In Chancellor, R.D. and B.-U. Meyburg, eds., *Raptors Worldwide*. World Working Group on Birds of Prey and Owls, Berlin, Germany.

Kenward, R.E. (2006). *The Northern Goshawk*. Poyser Series, A&C Black, London, UK.

Kenward, R.E., Clarke, R.T., Hodder, K.H. and Walls, S.S. (2001a). Density and linkage estimators of home range: nearest-neighbor clustering defines multi-nuclear cores. *Ecology*, **82**, 1905–1920.

Kenward, R.E., Hall, D.G., Walls, S.S., and Hodder, K.H. (2001b). Factors affecting predation by buzzards (*Buteo buteo*) on released pheasants (*Phasianus colchicus*). *Journal of Applied Ecology*, **38**, 813–822.

Kenward, R., Katzner, T., Wink, M., et al. (2007). Rapid sustainability modelling for raptors with radio-tags and DNA-fingerprints. *Journal of Wildlife Management*, **71**, 238–245.

Kenward, R.E. and Marcström, V. (1988). *How differential competence could sustain suppressive predation on birds*. In Proceedings of the XIX International Ornithological Congress, pp. 733–742.

Kenward, R.E., Marcström, V., and Karlbom, M. (1991). The goshawk (*Accipiter gentilis*) as predator and renewable resource. *Gibier Faune Sauvage*, **8**, 367–378.

Kenward, R.E., Marcström, V., and Karlbom, M. (1999). Demographic estimates from radio-tagging: models of age-specific survival and breeding in the goshawk. *Journal of Animal Ecology*, **68**, 1020–1033.

Kenward, R.E., Walls, S.S., and Hodder, K.H. (2001c). Life path analysis: scaling indicates priming effects of social and habitat factors on dispersal distances. *Journal of Animal Ecology*, **70**, 1–13.

Kenward, R.E., Walls, S.S., Hodder, K.H., Pahkala, M., Freeman, S.N., and Simpson, V.R. (2000). The prevalence of non-breeders in raptor populations: evidence from rings, radio-tags and transect surveys. *Oikos*, **91**, 271–279.

Kenward, R.E., Arraut, E.M., Robertson, P.A., Fuller, R.M., Walls, S.S., Casey, N.M., and Aebischer, N.J. (submitted). An assessment across home range estimators and core sizes: Resource-Area-Dependence Analysis. *Ecology*.

Lee, A.W., Shore, R.F., Turk, A., Pereira, M.G., and Best, J. (2008). The predatory bird monitoring scheme: identifying chemical risks to top predators in Britain. *Ambio*, **37**, 466–471.

Lees, A.C, Newton, I., and Balmford, A. (2013). Pheasants, buzzards, and trophic cascades. *Conservation Letters*, **6**, 141–144.

Levin, S.A. (1998). Ecosystems and the biosphere as complex adaptive systems. *Ecosystems*, **1**, 431–436.

Levin, S.A. (1999). *Fragile Dominion: Complexity and the Commons*. Helix Books, Perseus Publishing, 250p.

Lloyd, D.E.B. (1976). *Avian predation of reared pheasants*. Report for the British Field Sports Society, Game Conservancy, Royal Society for the Protection of Birds and the Wildfowlers Association of Great Britain and Ireland.

Macdonald, D.W. (1983). The ecology of carnivore social behaviour, *Nature*, **301**, 379–384.

Macdonald, D.W., Ball, F.G., and Hough, N.G. (1980), The evaluation of home range size and configuration using radio tracking data. In Amlaner, C.J. and Macdonald, D.W. *A Handbook on Biotelemetry and Radio tracking*, pp. 405–425. Pergamon Press, Oxford.

Macdonald, D.W. and Feber, R.E., eds. (2015). *Wildlife Conservation on Farmland. Managing for Nature on Lowland Farms*. Oxford University Press, Oxford.

Macdonald, D.W. and Rushton, S. (2003). Modelling space use and dispersal of mammals in real landscapes: a tool for conservation. *Journal of Biogeography*, **30**, 607–620.

Marcström, V. and Kenward, R.E. (1981). Movements of wintering goshawks in Sweden. *Swedish Game Research*, **12**, 1–35.

Martin, J. (2011). The transformation of lowland game shooting in England and Wales since the Second World War: the supply side revolution. *Rural History*, **22**, 207–226.

Moore, N.W. (1957). The past and present status of the buzzard in the British Isles. *British Birds*, **50**, 173–197.

Musgrove, A., Aebischer, N., Eaton, M. Hearn, R., Newson, S., Noble, D., Parsons, M., Risely, K., and Stroud, D. (2013). Population estimates of birds in Great Britain and the United Kingdom. *British Birds*, **106**, 64–100.

Newton, I. (1979). *Population Ecology of Raptors*. Poyser, Berkhamsted, UK.

Newton, I. (1998). *Population Studies of Birds*. Poyser, Berkhamsted, UK.

PACEC (2006). *Shooting Sports. Findings of an Economic and Environmental Survey*. Cambridge Public & Corporate Economic Consultants.

Park, K.J., Graham, K., Calladine, J., and Wernham, C.W. (2008). Impacts of birds of prey on gamebirds in the UK. *Ibis*, **150**, 9–26.

Post Office (2008). *Post Office counts the cost of wild animal accidents*. Press release. <http://www.royalmailgroup.com>, accessed March 2013.

Potts, G.R. (2012). *Partridges*. Collins, London.

Prytherch, R. (2013). The breeding biology of the Common Buzzard. *British Birds*, **106**, 264–279.

Railsback, S. and Grimm, V. (2012). *Agent-based and Individual-based Modeling: A Practical Introduction*. Princeton University Press, 329p.

Ratcliffe (1990). *The Peregrine Falcon*. Poyser, Berkhamsted.

Reynolds, C.W. (1987). Flocks, herds, and schools: a distributed behavioral model. *Computer Graphics*, **21**, 25–34.

Risely, K., Massimino, D., Newson, S.E., et al. (2013). *The Breeding Bird Survey 2012*. BTO Research Report 645. British Trust for Ornithology, Thetford.

Robertson, P.A. (1988). Survival of released pheasants, *Phasianus colchicus*, in Ireland. *Journal of Zoology*, **214**, 683–695.

RSPB (2008). *Birds of prey in the UK: on a wing and a prayer*.

RSPB (2013). <http://www.rspb.org.uk/>, accessed July 2013.

Rushton, S.P., Shirley, M.D.F., Macdonald, D.W., and Reynolds, J.C. (2006). Effects of culling fox populations at the landscape scale: a spatially explicit population modelling approach. *Journal of Wildlife Management*, **70**, 1102–1110.

Sim, I.M.W. (2003). Land use, common buzzards (*buteo buteo*) and rabbit (*Oryctolagus cuniculus*) in the Welsh Marches. In Thompson, D.B.A., Redpath, S.M., Fielding, A.H., Marquiss, M., and Galbraith, C.A., eds., *Birds of Prey in a Changing Environment*, pp. 351–371. The Stationery Office, Edinburgh.

Smart, J., Amar, A., Sim, I.M.W., Etheridge, B., Cameron, D., Christie, G., and Wilson, J.D. (2010). Illegal killing slows population recovery of a re-introduced raptor of high conservation concern—The red kite *Milvus milvus*. *Biological Conservation*, **143**, 1278–1286.

South, A.B. and Kenward, R.E. (2001). Mate finding and dispersal distances in invading species; a spatially explicit model. *Oikos*, **95**, 53–58.

South, A.B., Rushton, S.P., Kenward, R.E., and Macdonald, D.W. (2002). Modelling vertebrate dispersal and demography in real landscapes: how does uncertainty regarding dispersal behaviour influence predictions of spatial population dynamics? In Bullock, J.M., Kenward, R.E., and Hails, R., eds., *Dispersal: an ecological perspective*. Symposium of the British Ecological Society, pp. 327–349. Blackwell, Oxford.

Stahl, P., Ruette, S., and Gros, L. (2002). Predation on free-ranging poultry by mammalian and avian predators: field loss estimates in a French rural area. *Mammal Review*, **32**, 227–234.

Tubbs, C.R. (1974). *The Buzzard*. Collins.

Turner, C. and Sage, R. (2003). Fate of released pheasants. *The Game Conservancy Trust Review*, **35**, 74–75.

Valkama, J., Korpimaki, E., Arroyo, B., et al. (2005). Birds of prey as limiting factors of gamebird populations in Europe: a review. *Biological Reviews*, **80**, 171–203.

Walls, S.S. and Kenward, R.E. (1995). Movements of radio-tagged Common Buzzards *Buteo buteo* in their first year. *Ibis*, **137**, 177–182.

Walls, S.S. and Kenward, R.E. (1998). Movements of radio-tagged Buzzards *Buteo buteo* in early life. *Ibis*, **140**, 561–568.

Walls, S.S. and Kenward, R.E. (2001). Spatial consequences of relatedness and age in buzzards. *Animal Behaviour*, **61**, 1069–1078.

Walls, S., Kenward, R., and Holloway, G. (2004). How common is the Common Buzzard, *Buteo buteo*? *Biologist*, **51**, 23–26.

Walls, S.S., Kenward, R.E., and Holloway, G.J. (2005). Weather to disperse? Evidence that climatic conditions influence vertebrate dispersal. *Journal of Animal Ecology*, **74**, 190–197.

Watson, D. (1977). *The Hen Harrier*. T & AD Poyser, London.

Welsh Kite Trust (2011). Red Kite *Milvus milvus* distribution in Britain and Ireland. *Boda Wennol*, **26**, 24–25.

Wildlife and Countryside Act (1981). Act of Parliament of the United Kingdom. http://www.legislation.gov.uk/ukpga/1981/69.

CHAPTER 11

Brown rats on farmland: ecological citizens or subsidized carpet-baggers?

David W. Macdonald, Manuel Berdoy, and Joanne P. Webster

I wouldn't mind the rat race - if the rats would lose once in a while.

Tom Wilson

11.1 Background

The brown (or Norway) rat *Rattus norvegicus*[1] is an intriguing, if generally unwelcome, member of the British farmland fauna. It is a notoriously sociable and adaptable pest that belongs to a genus whose members' lives have been interwoven with those of humans for over a million years[2]. Rats have figured prominently in literature since medieval times, perhaps most famously in the epic poem *The Pied Piper of Hamlin*, adapted this century by Terry Pratchett in *The Amazing Maurice and his Educated Rodents*, a story in which a group of 'educated rats' gain intelligence from eating waste at the rubbish tip.

Having originated in South-east Asia, brown rats arrived in Europe in the early eighteenth century (Grzimek 1968; and see Barnett 2002). The question of when an introduced species can be said to have 'gone native' is always difficult—brown rats appeared only 100 years before grey squirrels *Sciurus carolinensis*, and several hundred years after feral ferrets *Mustela putorius furo* were introduced to hunt rabbits *Oryctolagus cuniculus*, themselves introduced by the Normans—and, as Macdonald and Burnham (2010) highlight, the question is particularly difficult for commensal[3] rodents. Macdonald and Burnham developed the notion of 'ecological citizenship', to describe the case of an introduced species that has become an integral part of the ecosystem, such that its removal would no longer rescue or restore the original natural state—and suggest that, in that case, the intruder's origins alone do not justify killing it. However, they also point out that it blurs a logical boundary to categorize rats and mice with most unintended invaders, in the sense that they evolved to travel with people, so it is entirely natural for them to do so (Macdonald et al. 2007). So, after almost three centuries in Britain, do brown rats qualify as ecological citizens, whether or not they are they still carpet-baggers, free-loading on the human enterprise? Does the answer affect the practicality, and the ethics, of managing rats? We have tackled these questions by researching their ecology on lowland farmland (in three contrasting circumstances with variable levels of human resource availability), their behaviour (partly in naturalistic enclosures) and their zoonoses[4] (including the extraordinary

[1] A different species from the black rat *Rattus rattus* associated (perhaps incorrectly) with the Black Death in fourteenth-century London.

[2] Bones of rodents of the genera *Mus* and *Rattus* are found alongside humans from mid-Pleistocene (2.5–1 mya) encampments (Macdonald et al. 2013).

[3] Meaning that they are usually found in association with people, 'sharing the table' (*mensa* in latin means 'a table'). Although since the word 'commensal' implies no damage to the host, rats (and mice) might be more precisely termed kleptoparasites (Macdonald et al. 2013)—an animal that takes prey or other food from another that has caught, collected, or otherwise prepared the food, including stored food (literally, 'parasitism by theft').

[4] Diseases that can be transmitted, directly or indirectly, from animals to humans.

Wildlife Conservation on Farmland. Conflict in the Countryside. Edited by David W. Macdonald and Ruth E. Feber.
© Oxford University Press 2015. Published 2015 by Oxford University Press.

phenomenon of parasite-altered behaviour, where our findings turn out to have implications not just for rats as agricultural pests, but to lay a foundation for potentially understanding human and animal behaviour in general, and even some cases of human psychoses).

Wild rat populations are typically associated with humans and anthropogenic food sources such as those found in warehouses, farm buildings, or refuse tips (Meehan 1984; Quy and Macdonald 2008). A number of studies on UK farmland suggest that hedges and fields are marginal habitats for rats, except when crops are available for food (e.g. Brodie 1981; Homolka 1983), and that rat populations in hedgerows are seasonally ephemeral (e.g. Middleton 1954). Males are more likely than females to leave a stable colony (Calhoun 1963; Kendall 1984) and enter hedgerow populations (Bishop and Hartley 1976), and farmland colonizers are mainly young animals (e.g. Farhang-Azad and Southwick 1979).

Given the close association between rats and humans, and the enormity of the brown rat's agronomic and public health impact (below, Section 11.3), surprisingly little is known of the ecology and behaviour of wild brown rats. In spite of pioneering post-war investigations (e.g. Chitty and Southern 1954), interest (and funding) waned, and by the 1980s a survey of the science citation index revealed that, of around 23,700 publications on rats each year, less than a dozen were on wild rats, and only a few of those were studies in the wild (Berdoy and Macdonald 1991). In an attempt to redress this balance, and with a view to seeking answers to practical problems (along with biological insight), we conducted a suite of field, enclosure, and laboratory studies of the ecology and behaviour of brown rats. We trapped and radio-tracked rats at field sites at Oxford University's University Farm at Wytham, Oxfordshire (the site of much of the WildCRU's research on small mammals, in Macdonald and Feber 2015: Chapter 4) (hereafter Wytham Farm) and at a nearby woodland managed for pheasant shooting. We also constructed naturalistic enclosures in Oxford University's Wytham woods, where semi-captive rats could be observed or filmed remotely, enabling detailed insights of rats and their behaviour not normally possible for a small, elusive, and nocturnal animal.

Through our field studies, we tested the hypothesis that rat population dynamics and demography are largely determined by the availability of anthropogenic foodstuffs. We predicted that (1) range sizes and movements would vary in response to resource availability, (2) changes in food availability would be reflected in sex and age ratios (with young males, indicative of an ephemeral population, predominating where resources were lowest), and (3) rats would not be able to breed year-round without access to such anthropogenic foodstuffs (and thus, that without supplemental food, farmland fields provide only marginal habitats). Before turning to our ecological findings we will first briefly review the behaviour and ecology of the brown rat, then give an overview of the problems that are associated with rats on British farmland (focusing specifically on zoonoses and zoonotic agents carried by rats on farmland), and describe a series of experiments to illustrate why rats are so difficult to control.

11.2 Rat behaviour and ecology

Brown rats are adapted to burrow, climb, and swim—as testament to their versatility, they can swim for 72 hours non-stop, enter houses through lavatory U-bends, and use kinaesthetic memory ('muscle memory'[5]) to run along pre-recorded paths at high speed (Berdoy 2002; Macdonald et al. 2013). In Britain, they are widely distributed in urban and rural areas, except in the most exposed mountainous regions and some smaller off-shore islands. Although commonly associated with buildings, rats also occur on unpopulated islands, such as the Hebrides[6].

Rats are omnivorous and opportunistic, and take a wide variety of foods including invertebrates (e.g. slugs, snails, and larvae), vertebrates (e.g. frogs, bird's eggs, and young small mammals), vegetable matter (e.g. agricultural cereals and seeds), and scavenged meat and bones (Macdonald and Tattersall 2001) (although they are exceptionally wary of unfamiliar food, Section 11.4.1). Rats are generally nocturnal but may adopt diurnal activity patterns to avoid predators (Fenn and Macdonald 1995[7]; see also Chapter 7, this

[5] A long-term muscle memory for a particular task, developed through repetition over time, eventually allowing it to be performed without conscious effort. An example of muscle memory in humans is riding a bicycle.

[6] Rats were recently successfully eradicated from the small Hebridean island of Canna (<http://www.guardian.co.uk/>, accessed October 2012), have been eradicated from Lundy Island (Lock 2006), and rat removal projects are currently underway on the Isles of Scilly (<http://www.ios-wildlifetrust.org.uk/>, accessed August 2013).

[7] Rats radio-tracked at a farm tip were diurnally active, and more strictly so in summer when foxes more often visited the tip to scavenge—the same rats reverted to nocturnal activity when they were moved to a fox-proof enclosure, suggesting that the rat's unusual behaviour at the tip was a strategy to reduce predation risk (Fenn and Macdonald 1995).

volume for similar behaviour in American mink *Neovison vison*). On English farmland, brown rats regularly travel several kilometres to reliable food sources (Taylor and Quy 1978; Hardy and Taylor 1979; Fenn and Macdonald 1987), although in farm buildings, where resources are plentiful, rats tend to be sedentary (Macdonald and Tattersall 2001).

Rats reach sexual maturity at two to three months of age and the females produce about five litters per year of seven to nine young (up to 15)[8] (Macdonald and Tattersall 2001). Theoretically, rats can breed year-round, producing up to 13 litters in a single year at 4- to 5-week intervals, and Fenn and Macdonald (1987) estimated that a pair of rats could, at a plausible extreme, produce 214 descendents in a single year. Rats may live, on average, up to three years in captivity, but longevity in the wild is probably less than 18 months (Calhoun 1963), and farm populations are likely to have high turnover rates.

Brown rats are highly social and gregarious; social organization is very flexible, ranging from large colonies with near-linear male dominance hierarchies, to single males with access to several females (Berdoy et al. 1995a; Macdonald and Fenn 1995; reviewed in Macdonald et al. 1999, Section 11.5.1; Berdoy and Drickammer 2007). Rats sometimes form colonies that are loose agglomerations of small family units or 'clans' (Calhoun 1963; Fenn and Macdonald 1987) of between five and 20 individuals. Clans defend territories and the number of clans that can be supported depends on resources (Macdonald and Fenn 1994). Movements within clans may be centred around a burrow that can function as an 'information centre'[9] where social transmission of information may take place (Galef 1991; Berdoy 1994; Macdonald et al. 1999).

As in other rodents, rats have a keen sense of smell (hearing, touch, and taste are also well-developed senses) and olfactory communication is of enormous social significance (reviewed in Brown and Macdonald 1985; Burn 2008; Wyatt 2013). Urine, faeces, and glandular secretions contain functional odours. At least 22 volatile components have been identified from the rat's haunch fur, proportions of which varied among individuals such that distinct odour profiles could be described according to sex, reproductive status, and age (Natynczuk and Macdonald 1994a; Natynczuk et al. 1995) (Box 11.1).

Box 11.1 The smell of sex

Distinct odours provide a basis for communication, but to receive a signal encoded in that odour, rats must both *smell* and *interpret* it. Signals may be encoded in odours in three ways: the presence of a single compound; ratios of a number of chemicals within an odour (White and Chambers 1989; White et al. 1989); or the relative abundance of a spectrum of odorous compounds, much like a chemical image (Albone 1984). Natynczuk and Macdonald (1994b), however, proposed a fourth model whereby odours might be interpreted by calibrating a fluctuating signal with a stable signal—self-calibration. Theoretically, this model might be particularly appropriate for signalling reproductive status, whereby a particular odour fluctuates through the reproductive cycle and is assessed (calibrated) against another 'non-sexual' stable odour. To explore whether this model might occur in nature, Natynczuk and Macdonald (1994b) used a combination of biochemical, behavioural, and histological techniques, and found that, consistent with their model, although males sniffed the females' haunches more than any other body zone[10] (irrespective of the rats' relatedness or reproductive condition), they sniffed the haunch *after* sniffing the forequarters as part of a sequence of sniffing along the female's body (Box Fig. 11.1.1).

Histology of the skin sebaceous glands indicated that the surface area-to-volume ratio of the sebaceous alveoli in the haunch, were significantly larger than those of the secretory glands in the forequarters, for five of six oestrous females, but not for four of five dioestrous females, and that

continued

[8] Gestation is two to three weeks and females are usually capable of *post-partum* oestrus; that is they can conceive immediately after giving birth, and so another litter can be produced as soon as the previous one is weaned (Kirkham 1910).

[9] *Sensu* Ward and Zahavi (1973) who initially proposed the existence of information centres in colonially breeding birds (see e.g. Bijleveld et al. 2010 for more recent work on this subject).

[10] 34–36% of sniffs in a ten-minute observation period were directed at the haunches, whereas all other body zones received < 15% sniffs, and the genitals received < 30% sniffs.

Box 11.1 *Continued*

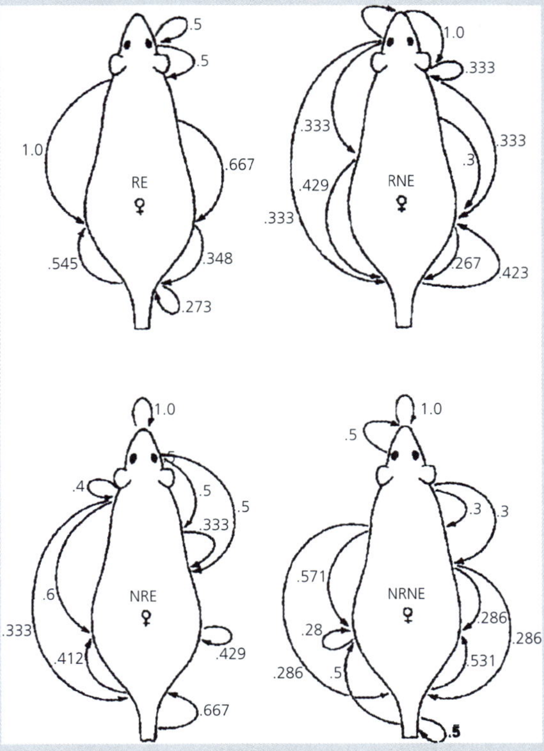

Box Figure 11.1.1 Transition frequencies for a sniff at one zone being followed by a sniff to another (for all interactions when male rats sniffed female rats twice in succession, depicting only those cases where probabilities exceeded 0.25). 10-minute behavioural observations of one male and one female rat in a 4.6 × 5.6 m rat-proof pen. RE = related oestrous female (n = 10), RNE = related dioestrous female (n = 21), NRE = non-related oestrous female (n = 9), NRNE = non-related dioestrous female (n = 22). There was no significant difference among categories of females in the absolute number of sniffs per 10-minute trial; males sniffed females an average of 23.38 times (SD = 1.3, n = 62). From Natynczuk and Macdonald (1994b) and reproduced with permission from Springer Science + Business Media.

surface area-to-volume ratios of sebaceous glands in the haunch region varied more than the ratios in the shoulder, suggesting a fluctuating production of sebum in the haunch only, in synchrony with the oestrous cycle. A male rat will therefore smell a discontinuity, or gradient, in scent along the body of an oestrous, but not a dioestrous, female. The self-calibration model has four distinct advantages: (a) it obviates the need for learning or inheriting responses to a pharmacopoeia of scents; (b) it minimizes the difficulty posed to the recipient by variation, in quality and quantity, between the odours of different signallers; (c) social odours are inevitably mixed with variable and copious background smells, and even one individual's scent may vary over only a few days; and (d) it could work not only with complex mixes of chemicals, but also with single compounds (Natynczuk and Macdonald 1994b).

11.3 Problems associated with rats

Rats are widely regarded as pests, due to the agricultural damage that they cause (predominantly through contamination of foodstuffs rather than direct consumption of food[11], but also through damage to property and materials[12]) and their risk to both human and farm animal health. Over half of the farmers participating in a questionnaire survey in the early 1980s in lowland England claimed that they had suffered damage due to rats, despite attempts to control them (Macdonald 1984). The damage on farms caused by rats was estimated to cost the UK farming industry £14 to £28 million in 2007 (Buckle 2007). There is some evidence that brown rat numbers in England are increasing (PTES 2008; Battersby 2005; TMP 2009) and speculation that climate change may exacerbate the situation if trends towards a warmer, wetter climate in the UK continue (Buckle 2009).

Despite the disease risks associated with rats, rat-borne infections in the UK had not, before the mid 1990s, received the same attention as had those of wild mice (e.g. Elton et al. 1931; Healing 1981). To identify the parasites and pathogens carried by wild rural rat populations in the UK, and to assess their prevalence, in 1995 we examined 510 wild brown rats in a series of surveys of 11 rural UK farms (nine within Oxfordshire, one in North Wales, and one in Hampshire), and nine homesteads or smallholdings around Somerset. At that time, we found that farmland rat populations carried at least 13 zoonotic and ten non-zoonotic parasitic species, few of which had been either detected, or investigated previously in UK rats (Table 11.1), and that individual rats simultaneously carried between two and nine potentially zoonotic parasites (Webster 1994a; Webster and Macdonald 1995a,b; Webster et al. 1995c).

11.3.1 Rat diseases and human health

All zoonotic parasites identified in wild rats can cause serious disease in humans, particularly those who are young, old, or immunosuppressed, and it is important that clinicians, veterinarians, and farmers are aware of the disease risks that wild rats pose to both humans and livestock. We found that rats were infected with bacteria that cause, for example, listeriosis, yersiniosis, and pasteurellosis in humans and/or domestic livestock. None of these bacteria had previously been recorded in wild rats in the UK. Listeriosis commonly causes encephalitis in ruminants, septicaemia and liver damage in other mammals, and in humans is particularly dangerous for pregnant women (causing miscarriage, premature delivery, or severe illness in newborn children). Yersiniosis causes symptoms similar to appendicitis in humans[13] (but does not affect livestock), and pasteurellosis causes pneumonia in farm animals. These bacteria may be passed by rats directly (e.g. via bites) or indirectly (e.g. by inhalation or via contamination of foodstuffs by droppings or urine). The role of rats in the spread of listeriosis is not clear (*Listeria* is carried by many animals and is found in water and silage), but the relatively high prevalence of *Listeria* in rats (11%) is of particular relevance to farmers, since rats, rather than the more frequently blamed moles (Atkinson et al. 1994; Chapter 12, this volume), could be responsible for *Listeria* contamination of silage (via contamination of stored silage).

The most prevalent protozoan parasite detected in wild rats was *Cryptosporidium parvum* (Table 11.1). *C. parvum* causes cryptosporidiosis, which can trigger enteritis and enterocolitis (and watery diarrhoea) in humans and other mammals (including cattle), and is particularly debilitating in the immunosuppressed (Perryman 1990). Transmission is usually due to contamination with infected faeces; humans may also be infected by swimming in contaminated lakes or rivers. Another protozoan, *Toxoplasma gondii*, was also detected at a much higher prevalence than the 10% previously reported by Jackson et al. (1986) (Table 11.1). Toxoplasmosis reputedly accounts for more human congenital abnormalities than rubella, syphilis, and herpes combined (Schmidt and Roberts 1989) and the loss of lambs due to toxoplasmosis has been estimated to cost the UK sheep industry £12.4 million a year (Brian 2008). Wild rats were previously dismissed as a negligible factor in the dissemination of this disease, partly because cats are the definitive hosts, and rats only an intermediate host; however, we have since discovered that *T. gondii* can persist in wild rat populations in the absence of cats via, at least in part, congenital transmission (Webster 1994a) and

[11] An enclosed rat population spoiled 70% of a store of wheat, but ate only 4.4% of it (Barnett 1950).

[12] 50% of fires on farms are thought to be caused by rats gnawing on electrical wires (Buckle 2007).

[13] Food-borne yersiniosis is usually caused by *Yersinia enterocolitica*; a related species, *Yersinia pestis*, is the causative agent of plague (but it is not currently found in the UK).

Table 11.1 Parasites detected in wild brown rats. ? = unconfirmed identification. $_v$ = vector for. Farms are the number of farms in which at least one rat was infected/number of farms sampled. From Webster and Macdonald (1995b), reproduced with permission from Cambridge University Press.

Parasites	Disease	Disease of Man	Disease of Livestock	n	Rats (% +ve)	Farms
Ectoparasites						
Fleas	$_v$Typhus	N_v	N_v	510	100	11/11
Mites	$_v$Typhus	N_v	N_v	510	67	11/11
Lice	$_v$Typhus	N_v	N_v	510	38	11/11
Ticks	$_v$Lyme disease	Y_v	Y_v	510	0	0/11
Helminths						
Syphacia muris	-	N	N	215	67	11/11
Nippostrongylus brasiliensis	-	N	N	112	23	11/11
Capillaria spp.	Capillariasis	Y	Y	44	23	3/3
Hymenolepis diminuta	Dwarf tapeworm	Y	N	243	22	9/11
Toxocara cati	Toxocariasis	Y	Y	112	15	4/11
Heterakis spp.	-	N	N_v	44	14	3/3
Hymenolepis nana	Rat tapeworm	Y	Y	243	11	9/11
Taenia taeniaeformis	Tapeworm	N	Y	147	11	4/11
Bacteria						
Leptospira spp.	Leptospirosis	Y	Y	259	14	9/11
Listeria spp.	Listeriosis	Y	Y	44	11	3/3
Yersinia enterocolitica	Yersiniosis	Y	N	44	11	3/3
Pasteurella spp.	Pasteurellosis	Y	Y	35	6	2/3
Pseudomonas spp.	Melioidosis	Y	Y	44	4	1/3
Borrelia burgdorferi	Lyme disease	Y	Y	30	0	0/3
Salmonella spp.	Salmonellosis	Y	Y	44	0	0/3
Protozoa						
Cryptosporidium parvum	Cryptosporidiosis	Y	Y	73	63	6/9
Toxoplasma gondii	Toxoplasmosis	Y	Y	235	35	7/7
Trypanosome lewisi	-	N	N	293	29	11/11
?Eimeria seperata	-	N	N	112	8	1/3
Babesia spp.	Babesiosis	Y	Y	293	0	0/11
Sarcocystis spp.	Sarcosporidiosis	Y	Y	210	0	0/11
Rickettsia (antibodies)						
Coxiella burnetii	Q-fever	Y	Y	127	34	5/5
Virus (antibodies)						
Hantavirus	Hantaan fever	Y	N	127	4	2/5
Cowpox	Cowpox	Y	Y	61	0	0/5

there is evidence that, to complete its life cycle, the parasite affects rats' behaviour in such a way as to increase predation by cats (Box 11.2). It is estimated that 7–34% of people in the UK have been infected with *T. gondii* but, because physical symptoms in healthy people are generally mild and non-specific, most infections probably go unnoticed (HPA 2012), with the main risks concerning foetal development. As with rats, though, there may be subtle effects on human behaviour, including links between *T. gondii* infection and mental health disturbances such as schizophrenia (e.g. Webster et al. 2013; Connor 2012; Webster et al. 2006, Box 11.2).

Antibodies to the rickettsian parasite of *Coxiella burnetii* and of *Hantavirus*, the causative agents of Q-fever and Hantaan fever (or haemorrhagic fever with renal syndrome), respectively, were also detected in wild rats for the first time in the UK and, indeed, continental Europe (Table 11.1). Both can be fatal in humans. In Britain, Q-fever has been described as the most serious, and potentially chronic, zoonosis transmissible to humans through aerosol from domestic animals (Geddes 1983). Although human infection is uncommon in the UK, Q-fever almost invariably causes illness in humans, ranging from mild influenza-like symptoms to endocarditis or hepatitis; outbreaks tend to be unpredictable and sporadic, with no identified common source (hence the name Q 'query' fever) (infection in livestock is usually asymptomatic). Cats appear to be important in the epidemiology of the disease, yet dogs, despite occupying a similar domestic niche, do not (Marrie et al. 1985; Marrie 1990)—rats (and their predation by cats) might be the missing link (Webster et al. 1995c).

Hantaan fever causes a harmless persistent infection in rodents, but in humans, is characterized by fever, headache, backache, haemorrhagic manifestations, and acute renal failure (Gajdusek 1962). Transmission from rats to humans most likely occurs via the inhalation of infectious aerosols of rat excreta or secreta (Mills 1999). The disease has been known in the Far East for some time, but its first recorded impact in western nations was the death of 5–10% of 3000 UN troops that developed the disease during the Korean war (Sheedy et al. 1954). Prevalence in the USA can be as high as 74% locally (Gibbs et al. 1982) and, although seroprevalence[14] in our survey was only 5%, now that the virus has reached the UK, it is possible that it will spread.

Undoubtedly, at the time of our study, the most publicised and feared rat-borne disease since bubonic plague was Weil's disease (caused by the bacteria *Leptospira icterohaemorrhagiae*)—a significant problem for people who regularly use rivers or other waterways in some capacity (Faine 1998) and said to be carried primarily by wild rats via their urine (e.g. Simpson 2002). Although other studies suggested prevalence rates of 70–90% (e.g. Waitkins 1991), we found only 14% of rats to be infected with *Leptospira* spp. (Table 11.1), and approximately half of those were infected with *Leptospira bratislava* (a bacterium that is much less serious for human health). The role of rats in the spread of leptospirosis to humans, at least on some farms, may thus be less important than commonly believed (Webster et al. 1995a,b) and other mammals are also likely to play a role. For example, 6% of wild water voles throughout the UK were found to be infected with *Leptospira* spp. (Gelling et al. 2012).

[14] The number of persons (or animals) in a population who test positive for a specific disease based on serology (blood serum) specimens.

Similarly, at 6%, the prevalence of *Pasteurella* (a bacteria that may cause pneumonia in farm animals, and reported to be the most common pathogen transmitted from pet bites, Chomel 1992), was much lower than the 41% reported by Curtis et al. (1980) for rats on UK farmland. Perhaps more surprisingly, although the general assumption is that many, if not all, rats carry *Salmonella* organisms (Chomel 1992) (the bacteria that cause salmonellosis—food poisoning, transmitted predominantly through eating food contaminated with *Salmonella*-infected droppings), we did not find any rats infected with this bacterium.

In short, wild brown rats on farmland may be less important in the spread of Weil's disease and common pathogens such as *Salmonella* than is commonly believed, but present a significant risk for the spread and transmission of a number of other zoonotic diseases. Q-fever and Hantaan virus are potentially of serious concern since both cause initial, non-specific, influenza-like symptoms in humans (as does Weil's disease), but can be fatal if untreated or misdiagnosed. There are risks for farmers, and for their livestock from rat-borne diseases. Management of disease risks is clearly helped by awareness, but understanding the risks and how best to respond to them also benefits from a better understanding of the ecology and population dynamics of the relevant parasites. We found, for example, that *C. parvum* prevalence in rats reached a peak (up to 90%) in spring and winter, when rats in the UK tend to migrate into farm buildings to feed on stored grain during cold weather. In contrast, prevalence dropped to 12% in summer when rats moved out into the fields to feed on growing crops. Thus, *C. parvum* prevalence in our study was highest at a time when rats were most likely to have contact with humans and livestock (Webster and Macdonald 1995a) (although Quy et al. (1999) report opposite seasonal trends for a population of rats at a rubbish tip on an arable farm, with peaks in infection in summer). *C. burnetii*, however, was detected in rats through the year, and showed no significant seasonal fluctuations, suggesting that rats constitute a persistent reservoir for this parasite in the UK and thus present a risk of Q-fever year-round (Webster et al. 1995c).

11.4 Why are rats so difficult to control?

Attempts to eradicate or to reduce the size of wild brown rat populations, to reduce damage or disease risk, have been on-going for thousands of years and continue today worldwide (Singleton et al. 1999).

Brown rats are, however, notoriously difficult to control (to trap, poison[15], or exclude). Brown rats are intelligent (indeed, they are the only species other than primates, including humans, for which it has been shown that they 'know what they do not know'[16]; Foote and Crystal 2007). They are capable of associative and social learning (reviewed in Galef 2012a), and are among the most innately neophobic[17] mammals known (Barnett and Cowan 1976). Rats' foraging behaviour (reviewed in Berdoy and Macdonald 1991; and Macdonald et al. 1999), an understanding of which is crucial to the development of effective baiting systems, is complex, and may be influenced by innate taste preference, experience, nutrient content, social status, and intra-group communication. Despite significant advances, new control methods are, over time, matched by new defence mechanisms, and the development of genetic resistance to some poisons demonstrates the power of natural selection (facilitated in rats by rapid population growth and high turnover of generations). Three features of the rat's biology and behaviour, which we explored through a series of enclosure and field experiments, are particularly relevant to control attempts: (1) learning, (2) physiological resistance, and (3) behavioural resistance.

11.4.1 Learning

A rat's capacity for making specific associations and for long-delay learning enables it to acquire information on the metabolic consequences of foods (Overman 1976; Rozin 1976): a rat that has been sublethally poisoned after ingesting a novel food will avoid that food for up to four months afterwards (Barnett 1958). Food aversion can also be mediated by other rats: laboratory rats experiencing a novel flavour will not consume it again if they are exposed to another sick rat (the 'poisoned partner effect'; Lavin et al. 1980) up to six hours afterwards (Coombes et al. 1980) or, in the case of mother rats, if it makes her pups sick (Gemberling 1984)[18]. In the laboratory, rats also develop a *preference* for a specific food after smelling that food on another rat's breath (see Galef 2012a,b). To test whether this 'demonstrator effect' also occurs in the wild, we recorded the foraging behaviour of wild rats in our enclosures (using an automated recording system described in Berdoy and Evans 1990) and were able to show that rats were more likely to eat cinnamon-scented, than peppermint-scented, wheat grain (previous experiments had shown these to be equally palatable to rats, but rats were neophobic to both) if they were exposed to other rats that had drops of cinnamon oil deposited on their neck, shoulders, and near their mouth (i.e. they appeared to have already eaten the cinnamon-scented grain) (Berdoy 1994; reviewed in Macdonald et al. 1999). The preference for cinnamon-scented grain persisted for two weeks after exposure to the cinnamon-scented 'informant' rats (Fig. 11.1). The informants themselves did not show any initial preference for the scents that were daubed on their fur, although they did show a delayed preference (perhaps due in turn to the influence of odours on the colony members that they had beguiled into eating cinnamon) (Macdonald et al. 1999).

11.4.2 Physiological resistance

The introduction of anticoagulant rodenticides[19], such as warfarin, in the 1950s, revolutionized rat control because symptoms of poisoning are delayed until several days after ingestion of a lethal dose, preventing associative learning and the development of

[15] This nineteenth-century recipe for rat poison demonstrates that people were well aware of these difficulties over 150 years ago: 'One pound of flour of malt, three drops of oil of rhodium, two ounces of loaf sugar, eight cloves, and a tablespoonful of caraway seeds, all beat fine in a mortar. Lay it in small parcels where they frequent, three or four nights, till they eat it freely; then add prepared arsenic, and set water in different places, with some infused into it. To prepare the arsenic, pour spirit of salt on it till it dissolves. When it is thus managed, it will not make them sneeze, which is the cause of their refusing to eat it.' John Mayor, Gamekeeper. 1845. *The sportsman's directory; and park and gamekeeper's companion: being a series of instructions, in ten parts, for the chase in its various classes ... with copious directions for trapping and destroying vermin, and detecting the operations of the poacher, to which is added an appendix, containing numerous valuable receipts, and useful abstracts of the game laws, and the laws of coursing*. Simpkin, Marshall and Co., London.

[16] This is termed metacognition—the ability to reflect on one's own mental processes, and is considered to be a defining feature of human existence (Foote and Crystal 2007 and references therein).

[17] A fear of novelty or anything new (operationally defined as the initial avoidance of novel objects in an otherwise familiar environment, Barnett 1958)—'fear' is not necessarily involved (Brigham and Sibly 1999).

[18] It was originally suggested that the poisoned conspecific functioned as an aversive unconditional stimuli in associative conditioning (Revusky et al. 1982), but more recent laboratory tests suggest that they are simply a salient stimulus that attracts the rat's attention and thus that aversion is caused by sensitization rather than association (Hishimura 2000).

[19] Chemicals that inhibit part of the vitamin K cycle involved in the blood clotting mechanism, causing animals that have consumed a lethal dose to die of internal bleeding.

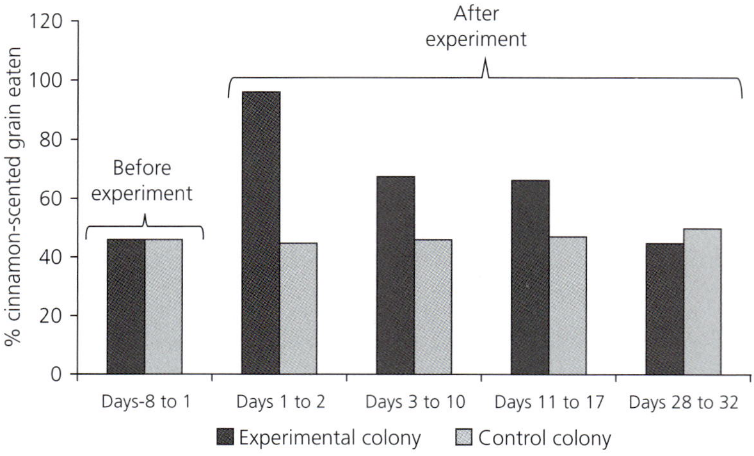

Figure 11.1 Change in preference for cinnamon-scented grain over peppermint-scented grain (shown as the percentage of cinnamon-scented grain consumed). 'Informants' (cinnamon-scented rats) were introduced to the experimental colony before the introduction of scented grain. The preference shown for cinnamon-scented grain in the experimental colony was statistically significant for two weeks; rats in the control colony showed no preference. From Macdonald et al. (1999).

aversion (bait shyness) (e.g. Nachman and Hartley 1975). However, by 1972, warfarin was largely ineffective in some regions due to the evolution of heritable physiological resistance (Greaves and Rennison 1973). Further, despite evidence that poison resistance can carry a fitness cost (warfarin-resistant individuals are prone to vitamin K deficiency[20] (Hermodsen et al. 1969) and are therefore at a disadvantage in a poison-free environment), we found an example where one type of warfarin-resistant individuals (those not suffering from vitamin K deficiency), in a warfarin-free environment, were significantly more abundant than expected, as well as heavier (i.e. were fitter) than were susceptible animals, suggesting that they retained a selective advantage over warfarin-susceptible individuals, *even in the absence of poison* (Smith et al. 1993).

Some rat populations later began to show resistance to two of the more potent second-generation anticoagulants (bromadiolone and difenacoum)[21] (reviewed in Buckle 2013), and recent research at Huddersfield University suggests that as many as 75% of rats in the west of England are now resistant to these rodenticides (Anon 2012)[22]. Unlike warfarin-resistant rats, however, difenacoum-resistant rats (rat populations for whom difenacoum treatment is ineffective) remain relatively susceptible to the compound (i.e. they succumb to treatment in the laboratory), and Greaves and Cullen-Ayers (1988) concluded that physiological resistance was not an adequate explanation for the high survival of some rat populations (in this case in Hampshire) subjected to anti-coagulant poisons (see also Quy et al. 1992). Two (not necessarily exclusive) explanations are that (1) there are differences in the ecology of the agricultural environment that affect the abundance and dispersion of rats in such a way that some rat populations are less susceptible to control (perhaps due to reduced access to baits) (Taylor 1978) (ecological effects), and (2) there are differences in the behaviour of individual rats (or rat populations) that lead to differing susceptibility to control by poison (Elkins 1973) (behavioural effects). There are a number of behavioural traits (collectively termed 'behavioural resistance')[23] that might vary,

[20] Warfarin resistance has been associated with impairment of the efficiency of the vitamin K cycle (Greaves and Ayres 1969), meaning that resistant rats require up to 20 times more vitamin K (Hermodsen et al. 1969). Vitamin K deficiency may hinder effective blood clotting and ultimately lead to haemorrhage (Suttie 1985) and death (Bishop et al. 1977).

[21] Second-generation anticoagulants only require one feed, whereas the first generation anticoagulants (such as warfarin) required multiple feeds to result in toxicity.

[22] Other more effective, resistance-breaking, anticoagulants exist (brodifacoum, flocoumafen, and difethialone), but their use is currently legally restricted to 'indoor only' in the UK because of the potential risks of these poisons to non-target wildlife. The European Commission regulations will now, however, permit the use of all five anticoagulants for use 'in and around buildings', and Buckle (2013) advises that the UK adopts the Commission's regulatory position. Currently practitioners in the UK can apply for a licence for emergency use of brodifacoum, flocoumafen, and difethialone to control rats at sites where the use of bromadiolone and difenacoum are ineffective.

[23] Defined here as behavioural traits which diminish a rat's tendency to eat palatable poison to which it otherwise has access to consume a lethal dose (Brunton et al. 1993).

Box 11.2 Kamikaze rats

T. gondii (the parasite responsible for the disease toxoplasmosis) is capable of infecting all endothermic vertebrates but has only one definitive host: the cat (Webster 2001). Parasites in an intermediate host (in this case, the rat) need to return to the cat to complete their life cycle (although infection can be perpetuated within the rat population by congenital transmission; Webster et al. 1994). Indeed, infected rats display abnormal behaviours that make them more susceptible to predation by cats, and also, incidentally, to poisoning (Berdoy et al. 1995b). *T. gondii*-infected rats showed statistically significantly higher levels of activity than uninfected rats, which may predispose them to be attractive to cats (which are attracted to moving objects, but show less interest in stationary ones) (Webster 1994b). No such increase in activity was observed for rats infected with parasites that do not require a definitive host to complete their life cycle (*Leptospira* spp., *Cryptosporidium parvum*, *Coxiella burnetii*, *Hymenolepis nana*, *Syphacia muris*) and for which predation would result in death of both host and parasite. Further, whereas uninfected rats showed an innate avoidance of cat-scented areas (tested using bedding material treated with a few drops of cat urine), infected rats not only did not avoid the cat-treated areas, but showed a significant (apparently suicidal) preference for them (Berdoy et al. 2000; Box Fig. 11.2.1). As reviewed in Webster et al. (2013), this behaviour is highly specific to cat hosts and no such 'fatal attraction' has been detected towards other predatory mammals such as mink (Lamberton et al. 2008) and dogs (Kannan et al. 2010), or to non-predatory mammals such as rabbits (Berdoy et al. 2000; Vyas et al. 2007; Webster et al. 2006). That these behavioural changes are due to the parasite altering its host's behaviour for its own benefit (i.e. enhancing its own transmission rate by increasing the likelihood of cat predation) is supported by Berdoy et al.'s (1995c) observations of 36 rats over 57 nights, that revealed no difference between infected and uninfected rats in other behavioural traits, such as social status or mating success—behaviours that might indicate generalized illness but have no obvious benefit to the parasite.

T. gondii-infected rats were also significantly less neophobic towards novel food-related items (Webster et al. 1994), were significantly more likely to be trapped first (Webster et al. 1994), and were significantly more likely to approach a mildly fearful object (in this case, a human observer)[24] than

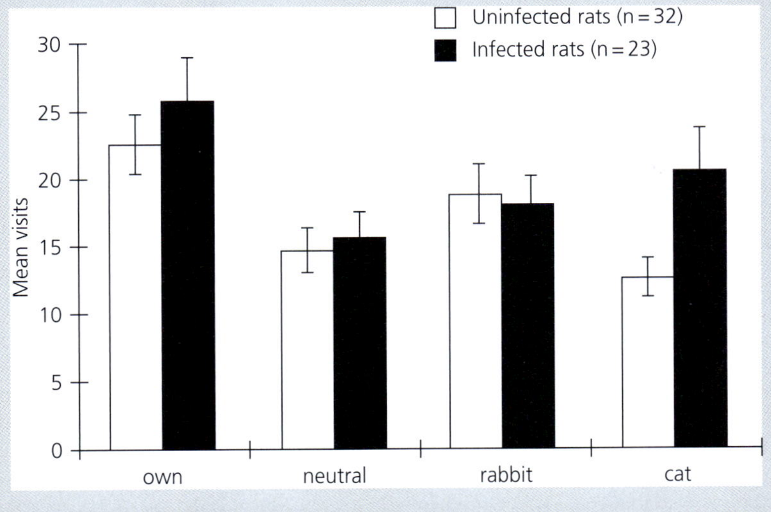

Box Figure 11.2.1 Mean number of visits (± SE, weighted by overall rat activity) to four scented areas in a 2 × 2 m outdoor pen. Uninfected and *T. gondii*-infected rats differ only in their response to areas associated with high predation risk (the cat-scented areas). From Berdoy et al. (2000).

continued

[24] Measured as the number of times a rat approached to ≤ 50 cm in 23 x 14 min sessions, over 16 non-consecutive nights: mean ± SE = 10. 9 ± 1.8 for infected animals versus 6.4 ± 1.2 for uninfected animals, n = 36 (Berdoy et al. 1995b).

Box 11.2 *Continued*

were their uninfected counterparts. Although most likely a side-effect of the parasite, these kamikaze tendencies render infected rats more susceptible to both poison and trapping (Webster et al. 1994). How the parasite is able to alter these behaviours is not fully understood, but appears to be associated with the parasite's interference with the brain's dopamine[25] system (Webster and McConkey 2010; Prandovszky et al. 2011; Kaushik et al. 2012).

Similar effects are possible in humans and further research has demonstrated links between *T. gondii* infection and personality, psychomotor performance (which may lead to increased risk of traffic accidents, e.g. Flegr 2007; Flegr et al. 2009), as well as a number of psychiatric disorders such as depression, anxiety and schizophrenia (e.g. Torrey and Yolken 2007), and probability of committing suicide (Ling et al. 2011).

with consequences for rat control, including capacity for conditioned or learned aversion, neophobia (which may also be affected by *T. gondii* infection; Box 11.2), and, more generally, food preferences, foraging tactics, and the ability to detect, recognize, and remember tastes.

11.4.3 Behavioural resistance

To examine whether behavioural effects were the most likely explanation for poor control with second generation anticoagulants in some rat populations, Brunton et al. (1993) treated three populations, at two farms in Hampshire, with a non-anticoagulant poison (calciferol[26,27], to which there was no known physiological resistance) for three to four months (Brunton et al. 1993). To rule out the most obvious ecological effects, a superabundance of bait points was used to ensure that all rats had access to the baits, and bait points were deployed around farm buildings and in the hedgerows bordering the neighbouring fields to eliminate the possibility that survivors might be immigrants.

None of the populations was successfully eradicated, and at two of the three sites bait-taken and track plate indices suggested that at least more than half the rats had survived. While it was possible that a super-abundance of desirable foods at these sites simply distracted a proportion of rats from the poison and so they ate none of it (or only sub-lethal quantities), this seemed unlikely and, although resources were not quantified, they did not seem exceptional. The most likely explanation, therefore, for the survival of a proportion of rats in this study was that 'behavioural resistance' in the form of learned food aversion, and potentially also neophobia, prevented them from eating a fatal dose of poison. A later study by Quy et al. (1995) concluded that a rat population in northwest Berkshire also demonstrated learned aversion to calciferol-treated bait. If learned aversion did occur, it is likely that neophobia played a role, insofar as the most neophobic individuals might be those that consumed a sub-lethal dose and thus were able to develop an aversion.

That neophobic behaviour varies among rat populations has been suggested by a number of authors (e.g. Mitchell 1976; Cowan 1977; Quy et al. 1992; see also Klemann and Pelz 2006), but disentangling the various effects that might contribute to such behavioural differences is not straightforward. As well as the possible effect of *T. gondii* infection status on neophobia (Box 11.2, Webster et al. 1994), apparent differences in neophobic behaviour could be the result of differences in: (a) inherited levels of neophobia; (b) experience; (c) environmental conditions, or any combination of these. In an attempt to distinguish between these, Brunton and Macdonald (1996) set up a series of laboratory experiments in which two groups of 30 rats, of different provenance and experience, were subjected to three novel food-related treatments. The wild-caught group (from a population that had recently been subjected to successful control by rodenticides) was overall more

[25] A neurotransmitter involved in mood, sociability, attention, motivation, and sleep patterns.

[26] A vitamin D-based poison that assists in the uptake of calcium (essential in minute doses but, like most other fat-soluble vitamins, toxic in larger doses) causing hypercalcemia, which leads to mineralization (crystal formation) in blood vessels, stomach, kidneys, and lungs, and eventually heart problems, bleeding, kidney failure, and death.

[27] Although there is debate over the efficiency of calciferol as a rat poison, Rennison (1974) achieved complete control on most of 23 farms within 18 days.

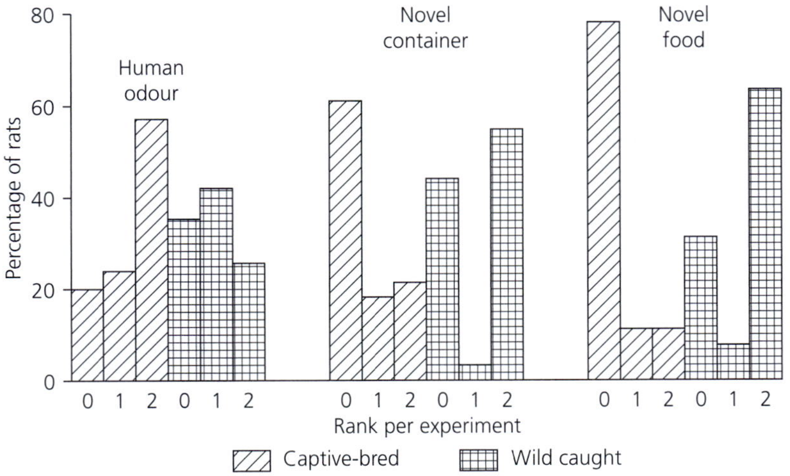

Figure 11.2 Percentage of rats from each group allocated a neophobia rank of 0, 1, or 2 in each experiment. 0 = no change in feeding (i.e. food consumed on the night of treatment), 1 = significant reduction in feeding for ≤ 5 days after treatment, 2 = significant reduction in feeding for > 5 days. From Brunton and Macdonald (1996).

neophobic (mean neophobic score[28] summed across all tests = 3.38 ± 0.33) than the captive-bred group (the first generation progeny of the survivors of a behaviourally resistant wild population)[29] (2.27 ± 0.26).

However, the two groups differed in the ranking of their response to the three treatments. The wild-caught group was most neophobic towards a novel container and novel food, but the captive-bred group was most neophobic towards a novel (human) odour (Fig. 11.2). This is consistent with the effect of experience rather than inheritance. Given the experience of the wild-caught group of toxic food, it is expected that they would be more neophobic towards food-associated novel stimuli (food or food containers) than captive-bred animals with no experience of harmful foods. Likewise, the former group had probably come into frequent contact with human odour before in harmless encounters, such as on bales of straw or farm machinery (i.e. humans for this group were neither novel nor threatening). Captive-bred rats, in contrast, were trapped from an enclosure where humans were absent (and so, for this group, humans were novel and potentially harmful), but they had never experienced a novel or unpleasant food before (so neophobia towards food, which was not perceived as harmful, was generally low).

Within both groups, though, there was considerable variation in reaction to potentially threatening or dangerous stimuli among individuals (even when the experience of those individuals was similar): 8% (n = 30) of the captive-bred rats and 10% (n = 30) of the wild-caught rats exhibited no neophobia, and 4% and 10%, respectively, showed extreme neophobia (neophobia score = 6). That the wild-caught rats were highly neophobic towards food-related cues under experimental conditions, but easily caught on the farms from which they were obtained, also suggests that environmental conditions play a role. Access to other 'safe' and plentiful foods (e.g. large grain stores) probably influenced the level of bait shyness exhibited (Quy et al. 1992). Indeed, rats housed in our outdoor enclosures with stable and *ad libitum* food sources showed very high

[28] For each test, individuals were assigned a neophobia score of 0, 1, or 2, according to the number of nights that elapsed before consumption of the food, such that a rat that reduced its feeding for only the first few days of the experiments received a lower score than a rat that significantly reduced its feeding for the duration of the experiment. Summed across all three tests, rats were scored on a scale from 0 to 6, where 6 indicated the most neophobic behaviour.

[29] These animals were offspring, born in a captive enclosure in Wytham woods, of five wild adult survivors (two males, three females) of a calciferol poisoning operation in Hampshire. The adults had been trapped at the Hampshire farm and released in the enclosure; they and their progeny were fed wheat *ad libitum* and had no subsequent experience of toxic foods. The 'captive-born' animals thus never had experience of toxic foods.

levels of neophobia (Berdoy 1994; see also Berdoy and Drickammer 2007). Individual variation in neophobia within a population remains unexplained but is presumably a personality (or behavioural type) trait[30].

11.5 Influence of human resources on rat populations

Given the significant challenges involved in effectively controlling rat populations, several authors have pointed to the need, first, for a better understanding of the rat's ecology and behaviour, to improve the effectiveness of management (e.g. Singleton et al. 1999; Cowan et al. 2003), and second, to focus on the factors limiting their populations, with a view to manipulating their populations (and their densities) without necessarily killing them (e.g. Krebs 1999; see also Lambert et al. 2008). We illustrate the first (i.e. how knowledge of rat movements and resource use enabled effective targeted management) in a small pilot study at Denman's Farm, Oxfordshire before addressing the second below.

At Denman's Farm, radio-tracking four rats (one male, three females) over six nights revealed that, although each differed in the time spent in hedgerows versus the farmyard, and in particular out-buildings, all four visited a central grain mill, which nightly observations confirmed was a favoured feeding site (up to 174 rat sightings per hour), despite being positioned in the centre of an otherwise bare farm building that offered little shelter for rats (Fenn et al. 1987). Exploiting the rat's use of this one favoured feeding site, we positioned a single poison bait (300 g of rodenticide, 0.025% warfarin and 0.1% calciferol) at the mill, which led to the elimination of 95% of the farm's rat population within ten days (at which point < 1 rat was seen per hour[31]) (Fenn et al. 1987). This extraordinary level of success with a single poison bait point was unusual (and we do not imply that it could be generally achieved), but given that the frozen ground (during the study) made other food less accessible, and that the lack of shelter meant that no single group of rats could monopolize the site, we suspect the baiting success, in this case, arose because *all* rats in the vicinity were at least partly dependent on the mill for food. This leads to the second point and the main thread of this chapter: the hypothesis that wild rat populations on lowland farmland, in the wider farmscape, are dependent on human resources.

To test this hypothesis, we trapped and radio-tracked three farmland rat populations with access to high, medium, and low levels of anthropogenic foodstuffs (Macdonald and Fenn 1995) as represented respectively by (1) an agricultural rubbish tip on Wytham farm (*c.* 0.3 ha, *c.* 500 m from farm buildings, surrounded by fields of winter wheat and hedgerows, supplied constantly mainly with chaff and cleanings from a grain-dryer and supplemented with pig and sheep carcasses) (hereafter 'the tip'), (2) a woodland pheasant-rearing area *c.* 1.13 ha of mixed deciduous woodland, several kilometres from the tip, where large quantities of low quality chaff were available throughout the study and grain intermittently (hereafter 'the hoppers'), and (3) a stream running through woodland and hedgerow *c.* 500 m of waterway, adjacent to the hopper site, where low amounts of grain and chaff were available for the first ten months of the study, but not thereafter (hereafter 'the stream'). We predicted that movements and home range sizes would be restricted at the tip (where resources were high), that young males (colonists) would predominate at the stream (where resources were lowest), and that rats would only be able to breed year-round at the tip.

11.5.1 Were rat movements and home range sizes restricted to where food resources were high (at the tip)?

In accordance with our predictions, both male and female rat home ranges were smallest at the tip (when the crop was absent) (Macdonald and Fenn 1995). Similarly, female rats at the hoppers occupied smaller linear ranges than did females dependent on natural food along the stream. Hardy and Taylor (1979) also reported smaller home ranges for rats near food sources compared with those in open farmland. Home range sizes of male rats at both the hoppers and the stream, however, were large. Similarly, at Wytham farm, when the crop was present, males (but not females) ranged widely, and males living in fields travelled more extensively than did tip residents. As might be expected, where home ranges were small, abundance was high.

[30] See for example Rödel and Meyer (2011) who show that distinct behavioural types exist in rats, and that personalities in young rats are shaped by early development (heavier individuals were bolder and more explorative) and early life experiences, such as litter size (young rats from small and large litters were more anxious than were rats from medium-sized litters).

[31] There was no evidence that rats had simply moved into the fields, and, in general, few field signs of rats were detected around any of the farm buildings or in the nearby hedgerows post-poisoning.

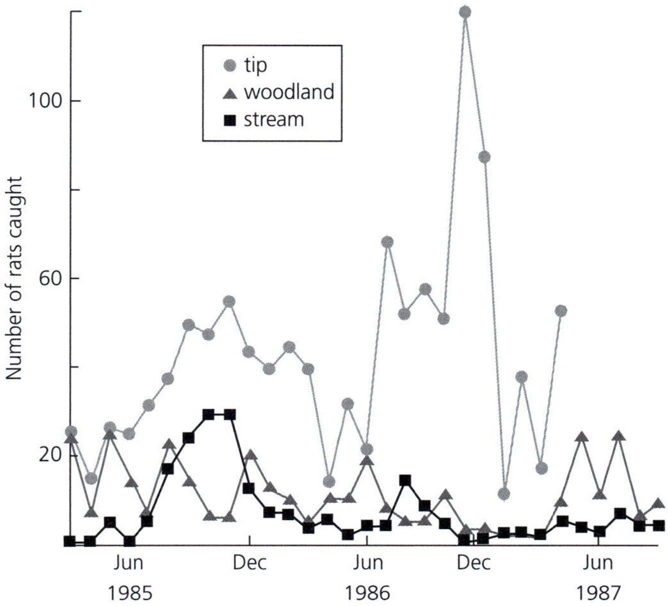

Figure 11.3 Variation in rat abundance (number of rats caught) across time in three habitats with different food resources. More rats were captured in the resource-rich environment of the tip than at the hoppers (the woodland pheasant rearing area, where resources were moderate) or stream (poor availability of resources). From Macdonald et al. (1999).

Thus, rat abundances were highest (mean monthly number of individuals per trap night = 0.19 ± 0.11, SD) at the tip (but fluctuated seasonally, with lowest numbers in spring and early summer), intermediate (0.05 ± 0.03) and unpredictable at the hoppers, and lowest (0.03 ± 0.04) at the stream (Fig. 11.3). Further, within all three sites, fluctuations in rat numbers appeared to reflect the vagaries of food provision (Fig. 11.3). Highest abundances at the tip occurred two months after a period of very high food availability. Similarly, although considerable month to month fluctuations at the hoppers were apparently unconnected with food availability, numbers were lowest at the end of a long period without supplementary food, and increased within three months of the reintroduction of feeding. At the stream, numbers were initially high but declined rapidly and remained very low after food provisioning stopped.

Differences in densities and ranging behaviour have implications for social organization insofar as population densities at food-rich sites (the tip and the hoppers) were likely to be too high for male rats to be able to hold individual defendable territories. At such sites, we envisaged a clan (sensu Fenn and Macdonald 1987) of rats whereby several adult males and a cluster of females occupy a single territory. Where resources were scattered or sparse, male movements were consistent with exclusive ranges and (at the stream) their large ranges gave access to clusters of females. The lack of females in the crop suggested that widely ranging males at Wytham farm were low ranking. In short, the flexibility of rat society means that, in resource-rich areas, rats might adopt multi-male systems, whereby several males mate with several females but, in impoverished areas, males appear to maintain single-male territories, embracing the overlapping ranges of several females with whom they presumably mate (Box 11.3).

11.5.2 Did young male rats predominate where food resources were low (at the stream)?

Contrary to our predictions, neither sex nor age ratios were related to relative food availability among the three study sites. Sex ratios did not differ significantly from equality at any site, and overall proportions of juveniles were greatest at the hoppers, especially in the first three quarters of the year (> 70% in April–June), and lowest at the stream (< 31%)—opposite to our prediction that young males would predominate where resources were lowest. There were, however, complex interactions between age (weight) and season. At all sites, for example, there were significantly more adult (> 150 g) females than males during April–June, suggesting either that population pressure forces males to emigrate or that males suffer higher juvenile mortality

Box 11.3 Multiple mating

Female rats mate with multiple males and, as for many other species, this behaviour has long been assumed to be a result of male coercion (see Chapter 6, this volume). Filming the behaviour of multi-male multi-female groups of rats over night in large naturalistic enclosures in Wytham woods, we were able to observe 'mating chases' during which an oestrous female was pursued by a string of males, and mated with several of them (Box Fig. 11.3.1, Berdoy 2002).

To test whether such behaviour was inflicted on the females, or sought by them, we set up an outdoor experimental arena that allowed the (smaller) female access to three (larger) males housed separately, but prevented the males from accessing the female's compartment. Replicating this trial with eight females, we showed that all eight females chose to mate with two or more of the three available males (average = 2.87) and made repeat visits to the male compartments (each male compartment was visited an average of 62 times by each female during each oestrous night). One way in which females might assess male fitness and thus select males for mating, might be to seek out conditions under which males are thrown into competition. However, although, when two males were housed together in a single compartment, the more dominant and active male of the two secured more matings, there was no evidence that females actively sought male–male competitive situations: females showed no preference for two males housed together versus a single male housed alone. Females did, however, display clear preferences among the three males in the days before oestrous, and those with which the female had spent most time before oestrous, were those allowed greater mating opportunities during oestrous (in contrast, house mice showed no preference for any particular male outside the mating period, but, during mating, preferred the dominant male and mated with them last; Rolland et al. 2003). Further, the 'preferred' males (the alpha males) appeared to take advantage of the greater mating opportunities offered and optimized the timing of their mating effort (Box Fig. 11.3.2); beta males attempted to mount the female whenever she entered his compartment continually through the night (as early as 11 hours before the peak of oestrous), but the alpha males only attempted to mate close to the peak of oestrous (despite repeat solicitations by the female). Mating opportunities thus appear to derive from a combination of male–male competition and female choice. There are 133 mammals for which there is evidence of multiple mating by females—the most general explanation appears to be to cause uncertainty about paternity and thereby diminish the risk of infanticide by males (Wolff and Macdonald 2004).

Box Figure 11.3.1 Mating chases—an oestrous female rat is pursued by a string of male rats. ©M. Berdoy.

continued

Box 11.3 *Continued*

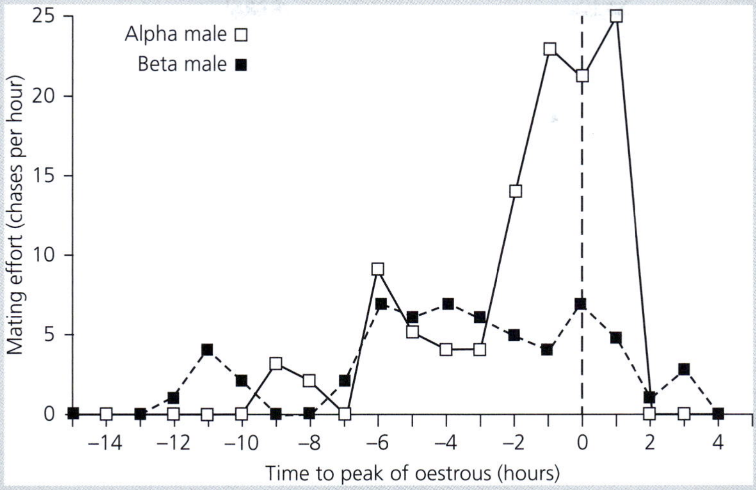

Box Figure 11.3.2 Mating effort (chases per hour) by an alpha and beta male rat, relative to time of the female's oestrous. M. Berdoy, unpublished data.

over winter[32]. Juveniles were present all year round (although animals < 100 g were rarely caught) and the proportion of juveniles at all sites was lower in late winter and spring, and higher in summer and autumn. Similarly, there was a relative excess of males at about 200–250 g, mainly in autumn, probably due to an influx of immigrant males (Macdonald and Fenn 1995). It would be interesting to test for differences in neophobia between this mobile and exploratory weight class and other rats.

11.5.3 Did year-round breeding occur at the tip (facilitated by anthropogenic food sources)?

Our third prediction was more difficult to assess. The occurrence of (palpable) pregnancies could not be readily interpreted in terms of presence or absence of food. However, relatively few young were caught in the second year of the study at the hoppers when there was little supplementary food (Fig. 11.3) and rats were probably subsisting, for the most part, on naturally-occurring foods, and at the stream there were even fewer young after food provisioning stopped.

In spite of the assertions of Davis and Jackson (1981) that, except in Alaska, rats breed throughout the year, Macdonald and Fenn's (1995) study, and others (Leslie et al. 1952; Andrews et al. 1972; Lattanzio and Chapman 1980), show that reproduction in female rats may cease with the onset of cold weather. At all three sites (i.e. regardless of anthropogenic food sources), pregnancy was broadly confined between March and October, and although some females did become pregnant outside this breeding season, no pregnant females were found in December (at any site)[33]. Human food, therefore, did not appear to influence either the number of the pregnancies or the time of year that they could occur.

There was, however, some evidence that reproduction in males was influenced by both food and season. Median weights of mature (scrotal) males were significantly heavier at the stream than at the hoppers in

[32] There was no evidence that the cull at the tip distorted the sex ratio because 52.3% of animals culled were males.

[33] Post-mortems of 152 rats culled at the tip (for the purposes of farm management), revealed that corpora lutea were present in the ovaries of females at a median body weight of 120.4 g (95% CI = 90.0–140.8) but pregnancies were not found in females < 200 g. In spring, 26.5% of females were pregnant (34.5% of females > 200 g), and in summer 58.6% of animals > 200 g, with 8.4 embryos on average.

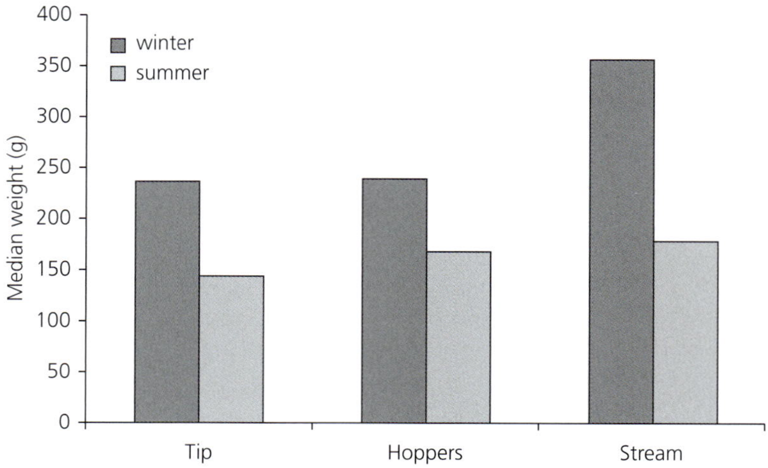

Figure 11.4 Median weight at which rat testes become scrotal in winter and in summer at sites that differed in resource availability (tip = resource-rich, hoppers, a woodland pheasant-rearing area = moderate, stream = poor). Males achieved sexual maturity at lower weights in summer than in winter, and also in resource-rich environments. From Macdonald et al. (1999).

winter, and heavier at the stream than at the tip in both summer and winter (Fig. 11.4) (but they did not differ between the tip and the hoppers), suggesting that food subsidies (at the tip and the hoppers) might allow earlier maturity in males (younger males are lighter, but were mature in these areas). Similarly, median weights of male rats with scrotal testes were significantly heavier during winter than summer at all three sites (Fig. 11.4). Post-mortems of males culled at the tip (for the purposes of farm management) revealed that, in spring, all males > 250 g (n = 32) were fertile[34] (median body weight = 353 g), whereas in winter only some of the males > 250 g were fertile. In particular, two individuals remained infertile in winter at body weights exceeding 300 g, suggesting that facultative cessation of spermatogenesis occurred in some individual rats.

To return to Macdonald and Burnham's (2010) question of ecological citizenship, in spite of their ubiquity on farmland (and remembering that their natural history in urban environments is even less well researched), the rat's foothold in rural Britain is surprisingly tentative, despite their weighty impact on agriculture and health. The numbers of rats at the stream (equivalent to unprovisioned woodland) were very low. At the hoppers, rat densities (supported by resources provided for pheasants) were higher than at the stream, but reproduction appeared to be low during periods when there was little supplemental food, and even at the tip, which was consistently provisioned, the rat's apparent ability to breed year-round was limited by the weather. These results from our three study sites emphasize the rat's dependence on human subsidies, and their brinkmanship in survival in the face of the British climate (although, importantly, this may be changing to suit them better).

[34] As evidenced by scrotal testes and the presence of visible tubules in the cauda epididymes.

11.6 Conclusions

The behaviour and ecology of rats is complex and variable, and interactions between their ecology, social systems, disease, and behaviour are intricate. Our argument is that better understanding of that complexity will facilitate better pest management, yet nearly three decades after we first began this work, little has changed in terms of ecological field studies of wild rats in the UK, which remain scant (and starkly in contrast to the abundance of laboratory studies). Of 99,000 publications on rats between 2008 and 2010, only 2056 of those were on wild rats and less than 15% of those were studies conducted in the wild (Macdonald et al. 2013).

Are brown rats truly naturalized in the UK as wildlife, or only as co-habitants (carpet-baggers)? With regard to whether we accept the rat as an ecological citizen, the answer doesn't really matter. Macdonald and Burnham (2010) argue that this decision instead hinges on the nature of the damage the species causes. They suggest that as long as that damage is judged intolerably undesirable then the intruder is considered a pest, and efforts to remove it, or otherwise mitigate its impact, are justified (presuming that there is an effective, and thus justifiable, method for doing that). But the answer does matter for the practicalities of controlling rats. As fellow travellers, brown rats seem equally adept at accepting our handouts, whether staying in the luxury of the metaphorical Hilton or squalor of a grubby hostel, and their apparent dependence on our handouts calls for ecological thinking. For any control method (ecological or not) there are three considerations: (1) effectiveness, (2) impact on other wildlife, and (3) humaneness. Twenty years ago we proposed an ecological ethic with regard to rat control (Macdonald and Fenn 1994), in which we argued that it would be preferable to explore the foraging tactics of rats in the hope of identifying more efficient and effective baiting strategies, than using more, and more toxic, poison. This idea remains relevant today; anticoagulant resistance remains a real problem and debates regarding the appropriate use of more potent, resistance-breaking products continue (see, for example, Buckle 2013). An even better (perhaps ideal) approach would be to reduce the rat's food supply, with a view to reducing rat populations without reverting to (or at least with reduced use of) poisons (which carry considerable risks to rat-eating wildlife on farmland through secondary poisoning, Cox and Smith 1992) or traps. The question is: if we eliminate these super-abundant accessible anthropogenic food sources, can we limit rat populations? Our field studies suggest yes. The difficulties of such an approach within the farmstead have been discussed widely before (e.g. Lambert et al. 2008), but the issues in the wider farmscape are different. The disposal of animal carcasses is now subject to EU regulations (Defra 2012), reducing one potentially rich source of food for rats, but the widespread provision of grain for pheasants remains a problem. The irony is that this similarly subsidized, deliberate invader provides a foothold for rats, and an evaluation of the impact of the pheasants, and particularly the food provided for them, in the English countryside on wildlife (including rats) is a major research topic that is high on our list of priorities (see Chapter 9, this volume).

Society may decide that killing rats is easier (or more acceptable) than tackling the pheasant-rearing industry (certainly, killing rats has never been controversial), but the pest status of the rat does not negate concern for unnecessary suffering. Many thousands of rats are trapped each year in the UK in spring traps that are currently exempt from formal approval and may vary considerably in their humaneness (Baker et al. 2012; Chapter 13, this volume). Poisoning with anticoagulants (when it is effective) probably involves an excruciating death (Sainsbury et al. 1995; Mason and Littin 2003). Rats may not be ecological citizens but it is hard today to think of a valid reason why their welfare should not be given similar attention to that afforded to other species that are.

Acknowledgements

Lauren Harrington worked with exceptional diligence and insight helping to bring together the strands of this chapter, and Sandra Baker, Ruth Feber, and Eva Raebel offered helpful comments. We warmly acknowledge the work of former members of our rat team, including Clair Brunton, Mike Fenn, and Stephan Natynczuk.

References

Albone, E.S. (1984). *Mammalian Semiochemistry*. Wiley Interscience, Chichester.

Andrews, R.V., Belknap, R.W., Southard, S., Lorincs, M., and Hess, S. (1972). Physiological, demographic and pathological changes in wild Norway rat populations over an annual cycle. *Comparative Biochemistry and Physiology*, **41**, 149–165.

Anon (2012). *Science versus the super rats*. University of Huddersfield Research News, 18 October 2012. <http://www.hud.ac.uk/news/allstories/scienceversusthesuperrats.php >, accessed March 2015.

Atkinson, R.P.D., Macdonald, D.W., and Johnson, P.J. (1994). The status of the European mole *Talpa europea* L. as an agricultural pest and its management. *Mammal Review*, **24**, 73–90.

Baker, S.E., Ellwood, S.A., Tagarielli, V., and Macdonald, D.W. (2012). Mechanical performance of rat, mouse and mole spring traps, and possible implications for welfare performance. *PLOS ONE*, **7**, e39334.

Barnett, S.A. (1950). Damage to wheat by enclosed populations of Rattus norvegicus. *Journal of Hygiene*, **49**, 22–25.

Barnett, S.A. (1958). Experiments on 'neophobia' in wild and laboratory rats. *British Journal of Psychology*, **49**, 195–201.

Barnett, S.A. (2002). *The Story of Rats: Their Impact on Us, and Our Impact on Them*. Allen and Unwin, Crows Nest, Australia.

Barnett, S.A. and Cowan, P.E. (1976). Activity, exploration, curiosity and fear: an ethological study. *Interdisciplinary Science Review*, **1**, 43–62.

Battersby, J. (ed.) and Tracking Mammals Partnership (2005). *UK Mammals: Species Status and Population Trends. First Report by the Tracking Mammals Partnership*. JNCC/Tracking Mammals Partnership, Peterborough.

Berdoy, M. (1994). Making decisions in the wild: constraints, conflicts and communication in foraging rats. In B.G. Jr. Galef and P. Valsechhi, eds., *Behavioural Aspects of Feeding, Ettore Majorana International Life Sciences Series, Vol 12*, pp. 289–313. Harwood Academic Publishers, Chur, Switzerland.

Berdoy, M. (2002). *The laboratory rat: A natural history*. (27 minute film). <http://www.ratlife.org>.

Berdoy, M. and Drickammer, L. (2007). Comparative social organisation and life history of Rattus and Mus. In J. Wolff and P. Sherman, eds., *Rodent Societies: An Ecological and Evolutionary Perspective*, pp. 380–392.University of Chicago Press.

Berdoy, M. and Evans, S. (1990). An automatic recording system for identifying individual small animals. *Animal Behaviour*, **39**, 998–1000.

Berdoy, M.L. and Macdonald, D.W. (1991). Factors affecting feeding in wild rats. *Acta Oecologica-International Journal of Ecology*, **12**, 261–279.

Berdoy, M.L., Smith, P., and Macdonald, D.W. (1995a). Stability of social-status in wild rats: age and the role of settled dominance. *Behaviour*, **132**, 193–212.

Berdoy, M.L., Webster, J.P., and Macdonald, D.W. (1995b).The manipulation of rat behaviour by Toxoplasma gondii. *Mammalia*, **59**, 605–613.

Berdoy, M.L., Webster, J.P., and Macdonald, D.W. (1995c). Parasite-altered behaviour: is the effect of Toxoplasma gondii on Rattus norvegicus specific? *Parasitology*, **111**, 403–409.

Berdoy, M.L., Webster, J.P., and Macdonald, D.W. (2000). Fatal attraction in *Toxoplasma*-infected rats: a case of parasite manipulation of its mammalian host. *Proceedings of the Royal Society B*, **267**, 1591–1594.

Bijleveld, A.I., Egas, M., van Gils, J.A., and Piersma, T. (2010). Beyond the information centre hypothesis: communal roosting for information on food, predators, travel companions and mates? *Oikos*, **119**, 277–285.

Bishop, J.A. and Hartley, D.J. (1976). The size and age structure of rural populations of Rattus norvegicus containing individuals resistant to the anticoagulate poison Warfarin. *Journal of Animal Ecology*, **45**, 623–646.

Bishop, J.A., Hartley, D.J., and Partridge, G.G. (1977). The population dynamics of genetically determined resistance to warfarin in *Rattus norvegicus* from mid-wales. *Heredity*, **39**, 389–398.

Brian, K. (2008). *Sheep Diseases Directory. A producer's guide to keeping sheep healthy and profitable*. EBLEX Ltd., Huntingdon.

Brigham, A.J. and Sibly, R.M. (1999). A review of the phenomenon of neophobia. In P.D. Cowan and C.J. Feare, eds., *Advances in Vertebrate Pest Management*, pp. 67–84. Filander-Verlag, Fürth.

Brodie, J. (1981). Norway rats (*Rattus norvegicus*) on cereal stubble. *Journal of Zoology*, **195**, 542–546.

Brown, R.E. and Macdonald, D.W. (1985). *Social Odours in Mammals*. Oxford University Press, Oxford.

Brunton, C.F.A. and Macdonald, D.W. (1996). Measuring the neophobia of individuals in different populations of wild brown rats. *Journal of Wildlife Research*, **1**, 7–14.

Brunton, C.F.A., Macdonald, D.W., and Buckle, A.P. (1993). Behavioral resistance towards poison baits in Brown rats, *Rattus norvegicus*. *Applied Animal Behaviour Science*, **38**, 159–174.

Buckle, A. (2007). Rat Control 1: Problems caused on farms. *Farmers Weekly*. <http://www.fwi.co.uk/>, accessed October 2012.

Buckle, A. (2009). Rat control 3: climate change. *Farmers Weekly*. <http://www.fwi.co.uk/>, accessed October 2012.

Buckle, A. (2013). Anticoagulant resistance in the United Kingdom and a new guideline for the management of resistant infestations of Norway rats (*Rattus norvegicus* Berk.). *Pest Management Science*, **69**, 334–341.

Burn, C. (2008). What is it like to be a rat? Rat sensory perception and its implications for experimental design and rat welfare. *Applied Animal Behaviour Science*, **112**, 1–32.

Calhoun, J.B. (1963). *The Ecology and Sociology of the Norway Rat*. US Public Health Service Publication 1008.

Chitty, D. and Southern, H.N. (1954). *Control of Rats and Mice*. Clarendon Press, Oxford.

Chomel, B.B. (1992). Zoonoses of house pets other than dogs, cats and birds. *Paediatric Infectious Diseases Journal*, **11**, 479–487.

Connor, S. (2012). Professor Joanne Webster: The scientist who uncovered 'fatal feline attraction'. *The Independent*. Tuesday 4th September 2012. <http://www.independent.co.uk/>, accessed October 2012.

Coombes, S., Revusky, S., and Lett, B.T. (1980). Long-delay taste aversion learning in an unpoisoned rat: exposure to a poisoned rat as the unconditioned stimulus. *Learning and Motivation*, **11**, 256–266.

Cowan, P.E. (1977). Neophobia and neophilia: new-object and new-place reactions in three Rattus species. *Journal of Comparative Physiology and Psychology*, **91**, 63–71.

Cowan, P.E., Quy, R.J., and Lambert, M.S. (2003). Ecological perspectives on the management of commensal rodents. In G.R. Singleton, L.A. Hinds, C.J. Krebs, and D. Sprat, eds., *Rats, mice and people: rodent biology and management*, ACIAR Monograph no. 96, pp. 433–439. Australian Centre for International Agricultural Research, Canberra.

Cox, P. and Smith, R.H. (1992). Rodenticide ecotoxicology: pre-lethal effects of anticoagulants on rat behaviour. In J.E. Borrecco and R.E. Marsh, eds., *Proceedings of the 15th Vertebrate Pest Conference 1992*. Paper 86.

Curtis, P.E., Ollerhead, G.E., and Ellis, C.E. (1980). Pasteurella multocida infection of poultry farm rats. *Veterinary Record*, **107**, 326–327.

Davis, D. and Jackson, W.B. (1981). Rat Control. In H.T. Coaker, ed. *Advances in Applied Biology*. Academic Press, London.

Defra (2012). *Animal by-products: collection, storage and disposal*. <https://www.gov.uk/>, accessed November 2012.

Elkins, R.L. (1973). Attenuation of drug-induced bait-shyness to a palatable solution as an increasing function of its availability prior to conditioning. *Behavioral Biology*, **9**, 221–226.

Elton, C., Ford, E.B., Baker, J.R., and Gardner, A.D. (1931). The health and parasites of a wild mouse population. *Proceedings of the Zoological Society of London*, **108**, 657–721.

Faine, S. (1998). Leptospirosis. In A.S. Evans and P. Brachman, eds., *Bacterial Infections of People: Epidemiology and Control*. Springer, London.

Farhang-Azad, A. and Southwick, C.H. (1979). Population ecology of Norway rats in the Baltimore Zoo and Druid

Hill Park, Baltimore, Maryland. *Annals of Zoology*, **15**, 1–42.

Fenn, M.G.P. and Macdonald, D.W. (1987). The contribution of field studies to stored product rodent control. *BCPC Monographs*, **37**, 107–113.

Fenn, M.G.P. and Macdonald, D.W. (1995). Use of middens by red foxes: risk reverses rhythms of rats. *Journal of Mammalogy*, **76**, 130–136.

Fenn, M.G.P., Tew, T.E., and Macdonald, D.W. (1987). Rat movements and control on an Oxfordshire farm. *Journal of Zoology*, **213**, 745–749.

Flegr, J. (2007). Effects of Toxoplasma on human behaviour. *Schizophrenia Bulletin*, **33**, 757–760.

Flegr, J., Klose, J., Novotna, M., Berenreitterová, M., and Havlíček, J. (2009). Increased incidence of traffic accidents in Toxoplasma-infected military drivers and protective effect RhD molecule revealed by a large-scale prospective cohort study. *BMC Infectious Diseases*, **9**, 72.

Foote, A.L. and Crystal, J.D. (2007). Metacognition in the rat. *Current Biology*, **17**, 551–555.

Gajdusek, D.C. (1962). Virus haemorrhagic fevers. Special reference to haemorrhagic fever with renal syndrome (epidemic haemorrhagic fever). *Journal of Pediatrics*, **60**, 841–857.

Galef, B.G. (1991). Information centres of Norway rats: sites for information exchange and information parasitism. *Animal Behaviour*, **41**, 295–301.

Galef, B.G. (2012a). Social learning in rats. Historical context and experimental findings. In T.R. Zentall and E.A. Wasserman, eds., *The Oxford Handbook of Comparative Cognition*, pp. 803–818. OUP, Oxford.

Galef, B.G. (2012b). A case study in behavioural analysis, synthesis and attention to detail: social learning of food preferences. *Behavioural Brain Research*, **231**, 266–271.

Geddes, A. (1983). Q-fever. *British Medical Journal*, **287**, 927–928.

Gelling, M., Macdonald, D.W., Telfer, S., et al. (2012). Parasites and pathogens in wild populations of water voles (*Arvicola amphibius*) in the UK. *European Journal of Wildlife Research*, **58**, 615–619.

Gemberling, G.A. (1984). Ingestion of a novel flavour before exposure to pups injected with lithium chloride produces a taste aversion in mother rats. *Journal of Comparative Psychology*, **98**, 285–301.

Gibbs, C.J.,Jr., Takaenak, G.A., Franko, M., et al. (1982). Seroepidemiology of Hantaan virus. *Lancet*, **18**, 1406–1407.

Greaves, J.H. and Ayres, P.B. (1969). Linkages between genes for coat colour and resistance to warfarin in *Rattus norvegicus*. *Nature*, **224**, 284–285.

Greaves, J.H. and Cullen-Ayres, P.B. (1988). Genetics of difenacoum resistance in the rat. In J.W. Suttie, ed. *Current Advances in Vitamin K Research*, pp. 389–397. Elsevier, Amsterdam.

Greaves, J.H. and Rennison, B.D. (1973). Population aspects of warfarin resistance in the brown rat. *Mammal Review*, **3**, 27–29.

Grzimek, B. (ed.) (1968). *Enzyklopädie des Tierreiches*, Vol. 11. Kindler Verlag, Zurich.

Hardy, A.R. and Taylor, K.D. (1979). Radio tracking of Rattus norvegicus on farms. In C.J. Amlaner and D.W. Macdonald, eds., *A Handbook of Biotelemetry and Radio Tracking*, pp. 657–665. Pergamon Press, Oxford.

Healing, T.D. (1981). Infections with blood parasites in the small British rodents *Apodemus sylvaticus*, *Clethrionomys glareolus* and *Microtus agrestis*. *Parasitology*, **83**, 179–189.

Hermodsen, M.A., Suttie, J.W., and Link, K.P. (1969). Warfarin metabolism and vitamin K requirement in the warfarin-resistant rat. *American Journal of Physiology*, **217**, 1316–1319.

Hishimura, Y. (2000). Re-examination of the poisoned-partner effect with the two-bottle testing method. *Behavioural Processes*, **50**, 95–99.

Homolka, M. (1983). On the problem of the exanthropic occurrence of *Rattus norvegicus*. *Folia Zoologica*, **32**, 202–211.

HPA (Health Protection Agency) (2012). *Infectious Diseases A-Z*. <http://www.hpa.org.uk/Topics/InfectiousDiseases/InfectionsAZ/>, accessed October 2012.

Jackson, M.H., Hutchison, W.M., and Siim, J.C. (1986). Toxoplasmosis in a wild rodent population in central Scotland and a possible explanation of the mode of transmission. *Journal of Zoology*, **209**, 549–557.

Kannan, G., Moldovan, K., Xiao, J.-C., Yolken, R.H., Jones-Brando, L., and Pletnikov, M.V. (2010). Parasite strain-dependent effects of *Toxoplasma gondii* on mouse behaviour. *Folia Parasitologica*, **57**, 151–155.

Kaushik, M., Lamberton, P.H.L., and Webster, J.P. (2012). The role of parasites and pathogens in influencing generalised anxiety and predation-related fear in the mammalian central nervous system. Invited (JPW) for 'The Neuroendocrine-Immune Axis in Health and Disease' special volume: *Hormones and Behaviour*, **62**, 191–201.

Kendall, P.B. (1984). Seasonal changes of sex ratio in Norway rat (Rattus norvegicus) infestation of agricultural premises. *Journal of Zoology*, **203**, 208–211.

Kirkham, W.B. (1910). Ovulation in mammals, with special reference to the mouse and the rat. *Biological Bulletins*, **18**, 245–251.

Klemann, N. and Pelz, H.J. (2006). The feeding pattern of the Norway rat (*Rattus norvegicus*) in two differently structured habitats on a farm. *Applied Animal Behaviour Science*, **97**, 293–302.

Krebs, C.J. (1999). Current paradigms of rodent population dynamics—what are we missing. In G.R. Singleton, L.A. Hinds, H. Leirs, and Z. Zhang, eds., *Ecologically-based management of rodent pests. ACIAR Monograph No. 59*, pp. 33–48. Australian Centre for International Agricultural Research, Canberra.

Lambert, M.S., Quy, R.J., Smith, R.H., and Cowan, D.P. (2008). The effect of habitat management on home-range size and survival of rural Norway rat populations. *Journal of Applied Ecology*, **45**, 1753–1761.

Lamberton, P.H.L., Donnelly, C.A., and Webster, J.P. (2008). Specificity of the *Toxoplasma gondii*-altered behaviour to definitive versus non-definitive host predation risk. *Parasitology* **135**, 1143–1150.

Lattanzio, R.M. and Chapman, J.A. (1980). Reproductive and physiological cycles in an island population of Norway rats. *Bulletin of the Chicago Academy of Science*, **12**, 1–68.

Lavin, M.J., Freise, B.,and Coombes, S. (1980). Transferred aversions in adult rats. *Behavioural and Neural Biology*, **28**, 15–33.

Leslie, P.H., Venables, U.M., and Venables, L.S. (1952). The fertility and populations structure of the brown rat (Rattus norvegicus) in corn ricks and some other habitats. *Proceedings of the Royal Society of London*, **122**, 187–238.

Ling, V.J., Lester, D., Mortensen, P.B., Langenberg, P.W., and Postolache, T.T. (2011). Toxoplasma gondii seropositivity and suicide rates in women. *Journal of Nervous and Mental Disease*, **199**, 440–444.

Lock, J. (2006). Eradication of brown rats *Rattus norvegicus* and black rats *Rattus rattus* to restore breeding seabird populations on Lundy Island, Devon, England. *Conservation Evidence*, **3**, 111–113.

Macdonald, D.W. (1984). A questionnaire survey of farmers' opinions and actions towards wildlife on farmlands. In D. Jenkins, ed. *Agriculture and the Environment, Institute of Terrestrial Ecology Symposium No. 13*, pp. 171–177. Institute of Terrestrial Ecology, Cambridge, UK.

Macdonald, D. and Burnham, D. (2010). *The State of Britain's Mammals—a Focus on Invasive Species*. PTES, London.

Macdonald, D.W. and Feber, R.E., eds. (2015). *Wildlife Conservation on Farmland. Managing for Nature on Lowland Farms*. Oxford University Press, Oxford.

Macdonald, D.W. and Fenn, M.G.P. (1994). The natural history of rodents: preadaptations to pestilence. In A.P. Buckle and R.H. Smith, eds., *Rodent Pests and Their Control*, pp. 1–21. CAB International, Wallingford.

Macdonald, D.W. and Fenn, M.G.P. (1995). Rat ranges in arable areas. *Journal of Zoology*, **236**, 349–353.

Macdonald, D.W., Fenn, M.P.G., and Gelling, M. (2013). The natural history of rodents: preadaptations to pestilence. In A.P. Buckle and R.H. Smith, eds., *Rodent Pests and their Control. 2nd Edition*. CABI Publishing, Wallingford.

Macdonald, D.W., King, C.M., and Strachan, R. (2007). Introduced species and the line between biodiversity conservation and naturalistic eugenics. In D. Macdonald and K. Service, eds., *Key Topics in Conservation Biology*, pp. 186–205. Blackwell Publishing, Oxford.

Macdonald, D.W., Mathews, F., and Berdoy, M.L. (1999). The behaviour and ecology of Rattus norvegicus: from opportunism to kamikaze tendencies. In G.R. Singleton, L.A. Hinds, H. Leirsand, and Z. Zhang, eds., *Ecologically-based management of rodent pests. ACIAR Monograph No. 59*, pp. 49–80. Australian Centre for International Agricultural Research, Canberra.

Macdonald, D.W. and Tattersall, F. (2001). *Britain's Mammals: the Challenge for Conservation*. PTES, London.

Marrie, T.J. (1990). Q fever—a review. *Canadian Veterinary Journal*, **3**, 555–564.

Marrie, T.J., Van Buuren, J., and Fraser, J. (1985). Seroepidemiology of Q-fever among domestic animals in Nova Scotia. *American Journal of Public Health*, **75**, 763–766.

Mason, G.M. and Littin, K.E. (2003). The humaneness of rodent pest control. *Animal Welfare*, **12**, 1–37.

Meehan, A.P. (1984). *Rats and Mice. Their Biology and Control*. East Grinstead: Rentokill.

Middleton, A.D. (1954). Rural rat control. In D. Chitty and H.N. Southern, eds., *Control of Rats and Mice*. Clarendon Press, Oxford.

Mills, J.N. (1999). The role of rodents in emerging human disease: examples from the hantaviruses and arenaviruses. In G.R. Singleton, L.A. Hinds, H. Leirs, and Z. Zhang, eds., *Ecologically-based management of rodent pests, ACIAR Monograph No. 59*, pp. 134–160. Australian Centre for International Agricultural Research, Canberra.

Mitchell, D. (1976). Experiments on neophobia in wild and laboratory rats: a re-evaluation. *Journal of Comparative Physiology and Psychology*, **90**, 190–197.

Nachman, M. and Hartley, P.L. (1975). Role of illness in producing taste aversion in rats: a comparison of several rodenticides. *Journal of Comparative Physiology and Psychology*, **89**, 1010–1018.

Natynczuk, S.E. and Macdonald, D.W. (1994a). Analysis of rat haunch odor using dynamic solvent effect and principal component analysis. *Journal of Chemical Ecology*, **20**, 1859–1866.

Natynczuk, S.E. and Macdonald, D.W. (1994b). Scent, sex, and the self-calibrating rat. *Journal of Chemical Ecology*, **20**, 1843–1857.

Natynczuk, S.E., Macdonald, D.W., and Tattersall, F.H. (1995). Morphology and chemistry of brown rat, Rattus norvegicus, preputial and clitoral glands. *Journal of Chemical Ecology*, **21**, 247–260.

Overman, S.R. (1976). Dietary self-selection by animals. *Psychological Bulletin*, **83**, 218–235.

Perryman, L.E. (1990). Cryptosporidium in rodents. In J.P. Dubey, C.A. Spper, and R. Fayer, eds., *Cryptosporidiosis of Man and Animals*, pp. 125–132. CRC Press.

Prandovszky, E., Gaskell, E., Dubey, J.P., Webster, J.P., and McConkey, G.A. (2011). The neurotropic parasite Toxoplasma gondii increases dopamine metabolism. *PLOS ONE*, **6**, e23866.

PTES (2008). *Living with mammals—survey update 2008*. <http://www.ptes.org/>, accessed October 2012.

Quy, R.J., Cowan, D.P., Haynes, P.J., et al. (1999). The Norway rat as a reservoir host of *Cryptosporidium parvum*. *Journal of Wildlife Diseases*, **35**, 660–670.

Quy, R.J., Cowan, D.P., Prescott, C.V., et al. (1995). Control of a population of Norway rats resistant to anticoagulant rodenticides. *Pesticide Science*, **45**, 247–256.

Quy, R.J. and Macdonald, D.W. (2008). Common Rat. In S. Harris and D.W. Yalden, eds., *Mammals of the British Isles: Handbook*. The Mammal Society.

Quy, R.J., Shepherd, D.S., and Inglis, I.R. (1992). Bait avoidance and effectiveness of anticoagulant rodenticides against warfarin- and difenacoum-resistant populations of Norway rats (*Rattus norvegicus*). *Crop Protection*, **11**, 14–20.

Rennison, B.D. (1974). Field trials of calciferol against warfarin resistant infestations of the Norway rat (*Rattus norvegicus*, Berk.). *Journal of Hygiene*, **73**, 361–367.

Revusky, S., Coombes, S., and Pohl, R.W. (1982). US preexposure: effects on favour aversions produced by pairing a poisoned partner with ingestion. *Animal Learning and Behavior*, **10**, 83–90.

Rödel, H.G. and Meyer, S. (2011). Early development influences ontogeny of personality types in young laboratory rats. *Developmental Psychobiology. Special Issue: The Emergence of Personality in Animals*, **53**, 601–613.

Rolland, C., Macdonald, D.W., De Fraipont, M., and Berdoy, M. (2003). Free female choice in house mice: leaving best for last. *Behaviour*, **140**, 1371–1388.

Rozin, P. (1976). The selection of food by rats, humans and other animals. *Advances in the Study of Behaviour*, **6**, 21–76.

Sainsbury, A.W., Bennett, P.M., and Kirkwood, J.K. (1995). The welfare of free-living wild animals in Europe: harm caused by human activities. *Animal Welfare*, **4**, 183–206.

Schmidt, G.D. and Roberts, L.S. (1989). *Foundations of Parasitology*. 4th Edition. Times Mirror/Mosby College Publishing, St. Louis, Missouri.

Sheedy, J.A., Froeb, H.F., Batson, H.A., et al. (1954). The clinical course of epidemic haemorrhagic fever. *American Journal of Medicine*, **16**, 619–628.

Simpson, V.R. (2002). Wild animals as reservoirs of infectious diseases in the UK. *Veterinary Journal*, **163**, 128–146.

Singleton, G.R., Leirs, H., Hinds, L.A., and Zhang, Z. (1999). Ecologically-based management of rodent pests—re-evaluating our approach to an old problem. In G.R. Singleton, L.A. Hinds, H Leirs, and Z. Zhang, eds., *Ecologically-based management of rodent pests, ACIAR Monograph No. 59*, pp. 17–29. Australian Centre for International Agricultural Research, Canberra.

Smith, P., Berdoy, M., Smith, H.R., and Macdonald, D.W. (1993). A new aspect of warfarin resistance in wild rats: benefits in the absence of poison. *Functional Ecology*, **7**, 190–194.

Suttie, J.W. (1985). Vitamin K. In A.T. Diplock, ed. *Fat soluble vitamins—their biochemistry and applications*, pp. 225–311. Heinemann, London.

Taylor, K.D. (1978). Range of movement and activity of common rats (*Rattus norvegicus*) on agricultural land. *Journal of Applied Ecology*, **15**, 663–677.

Taylor, K.D. and Quy, R.J. (1978). Long distance movements of a common rat (*Rattus norvegicus*) revealed by radio-tracking. *Mammalia*, **42**, 63–71.

TMP. (2009). *Tracking Mammals Partnership Update 2009*. <http://jncc.defra.gov.uk/page-5497>, accessed October 2012.

Torrey, E.F. and Yolken, R.H. (2007). Special Theme: Schizophrenia and Toxoplasmosis. *Schizophrenia Bulletin*, **33**, 727–728.

Vyas, A., Seon-Kyeong, K., Giacomini, N., Boothroyd, J.C., and Sapolsky, R.M. (2007). Behavioural changes induced by Toxoplasma infection of rodents are highly specific to aversion of cat odours. *Proceedings of the National Academy of Sciences of the United States of America*, **104**, 6442–6447.

Waitkins, S.A. (1991). Rats as a source of leptospirosis—Weil's disease. *Sorex Technical Publication No. 4*. Sorex Ltd., Cheshire.

Ward, P. and Zahavi, A. (1973). The importance of certain assemblages of birds as 'information centres' for food-finding. *Ibis*, **115**, 517–534.

Webster, J.P. (1994a). Prevalence and transmission of Toxoplasma gondii in wild brown rats, *Rattus norvegicus*. *Parasitology*, **108**, 407–411.

Webster, J.P. (1994b). The effect of Toxoplasma gondii and other parasites on activity levels in wild and hybrid *Rattus norvegicus*. *Parasitology*, **109**, 583–589.

Webster, J.P. (2001). Rats, cats, people and parasites: the impact of latent toxoplasmosis on behaviour. *Microbes and Infection*, **3**, 1037–1045.

Webster, J.P., Brunton, C.F.A., and Macdonald, D.W. (1994). Effect of Toxoplasma gondii upon neophobic behaviour in wild brown rats, *Rattus norvegicus*. *Parasitology*, **109**, 37–43.

Webster, J.P., Ellis, W.A., and Macdonald, D.W. (1995a). Prevalence of Leptospira and other zoonoses in wild brown rats on UK farms. *Mammalia*, **59**, 615–622.

Webster, J.P., Ellis, W.A., and Macdonald, D.W. (1995b). Prevalence of Leptospira spp in wild brown rats (Rattus norvegicus) on UK farms. *Epidemiology and Infection*, **114**, 195–201.

Webster, J.P., Kaushik, M., Bristow, G.C., and McConkey, G.A. (2013). *Toxoplasma gondii* infection, from predation to schizophrenia: can animal behaviour help us understand human behaviour? *Journal of Experimental Biology*, **216**, 99–112.

Webster, J.P., Lamberton, P.H.L., Donnelly, C.A., and Torrey, F. (2006). Parasites as causative agents of human affective disorders?: The impact of anti-psychotic and anti-protozoan medication on *T. gondii*'s ability to alter host behaviour. *Proceedings of the Royal Society* (*London*), *Series B*, **273**, 1023–1030.

Webster, J.P., Lloyd, G., and Macdonald, D.W. (1995c). Q fever (Coxiella burnetii) reservoir in wild brown rat (*Rattus norvegicus*) populations in the UK. *Parasitology*, **110**, 31–35.

Webster, J.P. and Macdonald, D.W. (1995a). Cryptosporidiosis reservoir in wild brown rats (*Rattus norvegicus*): first report in the UK. *Epidemiology and Infection*, **115**, 207–209.

Webster, J.P. and Macdonald, D.W. (1995b). Parasites of wild brown rats, *Rattus norvegicus*, on UK farms. *Parasitology*, **111**, 247–255.

Webster, J.P. and McConkey, G.A. (2010). *Toxoplasma gondii*-altered host behaviour: clues as to mechanism of action. *Folia Parasitologica*, **57**, 95–104.

White, P.R. and Chambers, J. (1989). Saw-toothed grain beetle *Oryzaephilus surinamensis* (L.) (Coleoptera: Silvanidae). Antennal and behavioural responses to individual components and blends of aggregation pheromone. *Journal of Chemical Ecology*, **15**, 1015–1031.

White, P.R., Chambers, J., Walter, C.M., Wilkins, J.P.G., and Millar, J.G. (1989). Saw-toothed grain beetle *Oryzaephilus surinamensis* (L.) (Coleoptera: Silvanidae): collection, identification and bioassay of attractive volatiles from beetles and oats. *Journal of Chemical Ecology*, **15**, 999–1013.

Wolff, J.O. and Macdonald, D.W. (2004). Promiscuous females protect their offspring. *Trends in Ecology and Evolution*, **19**, 127–134.

Wyatt, T.D. (2013). *Pheromones and Animal Behavior: Chemical Signals and Signature Mixes*, *2edn*. Cambridge: Cambridge University Press.

CHAPTER 12

Moles on farmland: making mountains out of molehills?

Sandra E. Baker and David W. Macdonald

Forget about likes and dislikes. They are of no consequence. Just do what must be done.

George Bernard Shaw

12.1 Introduction

European moles *Talpa europaea* (Fig. 12.1) occur throughout Britain (but not Ireland) and in most parts of mainland Europe (Gorman and Stone 1990). They are rarely seen alive, but evidence of their presence is often conspicuously widespread in the countryside. The enigmatic mole has a mixed reputation, loved by many as 'Moley' in Kenneth Grahame's *The Wind in the Willows*, and yet also regarded as 'one of the most universally troublesome of agricultural pests' (Lovegrove 2007). For centuries moles have been persecuted in Britain as pests and for other reasons (see Box 12.1), and even today they are subject to unequal treatment in terms of their management (Box 12.3).

12.1.1 Natural history

Moles are primarily solitary animals with largely exclusive home ranges (Macdonald et al. 1997). They are fossorial, living almost permanently underground in a network of chambers and feeding tunnels which they excavate as both a home and hunting ground. Moles are insectivores, feeding mainly on earthworms and other soil invertebrates, such as insect larvae and molluscs, which they locate in their tunnel systems (Gorman and Stone 1990). Generally, mole activity occurs in three cycles of eight hours a day, about half of each cycle being spent on feeding and other activity, and the other half resting (Stone 1986; Macdonald et al. 1997; see Box 12.2). Digging for a living is hard work (Gorman and Stone 1990) and moles have a high metabolic rate, eating an estimated 50–60 g (approximately half their bodyweight) of food a day (Mellanby 1967; Gorman and Stone 1990). Moles vary widely in bodyweight but males are generally larger than females, estimates of average adult bodyweight for the two sexes being 110 g and 87 g respectively (Atkinson 2013). Moles have relatively poor eyesight and, while they have acute hearing, can detect only low frequency sounds (0.1–15 kHz). They use touch and odour to navigate and defend their territories, and to find food. The mole's sense of smell functions only at short range (6–7 cm), but their sense of touch is particularly sensitive, aided by touch-sensitive Eimer's organs in the nose, and sensitive hairs on their chin, muzzle, sides of the face, and on their tail (Gorman and Stone 1990; Atkinson 2013). Moles have extremely powerful shoulders, and large front paws with long claws, to assist with digging (Atkinson 2013).

Moles breed in spring, although timings vary with latitude throughout their range (occurring earlier on the European continent and later in Scotland). A male mole in England leaves its solitary territory in February/March, in search of females with which to mate, and leaving a trail of new molehills in its wake as it attempts to encounter as many receptive females as possible. Mating generally occurs in March and females are pregnant for just over four weeks, most giving birth to three to four young in April. Young moles are suckled for a few weeks before being driven out of the maternal territory, from early June onwards, at which time they disperse above ground (and sometimes even over open water), to find and establish a territory of their own (Atkinson 2013). This is one of the most dangerous times of a mole's life as they are vulnerable to predation by birds and mammals, and to being run over on roads (Gorman and Stone 1990).

Wildlife Conservation on Farmland. Conflict in the Countryside. Edited by David W. Macdonald and Ruth E. Feber.
© Oxford University Press 2015. Published 2015 by Oxford University Press.

Figure 12.1 The European mole *Talpa europaea*. Photograph © R. Shaw.

Box 12.1 History, folklore, and trapping

From at least the 1300s, moles were known as *Moldewarps* (or 'earth throwers'), a term which has come to mean an underhand person, a sneak, a plotter, a spy, and a stupid or worthless person[1]. Historically, moles were associated with superstition; for example, the appearance of molehills around a house foretold of a death, wearing a dead mole around your neck cured toothache, and wearing a pair of the mole's large front digging feet (cut off while the mole was still alive), as a necklace, brought good luck. Moles have also been prized for their fur, which was used to make moleskin hats and waistcoats etc, particularly in the nineteenth century, and although they are no longer made of mole skins, we still refer to 'moleskin' garments today.

Evidence for mole trapping in Britain goes back to Roman times (Nicholls 2010) and traps have been made of earthenware, clay, wood and, since the 1800s, steel. In the sixteenth century, moles were among the many vermin listed in Queen Elizabeth I's 1566 *Acte for the Preservation of Grayne* (Lovegrove 2007). From the early 1600s onwards, there are records in church wardens' accounts of moles being killed as vermin in some parts of Britain, surviving records originating mainly in the East Midlands. Moles were trapped either by dedicated mole catchers, employed by parishes, or by individuals in return for bounty payments. Before that, a single parish record, from Bishop's Stortford (Hertfordshire) survives, relating to a vermin payment for 53 moles over the period 1569–1571. Private estates also employed mole catchers. Examples of payments include the following: 1d per mole was paid to individuals in North Hampshire between 1673–1703; a fixed annual rate of £3 and 18s was paid to a succession of mole catchers in Lincolnshire between 1742–1810; and in 1882, following an election in Shropshire, a Mr T. Jones was given a 21-year appointment as a mole catcher for ¾d per acre (Lovegrove 2007).

There have been plenty of reasons to catch moles over the years and, according to Lovegrove (2007), moles are second only to the sparrow, in terms of numbers, among the many mammals and birds on the old lists of vermin killed over the centuries (Lovegrove 2007). Many millions of moles must have been killed and mole-trapping has become something of a tradition in Britain (Mellanby 1971; Nicholls 2010). Even today there are molecatching organizations in Britain, including the Guild of British Molecatchers, the British Traditional Molecatchers Register, and the Association of Professional Molecatchers.

[1] Oxford English Dictionary, <http://www.oed.com/>, accessed February 2014.

Moles dig tunnels and produce new spoil heaps, well-known as molehills, only when necessary, for example, when setting up a new territory or when tunnels have collapsed, or been damaged. The soil tends to be damp and easier to work in autumn and winter, when tunnels are also most likely to be damaged by frost, flooding, and cultivation (Gorman and Stone 1990). In late winter, moles may move from wet areas to higher ground, extending their burrow systems as they go (Lund 1976). Moles dig less in the summer and move deeper underground in search of earthworms when the ground is dry (Atkinson 2013). Because digging activity is affected by so many factors, the number of molehills on the ground is not a good indicator of the number of moles beneath. Indeed, high density mole populations living in permanent pasture may produce no molehills for most of the year (Gorman and Stone 1990). Finding a way of discerning the number of moles present in an area of land could be a significant milestone on the road to their more effective and humane management.

Mole activity on British farmland is reported by farmers to peak in spring and, to a lesser extent, autumn (Fig. 12.2); the pattern is similar across Britain, although the spring peak in southern counties is less pronounced. These seasonal patterns in mole activity are likely to reflect periods of greater digging activity either during breeding, when tunnels need repairing, or when soil is more workable. Other factors in the reported seasonal variations in mole activity on farmland may be that moles are likely to be more conspicuous on some cultivation types, and their perceived activity is likely to vary with the time of year and the activity of the farmer reporting that of the mole (Atkinson et al. 1994). Seasonal patterns in reported mole control on farmland broadly echo reported mole activity (Fig. 12.2), with a peak between February and April, and another, less pronounced, peak between October and December (Atkinson et al. 1994; Baker et al., unpublished data).

Mole home range sizes vary widely with habitat quality. In order to establish sufficient length of feeding tunnel in poorer soils, moles need to make more molehills over a larger area. Moles established in food-rich areas need to tunnel less and make fewer molehills as a result (Atkinson 2013). For example, in a lush lowland meadow, with plentiful food, a single mole might use an area of 300 m^2, while on a barren Highland muir it might need up to 3000 m^2 to find sufficient food (Gorman and Stone 1990). Over three grassland sites in and around Oxford, Macdonald et al. (1997) reported an average home range size (excluding breeding males) of 2324 m^2, with no significant difference between the sexes. On Scottish pasture, Stone and Gorman (1985) recorded non-breeding home ranges of 2060 m^2 and 2072 m^2 for males and females respectively, and while female ranges remained fairly constant year-round, male ranges expanded to 6040 m^2 during the breeding season, when they make mating excursions outside their normal range (Lund 1976; Stone and Gorman 1985; Hartman 1995).

Mole tunnels are little wider in diameter (5 cm) than a mole itself (Gorman and Stone 1990; Atkinson 2013). Moles dig two types of tunnel: surface tunnels and deep tunnels. Surface tunnels can be seen breaking the soil surface and tend to occur in sandy or

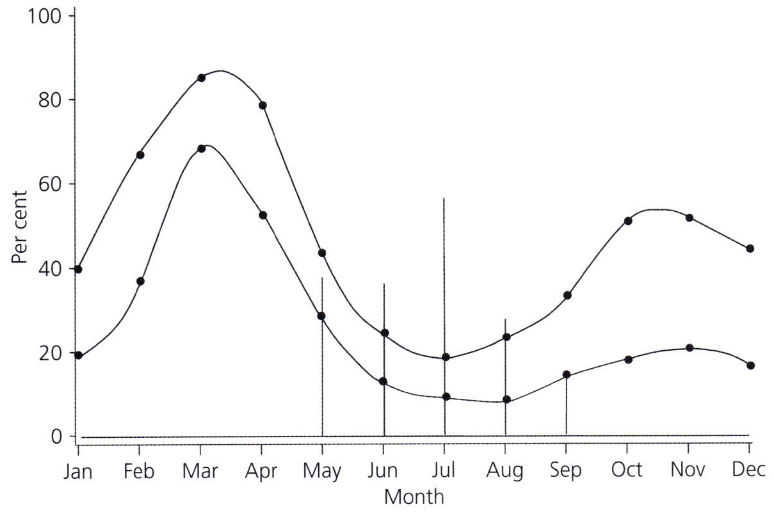

Figure 12.2 National pattern of mole activity (upper line, proportion of farms recording that moles were active) and control (lower line, proportion of farms where control was in effect in 1991 in the month in question). Smoothed curves are drawn through the points. Vertical lines denote national silage cutting activity: the proportion of farms where silage was made which were cutting this month. From Atkinson et al. (1994), reproduced with permission from John Wiley & Sons.

Box 12.2 Mole activity patterns

Mole activity patterns and movements are dictated by the temporal and spatial dispersion of food. Moles maintain largely exclusive home ranges with a small degree of overlap between neighbours. On permanent pasture and rough grassland in the Oxford area, Macdonald et al. (1997) radio-tracked 29 moles to discover that home ranges overlapped, on average, by only 13%, with non-breeding neighbours of opposite sexes tending to share more of their home ranges than neighbours of the same sex.

So, how do moles successfully share parts of their range with others? In their Oxfordshire study, Macdonald et al. (1997) found that each mole spent less than 1% of its time within 6 m of another mole. Only three of 46 mole dyads (a dyad here meaning two animals) appeared to be attracted to each other, but there was no evidence from simultaneous movements of neighbouring moles that they actively attempted to avoid each other. In their study, however, Stone and Gorman (1985) found that while the activity patterns of neighbouring moles were largely synchronized, they avoided direct confrontation by foraging in areas of range overlap during different activity periods.

In general, moles exhibit a tri-phasic pattern of activity (three 8-hour cycles in 24 hours) (Godfrey 1955; Macdonald et al. 1997), each 8-hour cycle comprising approximately 3.5 hours of resting and 4.5 hours of activity, consisting of foraging, digging, and moving about (Godfrey 1955). Macdonald et al. (1997) demonstrated that moles may switch to a tetra-phasic activity pattern (four 6-hour cycles), and more sleeping, under drought conditions. Because drought conditions are associated with reduced food availability for moles, these results suggest that under such conditions moles may be fatigued more quickly and then woken earlier by hunger. In their study on Scottish pasture, Stone and Gorman (1985) observed that, in September and October, males switched to just two periods of activity a day, while females maintained a tri-phasic pattern throughout the year, except when lactating in May and June, when they returned to the nest four or five times in a 24-hour period, presumably to feed their young.

newly cultivated soil, while deeper tunnels cannot be seen from above ground and may be anything up to 1.5 m deep (Gorman and Stone 1990; Atkinson 2013). In digging deeper tunnels, moles produce a series of molehills to remove soil as they go; this is not necessary when making surface tunnels. As well as providing a way of disposing of tunnel soil, molehills may also help to aerate the underground tunnels. The distance between molehills varies with the depth of the tunnel being dug, because the energy needed to push soil upwards (against gravity), to make a hill, is several times greater than that needed to push soil along a tunnel (Atkinson 2013). On average, molehills are 8 cm tall and 30 cm wide. However, occasionally, a mole will construct a much grander mound, known as a fortress, on average, an impressive 30 cm tall and 90 cm wide. Both sexes are known to build fortresses, but fewer than 5% of moles do this, and it is not known why. Fortresses usually contain nests and sometimes food stores containing earthworms, and they are more often found in areas prone to flooding and on thin soil. This has led to the ideas that they provide refuge from flooding (e.g. for young moles in a breeding nest, above ground level, within the fortress) and insulation for nests, built near the surface, in shallow soil (Atkinson 2013).

Molehills and tunnels are often viewed as damage on farmland and, as a consequence, moles have been considered agricultural vermin for centuries (Lovegrove 2007; Nicholls 2010; Box 12.1) and much effort is expended in an effort to control mole numbers and activity in agricultural settings (Atkinson et al. 1994). In this chapter we examine the impact of moles on agriculture, possible solutions, and the associated costs of both. We also highlight issues regarding the legality and welfare impacts of some mole control methods, and finally we discuss the way forward.

12.2 Moles as pests

12.2.1 Mole pest status

According to Lovegrove (2007), the mole was originally a woodland animal but, as forests were cleared for farming, it thrived on open pasture and arable land and became established as an agricultural pest (see Box 12.1). There is no information on historic mole populations (Tracking Mammals Partnership 2005) and, because mole bags (the number of moles killed, e.g. by gamekeepers) have not been recorded systematically (e.g. through the National Gamebag Census, Game and Wildlife Conservation Trust 2012), there is no information on long-term population trends. Even recent trends have not been established reliably (Tracking Mammals Partnership 2009), because data relate almost exclusively to the presence or absence of mole signs rather than live sightings and, as mentioned before, molehills and other mole signs are poor indicators of mole numbers. Nevertheless, there are claims that mole populations have 'exploded' recently as a result either of warmer, wetter winters that may have increased earthworm numbers[2], or of the 2006 ban on strychnine for poisoning moles[3], but these claims are unsubstantiated (see, e.g. Hartley and Campbell 2011).

Whatever has happened to mole numbers, since the sixteenth century, little has changed regarding mole pest status. This has been revealed by three studies, conducted in 1981, 1992, and 2007, of farmers' experiences and opinions of moles. Atkinson et al. (1994) describe one study on farmland wildlife in lowland England (in 1981) and another (in 1992) on moles on farms all over Britain. The 1981 questionnaire consisted of Parts A and B and was distributed by post to farm addresses provided by local fox-hunts. Part A was sent to 2280 farmers in 10 regions representing the major lowland landscapes in England. The survey asked questions about wildlife on respondents' farms and 867 (38%) responses were received. Part B was completed by a further 101 'Midland' farmers, from Oxfordshire, Buckinghamshire, Northamptonshire, and Warwickshire. Part B was a more comprehensive wildlife survey, including questions prompted by responses in Sample A. In 1992, Atkinson et al. (1994) distributed a further questionnaire at local NFU meetings to 460 farmers in England, Scotland, and Wales. The survey was designed to reassess farmers' perceptions of moles as pests and 157 (34%) responses were received. In the summer of 2007, we conducted a large-scale study of farmers' opinions regarding mole activity, damage, and control, on farms in England, Scotland, and Wales (Baker et al, unpublished data). This was the first survey of moles on British farms since strychnine was withdrawn. Questionnaires were posted to 1146 farmers and 720 (63%) responses were received.

In the 1992 study (Atkinson et al. 1994), moles were the third most often cited pest of farmland, after rats *Rattus norvegicus* (see Chapter 11, this volume) and magpies *Pica pica* (see Chapter 13, this volume), and almost three-quarters of farmers said they would prefer fewer moles on their land (Atkinson et al. 1994). According to the 1992 survey (n = 157), and our follow-up in 2007 (n = 685, conducted after the withdrawal of strychnine), more than 90% of British farmers reported moles on their farm, with about two-thirds of these farmers considering them to be pests, most of the rest feeling indifferent towards moles, and only a small minority thinking them desirable (Atkinson et al. 1994; Baker et al., unpublished data). Today, moles are considered farmland pests for a wide variety of reasons, but primarily because molehills can contaminate silage crops with soil bacteria, and molehill soil can cover pasture (making it unavailable to livestock), while mole activity can damage crops and drainage systems (see Section 12.2.2). Effects of mole digging activities and molehills on pasture, crops, and drainage systems are likely also to have been the cause of historical mole problems. For example, there are historical records, from the Lake District, showing that moles were captured, and molehill soil spread, to facilitate harvesting and haymaking by scythe (Lovegrove 2007).

Returning to farmers' opinions of moles as pests, these vary regionally across Britain. In 1992, Atkinson et al. (1994) found that fewer farmers regarded moles as pests in the Central and Eastern region (53%) (Cambridgeshire, West Midlands, Buckinghamshire, Oxfordshire, Suffolk, and Essex) compared to 71–76% of those in Southern, Northern (Scotland and northern England), and Wales and Western regions. In 2007, we found a somewhat similar pattern, in that farmers with moles in the South-west of England and Central and Eastern areas were less likely to consider them pests than those in Scotland, Wales, and the North of England. Atkinson et al. (1994) related regional variation in attitudes towards moles to regional differences in soil quality, with moles being less often considered pests in the Central and Eastern region, where soil quality was reported to be better overall than elsewhere. Moles were more often regarded as pests on soils classed by farmers as 'poor quality', while 'good quality' soils were more prevalent where moles were less of a problem. (Of course, it is not clear what farmers may have interpreted as 'good quality' soil, and farmers of different enterprises may value soil types differently.) One possible explanation for this association has been mentioned earlier—that moles on poorer soils need to dig more extensive tunnels to fulfil their feeding needs, and more tunnelling is likely to lead to greater mole pest status. There may also have been some circularity in farmers' responses to this question, and it is possible that they think of soil with moles as poorer than soil without.

[2] 'Unmaking a mountain', *The Daily Telegraph*, 10.09.12.

[3] McCarthy, M. (2010). 'Dark doings: the secret life of an enigmatic pest, *The Independent*, 19.6.2010, pp. 18–19.

Moles generally inhabit well-structured soils in which they can dig complex tunnel networks including shallow and deep tunnels (Gorman and Stone 1990). In 2007, they were most often reported on loamy soils and least often on clay. Loams are fertile, well-drained, and easily worked soils, comprising a mixture of clay, sand, and silt—and considered to be of good quality (Royal Horticultural Society 2014)—so it makes sense that moles should favour loams for soil workability. Soil type, as opposed to soil quality, was not related to mole pest status among farms with moles, but soil type did vary regionally, with chalk and clay being best represented in the South-west and Central and Eastern areas, where moles were least likely to be perceived as pests. Soil type also varied with farm enterprise type, with clay being best represented on arable farms. Given that arable farming predominates in eastern England (Countryside Survey 2009), arable farming may be more common in areas where soil is less suitable for moles.

In our 2007 survey, we also found that both reported mole presence on farms, and reported mole pest status, were related to farm enterprise type, with moles being reported as present, and as pests, most often on livestock and mixed farms. Of course, this may be because mole activity is more conspicuous, and more of a concern, on grassland (both grazed and cut for silage or hay) than in arable crops, rather than because of a real difference in mole presence and activity. In their 1992 survey, Atkinson et al. (1994) found that farmers who were indifferent towards moles had a greater proportion of cereal/arable farming on their land, while farmers who considered moles to be pests devoted a greater proportion of their land to silage production. These findings are consistent with the geographic bias discovered in soil type and the relationship between soil type and farm enterprise discussed above.

12.2.2 Mole damage

Mole activity is blamed for a bewildering array of agricultural impacts. These are reviewed by Atkinson et al. (1994). The catalogue of mole-related problems reported on farms includes the following: contamination of silage, covering pasture with molehill soil (which reduces the grazing area available (see Fig. 12.3)), covering young cereal plants, contamination of crops with soil at harvest, injury to livestock on pasture, damage to agricultural machinery, damage to drainage systems and watercourses, damage to young plants by cutting through tap roots, wilting of young crops caused by drying of soil above tunnels, weed invasion followed by degeneration of pasture on molehills, consolidation of molehills by ants, bringing stones to the soil surface, incidence of TB and tick-borne encephalitis, and molehills eroding down to bare rock (Atkinson et al. 1994; Lodal 1999; Quy and Poole 2004; Katzerke et al. 2010).

Potential contamination of silage is one of farmers' primary concerns regarding moles on their land. In Atkinson et al. (1994)'s 1992 survey, 64% of 157 farmers made silage. Of the silage-producing farmers, 15% claimed that moles adversely affected the quality of their silage and 11% said that silage was actually lost as a result of mole activity, although loss rarely exceeded 5% of the crop. Of silage producers who reported no silage damage by moles in the previous year, 40% considered them great pests of silage nevertheless (Atkinson et al. 1994).

Silage is produced as winter fodder for cattle and sheep. Grass is cut for silage in spring or late summer, and usually stored in plastic-wrapped bales where it

Figure 12.3 Molehills on pasture. Photograph ©R. Shaw.

is fermented by bacteria; this creates lactic acid, which preserves it as silage. However, if *Clostridium* or *Listeria* bacteria are present in soil they may be gathered up with grass during silage cutting, especially where there are molehills or other uneven soil (e.g. caused by poaching by livestock or by poor cultivation before reseeding). Sometimes, silage may be contaminated by these same bacteria if present in farmyard soil or manure (Atkinson et al. 1994). When these soil bacteria, particularly *Clostridium* spp., make their way into silage they suppress lactic acid formation, preventing the proper preservation of silage. *Clostridium* spp. also digest the proteins in grass and excrete ammonia. The resulting silage is less nutritious and either unpalatable or inedible (MAFF 1977, 1986). Contamination of silage with *Listeria monocytogenes* can cause listeriosis, especially in sheep, and may result in septicaemia, abortion, and encephalitis (Barlow 1971). Listeric abortion rates can be up to 20% in some sheep flocks and while an outbreak of listeric encephalitis usually affects fewer than 2% of animals, 10–30% of a flock may be affected in exceptional circumstances (Merck 2012). Of course, in thinking about this, it is important to remember that moles are not responsible for the bacteria, which are present naturally in soil.

Farmers have various options for protecting their silage from contamination by soil bacteria. Options include raising the height of grass cutting blades (which has the downside of reducing yield), allowing grass cut for silage to wilt in the field before baling (leaving the cut grass to dry out before baling allows soil on the grass to dry and fall out, reduces silage effluent production, and improves silage fermentation quality), and using silage additives to reduce the effect of any harmful bacteria that do find their way into silage bags (Atkinson et al. 1994; Lodal 1999; Bouriako et al. 2001). Harrowing and/or rolling fields early in the season, and as necessary after that, to remove molehills and other uneven areas, can considerably reduce silage contamination with soil bacteria (MAFF 1986).

All farmers producing silage in the 1992 survey, and most in 2007, wilted the cut grass before baling, but most did not raise the blades when cutting silage, or use a silage additive to protect it from bacterial contamination. However, the farmers who did report mole damage were more likely to have raised cutting blades or used a silage additive, whereas those without mole damage were more likely to have wilted the grass or taken no mitigation precautions. While farmers who raised cutting blades claimed this reduced their silage yield by 5% (Atkinson et al. 1994), wilting cut grass before baling is likely to be an easy, no/low cost, option, with the potential benefit of avoiding other sources of silage contamination as well as those resulting from mole activity.

Moles are not all bad for farming, although this is not necessarily recognized by the farmer. Moles can help to improve drainage and they eat undesirable invertebrate soil pests, aerate soils, and turn over the soil which contributes to soil nutrient levels. Old molehills also provide a suitable bed for the germination of rare herbs, and molehill soil is prized by gardeners as a source of fine topsoil or for use as a seedbed because of its fine tilth (Atkinson et al. 1994; Lovegrove 2007). It may be that the benefits brought by moles are less conspicuous to farmers than the problems—and farmers may have entrenched ideas that are difficult to change.

12.2.3 Mole control

Moles are still widely killed or otherwise managed in Britain (the methods used are described in Section 12.2.4). Our 2007 survey indicated that mole control had taken place at some time on two-thirds of 688 British farms. In Atkinson et al. (1994)'s 1992 study, half of 152 farms with moles conducted some form of mole control, with moles being controlled on 61% of farms where they were considered a pest, and even on 31% of farms where the farmer claimed indifference to them. Control varied by region, mirroring the pattern of reported pest status, with most control occurring in the North (Scotland and northern England; 77%), then Wales and the West (64%), and the South (40%). Least control was reported in the Central and Eastern region (20%), where moles were least often considered pests. In 2007, reported past mole control was similarly most prevalent in Wales, Scotland, and the North of England. Mole control was also more likely to occur on loamy, peaty, or sandy soils, and was most common on livestock farms. Again, this pattern of control reflected reported mole pest status.

However, the more localized study of wildlife, on lowland English farms (in 1981), revealed that the relationship between mole damage and mole control may not be straightforward. While the overall number of farmers reporting mole control (40%) was similar to that reporting damage (41%), there were big differences in some areas between the number of farmers reporting mole damage and those reporting mole control (Atkinson et al. 1994). In six areas (Dorset, Exmoor, Leicestershire, Oxfordshire and Buckinghamshire, Shropshire, and Sussex), more farmers reported damage than reported control, indicating that at least some were tolerating perceived damage without taking action to reduce it. In the other four areas (Cornwall, Suffolk, Warwickshire, and Yorkshire), however, more farmers reported controlling moles than reported mole damage. These differences between reported damage and control are summarized in Fig. 12.4, as a control/damage ratio (where a value > 100 means control > damage, and vice versa); ratios ranged between 73 (Oxfordshire and Buckinghamshire) and 163 (Cornwall). However, ratios exceeding 100 (i.e. where control > damage) are not easily interpreted because they could represent two quite different circumstances. In the first of these, farmers may have applied mole control even when they did not consider mole activity to constitute damage; alternatively, farmers may have controlled moles so effectively that they did not cause any damage. Overall, mole control on lowland farms varied widely among areas and the control/damage ratio did not seem to be related to reported damage (Atkinson et al. 1994).

Where mole control is deemed necessary, the timing is crucial and depends on the land manager's goal. Control may be approached with one, or both, of two aims in mind: first, to minimize mole damage in the short term, e.g. for cutting silage or hay and, second, to reduce the mole population in the longer term, e.g. to protect pasture on an ongoing basis. If the goal is to reduce short-term damage, then control may be conducted as and when thought necessary. However, new molehills often appear overnight, so this may not be worthwhile. Longer-term population control is a different matter. Male mole activity increases between January and March, when they leave their home ranges to seek mating opportunities (Hartman 1995). This activity probably constitutes much of the reported spring peak in mole activity on farms, and the corresponding peak in reported mole control is likely to be a response to this. However, there is good evidence that trapping during the January–March peak in male breeding activity results in a significant male bias among trap captures (2:1, Hartman 1995; 7.7:1, Baker et al. 2015). Trapping moles at this time may help to reduce immediate damage, but doing this in the hope of reducing mole populations, and therefore longer-term damage, may prove futile. These observations demonstrate the need to understand the fundamental biology of a species in order to manage it effectively (Baker et al. 2006). Where moles are to be removed it might be most productively done in April/May, when females are pregnant or lactating, but trapping at this time would raise welfare issues regarding the fate of dependent young (Baker et al. 2015). (There are also more general concerns regarding the humaneness of some mole traps and these are described in Section 12.2.4 and Box 12.3.) Phosphine gassing during April/May ought to kill any dependent young along with their parents (thus removing the issue of young starving to death), but there are issues regarding the humaneness of gassing moles *per se*. (These are outlined in Section 12.2.4.)

12.2.4 Mole control methods

Over the years, people have employed an enormous variety of practices for killing moles, or otherwise controlling their activities (Atkinson et al. 1994; Quy and Poole 2004). Efforts to kill them have included poisoning (e.g. with strychnine), kill-trapping, gassing (e.g. with phosphine), shooting (mainly into moving molehills or tunnels), bludgeoning, killing with dogs, and exploding devices (e.g. Rodenator^{TM}) (Gorman and Stone 1990; Natural England 2008; Nicholls 2010). Moles have been live-trapped and moved to different locations (Shaw et al. submitted) and various methods have been tried for persuading moles to move elsewhere. These have included habitat manipulation to

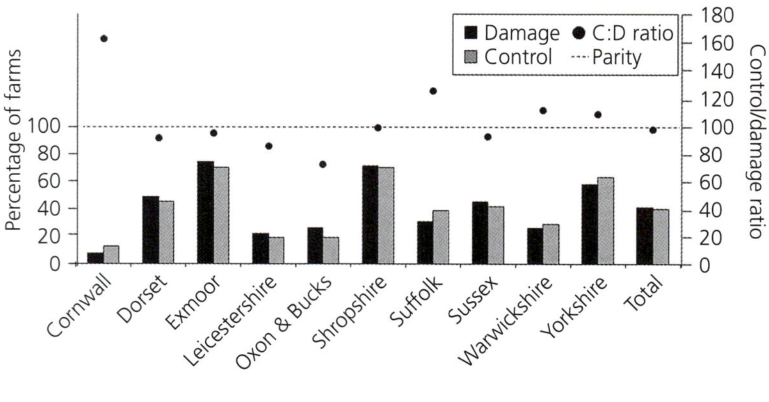

Figure 12.4 Percentage of farmers reporting mole damage and mole control in the 1981 survey, together with the control/damage ratio. Drawn from data in Atkinson et al. (1994), reproduced with permission from John Wiley & Sons.

reduce earthworm numbers in the soil, poking prickly vegetation (such as rose prunings or brambles) into mole tunnels, burying bottles in the ground (with the hope that wind blowing over the bottle mouths will produce unpleasant low frequency vibrations in the soil), commercially available sonic devices for producing the same effect, mole smokes, chemical repellents (such as Renardine™, which is no longer available), and introducing all manner of substances into their burrow systems, for example, vehicle exhaust fumes, naphthalene (moth balls), carbide, Jeyes fluid, rotten fish, tar, diesel oil, or creosote (Atkinson and Macdonald 1994; Gorman and Lamb 1994; Edwards et al. 1999; Lodal 1999; Nicholls 2010).

The most popular mole control method overall among farmers in Atkinson et al. (1994)'s 1992 survey was strychnine poison (41%), followed by trapping (37%), gassing (14%), shooting (7%), and harrowing (1%), and methods were sometimes combined. The methods used varied regionally, with strychnine being the most popular in Wales and the West (63%), gassing in the South (40%), and trapping in both the Central and Eastern (46%) and Northern (44%) regions. Strychnine came a close second in the North (41%), and overall strychnine was used more often in the North and in the Wales and Western region than elsewhere (Atkinson et al. 1994). There seems to have been a tendency to use strychnine (which required training and Defra approval and was often applied by a professional mole controller) in areas where moles were most likely to be considered pests, and trapping where this was least likely. The perceived advantages of strychnine were its effectiveness, ease of application, and low cost (see also Mead-Briggs and Trout 1975), while the main disadvantages reported were the danger and bureaucracy involved in its use. When asked which mole control methods were effective, all farmers who answered the question said strychnine (100%), while fewer thought other methods were effective: trapping (83%), gassing (66%), and harrowing (26%). Farmers that did not use strychnine were more likely to say it was inhumane (Atkinson et al. 1994). There are apparent differences in the seasonal pattern of use of different control methods. In 1992, Atkinson et al. (1994) found that strychnine poison was used primarily in spring (particularly in March/April), phosphine gassing was used between March and September, but most often in May, and trapping was used year-round, peaking in March (Atkinson et al. 1994). In a post-mortem study, conducted in 2008/2009 in Oxfordshire, we found that land managers trapped more moles, per month, between January and March than at other times of the year. We have already discussed the likely effectiveness of trapping at this time of year (see above).

Strychnine is an alkaloid poison, which causes death by asphyxia due to paralysis of the respiratory muscles (Neville 1985). It was administered by inserting strychnine-treated earthworms into mole tunnels at intervals. Because of the potential for mis-use of strychnine, Defra approval was required before it could be purchased or used (Quy and Poole 2004). Until 2006, strychnine poison was uniquely legal for controlling moles in the UK (see Box 12.3), but nowhere else in the EU (Neville 1985). A 2004 review of mole management options concluded that strychnine was by far the most cost-effective method of mole control available in the EU (Quy and Poole 2004). However, strychnine posed risks for users, pets, and predators (through secondary poisoning) (Nicholls 2010) and, in September 2006, it was withdrawn from use for mole control in the UK (Natural England 2011). Alphachloralose is a poison used for controlling moles in France but it has not proved effective in field trials in the UK, and is not registered for use outside France (Quy and Poole 2004). So, in 2014, no poisons are available for controlling moles in the UK, and—apart from alphachloralose in France—no other poisons are registered for use with moles elsewhere in Europe (Quy and Poole 2004). Strychnine use declined as its withdrawal in 2006 approached, and over the five years prior to our 2007 survey, strychnine was used on only one-fifth of farms, approximately half of those reporting strychnine use 15 years before. However, most farmers who had used strychnine in the previous five years were keen to use the poison again if it became legally available in the future.

Trapping and gassing are the two main legal mole control methods in use today. In 2007, one year after strychnine was withdrawn, kill-trapping with spring traps had become the most popular mole control method among farmers, followed by phosphine gassing, and harrowing. Mole-trapping has been something of a tradition in Britain (see Box 12.1). Traps are set beneath ground in mole tunnels, and trapping may be conducted year-round (and 'in-house') on farms as maintenance/prophylactic control. Kill-trapping, using metal spring traps such as scissors and Duffus traps, is widely used in Britain, Denmark, and France, and is thought to be one of the most effective methods of controlling moles, when conducted by an experienced operator in small-scale applications (Quy and Poole 2004). Mole traps are not required to meet welfare standards that apply to other spring traps in Britain, because they are exempt from The Pests Act 1954

Box 12.3 Inequality for moles

Not only have moles been persecuted in Britain (see Box 12.1), but legislation on the management of moles (and rats and mice) has been unfairly biased. For example, under the Protection of Animals Act 1911 and the Protection of Animals (Scotland) Act 1912 it was made illegal to poison any mammal in Britain except, under certain circumstances, where it was a defence that the poison was used to destroy invertebrates, rats, mice, or other small ground vermin (which would include moles). In the 1960s, the Animals (Cruel Poisons) Act 1962 and the Animals (Cruel Poisons) Regulations 1963 went further, specifically banning the use of phosphorus for destroying any mammal, and the use of strychnine for destroying any mammal *except moles.* Strychnine causes death through asphyxia due to paralysis of the respiratory muscles and is considered to be an extremely dangerous and inhumane alkaloid poison (Atkinson et al. 1994). The exception for continuing to allow the use of strychnine with moles was probably made on the grounds that no other suitable methods of destroying them existed and that when used underground, for killing moles, strychnine did not pose a risk to other species (e.g. see the later *Hansard* 23 June 1976 col 297–299). Other reasons suggested by Atkinson et al. (1994) for this exception are that: there was a perceived need to manage moles; strychnine was the only method cost-effective on an agricultural scale; and moles die underground where their painful deaths cannot be seen.

However, under Directive 91/414/EEC, the EU conducted a review of all pesticides used within member states. As part of this review, manufacturers were required to provide health and safety data to support continued registration of their products, but the manufacturers of strychnine failed to do this and, despite complaints from users, strychnine was finally withdrawn from use from 1st September 2006 (Davies 2006). The only methods legally available for killing moles in Britain today are spring trapping and phosphine gassing (Natural England 2011).

However, the spring trapping legislation is also unfairly biased against moles (and rats and mice). In 1951, The Committee on Cruelty to Wild Animals recommended that spring traps should require welfare approval. However they also commented that there was no evidence that mole traps or break-back traps used for rats or mice involved any unnecessary suffering. In 1954, the Pests Act gave effect to the Committee's recommendations, making it an offence to use any spring trap that had not been approved, and the 1958 Small

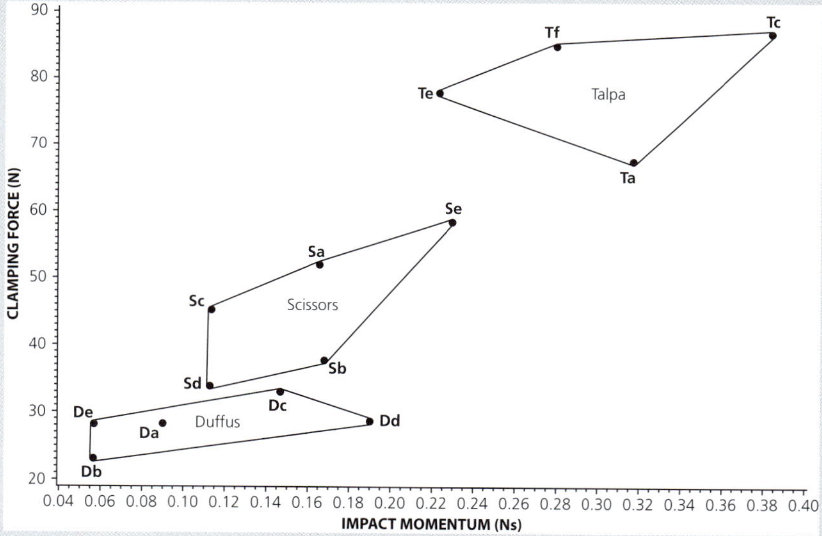

Box Figure 12.3.1 Mechanical performance of mole traps: impact momentum against clamping force for different types and manufacturers of mole traps. Each polygon represents a trap type (Scissors, Duffus, or Talpa). Each point within a polygon is the mean of measurements on ten individual traps for impact momentum and 20 traps for clamping force. Letters a–f represent different trap manufacturers. Points Da–e represent Duffus traps (n = 5), points Sa–e represent Scissors traps (n = 5), and points Ta,c,e,f represent Talpa-style traps (n = 4). From Baker et al. (2012).

continued

Box 12.3 *Continued*

Ground Vermin Traps Order implemented a provision in the Pests Act to exempt from the approval process mole traps and break-back traps. When strychnine was withdrawn in 2006, molecatchers predicted an increase in the use of poor quality traps by untrained operators (Davies 2006).

In 2010, we observed that there had been a proliferation in the brands of traps available for killing moles and in the types and brands of break-back traps for killing rats and mice. As a result we decided to examine the evidence for scope to improve the welfare standards of these traps. We measured the mechanical performance of 23 types of mouse trap, 18 types of rat trap, and three types of mole trap—scissor, duffus, and Talpa (each of these mole traps made by several different brands) (Baker et al. 2012). Impact momentum varied 6–8 fold, and clamping force values varied 4–5.5 fold among traps for killing each species (see Box Figure 12.3.1 regarding mole traps), and there was significant overlap in the performance of rat and mouse traps. There was no relationship between the price and the performance of traps except for Talpa mole traps where traps with greater clamping force were more expensive. The wide variation in mechanical performance in traps for each species, overlap in performance between rat and mouse traps, and increasing availability of weaker plastic rodent traps indicated considerable scope for improving the humaneness of spring traps for rats, mice, and moles. We concluded that all spring traps should be subject to the UK approval process. Our results might also assist consumers in selecting more powerful traps, and many thousands of rats, mice, and moles might benefit (Baker et al. 2012).

under The Small Ground Vermin Traps Order 1958. And there is concern regarding the welfare performance of at least some of the traps available for killing moles in Britain today, and clearly room for improvement in terms of reducing their welfare impact (see Box 12.3 and Box Fig. 12.3.1).

Phosphine is a systemic poison which depresses nervous system and respiratory function (Pesticides Safety Directorate 1997, section 5.2). Phosphine gassing is probably the most widely used mole control method in Europe (Quy and Poole 2004), and two products, Luxan Talunex™ and Phostoxin™, which produce phosphine gas, are approved for mole control in the UK (Pesticide Safety Directorate, pers. comm.). No other gases, such as cyanide, or those found in exhaust fumes, may be used to kill moles, legally, in Britain. Phosphine gassing involves placing aluminium phosphide tablets or pellets in mole tunnels at intervals, and once in contact with moisture from the air or soil they produce phosphine gas in the tunnel system. As was the case with using strychnine, phosphine gassing is probably a larger undertaking and so perhaps more likely to be conducted routinely once or twice a year by a professional mole controller. The humaneness of phosphine gassing is questionable. There has been almost no research on the effects of phosphine gassing on moles, but one study on captive individuals found that constant exposure to phosphine concentrations ranging between 1500 ppm and 4.5 ppm caused moles to die after between 9 minutes and 5.5 hours (MAFF 1983). Visible signs of illness included uncoordinated movements, rocking/over-balancing, shallow and gasping breath. Moles may also suffer the invisible symptoms reported by human victims of phosphine gassing, e.g. strong chest pains, stomach and head ache and double-vision. When captive moles were exposed to phosphine for limited time periods (as might occur in the field), moles were ill for protracted periods (with more severe symptoms, such as convulsions and respiratory spasms) and died much later; one mole took 3–4 days to die. The gas therefore needs to be distributed quickly and evenly throughout the tunnel system at a concentration sufficient to cause rapid insensibility for all moles. However there have been no studies to determine the treatment protocol needed to achieve this outcome.

Currently, there are no commercially available chemical repellents approved for use with moles under the Control of Pesticides Regulations 1986. A bone oil product, Renardine™, was previously approved as a mole repellent in Britain. While it deterred invasion of cleared areas for short periods, its application raised welfare issues and was far too labour-intensive for use on an agricultural scale (Atkinson and Macdonald 1994). Renardine™ is no longer approved and it would also be illegal to use any other chemicals, e.g. creosote, and any other substances such as those reported to have been used as pesticides against moles (see above). Acoustic repellers for deterring moles are available from garden centres and online. These battery or solar-powered, spike-shaped devices are designed to be pushed into the soil, from where they produce vibrations that are supposed to scare moles away. However, a British study that tested three acoustic mole deterrents concluded that they were not effective because the vibrations produced by the repellers were rapidly attenuated in soil and could not be detected even a few centimetres away (Gorman and Lamb 1994). Nevertheless, sonic/ultrasonic devices remain widely available in Britain (Quy and Poole 2004). Another type of gadget sold for controlling other mole species in parts of the United States is an exploding device, known as the Rodenator™. Under the Wildlife and Countryside Act 1981, it is illegal to use any explosive device, other than ammunition, to kill a wild animal in Britain. Therefore the Rodenator™ could only be used here legally for destroying empty tunnel/burrow systems. While Defra strongly advises that all reasonable means are taken to ensure that no animals remain in burrows before using the device to collapse them (Defra 2007), Natural England has serious concerns about the practical difficulties of doing this and therefore of using these devices at all (Natural England 2008).

Live-trapping and translocating moles could potentially be a mole management option for small-scale applications (Shaw et al. submitted). However, there are concerns about the welfare impact of catching and moving live moles (Natural England 2011; Baker and Macdonald 2012). Releasing a mole into another mole's territory may result in fighting (moles are known to fight to the death, Atkinson and Macdonald 1994), while releasing a mole into an area with no existing run system could put unreasonable pressure on the mole to find sufficient food. Natural England (2011) suggest that this could potentially constitute an offence under the Animal Welfare Act 2006 (which *inter alia* places a duty of care on anyone holding a live wild animal in a trap), and they specifically do not recommend relocating a live-trapped mole, instead suggesting that if a mole is captured alive it should be dispatched humanely. Furthermore, the only commercially available mole live trap, the plastic tube trap, is particularly unsuitable (in that the space inside is extremely restricted such that there is no room to supply bedding, and because of the possibility of more than one mole entering the trap and fighting as a result) and we have called for it to be withdrawn from sale on welfare grounds (Baker and Macdonald 2012).

12.2.5 Costs of mole damage and control

A useful index of the scale of a perceived 'pest' problem can be to examine the financial cost of damage suffered and control administered, as reported by the injured parties. In Atkinson et al. (1994)'s 1992 survey, those farmers reporting either mole damage or control in the previous year recorded a mean overall loss (damage and control combined) of £126 (Atkinson et al. 1994), or £221 at 2012 prices (Bank of England 2013). However, this figure could not be separated into the amount lost to mole damage and the amount spent on mole control. In our 2007 survey, we asked farmers to report the cost of mole damage and mole control separately, with control costs further split into the costs of professional services and non-professional control. Surprisingly, many farmers who reported suffering 'financial damage' from moles were unable to put a value on this damage. Overall, those farmers reporting mole damage or control (and who did report the cost) sustained damage worth a mean of £568 (£667 at 2012 prices) and spent a mean of £298 (£350 in 2012) on mole control. The control costs comprised professional control costs of £51 (£60 in 2012) and non-professional control costs of £197 (£231 in 2012). The mean damage and control costs combined would be £866 (£1017 in 2012), more than 4.5 times the amount reported in the 1992 survey (allowing for inflation). When converted to costs per hectare, the 2007 costs (and 2012 prices in parentheses) were: damage, £4.75 (£5.58); total control, £3.32 (£3.90); professional control, £0.45 (£0.53); and non-professional control, £2.38 (£2.79). While reported costs per hectare were greatest for livestock farmers, followed by mixed and then arable, only the total costs of control differed significantly among farm enterprises.

Even where costs are reported separately for damage and control, it remains impossible to account for the fact that the amount of control may affect the amount of future damage. Hence the cost of potential unmanaged damage could be disguised by the control conducted, and this will be related to the amount spent on control. Farmers of larger establishments may tend to under-report mole damage costs per hectare (simply because they are not aware of all mole activity everywhere on their property), with the effect that smaller units are perceived to suffer greater losses per hectare. Also, livestock farms tend to be smaller than arable enterprises, so the greater reported impact, and control, of moles on livestock farms may be a result of a combination of effects, i.e. a greater chance per hectare of noticing mole activity on a smaller farm, a greater chance of noticing mole activity on pasture than on cropped land, and a greater concern regarding mole damage on silage-producing establishments.

The apparent 4.5-fold increase in the combined cost of mole damage and control, between Atkinson et al. (1994)'s 1992 survey and our 2007 survey, could be explained in two different ways. The first is that there may have been an increase in real *or perceived* mole problems on British farms between 1992 and 2007. This could potentially be a result of the loss of strychnine poisoning as a mole management option, if farmers are either choosing to use management methods that are more time consuming and therefore more expensive than strychnine, e.g. trapping, or they are simply not bothering to manage moles at all and are instead accepting an increase in mole damage. Alternatively, farmers may perceive that there are now more moles, and more mole damage, as a result of changes in British weather patterns. Indeed, in 2010, some pest controllers reported an increase in mole numbers, putting this down to recent warm and wet conditions and the withdrawal of strychnine[4]. Also, in 2009, the Tracking Mammals Partnership reported that mole numbers may have increased in the previous ten years, based on molehill reports from volunteers (Tracking Mammals Partnership 2009). However, any observed increase in molehill numbers may simply have been a result of wetter weather (increasing damage to tunnel systems) leading to the production of more molehills, and as mentioned earlier, claims of an increase in the number of moles themselves have not been substantiated (see Hartley and Campbell 2011). The second possible explanation for the large increase in reported mole-related costs between the two surveys could be that the surveys asked farmers about costs in different ways. Atkinson et al. (1994) asked farmers about their overall financial loss to moles, whereas we asked about the costs of damage and control separately. Our approach may have encouraged farmers to think more fully about their response and to include costs perhaps omitted in their responses to the 1992 survey. It is therefore difficult to discern the reason for the apparent increase in mole-related costs to farmers.

12.3 Conclusions

Moles are reported on the majority of British farms and widely regarded as farmland pests. However, our review reveals the contradictory nature of farmers' opinions and actions regarding moles, the damage they cause, and mole control. Some farmers may have entrenched ideas about moles, regardless of their own experiences. For example, there seems to be no direct relationship between whether farmers report mole damage and whether they attempt to control moles. In 1992, farmers controlled moles on less than two-thirds of farms where they were considered pests and on about a third of farms where they were not (presumably this was done on a prophylactic basis). Further, while contamination of silage is one of farmers' main stated concerns about moles as pests, 40% of the silage producers that reported no silage damage, in the 1992 farm survey, nevertheless regarded moles as great pests of silage (Atkinson et al. 1994), perhaps because they are aware of the issue within the farming community. Farmers' efforts to limit mole damage are not always related clearly to their individual circumstances. For example, most farmers producing silage wilt the cut grass before baling it, a relatively simple and no cost option, which can help to reduce the amount of soil bagged with the silage. However, in 1992 and in 2007, most silage producers did not seem to take the precautions of raising the cutting blades, when cutting silage, or using a silage additive to protect it from bacterial contamination. The truth of course may be that the cost of conducting these measures (e.g. cost of additives, or the 5% of yield reported to be lost through raising silage cutters; Atkinson et al. 1994) is greater than the perceived cost of any potential damage. This idea is supported by the fact that while many farmers in our 2007 survey reported that they suffered mole damage 'of financial value' they were often unable to attribute a value to it.

And this may be the nub of the matter. Farmers may be more annoyed by the presence of moles than they are damaged by it. This may explain why mole control responds to seasonal patterns in activity rather than being targeted necessarily at the most effective time of year. In addition, different control methods follow different seasonal patterns (Atkinson et al. 1994), which seem to be dictated perhaps more by circumstances or routine than by need. For example, trapping is likely to be done 'in-house', and in 1992 occurred primarily in March (Atkinson et al. 1994), when mole activity is most conspicuous on farmland and when mole trapping produces strongly male-biased captures (Hartman 1995; Baker et al. 2015). Trapping moles at that time may therefore be least efficient for population management purposes. Where a farmer reported financial costs of both mole damage and mole control, the costs of control often exceed the costs of damage. And while it could be that prophylactic mole control

[4] *Mole Population Booms Across Sussex*, <http://www.sussexeps.co.uk/>; *Mole population soars due to poison ban*, <http://www.telegraph.co.uk/>; *Mole man who dug himself out of a hole*, <http://www.yorkshirepost.co.uk/> (all accessed 06/10/14).

reduces damage (thus disguising its true potential cost), it could simply also be that more mole control is conducted than is worthwhile.

Killing moles has a long history in Britain. The behaviour and lifestyle of the mole present a particular challenge for any would-be pest controller, and people have risen to that challenge, devising all manner of devices over the centuries for trapping and killing moles (Nicholls 2010). Perhaps, because of the difficulty of catching moles, and the likely associated frustration, people have been prepared to resort to methods they might not have considered for other species. For example, it has been suggested that the poisoning of moles using strychnine was considered acceptable only because moles died underground, where their (doubtless painful and protracted) deaths could not be seen (Atkinson et al. 1994). Another inconsistency in mole control remains in place today; in contrast to spring traps for use with other species, those for moles (and certain traps for rats and mice) are not required to pass welfare approval (see Box 12.3; Baker et al. 2012), perhaps because the welfare of these particular species attracts indifference.

12.4 Recommendations

In summary, moles may be killed more readily than is worthwhile, at less efficient times of year, and sometimes routinely rather than in considered response to need. More use might be made of managing or preventing damage. Perhaps farmers should re-evaluate the relative costs and benefits of controlling moles on their land, and as a result resources might be saved and potentially fewer moles killed. This chimes with the conclusions of Quy and Poole (2004) that the withdrawal of strychnine would encourage land managers to assess more carefully whether moles need to controlled or whether they should be left alone, to avoid applying measures that are unnecessary or unlikely to reduce damage levels in the long term. Where lethal control really is considered necessary, to reduce mole populations in the longer term, they may be legally trapped or gassed. Trapping should be done outside the January–March period when there is a significant male bias among trapped moles. Both gassing and trapping may be most effectively targeted at breeding females in April/May. However, trapping at this time raises particular welfare issues regarding the fate of dependent young, and both gassing and trapping raise separate welfare concerns in general. Now that strychnine is no longer available, and farmers tend to kill moles using spring traps, the welfare of those moles that are killed could be safeguarded and potentially improved if mole traps were required to meet the same welfare standards as other spring traps in Britain (Baker et al. 2012).

Acknowledgements

WildCRU mole research was funded by the Royal Society for the Prevention of Cruelty to Animals. SB was supported by a Fellowship from the Humane Society International and the Baker Trust during preparation of this chapter. We are very grateful to Rob Atkinson for his contribution to the work described in this chapter. We thank Ruth Feber and Eva Raebel for their comments and input to the chapter.

References

Atkinson, R.P.D. (2013). *Moles*, 135 pp. Whittet Books, Stansted, Essex, UK.

Atkinson, R.P.D. and Macdonald, D.W. (1994). Can repellents function as a non-lethal means of controlling moles (*Talpa europea*)? *Journal of Applied Ecology*, **31**, 731–736.

Atkinson, R.P.D., Macdonald, D.W., and Johnson, P.J. (1994). The status of the European mole *Talpa europea L.* as an agricultural pest and its management. *Mammal Review*, **24**, 73–90.

Baker, S.E., Ellwood, S.A., Tagarielli, V.L., and Macdonald, D.W. (2012). Mechanical performance of rat, mouse and mole spring traps, and possible implications for welfare performance. *PLOS ONE*, **7**, e39334.

Baker, S.E. and Macdonald, D.W. (2012). Not so humane mole tube traps. *Animal Welfare*, **21**(4), 613–615.

Baker, S.E., Shaw, R.F., Atkinson, R.P.D., West, P., and Macdonald, D.W. (2015). Potential welfare impacts of kill-trapping European moles *(Talpa europaea)* using scissor traps and Duffus traps: a post-mortem examination study. *Animal Welfare*, **24**, 1–14.

Baker, S., Singleton, G., and Smith, R. (2006). The nature of the beast: using biological processes in vertebrate pest management. In D.W. Macdonald and K. Service, eds., *Key Topics in Conservation Biology*, pp. 173–185. Blackwell Publishing, Oxford.

Bank of England (2013). *Inflation calculator*. <http://www.bankofengland.co.uk>, accessed February 2014.

Barlow, R.M. (1971). Listerial Encephalitis ('circling disease') of sheep and cattle. *Animal Diseases Research Association Newsheet*, **11**.

Bouriako, I.A., Shihab, H., Kuri, V., and Margerison, J.K. (2001). *Influence of wilting time on silage compositional quality and microbiology*. Available from: <http://www.bsas.org.uk>.

Countryside Survey (2009). *Countryside Survey: England Results from 2007*. NERC/Centre for Ecology & Hydrology, Department for Environment, Food and Rural Affairs, Natural England (CEH Project Number: C03259).

Davies, C. (2006). The death of strychnine. *Pesticide News*, **72**, 7.

Defra (2007). *Defra position statement: May 2007. Rodenator and similar devices*. Available from: <http://defra.gov.uk>.

Edwards, G.R., Crawley, M.J., and Heard, M.S. (1999). Influencing molehill distribution in grassland: implications for controlling the damage caused by molehills. *Journal of Applied Ecology*, **36**, 434–442.

Game and Wildlife Conservation Trust (2012). *The National Gamebag Census*. Available from: <http://www.gwct.org.uk>.

Godfrey, G.K. (1955). A field study of the activity of the mole (*Talpa europaea*). *Ecology*, **36**, 678–685.

Gorman, M. and Lamb, A. (1994). An investigation into the efficacy of mechanical mole scarers. *Animal Welfare*, **3**, 3–12.

Gorman, M.L. and Stone, R.D. (1990). *The Natural History of Moles*, 138 pp. Christopher Helm, London, UK.

Hansard HL Deb (1976). *Mole control: use of strychnine, vol. 372 col. 297–299 (23 June 1976)*. Available at: <http://hansard.millbanksystems.com/>.

Hartley, G. and Campbell, S. (2011). Surveys of Scottish farmers and their vertebrate pests—case study from a long running dataset. In Book of Abstracts of the *8th European Vertebrate Pest Management Conference*, eds., J. Jacob, A. Esther, Berlin, Germany.

Hartman, G.D. (1995). Seasonal effects on sex ratios in moles collected by trapping. *American Midland Naturalist*, **133**, 298–303.

Katzerke, A., Bliss, P., and Moritz, R.F.A. (2010). Size matters: mole (*Talpa europaea*) hills and nest-site selection of the ant Formica exsecta. *Population Ecology*, **52**, 271–277.

Lodal, J. (1999). The mole: a matter of control or conservation. In D.P. Cowan and C.J. Feare, eds., *Advances in Vertebrate Pest Management*, pp. 43–55. Filander-Verlag, Furth.

Lovegrove, R. (2007). *Silent Fields; The Long Decline of a Nation's Wildlife*, 404 pp. Oxford University Press, Oxford, UK.

Lund, M. (1976). Control of the European mole, *Talpa europaea*, in *Proceedings of the 7th Vertebrate Pest Conference*, ed. C.C. Siebe, Monterey, California, USA.

Macdonald, D.W., Atkinson, R.P.D., and Blanchard, G. (1997). Spatial and temporal patterns in the activity of European moles. *Oecologia*, **109**, 88–97.

MAFF (1977). *Silage. ADAS Bulletin 37*. HMSO, London, UK.

MAFF (1983). *The toxicity of phosphine to the mole* (Talpa europaea). Ministry of Agriculture, Fisheries and Food, unpublished report.

MAFF (1986). *Ensiling grass. ADAS pamphlet 482*. HMSO, London, UK.

Mead-Briggs, A.R. and Trout, R.C. (1975). A field evaluation of the effectiveness of a phosphine fumigant for mole control. *Agro-Ecosystems*, **2**, 1–13.

Mellanby, K. (1967). Food and activity in the mole *Talpa europaea*. *Nature*, **215**.

Mellanby, K. (1971). *The Mole*, 159 pp. Collins, Glasgow, UK.

Merck (2012). Overview of Listeriosis. *The Merck Veterinary Manual*. <http://www.merckmanuals.com>.

Natural England (2008). *Natural England calls on Rodenator users to heed the law*. Natural England Wildlife Management and Licensing. Available from: <http://www.naturalengland.org.uk>.

Natural England (2011). *Moles: options for management and control*. Natural England Technical Information Note TIN033. Available from: <http://publications.naturalengland.org.uk>.

Neville, P. (1985). Humane Control of Moles. In D.P. Britt, eds., *Humane Control of Land Mammals and Birds*. Symposium of UFAW.

Nicholls, J. (2010). *Mole Catching; A Practical Guide*, 112 pp. Crowood, Marlborough, UK.

Pesticides Safety Directorate (1997). *Assessment of humaneness of vertebrate control agents, section 6.2.4 (p. 19). A report by the Department for Environment, Food and Rural Affairs and the Pesticide Safety Directorate*. Defra and PSD.

Quy, R. and Poole, D. (2004). *A review of methods used within the European Union to control the European mole*, Talpa europea, 31 pp. Defra.

Royal Horticultural Society (2014). *Soil types*. <http://www.rhs.org.uk>.

Shaw, R.F., Baker, S.E., Harrington, A.L., Macdonald, D.W., and Atkinson, R.P.D. (submitted). Preliminary observations on translocation as a method of non-lethal mole (*Talpa europaea*) management. *Animal Welfare*.

Stone, R.D. (1986). *The social ecology of the European mole* Talpa europaea *L, and the Pyrenean desman* Galemys pyrenaicus *G: a comparative study*, PhD Thesis, University of Aberdeen.

Stone, R.D. and Gorman, M.L. (1985). Social organization of the European mole (*Talpa europaea*) and the Pyrenean desman (*Galemys pyrenaicus*). *Mammal Review*, **15**, 35–42.

Tracking Mammals Partnership (2005). *UK Mammals: Species Status and Population Trends; First Report by the Tracking Mammals Partnership*. Available from: <http://jncc.defra.gov.uk>.

Tracking Mammals Partnership (2009). *UK Mammals Update 2009*, JNCC. Available from: <http://jncc.defra.gov.uk>.

CHAPTER 13

Managing wildlife humanely with learned food aversions

Sandra E. Baker and David W. Macdonald

Persuasion is often more effectual than force.
Aesop

13.1 Introduction

13.1.1 Human–wildlife conflict

A wide range of wild vertebrates conflicts with people worldwide over agricultural, economic, conservation, recreation, and health interests (Baker and Macdonald 1999; Baker et al. 2007b). Where the animal concerned is large (e.g. big cats taking livestock), interactions are usually referred to as human–wildlife conflict, but where the animal is small, and usually acting in greater numbers on an ongoing basis, it is often termed a pest (Macdonald et al. 2013). However, many of the issues are really the same and the range of options for resolving them may be similar.

People have tried many ways of reducing conflict with wildlife, most of which involve killing the animals thought responsible for the damage or loss in question (Baker et al. 2007b). Culling methods used in the UK (and most other places) have included shooting, trapping, poisoning, gassing, hunting with dogs, exploding, electrocuting, and drowning, not all of which are legal, depending on the species and circumstances involved. Hundreds of thousands of vertebrates are killed each year in the UK alone because they are, or are perceived to be, in conflict with land management objectives. For example, Game Bag Census figures from the Game and Wildlife Conservation Trust (GWCT) indicate the scale of killing. Estimated numbers of animals killed as pests on shooting estates in 1990 were: grey squirrels *Sciurus carolinensis* 40 km^{-2}, European rabbits *Oryctolagus cuniculus* 23 km^{-2}, brown rats *Rattus norvegicus* 8 km^{-2}, carrion crows *Corvus corone* 3.1 km^{-2}, magpies *Pica pica* 2.2 km^{-2}, red foxes *Vulpes vulpes* 2.1 km^{-2}, stoats *Mustela erminea* 1.7 km^{-2}, and weasels *Mustela nivalis* 1.2 km^{-2} (Tapper 1992). Trends for these species indicate that since then, the numbers killed have either increased or stayed the same (GWCT 2012).

Pest management of any kind, though, is only rational if three tests are met: (1) the manager suffers losses to the pest; (2) the action taken reduces those losses; and (3) management is cost effective (see also Chapters 3 and 10, this volume). Culling may prove ineffective or even counter-productive at reducing losses (Baker and Macdonald 1999), thus failing the second of these tests. For example, Reynolds et al. (1993) showed that intensive shooting of foxes benefited gamebird production only very locally, and that vacated fox territories were subject to immigration and recolonization by new foxes later the same year through the vacuum effect. In addition, Macdonald et al. (2006) demonstrated that badger *Meles meles* removal operations, intended to reduce transmission of bovine tuberculosis (bTB) to cattle, produced a perturbation effect (e.g. changing spatial organization, demography, reproductive biology, and the frequency and type of social interactions) among surviving animals, and this might oppose the hoped-for reduction of disease transmission to cattle, or even increase it (Riordan et al. 2011; Chapter 5, this volume). (See Tuyttens and Macdonald's (2000) review of perturbation).

13.1.2 Non-lethal management

In contrast, non-lethal management methods may offer advantages over killing, in terms of animal

Wildlife Conservation on Farmland. Conflict in the Countryside. Edited by David W. Macdonald and Ruth E. Feber.
© Oxford University Press 2015. Published 2015 by Oxford University Press.

welfare, ethics, conservation, and the law (Baker and Macdonald 1999; Baker et al. 2007a). Of course, non-lethal wildlife management interventions cannot be assumed automatically to have a lower welfare impact than lethal methods, and we are currently examining the relative humaneness of different lethal and non-lethal interventions (e.g. see Baker et al. in prep.). Crucially, targeting problem animals using non-lethal methods could prove more effective than lethal control, because the target animal can maintain its territory and social position, and population perturbation is minimized as a result (Baker et al. 2007a). This avoids the density dependent population responses and immigration which often follow culling, while the animal continues to exert whatever effect it has on limiting other resources or prey, or excluding conspecifics (Baker and Macdonald 1999). Non-lethal approaches to wildlife management include fencing, scaring devices, fertility control, translocation, repellents, and food aversions (Baker and Macdonald 1999). Other examples of non-lethal control, shown recently to be worthwhile outside the UK, include lion guardians in Kenya (Hazzah et al. 2014) and livestock-guarding dogs in Namibia (Marker et al. 2005). Given the many thousands of animals that are killed in the name of pest management each year in the UK, and the clearly uncertain benefits of this, there is at the least a moral and financial imperative to explore non-lethal approaches to managing human–wildlife conflict in agricultural and other contexts.

13.1.3 Food aversions

Conflicts between people and wildlife often stem from unwanted feeding behaviour, such as animals eating and damaging growing crops (cereals, fruit), eating stored crops, eating animal feed from storage or feeding areas, or eating eggs intended either for hatching or food. One type of non-lethal wildlife management that may offer a targeted way of controlling such behaviour is food aversion (Gill et al. 1999; Cowan et al. 2000). Food aversions include a range of natural behavioural mechanisms evolved by animals for avoiding toxins (Garcia and Hankins 1975, 1977). Chemical feeding repellents (or aversive agents) can be used to exploit these mechanisms for the management of wildlife (Gill et al. 1999). We use the word 'repellent' here as a general term to include substances that cause an animal to avoid a target food, either through innate (or primary) aversion towards unpleasant tastes or odours, or through learned (or secondary) aversions of various kinds. Innate aversions are inborn and learned aversions are acquired as a result of experience. A learned food aversion may be caused using a secondary repellent with emetic (inducing nausea/vomiting), or other negative (e.g. irritant), properties. Some repellents cause both innate and learned food aversions (Gill et al. 1999; Macdonald and Baker 2004).

Conditioned taste aversion (CTA) is a distinct form of learned food aversion, widespread across taxa from molluscs to humans, and with potential for use in wildlife management. In CTA, the taste of a food becomes associated with feeling ill (an adverse post-ingestional effect such as illness, nausea, or malaise) and so the animal is less inclined to eat that particular food in future (Garcia et al. 1955). A CTA is formed when an animal ingests a toxic food and relates the resulting illness to the taste of that food after one or a few exposures (Nicolaus and Nellis 1987). Such aversions are thought to be acquired subconsciously, the mechanism being that the animal makes an association between the food taste (conditioned stimulus), and the illness (unconditioned stimulus), and avoids the taste of that food as a result (Cowan et al. 2000). Taste is more likely than non-taste cues to become associated with illness (Garcia and Rusiniak 1980), because taste and visceral information are processed in the same part (*nucleus solitarius*) of the brain (Nachman et al. 1977), while information relating to non-taste cues is not (Garcia et al. 1973). However both visual and odour cues may also be conditioned.

The most likely goal of a wildlife manager who wants to control feeding behaviour is to protect foods that are not chemically treated, because it is often neither desirable nor practical to treat the foods concerned, e.g. those destined for human consumption (Baker et al. 2007b) or which are live prey items such as eggs (Conover 1989). A learned food aversion based entirely on avoiding the taste of a food (as in classical CTA) could not be used to protect untreated foods, because a target animal would need to re-experience the aversive taste each time before being averted (see Cowan et al. 2000) and the target food would most likely be damaged as a result of sampling (Nicolaus and Nellis 1987). CTA has been used successfully to protect *treated* seeds, lawns, and even ripening fruit from wild bird damage (Conover 1989). For example, Conover (1985) treated blueberries with methiocarb or trimethacarb to protect them from damage by northern mockingbirds *Mimus polyglottos*, European starlings *Sterna vulgaris*, and northern orioles *Icterus galbula galbula*, until the repellent washed off. Such treatment is acceptable provided the repellent has reduced to safe levels by harvest time (Guarino et al. 1974). Untreated foods however have not yet been protected successfully in real-life applications.

Early attempts to protect untreated food items from wild animals by using CTA aimed to protect livestock from attack by coyotes. In 1974, Gustavson et al. (1974) fed captive coyotes with lamb baits treated with lithium chloride, and to some coyotes, he also administered an intraperitoneal injection of lithium chloride after consumption of the poisoned meal (lithium chloride has a strong salty taste which is difficult to conceal in baits (Reynolds 1999), so injection allows a larger dose to be administered without detection). Coyotes subsequently avoided attacking or eating a live lamb in captivity, but retained their appetite for rabbits. Researchers moved rapidly to large-scale field trials that aimed to deter wild coyotes from attacking free-ranging sheep, but some trials were poorly designed and results inconclusive. In the early 1980s, Lowell Nicolaus and colleagues established two key concepts. First, Nicolaus et al. (1982) demonstrated explicitly that it was possible to deter the killing of live prey under field conditions: raccoons *Procyon lotor* feeding on chicken carcasses treated with lithium chloride, subsequently avoided live chickens. Second, Nicolaus et al. (1983) showed that conditioned aversion towards model baits could also protect mimics: wild American crows *Corvus brachyrynchos* that ate green-painted eggs containing an emetic (UC 27867) subsequently avoided green eggs, whether or not they were treated with the emetic, but continued eating unpainted, untreated eggs. Nicolaus and co-authors have since demonstrated avoidance of eggs through CTA by corvids (*Corvus* spp.), raccoons *Procyon lotor*, mongooses *Herpestes auropunctatus*, and guilds of mammalian predators. In 2000, Cowan et al. concluded that research on learned aversions should progress on two fronts: (1) fundamental studies on learned food aversion; and (2) field research on target animals' behaviour and field logistics. The research described in this chapter tackles both, using data on individual behavioural responses among captive and free-ranging wild animals. We also demonstrate the importance of understanding behaviour in devising wildlife management methods.

13.2 Conditioned taste aversion

13.2.1 CTA with captive wild corvids

Members of the crow family, from carrion crows to magpies (corvids), are blamed for taking the eggs and chicks of other bird species, e.g. terns, songbirds, or gamebirds. Carrion crows and magpies are of particular interest as pests in the UK. In a 1992 questionnaire survey of farmer attitudes in Britain, magpies were the species second (only to rats) most frequently cited as a pest (Atkinson et al. 1994), and in a 1981 questionnaire survey of farmers in lowland England (n = 795 farmers), crows were the species fourth most frequently cited as a pest, and as being controlled (Atkinson et al. 1994). Corvids are often shot, or trapped and killed, as pests.

CTA could potentially provide a more effective and acceptable alternative to culling, for controlling corvid predation on eggs with financial or conservation value. While there had been no research on CTA in either carrion crows or magpies, studies outside the UK had explored CTA for reducing predation by other corvids on the eggs of endangered birds or ground-nesting gamebirds. These corvids included American crows (Nicolaus et al. 1983; Nicolaus et al. 1989a; Dimmick and Nicolaus 1990), hooded crows *Corvus corone cornix* (Bogliani and Bellinato 1998), and ravens *Corvus corax* (Avery et al. 1995). These and other studies demonstrated that, while there is clear potential to manage corvids using CTA, the development and duration of CTAs are complex and influenced by many things, including social and environmental factors, individual variation, prey recognition cues, stimulus reinforcement, length of training period, and choice of secondary repellent (Nicolaus et al. 1983; Gustavson and Gustavson 1985; Dimmick and Nicolaus 1990; Semel and Nicolaus 1992; Avery and Decker 1994).

We conducted the first CTA studies on magpies and crows, with individually housed, wild-caught birds in captivity (Prescott et al. 1997; Cox et al. 2004). This allowed us to examine detailed individual responses to controlled doses of repellents and to establish and manipulate existing individual feeding preferences. We aimed to identify suitable secondary repellents and doses that were safe and effective at manipulating the birds' feeding behaviour. Identifying an optimal dose is important, because doses need to be safe but they also need to be sufficient. Avery and Decker (1994) showed, in captive fish crows *Corvus ossifragus*, that insufficient doses causing only slight illness resulted in crows sampling treated and untreated eggs repeatedly, while doses large enough to cause complete avoidance of treated eggs extended their avoidance to untreated eggs for longer. Birds are primarily visual animals and the question we tackled was whether colour-specific learned aversions might be used to protect a particular type of egg prey, while leaving other feeding behaviour intact (as demonstrated by Nicolaus et al.'s

(1983) work with American crows and green-painted eggs, described earlier). The most likely mechanism for this process was second-order conditioning, a two-stage process in which first, the taste of a food becomes aversive, and then, a non-taste cue, such as colour, becomes associated with the aversive taste (Gustavson et al. 1974).

13.2.1.1 Magpie trials

We began by attempting to create aversions among magpies towards quail *Coturnix coturnix* eggs. We chose quail eggs as models for the small songbird and gamebird eggs on which magpies prey. Using a realistic model food is important because features of a poor model may interfere with the development of CTA to the target food in need of protection. We tested two secondary repellents, lithium chloride and carbachol, both of which are known to produce learned food aversions (Prescott et al. 1997). While both chemicals are recognized standards for producing feeding aversions, both have a narrow margin of safety between aversive and lethal doses (Reynolds 1999). Carbachol (carbamyl choline chloride) is an emetic that is water soluble, odourless, and tasteless at doses capable of producing CTA in other species (L.K. Nicolaus, unpublished data, 1986), including mongooses (Nicolaus and Nellis 1987), American crows (Nicolaus et al. 1989a), and hooded crows (Bogliani and Bellinato 1998). Lithium chloride is also an emetic, with a strong salty taste (Reynolds 1999). To avoid detection of lithium chloride in CTA trials, it is sometimes injected (intraperitoneally) into experimental animals after they have eaten an example of the target food.

In our trials, we first conducted a pre-conditioning phase to determine each bird's existing egg colour preference. On each day of pre-conditioning we offered each magpie a choice of three yellow and three green eggs (dyed using domestic food colouring). Pre-conditioning continued for each bird for several days until it had demonstrated a clear colour preference by attacking and eating the eggs. Birds were divided into two groups, half allocated to the carbachol treatment and half to the lithium chloride treatment. During the conditioning phase we presented each bird every day with one egg of its preferred colour, treated with either lithium chloride or carbachol at one of a range of doses (lithium chloride: 4, 8, or 13.5 g kg^{-1} body weight; carbachol: 16, 32, or 50 mg per egg; doses were based on previous research with other species). We refer to each bird's preferred egg colour as 'toxic' and its less preferred colour as 'non-toxic' in all phases regardless of whether an egg was treated or not. During the post-conditioning phase, we offered magpies a choice of untreated eggs of both 'toxic' and 'non-toxic' colours to see whether they had switched their egg colour preferences.

In the lithium chloride trial none of the doses had a significant effect (that lasted throughout post-conditioning) on which colour of egg was taken first, on which colour of egg they preferred overall to attack or consume, or on the total number of eggs attacked in 24 hours. However, magpies given the largest lithium chloride dose (13.5 g/kg body weight) consumed less egg overall (consumption halved between pre-conditioning [37%] and the end of a 36-day post-conditioning period [18%]). Magpies treated with this dose also attacked fewer eggs of both colours during post-conditioning. This suggested that a weak CTA to the taste of raw egg had developed slowly in birds receiving the greatest lithium chloride dose, thus encouraging us that the technique could have potential although, clearly, lithium chloride itself was not a promising aversive agent at the doses tested.

The effects of carbachol conditioning were much more convincing. All doses caused magpies to switch the colour of egg that they preferred to attack first during post-conditioning and, in the short-term at least, induced colour-specific aversion to eggs. Magpies receiving the largest carbachol dose (50 mg per egg) exhibited the longest-lasting effect, switching egg colour preference for 36 days of post-conditioning. All magpies conditioned with carbachol consumed less egg in total during post-conditioning; those receiving the greatest dose developed a significant aversion after 12 days of post-conditioning that lasted until the 48th day of post-conditioning (consumption declined from 35% in pre-conditioning to 16%). This dose had no conspicuous ill-effects on magpies and equated to a dose of 244 mg of carbachol/kg body weight (mean body weight of magpies was 216 g). We concluded that while magpies receiving all carbachol doses developed CTA to the taste of egg per se, the largest dose of 50 mg produced a stronger aversion to the 'toxic' or previously preferred colour, most likely through second-order conditioning of the colour cue. It was clear that, under our study conditions, and at the doses tested, carbachol was a better secondary repellent than lithium chloride (Prescott et al. 1997). While the aim of these trials was to explore the potential of the technique in principle, we obviously had in mind whether these particular agents had a future in practice, e.g. in conferring protection on untreated eggs of a particular species or to eggs in general. We will revisit this practical question further below.

13.2.1.2 Carrion crow trials

Our carbachol conditioning trials with magpies were sufficiently encouraging that we sought to develop them by devising CTA trials with carrion crows (Cox et al. 2004). As before, our main aim was to establish a (carbachol) dose that was safe for the birds and effective at protecting untreated eggs of a preferred colour. When administering drugs to wild animals it is normal practice to extrapolate doses from other species, using allometric scaling, which takes account of differences in body size (Wolfensohn and Lloyd 1998). We began work with crows using a dose of 190.5 mg/kg body weight (dose 1) which (using allometric scaling), was equivalent to the highest dose used with magpies (which equated to 244 mg/kg body weight). We prepared individually dosed eggs for each bird based on its own body weight, as measured no more than two days before conditioning started.

Trials with crows differed from those with magpies in only two respects. First, we used small chicken eggs as models for natural crow prey such as lapwing *Vanellus vanellus*, curlew *Numenius arquata*, and other wading bird eggs. Second, we limited birds' access to trial eggs to four hours each day, to encourage them to feed quickly while still allowing time for all birds to feed (other food was provided *ad libitum* at all other times). We had seen that the captive magpies sampled untreated food during post-conditioning as a result of being kept in close proximity with it for 24 h a day (Prescott et al. 1997). Under natural conditions, wild predators are unlikely to be faced with potential prey for very long, being more likely either to eat it or leave the area.

We recorded an attack score for each egg, depending on whether it was 'untouched', 'moved', 'attacked', and/or 'consumed', after 1, 2, and 4 h. An egg consumption score was also recorded for each egg after 4 h. We recorded on video the responses of six crows to gather corroborative information on their behavioural responses. Having assessed the results of dose 1 on six birds, we tested a 50% larger carbachol dose (dose 2—285.75 mg/kg body weight) on five other birds. Because doses 1 and 2 had little aversive effect we doubled dose 1 (dose 3—381 mg/kg body weight) and tested this on three more birds.

When we combined the results for all carbachol doses, significantly fewer 'toxic'-coloured (but untreated) eggs were attacked in the first hour and first two hours after egg presentation, but this effect was no longer detectable after four hours. In other words, crows delayed their attacks on eggs of their previously preferred (or 'toxic') colour. There was no change in attacks on 'non-toxic' eggs in the first two hours, but after four hours, there was a significant increase in the number of 'non-toxic' eggs attacked and consumed. Results for the greatest dose of carbachol (dose 3) reflected those for the three doses combined, in that crows delayed their attacks on untreated eggs of their previously preferred colour (Fig. 13.1).

We used CTA successfully to train captive carrion crows to delay their attack on a previously favoured egg colour, even when this was subsequently untreated (Cox et al. 2004). We also established that captive crows did not generalize their aversion to eggs of a different colour (i.e. blanket protection for eggs of both colours did not occur). This is a potentially important consideration when designing realistic wildlife management strategies, because it would not be desirable to try to deter corvids from eating eggs of all types.

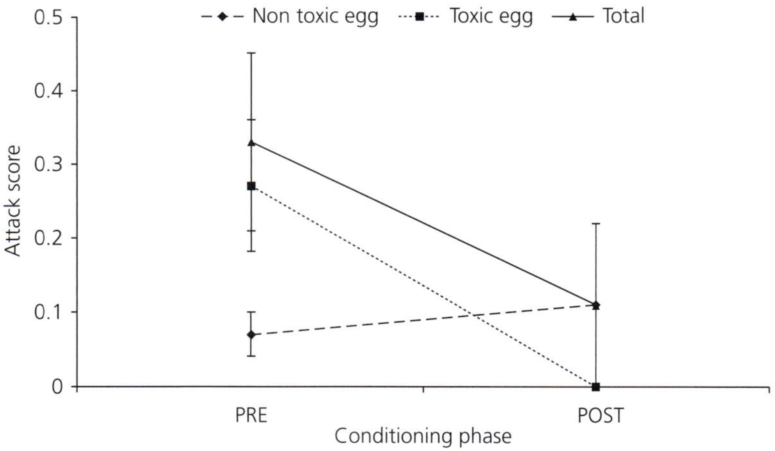

Figure 13.1 Mean (±S.E.) egg attack scores after 1 h of presentation during pre- and post-conditioning (birds exposed to dose 3: 381 mg kg^{-1} body weight of carbachol) ($n = 3$ birds). From Cox et al. (2004), reproduced with permission from Elsevier.

Our research suggests that CTA could potentially provide a powerful tool for managing egg predation by corvids. However, the over-riding problem seems to be in establishing a safe and effective emetic dose, with difficulties arising both in extrapolating effective doses between species and between individuals of the same species. It may be that wild bird physiologies and behaviours vary so widely that it is not possible to identify a suitable working oral dose that can be generalized either among species or conspecifics. However, taking this forward with crows would require two main things. First, rather than using a carbachol dose based on birds' individual bodyweights (as we have done here), it would be necessary to establish, in captivity, a dose which is safe and effective for the likely range of crow bodyweights encountered in the wild (perhaps at specific times of the year). Second, it would be necessary to conduct field trials on a larger sample of birds.

13.2.2 CTA with free-ranging wild badgers

Clearly, vertebrate pests are a major facet of wildlife conservation on farmland, and our pilot studies of captive magpies and crows encouraged us to explore further the potential of CTA as a non-lethal solution to other wildlife conflicts. Because our corvid CTA studies raised questions about the effects of captivity on the motivation for study animals to feed (and so their likelihood eventually of attacking untreated 'toxic' foods after conditioning), we decided that the next step should be to study CTA in free-ranging wild animals under field conditions.

We chose badgers *Meles meles* as models for this research because they are opportunistic omnivores, widely reported as crop pests, and distributed across much of the northern hemisphere (Neal and Cheeseman 1996; Moore et al. 1999). Our stable, high-density, long-term study population of badgers at Wytham Woods (Macdonald and Newman 2002) was ideal for this work (see Chapter 4, this volume). The complex mechanisms in animal behaviour and psychology that are involved in the development of CTA are so intricate that investigating them requires a carefully designed experiment, such as those usually conducted under strictly controlled laboratory conditions. However, our purpose was to unearth behaviour and to test interventions under realistic circumstances, and our challenge was to create sophisticated experiments under naturalistic field conditions. So we devised a series of iterative 'lab-style' field experiments and observed the behavioural responses of individually marked wild badgers (Baker et al. 2005a,b, 2007a, 2008) using CCTV surveillance video (Stewart et al. 1997). This work began in 1996 and was the first known CTA research on badgers. Further research on badger CTA has been conducted since (Cagnacci et al. 2005).

13.2.2.1 Testing three repellents

First, we used a 'cafeteria-style' experiment to ask whether any of three food-based chemical repellents had potential for manipulating badger feeding behaviour (Baker et al. 2005b). The repellents we tested were cinnamamide, capsaicin, and ziram, all known to produce both innate and learned aversive responses. Cinnamamide, an irritant and broad-spectrum repellent for herbivores including birds, rodents, and invertebrates (Watkins et al. 1998), is a synthetic derivative of plant defence compounds (Watkins 1996). Capsaicin, an irritant considered universally repellent to mammals (Mason et al. 1991), is the pungent principal component of capsicum peppers. 'Hot sauces' containing capsaicin are used in the USA as repellents, for example, to deter browsing by deer (Andelt et al. 1994). Ziram (zinc dimethyl dithio-carbamate) is a fungicide registered in the UK as a repellent against birds, deer, hares *Lepus europaeus*, and rabbits *Oryctolagus cuniculus*; it is also an irritant. We used a form of ziram, called AAprotect™, that is commercially available as a fungicide (Dalgety Agriculture Ltd, Reading, UK). AAprotect™ incorporates a 'sticking agent' which has an odour that is detectable to humans, but this odour wanes as the formulate dries.

We applied the three repellents to biscuit baits, which were presented to free-ranging badgers, on six paving slab feeders, at a feeding site established on rough pastureland. Each paving slab was divided into four quadrants ('bait patches') by a low wooden wall, supplying four bait portions (one in each quadrant) on each treatment night and offering badgers a choice of baits treated with each of the three repellents, and untreated baits. Treatment nights were alternated with control nights (on which all baits were untreated) to check for seasonal effects, such as badgers' interest in baits waning over time. We analysed data on the fate of baits, as well as on the detailed behavioural responses of five marked individual badgers.

All untreated baits were eaten on all control nights. On all treatment nights, all untreated baits and all baits treated with capsaicin or cinnamamide were also eaten. However, while all ziram treated baits were eaten on treatment nights 1 and 2, between treatment nights 3 and 9 consumption of ziram-treated baits declined to zero. This coincided with a sharp rise in

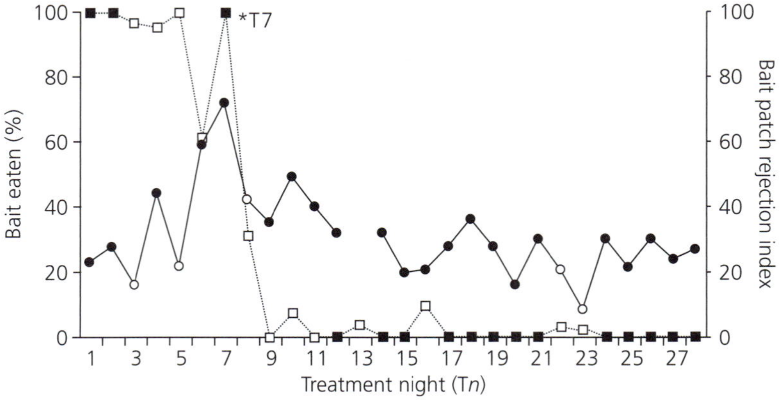

Figure 13.2 Ziram bait consumption, and total bait patch rejection associated with ziram, on each treatment night, for all badgers. Percentage of bait eaten: black squares indicate values based on video evidence; white squares indicate estimates calculated from retrieved pellets. On T7*, all baits were eaten by unmarked badger(s). Black circles indicate ziram bait patch rejection; white circles indicate the minimum value for incomplete nights. Bait patch rejection data were included to illustrate the association between the decline in ziram baits eaten and the peak in ziram bait patch rejection. From Baker et al. (2005b), reproduced with permission from John Wiley & Sons.

bait patch rejection (sniffing or tasting a particular bait patch which did not lead to feeding) (Fig. 13.2). This 'learning curve' peaked at treatment night 7 and we concluded that it was at this point that badgers developed CTA towards ziram treated baits. Consumption of ziram treated bait was practically zero over the last 20 treatment nights (spread over 40 trial nights) and individual badgers avoided ziram baits, without sampling, for the last 12–22 treatment nights (spread over 24–44 trial nights). The changes seen in badger behaviour suggested that they used odour cues to avoid ziram baits at a distance (anything up to a few metres away)—this most likely occurred through second-order conditioning of an odour cue. Because badgers were not deterred initially from eating ziram (AAprotect™) treated baits at the beginning of the treatment phase, we knew that any detectable odour from the formulation was not sufficiently deterrent in its own right to prevent feeding though innate aversion. Odour is an appropriate cue for mammals because they are primarily olfactory animals. The important next question arising from this trial was whether badgers would have continued to avoid baits that were not treated but which produced the same odour cue.

13.2.2.2 Ziram and clove odour

Our next trial was designed to test whether untreated foods could be protected by aversion to an odour (learnt through CTA), without being damaged through sampling (tasting) (Baker et al. 2007a). To answer this question we used the pure odourless form of ziram and a completely novel odour cue (clove oil). We used a three-phase trial (comprising pre-conditioning, conditioning, and post-conditioning phases) and applied three treatments at different sites. The treatments were: (a) ziram (food-based repellent) and clove oil (novel odour) combined; (b) ziram; (c) clove oil. Again, we applied treatments to biscuit bait and these were presented on paving slab feeders. Baits and feeders were treated as shown in Table 13.1.

Table 13.1 Bait and feeder treatments applied during the ziram–clove trials. From Baker et al. (2007a), reproduced with permission from Elsevier.

Treatment	Pre-conditioning phase		Conditioning phase		Post-conditioning phase	
	Baits	Feeders	Baits	Feeders	Baits	Feeders
(a) Ziram–clove	Untreated	Untreated	Ziram and clove	Clove	Untreated	Clove
(b) Ziram	Untreated	Untreated	Ziram	Untreated	Untreated	Untreated
(c) Clove	Untreated	Untreated	Clove	Clove	Untreated	Clove

We tested two predictions: (1) ziram would produce learned aversion towards untreated foods; and (2) training with the ziram–clove combination would produce greater aversion towards untreated foods in the presence of a clove odour cue. Clove oil was selected as a potential odour cue because it is non-toxic, inexpensive, pungent, persistent, and effusive, and therefore ought to be detectable at a distance; clove oil was novel to badgers and pilot trials indicated that it was neither aversive, nor attractive, to badgers. We found that ziram did not produce a significant learned aversion to untreated foods, but conditioning badgers with the ziram–clove combination produced learned aversion towards untreated foods in the presence of a clove odour cue. The clove control, on its own, did not affect badger behaviour (Fig. 13.3). This finding might have practical application for protecting untreated foods, e.g. cereal crops, from badger damage using an odour cue nearby.

While laboratory rats (*Rattus* sp.) can use odour cues to avoid potentially toxic food (Braun and Ryugo 1974; Nachman et al. 1977), nobody knew whether this phenomenon occurred among wild mammals. CTA studies with captive and wild mammals, other than rats, have demonstrated avoidance of untreated foods at a distance, and attributed this to conditioning of non-taste cues (e.g. auditory, visual, and olfactory; Gustavson et al. 1974), but none has investigated which cues were at work. Nicolaus and Nellis (1987) attempted, but failed, to train mongooses to associate one of two odours with an aversion to eggs. However, captive animals continued to attack eggs regardless of the odour cue provided while, in contrast, free-ranging animals generalized their aversion to eggs treated with either odour. The authors of that work attributed their findings to the captive conditions of the study, and to avoidance at a great distance, respectively. As far as we know, ours is the only research to date to successfully provide a natural mammal population with an odour cue for aversive foods. We concluded that conditioning badgers with a combined formulation of a secondary repellent and an odour cue (rather than the repellent on its own) produced enhanced aversion towards untreated foods in the presence of the odour cue, and that aversion was associated with the odour cue (Baker et al. 2007a).

13.2.2.3 Protecting maize cobs with ziram and clove odour

Our next step was to simulate a more realistic conflict situation, by testing the combined ziram and clove odour formulation for the protection of a model crop food eaten by wild badgers—we chose maize cobs. Cobs were likely to differ from the biscuit bait used in previous trials in terms of their consistency, their familiarity to badgers, and badgers' motivation to feed. Because of this we needed to determine whether coating cobs with ziram would deter badgers from eating them, and if so, what ziram concentration would be sufficient to achieve this. To find out, we conducted a multi-choice, dose–response trial to test the acceptability to free-ranging wild badgers of maize cobs treated topically with a range of ziram concentrations (Baker et al. 2005a). We successfully identified a minimum dose (1.3–1.5% w/w) that created either learned or innate aversion to treated baits among all participating badgers. We believe this to be the first investigation of the dose-dependent effects of a secondary repellent on free-ranging, wild mammals.

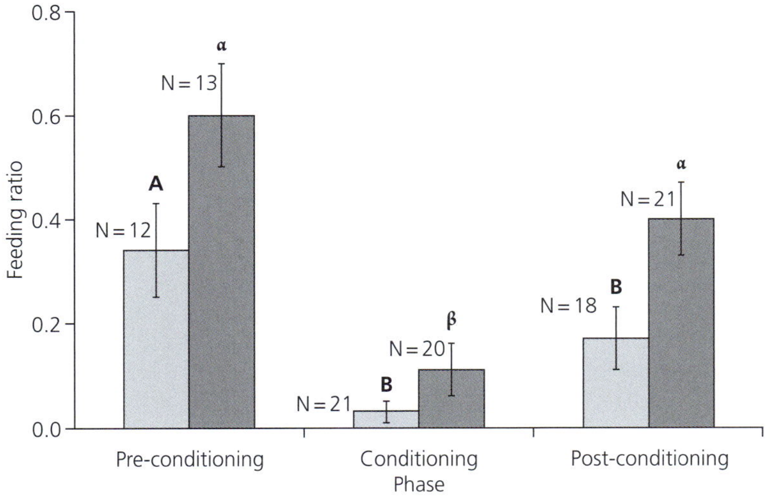

Figure 13.3 Mean individual badger feeding ratio (untransformed data) for the ziram–clove and ziram experiments (N = badgers). Feeding ratio = total 'feed' encounters/(total 'feed' + total 'reject' encounters). Error bars represent standard errors of the mean. Ziram–clove = pale grey; ziram = dark grey. Phases within an experiment that are denoted by the same symbol did not differ significantly (Tukey test). From Baker et al. (2007a), reproduced with permission from Elsevier.

Our final experiment in this series tested CTA with a clove odour cue as a paradigm for protecting untreated maize cobs (Baker et al. 2008). We used a three-phase conditioning trial (pre-conditioning, conditioning, and post-conditioning phases) using two treatments: (a) ziram (1.5% w/w) and clove oil combined; and (b) clove oil. Treatment of cobs and feeders in the three phases was as for treatments 'a' and 'c' in the ziram–clove trial described above (see Table 13.1). Following conditioning with the combined treatment, badgers avoided untreated maize cobs in the presence of a clove odour cue, whereas badgers were not conditioned by the clove oil control on its own.

We thus demonstrated for the first time that a natural mammal population can be conditioned in the wild to avoid potentially aversive foods using an odour cue. This type of field-based research on wild animals is a crucial step before proceeding to full-scale field trials, e.g. for protecting crops on a wider scale. By using CCTV video and detailed behavioural observations of individually marked wild animals we were able to examine detailed individual behaviour and the underlying mechanisms at work, and so tailor each subsequent research stage accordingly. This iterative process enabled us to move quickly, in four stages, from a relatively broad initial remit, of testing the potential of CTA in the wild, to a potential application in the form of second-order conditioning of odour cues for the protection of untreated foods. Of course there would be several hurdles to be crossed before this could become a practical reality.

13.3 Generalized aversion with captive foxes

A large number of chemicals are known to have aversive properties, but few meet the necessary requirements for producing CTA. These include causing short-term illness in target animals that consume them, being undetectable in referent food at the concentration needed to cause CTA, having a wide safety margin between aversive and lethal doses in members of the target population and potential non-target consumers, having a short time delay between ingestion and onset of nausea, and being stable in the environment (Reynolds 1999). For example, both lithium chloride and carbachol can be detected in some baits and both have a narrow margin of safety, making them less suitable as CTA agents. Oral oestrogen is a secondary repellent that is undetectable in eggs (Nicolaus et al. 1989b,c), but aversive doses may have abortifacient and teratogenic effects (Reynolds 1999). Finding a suitable emetic for real-life wildlife management applications might therefore present a problem. In pursuit of a potentially more expedient goal, we explored the use of a bitter taste in place of an emetic. This involved a learned food aversion process with features similar to both innate repellency and classic CTA, but which is different from both; we term this process 'generalized aversion' (Macdonald and Baker 2004; see Box 13.1).

In generalized aversion, an innately repellent substance (a primary repellent, in this experiment a bitter-tasting substance called Bitrex™) is used to create an aversion to untreated target food. There are similarities in the processes involved in generalized aversion and CTA, and generalized aversion may be viewed as a sub-set of CTA (Macdonald and Baker 2004). In both approaches, learning takes place through classical conditioning (where an association is formed between two stimuli—in this case, the bitter taste and cues related to consumption of the target food, and in the case of classic CTA, illness and the taste of the target food). The main distinguishing feature between generalized aversion and other CTAs is the form of the unconditioned stimulus. In classic CTA the unconditioned stimulus is the consumer's physiological response of nausea/vomiting, while in the context of generalized aversion it is the animal's reaction to an unpleasant taste. Generalized aversion has potential advantages over both classic taste-based CTA and innate aversion; on one hand there is no doubt regarding the source of the

Box 13.1 Classical conditioning in CTA and generalized aversions

In classical conditioning, such as occurs in conditioned taste aversion (CTA) and in generalized aversion, the pairing of an unconditioned stimulus (US) and a conditioned stimulus (CS) produces a conditioned response (CR) to the CS. For example, in CTA, the US might be an animal's physiological response (nausea, vomiting, malaise) to eating food treated with an emetic. The US becomes paired with the CS (the taste of the food), and the CS subsequently produces a CR, in this case the avoidance of the particular food taste. In a generalized aversion, created using Bitrex™, the US will be the animal's response to the bitter taste of Bitrex™-treated food. Aspects of the food, such as its smell or appearance (CS), will become paired with the US and the smell or appearance of the same type of food (the CS), whether or not it is treated with Bitrex™, will produce the avoidance of that food (CR).

unconditioned stimulus (here, the unpleasant taste), while on the other hand it may confer protection on untreated foods after a conditioning period. The disadvantage of generalized aversion (and of innate aversion, but not of classic CTA), is that the animal only needs to sample untreated target food to discover that it is not treated—the hope is that they will be discouraged from sampling because the treatment, if present, is so unpleasant.

Innate aversion to compounds which taste bitter to humans is known in many taxa, and thought to result from natural selection for avoidance of toxic alkaloids and glycosides which are bitter to humans (Garcia and Hankins 1975). Bitrex™ (Macfarlan Smith Ltd, Edinburgh) is the trade name for denatonium benzoate, an intensely bitter-tasting N-substituted lignocaine derivative, commonly used in everyday products such as shampoos and cleaning materials, to deter humans from consuming otherwise harmful substances, and in anti-nail biting products. Bitrex™ is repulsive to various mammals at particular concentrations (e.g. woodmice; Kleinkauf et al. 1999), and preliminary trials by the manufacturers had suggested that it may cause generalized aversion in dogs (Macfarlan Smith Ltd., unpublished data).

We used red foxes *Vulpes vulpes* as models for this work because they are opportunistic omnivores (Macdonald 1987), common across much of the northern hemisphere, and notoriously problematic in Australia where they have been introduced. Foxes are widely reported as pests in Britain (Burns et al. 2000), frequently conflicting with game-bird management through consumption of eggs, chicks, and adult birds (Reynolds et al. 1988), and they also kill threatened native species. For example, on nature reserves in the UK, red foxes predate fragmented populations of red grouse *Lagopus lagopus* (Macdonald et al. 1999) and terns *Sterna* spp. (see Baker and Macdonald 1999). Finding ways to discourage foxes from attacking and consuming eggs would be an ultimate goal; here, we describe a scoping trial on captive foxes in which we examined the potential of generalized aversion for conflict management, as the first step towards such a goal (Macdonald and Baker 2004).

Our aim was to test whether foxes could form a conscious association between an unpleasant taste, Bitrex™, and food cues other than taste (e.g. odour, appearance), such that they developed an aversion that was generalized to similar, but untreated, food. In other words, could they learn to avoid untreated food following consumption of the same type of food treated with Bitrex™? We ran trials on a litter of three captive, 6-month-old foxes, fed (in addition to trial food) on rabbits and other naturally taken food. The foxes were individually marked and we observed their behavioural responses using CCTV surveillance video. Because they were housed together their responses may not have been independent. Milk was selected as an appropriate test medium for four reasons. First, it was unlike the other foods that the foxes were receiving at the time (their basic diet consisted of rabbits, fowl, and occasionally canned dogfood), so they should be less likely to confuse the effects of conditioning between foods. Second, the Bitrex™ could be thoroughly dissolved in the liquid, facilitating even distribution throughout the bait. Third, they appeared to like untreated milk, and initially drank it very quickly, making it more straightforward to detect a change in their behaviour. Finally, it was impossible for them to carry it off or to play with it, so it was easy to determine clearly whether it had been consumed or left in the bowl.

We asked two questions: first, "what concentration of Bitrex™ in milk would deter foxes from drinking treated milk?" and second, "would foxes drinking Bitrex™-treated milk at this concentration then avoid untreated milk?". We addressed both questions using a continuous trial lasting ten weeks and consisting of pre-conditioning, dosage trial, conditioning, partial reinforcement, and post-conditioning phases (see Box 13.2). Each evening, we presented the foxes with

Box 13.2 Conditioning: continuous and partial reinforcement

In learned food aversions (including conditioned taste aversions) conditioning normally takes place using 'continuous reinforcement', where the aversive substance is present in all examples of the referent food presented and the desired behaviour is reinforced each time it occurs. In partial reinforcement, the aversive substance is present in only some examples of the referent food and the response is reinforced on only some occasions. Learned behaviours are acquired more quickly under continuous reinforcement but are less robust to extinction facilitated through sampling. Under partial reinforcement learned behaviours are acquired more slowly but are more resistant to extinction. Continuous reinforcement is sometimes used first to produce a strong association and then followed by partial reinforcement to deter animals from sampling foods.

two bowls of trial food, containing 0.5 pint (568 ml) of milk in one, and an equivalent mass of untreated canned dog food in the other, except during partial reinforcement, when both bowls contained milk. The milk was treated as follows in the various phases: pre-conditioning (milk untreated); dosage trial (milk treated with Bitrex™ for several days at each of 400 ppm, 1000 ppm, 2000 ppm, and 4000 ppm); conditioning (milk treated with Bitrex™ at 4000 ppm); partial reinforcement (untreated milk and Bitrex™-treated milk at 4000 ppm in the two bowls); and post-conditioning (milk untreated).

During the dosage trial, foxes took significantly longer to finish their Bitrex™-treated milk than they had taken previously to drink the same volume of untreated milk during pre-conditioning. They finished untreated milk in just over 2 min and milk treated with 400 ppm, 1000 ppm, or 2000 ppm of Bitrex™ in more than 7 h. When milk was treated at 4000 ppm the foxes failed to finish it within 24 h, on six of eight nights at that dose. Foxes took longer to approach, sample, and drink treated milk than untreated milk. The time lag between approaching and sampling, and between sampling and drinking milk, was significantly greater for Bitrex™-treated milk at most doses than for untreated milk during pre-conditioning (Fig. 13.4). Foxes spent a smaller proportion of their time drinking Bitrex™-treated than untreated milk, and performed fewer drinking bouts. Individual fox data for these two measures followed broadly similar patterns.

The dosage trial indicated that 4000 ppm Bitrex™ was a suitable dose for the conditioning trial. The next stage was to ask whether training with milk containing Bitrex™ at this concentration would persuade foxes subsequently to avoid untreated milk. To assess the effectiveness of conditioning foxes with Bitrex™-treated milk, we compared their responses towards untreated milk before (in pre-conditioning, n = 8 nights) and after conditioning (during partial reinforcement, n = 2 nights; and post-conditioning, n = 20 nights). During both partial reinforcement and post-conditioning, foxes conducted more pre-sampling activities (i.e. sniffing or burying the bowl, if performed before the first sampling event—drink, eat, take, or taste)—a sign that they were less eager to drink milk—and took longer to finish their untreated milk, than they had during pre-conditioning (finishing the untreated milk took just over 2 min during pre-conditioning, and a minimum of 10 h and 2.5 h, in partial reinforcement and post-conditioning, respectively). It took longer for the first fox to approach untreated milk during post-conditioning (15 s) than during pre-conditioning (1 s), and longer to sample and drink untreated milk during both partial reinforcement and post-conditioning (sample, ~2 h, and 45 s, respectively, compared to 1 s during pre-conditioning; drink, ~2 h, and ~2 min, respectively, compared to 1 s during pre-conditioning). There was a greater time lag between foxes approaching and sampling untreated milk in both partial reinforcement (~2 h) and post-conditioning (26 s) than during pre-conditioning (no lag).

In contrast, while the time lag between sampling and drinking Bitrex™-treated milk during conditioning had ranged between 1–10 h, there was no time lag between sampling and drinking untreated milk in any of the pre-conditioning, partial reinforcement,

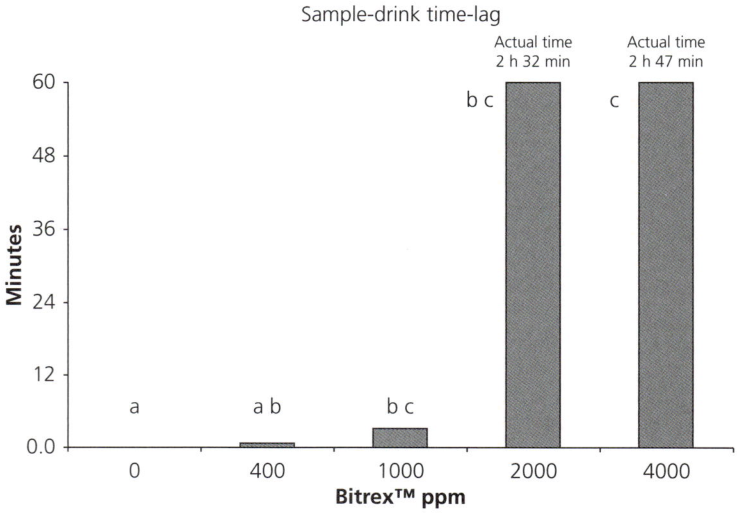

Figure 13.4 Median time lag between foxes sampling and drinking milk in the Bitrex™ dosage trial. Milk was untreated or treated with Bitrex™ as shown in ppm. Medians annotated with the same letter are not significantly different (at $P < 0.05$) for pairwise Kruskal–Wallis tests. df = 1 for all comparisons. Interquartile ranges were: 0 ppm, 0–1 s (n = 8); 400 ppm, 0–18 min (n = 3); 1000 ppm, 34 s–5 min (n = 2); 2000 ppm, 12 s–5 h (n = 2); 4000 ppm, 1–10 h (n = 5). Redrawn from Macdonald and Baker (2004). Copyright Universities Federation for Animal Welfare (UFAW).

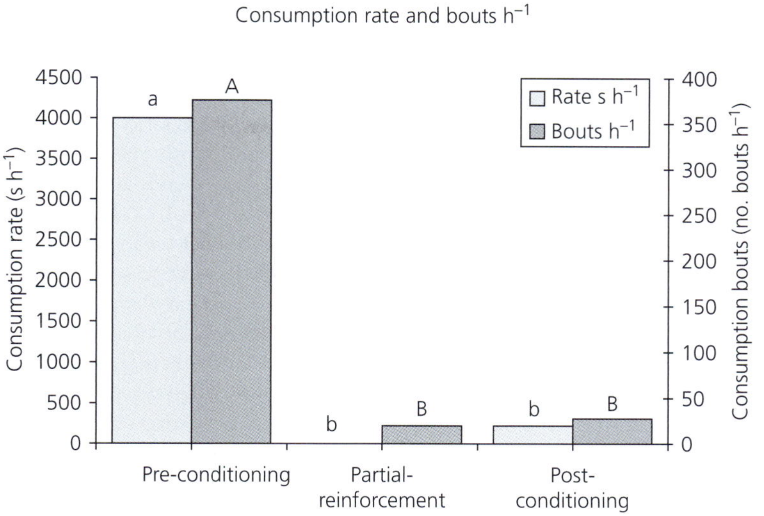

Figure 13.5 a : Median consumption rate (s/h); and b : median number of consumption bouts (bouts/h), performed by foxes (all foxes combined) to untreated milk in the pre-conditioning, and during partial reinforcement and post-conditioning. Medians with the same subscript letter are not significantly different (at $P < 0.05$) for pairwise Kruskal–Wallis tests. df = 1 for all comparisons. Interquartile ranges for consumption rate were: pre-conditioning, 58–75 min/h (n = 8); partial reinforcement, 0–5 min/h (n = 2); post-conditioning, 2–5 min/h (n = 13). Interquartile ranges for consumption bouts were: pre-conditioning, 318–423 bouts/h (n = 8); partial reinforcement, 0–33 bouts/h (n = 2); post-conditioning, 19–31 bouts/h (n = 13). Redrawn from Macdonald and Baker (2004). Copyright Universities Federation for Animal Welfare (UFAW).

and post-conditioning phases. This suggests that foxes could not detect whether milk was treated or untreated, without sampling, but that once they sampled, and found it to be untreated, they began drinking straight away. Foxes spent a smaller proportion of their time drinking untreated milk, and performed fewer drinking bouts, during partial reinforcement and post-conditioning than during pre-conditioning (Fig. 13.5). Individual fox data for these two measures followed broadly similar patterns.

From our experiments we determined a deterrent concentration of Bitrex^{TM} in milk and used this to create a generalized aversion in foxes to untreated milk. Wild and captive foxes may differ in their motivations to feed and responses to food. For example, wild foxes may experience varying and possibly extreme hunger levels, encouraging them to take greater risks, such as eating potentially unpalatable food. However, in some ways, the captive conditions of our study may have provided a more severe test than would have been the case for a fox in the wild, which would be free to move on to alternative food sources and with prey having opportunity to escape. Foxes in our study had access to milk for up to 24 hours, and untreated milk probably provided a great and constant temptation (a similar issue to that of the corvids and eggs described earlier). Keeping the animals together during the trials, as we did, may also have acted to enhance direct competition for milk, and created opportunities for social facilitation of food aversion, as well as reducing independence among their responses.

Generalized aversion is unlikely to last very long if conditioned animals sample the untreated foods under protection, and so may be suitable only for very short-term applications where extinction of aversions is less of an issue. For example, protection of pen-reared game birds (Hill and Robertson 1988; Leif 1994), and some rare ground-nesting birds, which are at risk only during the breeding season. If prey are mobile, and even if generalized aversion were sufficient only to delay rather than prevent approach or attack by a predator, this hesitation may be sufficient to allow prey to escape, and to prevent averted predators from discovering, by sampling, that prey are untreated.

13.4 Conclusions and applied recommendations

As we began, huge numbers of vertebrates are killed on farmland in an attempt to reduce pest problems, often routinely without proper consideration of the costs and benefits. McManus (2009) reveals how non-lethal management measures can reduce predation, e.g. of livestock by leopards, and make better financial sense than culling. While this example may seem poles away from the management of wildlife on UK farms, it is not, because the issues are the same. Even where killing is rational, it may involve suffering, which is an important cause for concern. It is therefore worth exploring the possibility of non-lethal options, as we have done, for a range of reasons (including economics, welfare, and ethics).

We showed that, in principle, learned food aversions, such as CTAs and generalized aversions, could provide a targeted, non-lethal approach for managing undesirable wildlife feeding on inanimate foods of either commercial importance or conservation concern, e.g. cereal crops, or the eggs of ground-nesting birds, respectively. We also demonstrated the importance of understanding behaviour in devising wildlife management methods (see discussion of this in Baker et al. 2007b), the benefits of video surveillance for capturing detailed behavioural observations from individually known animals (Stewart et al. 1997), both in captivity and in the wild, and that well-designed, laboratory-style experiments can be conducted under natural conditions on individually known free-ranging wild animals, e.g. Baker et al. (2005b).

On the face of it, learned food aversions could potentially have widespread application to wildlife management situations across a range of animal species worldwide. Carefully chosen non-taste cues, such as colours for birds, or odours for mammals, could help to improve the establishment and durability of aversions to untreated foods through second-order conditioning of CTA. Non-taste cues act to protect target foods from sampling and facilitate the creation of very specific aversions that are not generalized to other similar foods, e.g. aversion to one egg type only, an important consideration when designing realistic wildlife management strategies. Non-taste cues could potentially also be used to extend protection to different foods or to foods in different locations.

However, we are mindful of the many and varied practical difficulties of applying these ideas in practice, and these are summarized nicely by Reynolds (1999). The main issues relate to identifying a suitable repellent which meets both efficacy and safety requirements, in that it has sufficient margin between aversive and dangerous doses for the target species (and the likely range in individual sizes), is undetectable to the target animal in referent food at the required dose, is harmless to likely non-target consumers, meets human safety requirements, and does not cause environmental contamination (Reynolds 1999).

In spite of these serious drawbacks we think our findings are sufficiently promising that they merit further exploration as we nudge towards practicality with food aversions and with other possible non-lethal management methods. It may be that using learned food aversion to protect inanimate foods, such as crops or eggs, would be a more realistic goal than protecting live prey from predation, because live prey produce irresistible attack stimuli, whether or not the predator goes on to eat the prey. Alternatively it may be that learned food aversions can cause sufficient hesitation by a target predator to allow the prey to escape, whereas an inanimate food item remains available until the target animal has overcome any nervousness about approaching it. Finally, while there are difficulties in identifying CTA agents that are suitable for practical wildlife management, an alternative approach might be generalized aversion, which could be suitable where the window of vulnerability for target foods is very short.

Acknowledgements

Corvid research was funded by the RSPB. Research on badgers was conducted at Oxford University's Wildlife Conservation Research Unit in collaboration with the Central Science Laboratory, UK, and was funded first by the UK Government's Ministry for Agriculture, Fisheries and Food (MAFF), and by its successor, the Department for Environment, Food and Rural Affairs (Defra). Macfarlan Smith Ltd. provided Bitrex™ for the generalized aversion study. During the original food aversion research SB was supported by a studentship from the Royal Society for the Prevention of Cruelty to Animals, funding from the Ministry of Agriculture, Fisheries and Food, and subsequently the Department for Environment, Food and Rural Affairs. The People's Trust for Endangered Species provided additional support. SB was supported by a Fellowship from the Humane Society International and the Baker Trust during preparation of this chapter. We thank Ruth Feber and Eva Raebel for their helpful input to the chapter.

References

Andelt, W.F., Burnham, K.P., and Baker, D.L. (1994). Effectiveness of capsaicin and bitrex repellents for deterring browsing by captive mule deer. *Journal of Wildlife Management*, **58**, 330–334.

Atkinson, R.P.D., Macdonald, D.W., and Johnson, P.J. (1994). The status of the European mole Talpa europea L. as an agricultural pest and its management. *Mammal Review*, **24**, 73–90.

Avery, M.L. and Decker, D.G. (1994). Responses of captive fish crows to eggs treated with chemical repellents. *Journal of Wildlife Management*, **58**, 261–266.

Avery, M.L., Pavelka, M.A., Bergman, D.L., Decker, D.G., Knittle, C.E., and Linz, G.M. (1995). Aversive conditioning to reduce raven predation on California least tern eggs. *Colonial Waterbirds*, **18**, 131–245.

Baker, S.E., Ellwood, S.A., Slater, D., Watkins, R., and Macdonald, D.W. (2008). Learned food aversion plus odor cue protects crop from mammalian wildlife foraging. *Journal of Wildlife Management*, **72**, 785–791.

Baker, S.E., Ellwood, S.A., Watkins, R.W., and Macdonald, D.W. (2005a). A dose-response trial with ziram-treated maize and free-ranging European Badgers *Meles meles*. *Applied Animal Behaviour Science*, **93**, 309–321.

Baker, S.E., Ellwood, S.A., Watkins, R.W., and Macdonald, D.W. (2005b). Non-lethal control of wildlife: using chemical repellents as feeding deterrents for the European badger. *Journal of Applied Ecology*, **42**, 921–931.

Baker, S.E., Johnson, P.J., Slater, D., Watkins, R.W., and Macdonald, D.W. (2007a). Learned food aversion with and without an odour cue for protecting untreated baits from wild mammal foraging. In R. Swaisgood, ed. *Conservation Enrichment and Animal Behaviour; Applied Animal Behaviour Science, special edition*, pp. 410–428. Elsevier Science, Amsterdam, Netherlands .

Baker, S.E. and Macdonald, D.W. (1999). Non-lethal predator control: exploring the options. In D.P. Cowan and C.J. Feare, eds., *Advances in Vertebrate Pest Management*, pp. 251–266. Filander Verlag, Fürth.

Baker, S.E., Sharp, T., and Macdonald, D.W. (in prep). Assessing animal welfare impacts in wildlife management using rabbit, mole and crow interventions as case studies.

Baker, S.E., Singleton, G., and Smith, R.H. (2007b). The nature of the beast: using biological processes in vertebrate pest management. In D.W. Macdonald and K. Service, eds., *Key Topics in Conservation Biology*, pp. 173–185. Blackwell Publishing, Oxford.

Bogliani, G. and Bellinato, F. (1998). Conditioned aversion as a tool to protect eggs from avian predators in heron colonies. *Colonial Waterbirds*, **21**, 69–72.

Braun, J.J. and Ryugo, R.A. (1974). Taste facilitation of a learned odor aversion. *Bulletin of the Psychonomic Society*, **4**, 253.

Burns, T., Edwards, V., Marsh, J., Soulsby, L., and Winter, M. (2000). *Report of the Committee of Inquiry into Hunting with Dogs in England and Wales. CM 4763*. The Stationery Office, Norwich.

Cagnacci, F., Massei, G., Cowan, D.P., and Delahay, R.J. (2005). Can learned aversion be used to control bait uptake by Eurasian badgers? *Applied Animal Behaviour Science*, **92**, 159–168.

Conover, M.R. (1985). Using conditioned food aversions to protect blueberries from birds: comparison of two carbamate repellents. *Applied Animal Behaviour Science*, **13**, 383–386.

Conover, M.R. (1989). Potential compounds for establishing conditioned food aversions in raccoons. *Wildlife Society Bulletin*, **17**, 430–435.

Cowan, D.P., Reynolds, J.C., and Gill, E.L. (2000). Reducing predation through conditioned aversion. In L.M. Gosling and W.J. Sutherland, eds., *Behaviour and Conservation*, pp. 281–299. Cambridge University Press, Cambridge.

Cox, R., Baker, S., Macdonald, D.W., and Berdoy, M. (2004). Protecting egg prey from Carrion Crows: the potential of aversive conditioning. *Applied Animal Behaviour Science*, **87**, 325–342.

Dimmick, C.R. and Nicolaus, L.K. (1990). Efficiency of conditioned aversion in reducing depredations by crows. *Journal of Applied Ecology*, **27**, 200–209.

Garcia, J., Clarke, J.C., and Hankins, W.G. (1973). Natural responses to scheduled rewards. In P.P.G. Bateson and P.H. Klopfer, eds., *Perspectives in Ethology*, pp. 1–41. Plenum Press, New York.

Garcia, J. and Hankins, W.G. (1975). The evolution of bitter and the acquisition of toxophobia. In D.A. Denton and J.P. Coghlan, eds., *Olfaction and Taste V*, pp. 39–45. Academic Press, New York, NY.

Garcia, J. and Hankins, W.G. (1977). On the origin of taste aversion paradigms. In L.M. Barlor, M.R. Best, and M. Domjan, eds., *Learning Mechanisms in Food Selection*, pp. 3–22. Baylor University Press, Waco, TX.

Garcia, J., Kimmeldorf, D.J., and Koelling, R.A. (1955). Conditioned aversion to saccharin resulting from exposure to gamma radiation. *Science*, **112**, 157–158.

Garcia, J. and Rusiniak, K.W. (1980). What the nose learns from the mouth. In D. Muller-Schwarze and R. Silverstein, eds., *Chemical Signals*, pp. 141–156. Plenum Press, New York.

Gill, E.L., Cotterill, J.V., Cowan, D.P., et al. (1999). All in the worst possible taste: chemical repellents in vertebrate pest management. In P.D. Cowan and C.J. Feare, eds., *Advances in Vertebrate Pest Management*, pp. 283–296. Filander Verlag, Furth.

Guarino, J.L., Shake, W.F., and Schafer, E.W. (1974). Reducing bird damage to ripening cherries with Methiocarb. *Journal of Wildlife Management*, **38**, 338–342.

Gustavson, C.R., Garcia, J., Hankins, W.G., and Rusiniak, K.W. (1974). Coyote predation control by aversive conditioning. *Science*, **184**, 581–583.

Gustavson, C.R. and Gustavson, J.C. (1985). Predation control using conditioned food aversion methodology—theory, practice, and implications. *Annals of the New York Academy of Sciences*, **443**, 348–356.

GWCT (2012). The Game and Wildlife Conservation Trust, *National Gamebag Census*. Available from: <http://www.gwct.org.uk/>, accessed November 2013.

Hazzah, L., Dolrenry, S., Naughton, L., et al. (2014). Efficacy of two lion conservation programs in Maasailand, Kenya. *Conservation Biology*, **28**, 851–860.

Hill, D.A. and Robertson, P.A. (1988). Breeding success of wild and hand-reared ring-necked pheasants. *Journal of Wildlife Management*, **52**, 446–450.

Kleinkauf, A., Macdonald, D.W., and Tattersall, F. (1999). A bitter attempt to prevent non-target poisoning of small mammals. *Mammal Review*, **29**, 201–204.

Leif, A.P. (1994). Survival and reproduction of wild and pen-reared ring-necked pheasant hens. *Journal of Wildlife Management*, **58**, 501–506.

Macdonald, D.W. (1987). *Running with the Fox*, 224 pp. Unwin Hyman, London.

Macdonald, D.W. and Baker, S.E. (2004). Non-lethal control of fox predation: the potential of generalised aversion. *Animal Welfare*, **13**, 77–85.

Macdonald, D.W., Boitani, L., Dinerstein, E., Fritz, H., and Wrangham, R. (2013). Conserving large mammals: are they a special case? In D.W. Macdonald and K.J. Willis, eds., *Key Topics in Conservation Biology 2*, pp. 277–312. Wiley-Blackwell, John Wiley & Sons Ltd., Oxford.

Macdonald, D.W., Mace, G.M., and Barretto, G.R. (1999). The effects of predators on fragmented prey populations: a case study for the conservation of endangered prey. *Journal of Zoology*, **247**, 487–506.

Macdonald, D.W. and Newman, C. (2002). Population dynamics of badgers (Meles meles) in Oxfordshire, UK: numbers, density and cohort life histories, and a possible role of climate change in population growth. *Journal of Zoology*, **256**, 121–138.

Macdonald, D.W., Riordan, P., and Mathews, F. (2006). Biological hurdles to the control of TB in cattle: a test of two hypotheses concerning wildlife to explain the failure of control. *Biological Conservation*, **131**, 268–286.

Marker, L.L., Dickman, A.J., and Macdonald, D.W. (2005). Perceived effectiveness of livestock guarding dogs placed on Namibian farms. *Rangeland Ecology and Management*, **58** (**4**), 329–336.

Mason, J.R., Bean, N.J., Shah, P.S., and Clark, L. (1991). Taxon specific differences in responsiveness to capsaicin and several analogues: correlates between chemical structure and behavioural aversiveness. *Journal of Chemical Ecology*, **17**, 2539–2552.

McManus, J.S. (2009). *The spatial ecology and activity patterns of leopard* (*Panthera leopardus*) *in the Baiaanskloof and Greater Addo Elephant National Park.*, thesis, Rhodes University.

Moore, N., Whiterow, A., Kelly, P., et al. (1999). Survey of badger, *Meles meles*, damage to agriculture in England and Wales. *Journal of Applied Ecology*, **36**, 974–988.

Nachman, M., Rauschenberger, J., and Ashe, J.H. (1977). Stimulus characteristics in food aversion learning. In N.W. Milgram, L. Krames, and T.M. Alloway, eds., *Food Aversion Learning*, pp. 105–131. Plenum Press, New York.

Neal, E. and Cheeseman, C. (1996). *Badgers*, 271 pp. Poyser, London.

Nicolaus, L.K., Cassel, J.F., Carlson, R.B., and Gustavson, C.R. (1983). Taste aversion conditioning of crows to control predation on eggs. *Science*, **220**, 212–214.

Nicolaus, L.K., Farmer, P.V., Gustavson, C.R., and Gustavson, J.C. (1989b). The potential of estrogen-based conditioned aversion in controlling depradation: a step closer toward the 'magic bullet'. *Applied Animal Behaviour Science*, **23**, 1–14.

Nicolaus, L.K., Herrera, J., Nicolaus, J.C., and Dimmick, C.R. (1989a). Carbachol as a conditioned taste aversion agent to control avian predation. *Agriculture, Ecosystem and Environment*, **26**, 13–21.

Nicolaus, L.K., Herrera, J., Nicolaus, J.C., and Gustavson, C.R. (1989c). Ethinyl estradiol and generalized aversions to eggs among free-ranging predators. *Applied Animal Behaviour Science*, **24**, 313–324.

Nicolaus, L.K., Hoffman, T.E., and Gustavson, C.R. (1982). Taste aversion conditioning in free-ranging raccoons, *Procyon lotor. Northwest Science*, **56**, 165–169.

Nicolaus, L.K. and Nellis, D.W. (1987). The first evaluation of the use of conditioned taste aversion to control predation by mongooses on eggs. *Applied Animal Behaviour Science*, **17**, 329–346.

Prescott, M.N., Berdoy, M., and Macdonald, D.W. (1997). *Protecting egg prey from magpies (Pica pica): the potential of aversive conditioning. A report for the RSPB, Bedfordshire, UK.* WildCRU, Department of Zoology and Veterinary Services, University of Oxford.

Reynolds, J.C. (1999). The potential for exploiting conditioned taste aversion (CTA) in wildlife management. In D.P. Cowan and C.J. Feare, eds., *Advances in Vertebrate Pest Management*, pp. 267–282. Filander Verlag, Furth.

Reynolds, J.C., Angelstam, P., and Redpath, S. (1988). Predators, their ecology and impact on gamebird populations. In P.J. Hudson and M.R.W. Rands, eds., *Ecology and Management of Gamebirds*, pp. 72–97. Blackwell Scientific Publications, Oxford.

Reynolds, J.C., Goddard, H.N., and Brockless, M.H. (1993). The impact of local fox Vulpes vulpes removal on fox populations at two sites in southern England. *Gibier Faune Sauvage*, **10**, 319–334.

Riordan, P., Delahay, R.J., Cheeseman, C., Johnson, P.J., and Macdonald, D.W. (2011). Culling-induced changes in badger (Meles meles) behaviour, social organisation and the epidemiology of bovine tuberculosis. *PLOS ONE*, **6**, e28904. doi:28910.21371/journal.pone.0028904.

Semel, B. and Nicolaus, L.K. (1992). Estrogen-based aversion to eggs among free-ranging raccoons. *Ecological Applications*, **2**, 439–449.

Stewart, P.D., Ellwood, S.A., and Macdonald, D.W. (1997). Remote video-surveillance of wildlife—an introduction from experience with the European badger *Meles meles*. *Mammal Review*, **27**, 185–204.

Tapper, S. (1992). *Game Heritage: An Ecological Review from Shooting and Gamekeeping Records*. Game Conservancy Ltd, Fordingbridge, Hampshire, UK.

Tuyttens, F. and Macdonald, D.W. (2000). Consequences of social perturbation for wildlife management and conservation. In M. Gosling and W. Sutherland, eds., *Behaviour and Conservation*. Cambridge University Press, Cambridge, UK.

Watkins, R.W. (1996). Efficacy of cinnamamide as a repellent for vertebrate and invertebrate pests. *Pesticide Outlook*, **7**, 21–24.

Watkins, R.W., Gurney, J.E., and Cowan, D.P. (1998). Taste aversion conditioning of house mice (*Mus domesticus*) using the non-lethal repellent, cinnamamide. *Applied Animal Behaviour Science*, **57**, 171–177.

Wolfensohn, S. and Lloyd, M. (1998). *Handbook of Laboratory Animal Management and Welfare*. Blackwell Science, Oxford, UK.

CHAPTER 14

From science to practice: delivering conservation across the landscape

Philip Riordan, Ruth E. Feber, Eva M. Raebel, Paul J. Johnson, and David W. Macdonald

Realize that everything connects to everything else.

—Leonardo da Vinci

14.1 Farmers as custodians of the countryside

Most of the chapters in this volume and in Macdonald and Feber (2015) have focused on the scientific research projects designed to inform measures for conserving wildlife on farmland. But, whatever the research outcomes and recommendations, farmers are ultimately the key delivery agents for food production and wildlife conservation across lowland landscapes. Farmer actions are driven by changing financial support structures, policies, and environmental and societal concerns (Macdonald and Johnson 2000). For example, over the last 15 years, removal of the link between agricultural subsidies and production and the introduction of agri-environment schemes (AES) have enabled farmers, at the farm scale, to support initiatives aimed at reversing the habitat losses (the 'locust years'; Rackham 2003) of the earlier post-war decades (Macdonald and Johnson 2000, 2003; Macdonald and Feber 2015: Chapter 1). In this chapter we turn to our engagement with farmers and explore, using two case studies, how the delivery of wildlife conservation on farmland might be facilitated, particularly at the scale of the landscape.

So how have farmers viewed their purpose and responsibilities through the transformation in society's priorities for farmland? We can trace some robust trends and constancies in farming attitudes dating from the 1970s. The Countryside Commission[1] (cited by Mabey 1980) surveyed farmers at a time when habitat destruction was in full swing. While farmers were aware that the landscapes they were creating were not liked, they often felt this was no one else's business. Nonetheless, most recognized a duty towards the landscape and society beyond the production of food. Interestingly, attitudes as to whether the form of the landscape had to be accepted as a by-product of food production were related to how degraded the existing landscape was; farmers of mixed farms in the Midlands were more likely to find this degradation unacceptable than were those of monocultures in East Anglia. The Countryside Commission concluded that further 'large-scale changes' were inevitable, stating an aspiration to encourage the creation of 'new' agricultural landscapes.

Realizing that farmers' views on wildlife and non-agricultural land were important and largely unknown, in 1981 we carried out a survey of farmers in different regions of the UK by means of questionnaires concerning varied aspects of farming practices relevant to wildlife, hoping the results would inform farmland conservation polices (Macdonald 1984). The survey circulated 2288 forms, of which 38% were returned. In 1998, we repeated this survey in the same areas, circulating 1000 forms of which 45% were returned. We reported in detail on our findings in Macdonald and Johnson (2000), but add some original observations here.

Farmers reported widespread habitat removal in the years preceding the 1981 survey—the 'locust years' were coming to a close, but were not over. For

[1] A forerunner of what is Natural England in 2014.

Wildlife Conservation on Farmland. Conflict in the Countryside. Edited by David W. Macdonald and Ruth E. Feber.
© Oxford University Press 2015. Published 2015 by Oxford University Press.

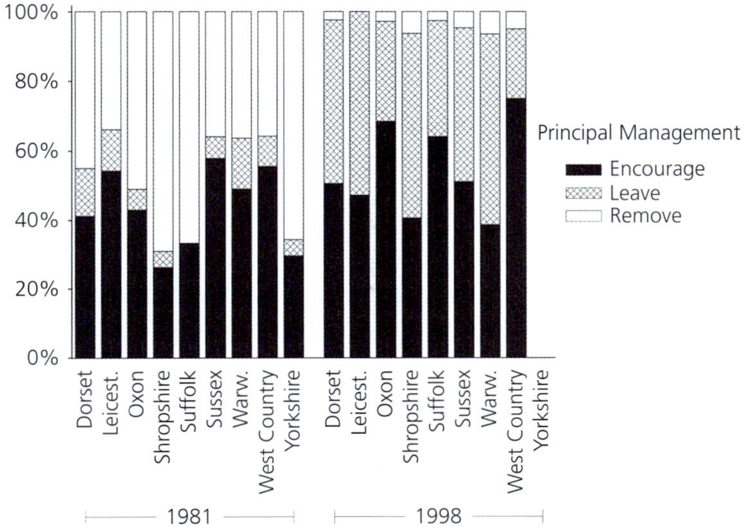

Figure 14.1 Hedgerow management strategy, UK farmers in 1981 and 1998. Reported principal management strategies. Adapted from Macdonald and Johnson (2000).

example, most farmers reported that their dominant strategy with respect to hedgerows was removal (Fig. 14.1). Reported rates of loss (in the preceding decade), were as high as 8.1 m/ha in Shropshire. Almost half (47.0%) of respondents reported some hedgerow removal in the 1981 survey and, although this varied significantly from region to region, nowhere was the rate lower than 38.5% (Devon); the highest rates were reported in Shropshire, Suffolk and Yorkshire (at 58.0%). In the Upper Thames region, the figure was close to the national average (Oxfordshire, 49.6%). Some other non-productive habitats, including scrubland, ponds, and ditches were also faring badly at this time (nationally, 26.7% of farmers said removal was their principal strategy for scrubland, and 22.1% for open water, ponds, and ditches; almost 10% of farms reported removal of trees as their dominant policy). In Oxfordshire, these habitats were removed by more farmers than the national average, markedly more for ponds (27.1% for scrub, 13.5% for trees, and 28.7% for ponds).

The most cited reason for hedgerow removal was 'efficiency' (87.3%), and this varied little regionally. Stock-proofing was also widely cited (57%), presumably where a hedge was replaced with a fence, and unsurprisingly this did vary substantially among regions, being highest in the West Country (68.9%; where livestock farming predominates) and lowest in Suffolk (21.8%; where cereal farming predominates). Oxfordshire farmers cited stock-proofing for hedgerow removal about as frequently as the national average (58.2%). 'Tidiness' was cited by a substantial minority of farmers everywhere (23.9% overall, and 21.0% of Oxfordshire farmers).

However, when we repeated the survey seventeen years later, in 1998, the differences in responses were dramatic. Only 23.0% of farmers reported that their principal involvement regarding hedgerows was to remove them, and the proportion leaving them alone had increased accordingly (Fig. 14.1). Rackham, commenting in 2003, noted that he saw less hedgerow removed in a year than in one week of the 1960s (Rackham 2003). The Countryside Surveys of 1990 and 2000 reflected on the ground what our questionnaire surveys had suggested—there was a net loss of hedgerow between 1984 and 1990, which had been halted in the following decade. The introduction of limited legal protection for hedgerows in 1997[2] may have been partly responsible, but similar changes were apparent for other non-productive habitats; scrubland removal as a principal strategy declined to 9.3%; ponds and ditches to 0.3% (n = 1), and not one farmer reported 'removal' as a principal strategy with respect to trees. It appeared that motives had changed: more farmers in the second survey compared to the first cited 'wildlife' as a motive

[2] Under the Hedgerows Regulations 1997 it is against the law to remove or destroy hedgerows that are at least 20 metres in length, more than 30 years old, and contain certain plant species, without permission from the local planning authority (NE 2010).

for keeping hedgerows (75.7% versus 61.6% in 1981). Economic motivation for maintaining hedgerows was conspicuous in both surveys: 78.4% of farmers cited stock-proofing as a reason for hedgerow retention in the earlier survey, and 70.2% in the second. In Oxfordshire, this percentage increased from 68.1% in the first survey to 83.9% in the second.

The altered emphasis from production to environmental protection allowed farmers who valued wildlife to express that in their management in the 1990s. In the first survey, farmers who said they were interested in wildlife were no less likely to remove hedgerow than those who were not, while, in the second survey, this category of farmers was less likely to remove hedgerows. The changing economic pressures apparently allowed a latent enthusiasm to be revealed. Morris and Potter (1995) identified a 'participation spectrum' among farmers engaging in conservation initiatives—younger farmers running large businesses were more likely to participate.

Reported uptake of advice (e.g. from government agencies or wildlife trusts) on wildlife and conservation was much more prevalent in the second survey—up considerably from 10.4% to 40.7%. Most farmers in the 1998 survey were aware of the commonly adopted agri-environment schemes (AES) subsidizing sensitive management of wildlife habitats. In the first survey we had asked farmers, at the time a hypothetical question, of whether they would be willing to create wildlife refuges if subsidized—66.2% said they would. By 1998, this had risen to 91.9% and was no longer hypothetical. The outcomes associated with these attitudes were encouraging: 35% of farmers reported hedgerow creation in 1998 compared with only 6.6% in 1981, and 46.3% said they had created woodland—'for wildlife' was the most conspicuously cited motive. There were also clear links between farmers' interest in field sports and habitat management: 'hunting' and game-rearing farmers had removed less hedgerow and woodland, while farmers who were principally interested in shooting had removed more. These effects were undetectable in the second survey when far less hedgerow was being removed.

While the political backdrop to the agricultural landscape had brought about profound changes in these patterns of habitat management there was one conspicuous constant. Almost all respondents in both surveys reported that they believed that they, the farmers, were the 'principal custodians of the countryside'—95.6% chose this option in 1981 and 95.3% in 1998—'the state' and 'local conservation bodies' were the alternatives offered, but few farmers ticked these. This echoes both the reports in the Countryside Commission survey of the early 1970s and anecdotal encounters with farmers: they report resentment at perceived 'interference'. No conservation initiative on agricultural land should ignore the farmers' perspective, their values being crucially important for conservation success (Feber et al. 2007).

14.2 Landscape-scale conservation

Farmer attitudes, always important, become particularly relevant when considering landscape-scale approaches to wildlife conservation. The lessons learnt from ecological theory and practice are that, for biodiversity conservation, land management is best designed at a landscape scale. Scale and spatial configuration of habitat patches is important for many aspects of conservation on farmland (Macdonald et al. 2000). For example, different taxonomic groups are affected by habitat heterogeneity at different scales (Macdonald et al. 2000); for many more mobile groups, such as mammals, conservation action is most effective when coordinated at the scale of a landscape, not on single unconnected farms (Macdonald et al. 2007; Macdonald and Feber 2015: Chapters 8 and 10; Box 14.1). Such landscape-scale conservation can best be achieved by joined-up action to form an ecological network expanding over a large geographical area; such networks may also be more resilient to changing pressures (e.g. climate change; Natural Environment White Paper (HM Government 2011)). Yet, due to the voluntary nature of most AESs (currently the most important instruments for delivering biodiversity gains on farmland), farmer participation, although widespread, is fragmented across the landscape. The result is that conservation action is often limited to the field or farm scale, leading some to question the value of such schemes in achieving their objectives for biodiversity (Butler et al. 2007). An obvious approach to this challenge is to try to recruit landowners to AES in a way that facilitates a contiguous and integrated farmland management plan across a landscape. But is it possible to engage landowners in this way, is it administratively manageable, and does it indeed lead to positive ecological outcomes? In the remainder of this chapter we describe two case studies which explore how locally sensitive, landscape-scale approaches to land-use planning and farmer engagement might be undertaken, and whether their conservation objectives can be achieved.

Box 14.1 Hugging the hedges: might agri-environment manipulations affect landscape permeability for hedgehogs?

In Britain, the number of European hedgehogs *Erinaceus europaeus* fell from an estimated 30 million in the 1950s to only 1.5 million in 1995 (i.e. most recent estimate), and is believed to continue to decline by several per cent each year (Wembridge 2011). As hedgehogs are distributed discontinuously, influenced by the distribution of predators and food, they can provide an illuminating model—relevant to other species which disperse through agricultural landscapes—for studying how the connectivity, and therefore permeability, of agricultural landscapes may vary with the availability of semi-natural habitats. Urban habitats often suit this species (Hof and Bright 2010), but suitable habitat in the countryside is increasingly fragmented within a matrix of cultivation. Consequently, the dispersion between suitable habitats on farmland is likely to be critical to hedgehog survival. Hedgehogs avoid cropped areas (Hof and Bright 2010) and disperse along linear features (Doncaster et al. 2001), and so edge habitats such as hedges, boundaries of woodlands, field margins, and road verges represent movement corridors (Doncaster et al. 2001). These features may therefore be crucial for linking separate populations and ensuring their stability (e.g. Hanski 1999; Baguette et al. 2013). Agri-environment scheme options to restore lost semi-natural agricultural habitats might be expected to increase connectivity, and therefore population stability for hedgehogs in farmscapes, but the crucial questions are which habitats, and what quantity of these habitats, would provide the best increase in landscape permeability for this species?

Working with colleagues in the University of Aberdeen, WildCRU's Tom Moorhouse addressed these questions using a model called the Stochastic Movement Simulator (SMS3), an individual-based, 'least-cost' model which allows virtual 'hedgehogs' to move through a number of virtual landscapes (Moorhouse et al. 2014). In the model, virtual landscapes are represented as a grid of 'costs', which they derived from radio-tracking studies of hedgehog preference for given habitats. For example, a real-life 20 × 20 m garden with a hedge running north–south down the middle would be represented as one hundred 2 × 2 m squares where the squares representing the hedge are given a low cost of 1 unit (because in similar situations radio-tracked individuals selected this habitat most often) and the squares in the lawn are given a high cost of 14 units (because radio-tracked hedgehogs spent 14 times less time on the lawn than in the hedge). In SMS3 the 'hedgehog' assesses the grid-cells around its current cell and moves in the direction with the lowest overall cost. It then repeats the process for a given number of steps (e.g. 50 in our garden). During this process, the animal's choices are limited by parameters such as its 'perceptual range' (how many cells it can 'see') and its inclination to keep going in a straight line. A virtual 'hedgehog' released in the middle of our example garden would thus probably follow the hedge rather than venture into the lawn. With enough steps, it would most likely end up at the north or south edge of the virtual garden.

The same process underlies the choices of 'model hedgehogs' in our virtual landscape, which accurately represents a real landscape in south Oxfordshire (a rectangular area of land, 11.2 km long by 8.47 km) in which every relevant feature (hedges, field margins, houses, rivers, roads, grass verges, fields, etc.) is represented to a 2 m accuracy. This allowed us to produce a set of manipulated virtual landscapes, by increasing the percentage of fields that had hedgerows, field margins, or which contained pasture or arable crops, which we could then use to examine landscape permeability using SMS3. Moorhouse and colleagues discovered that doubling the percentage of hedges from 25% to 50% resulted in an additional 13% of model 'hedgehogs' making movements of 500 m, while 25% more hedgehogs travelled 1000 m, 35% more travelled 1500 m, and 51% more travelled 2000 m. Their findings indicated that actions to reinstate or repair hedges in order to double the percentage of hedgerow could enhance connectivity between hedgehog populations, particularly in landscapes where these populations are distributed discontinuously.

A key factor affecting hedgehog numbers is the presence of badgers *Meles meles* (Hof and Bright 2010), which have increased by 25% nationally over the past few decades (see Chapter 4, this volume). Both species forage for earthworms in similar habitats (Doncaster et al. 2001), leading to intra-guild competition, and badgers also predate hedgehogs (Doncaster et al. 2001). An experimental translocation of wild hedgehogs into two woodland sites revealed that, in areas with high badger density, mortality and dispersal rates of hedgehogs were higher than in low badger density areas (Doncaster 1992). Hedgehogs have been shown to recognize the odour of badger faeces (Ward et al. 1996), and in the presence of badger odour in the wild, reduce their foraging effort (Ward et al. 1997; see also Ward et al. 2000), and remain closer to linear features (Hof and Bright 2012). Given the protection from predators, and access to nesting sites, that hedges provide for hedgehogs, and recent calls to reinstate hedgerows both to mitigate badger predation and to increase the connectivity between suitable hedgehog habitats (Hof and Bright 2012), restoration of hedgerows at a landscape scale should benefit hedgehogs and many other species, such as small mammals (Fitzgibbon 1997; Gelling et al. 2007), moths (Slade et al. 2013), butterflies (Merckx and Berwaerts 2010), birds (Green et al. 1994; Macdonald and Johnson 1995; Cornulier et al. 2011), and bats (Verboom and Huitema 1997).

14.3 Case study: The Chichester coastal plain sustainable farming partnership project

In 2000, we established a project on the Chichester coastal plain in West Sussex, UK, the aim of which was to restore habitat connectivity across the richly productive and diverse floodplain—an area criss-crossed by watercourses—and in doing so bring about increases in local wildlife populations, particularly the severely declining water vole *Arvicola amphibius*. Instead of viewing farms as individual entities, the project considered the whole landscape of the coastal plain in terms of its land use and river catchment management. This entailed working with a number of partners, including the Farming and Wildlife Advisory Group (FWAG), the Environment Agency, English Nature, and Sussex Wildlife Trust. Through targeted advice and delivery of whole farm conservation plans that addressed the need for habitat creation and enhancement on individual farms, the partnership's aim was to create linked habitat corridors across and between neighbouring farms. The method for achieving these outcomes therefore differed from conventional AES methodology, in that farmers were approached to join AES on the basis of the contiguous nature of their farmland, the success of the project hinging on being able to join up the activities of neighbouring farmers.

We offered farmers in the study area a free biodiversity audit and a Whole Farm Conservation Plan (WFCP) of their holdings through FWAG's Landwise package. The WFCP incorporated advice on best practice management of watercourses, ditches, wetlands, and the targeting of waterside land. Farmers were encouraged to apply for AES payments and much of the enhancement work for improving water vole relevant habitats was met by acceptance into the Countryside Stewardship Scheme (CSS, the AES operational at the time), for which we guided farmers through the funding application process.

In the end, the partnership provided WFCPs to 42 different farm holdings, covering an area of 8400 ha, and the project was an outstanding success. The plans included advice on livestock and grassland management, hedge management and creation, and wetland advice, and resulted in a network of AES (Stewardship) agreements that created over 61 km of field margins, restored 27 ponds, created 28 ditch ponds, and maintained over 42 km of ditches. Together with mink *Neovison vison* removal, these measures had a dramatic effect on water voles (Box 14.2), and other wildlife, including hares *Lepus europaeus*, farmland birds, and butterflies (Strachan and Holmes-Ling 2003).

Box 14.2 Reversing the decline of the water vole

A national survey in 1989–1990 indicated that water voles *Arvicola amphibius* had declined by 67.7% over 50 years due to a combination of habitat loss and the arrival of the American mink (Strachan and Jefferies 1993, see Macdonald and Feber 2015: Chapter 15 for full details). Our overall aim in the Chichester plain study was to ask if a joined-up approach of targeting contiguous farmers and landowners to undertake habitat management, often through agri-environment schemes, would be effective in delivering biodiversity gains, in particular, reversing the decline of the water vole (see Macdonald and Feber 2015: Chapters 14 and 15). The answer was an unequivocal yes. Initial water vole surveys in 2000 estimated that in the overall farms, there were only 103 individuals (Box Fig. 14.2.1a; Strachan and Holmes-Ling 2003). In 2002, ten pairs of captive bred voles were released into one watercourse. Mink trapping was undertaken over three years, as well as pond creation, raising summer water levels in ditches, and we erected 7200 m of fencing to protect watercourses from livestock. We found that water voles were strongly associated with these recently fenced banks, which supported dense bankside vegetation: from all occupied stretches, 88% were protected from cattle grazing and trampling by fences, while on unfenced sites, only 16% had evidence of water vole presence (Macdonald and Feber 2015: Chapter 15). By 2003, 4556 ha (of the 8400 ha area) had been entered into the Countryside Stewardship Scheme (with a further 1328 ha having had improvements). Overall, water vole numbers in the Chichester plain responded to habitat improvements and mink control by more than tripling in three years to an estimated 348 individuals (Box Fig. 14.2.1b; see Macdonald and Feber 2015: Chapter 15).

continued

Box 14.2 *Continued*

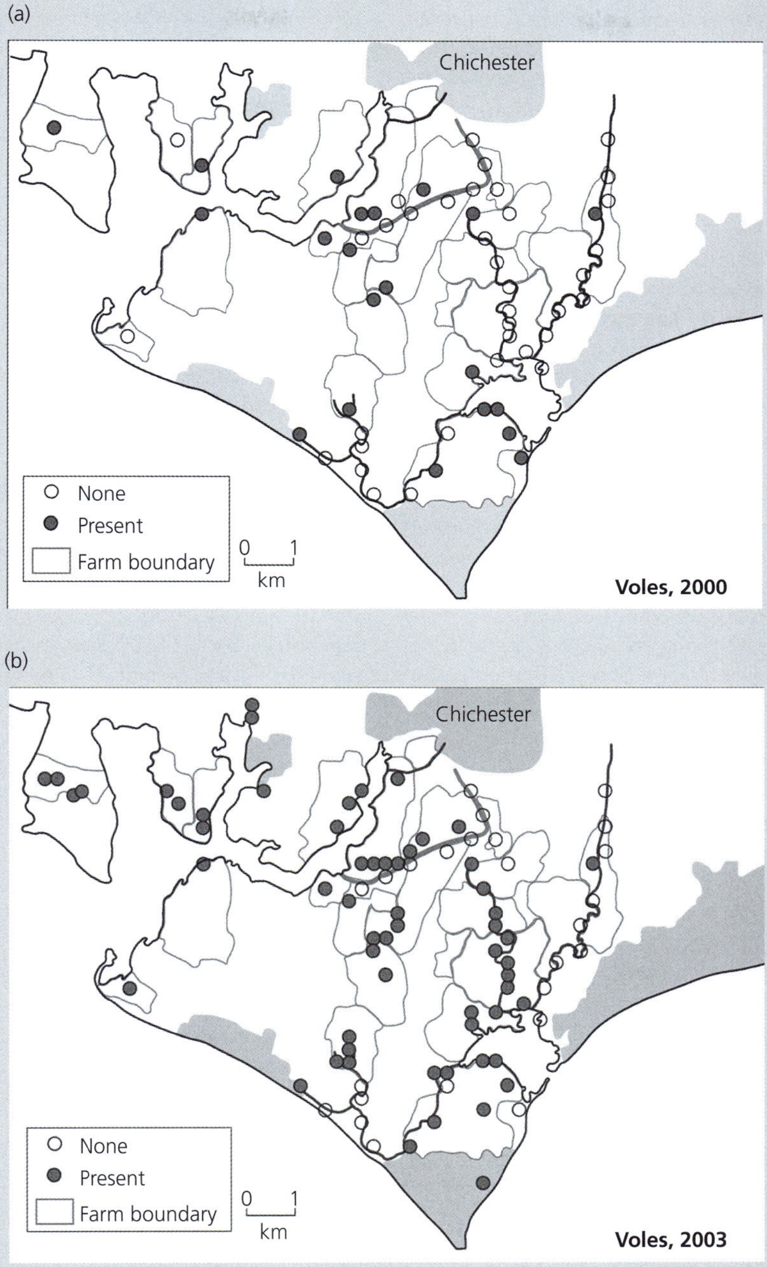

Box Figure 14.2.1 Relative distribution of water vole colonies (none/present) across the Chichester coastal plain farms in (a) 2000–2001 and (b) following habitat restoration in 2002–2003. The most easterly watercourse was a control area receiving little/no proactive management for water voles apart from mink removal. From Strachan and Holmes-Ling (2003) and Dutton et al. (2008), reproduced with permission of John Wiley & Sons.

Overall, we found that the process of engaging farmers in the Chichester coastal plain project was personalized and decentralized, which increased farmer participation and enthusiasm (Dutton et al. 2008). The decision-making process reduced farmers' concerns about 'independence': they were consulted about their individual needs and interests, and thereby allowed to take individual responsibility for conservation.

14.4 Case study: The Upper Thames project (UTP)

The Chichester coastal plain project was hailed as a 'demonstration area of best practice for sustainable farming and biodiversity enhancement, where land managers, farmers, and practitioners could experience sympathetic management first-hand' (Strachan and Holmes-Ling 2003). Importantly, the project had shown that it was possible to plan and implement conservation measures at the landscape scale, and that this approach can lead to benefits for biodiversity. But could this partnership approach to farmer engagement be achieved on a larger scale, across catchments, and could the biodiversity benefits be more rigorously assessed within a landscape-scale experimental design? Using the lessons learnt from the Chichester project, we embarked upon an ambitious five-year programme, the Upper Thames Project (UTP). The idea was to join up clusters of farms, and empower communities of individual farmers through a bottom-up approach stimulating a 'word-to-mouth' effect to encourage neighbours to participate and increase the connectivity of areas under AES schemes.

A key driver for UTP was a change in AES structure in the UK. As described in Macdonald and Feber 2015: Chapter 1, previous AES had focused on key existing areas of unique environmental features but, in 2005, in a major change to AES in England, the Environmental Stewardship Scheme (ESS) was launched. This was a two tier scheme—comprising a 'broad and shallow' tier (Entry Level Scheme, ELS) which encouraged extensive participation by being open to all farmers, who were required to undertake simple management options, and a 'narrow and deep' tier (Higher Level Scheme, HLS) requiring more complex management and which was not open to all (see Macdonald and Feber 2015: Chapter 1). Both tiers of the scheme, however, relied upon voluntary participation by farmers, leading to a scatter-gun uptake, and were not designed to address the challenge of joined-up habitat management.

Our approach was designed to deliver changes to habitat management, enhancement, and creation at a landscape scale through practical conservation while the research results unfolded. In this way, we could develop adaptive management, identifying critical linkages across the area, and refining and improving as the evidence base developed. Through discussion with partners and funders, we set targets for habitat delivery at the start of the project (Table 14.1) within an experimental framework (the project 'backbone') which would allow us to study habitat and biodiversity impacts in areas where farmers were proactively targeted to join agri-environment schemes (using the methods described above), and compare them with areas in which we did not influence AES applications and scheme uptake remained *laissez faire*. The overall study area comprised catchments of five of the tributaries of the Thames (Fig. 14.2), lying within Oxfordshire, Gloucestershire, Buckinghamshire, and south of Northamptonshire. Two of these catchments, the Rivers Ray and Windrush, were actively targeted (Fig. 14.2), while in three other catchments, the Thames, Cherwell, and Evenlode (Fig. 14.2), there was no targeting. Using moths as our biodiversity indicator group, we found that the approach of targeting farmers to join up whole landscapes for AES uptake, rather than the classic scattered approach focused on single sites/individual fields, resulted in biodiversity gains (Macdonald and Feber 2015: Chapter 8; Box 14.3).

14.4.1 A partnership approach

A crucial factor in the success of the Chichester coastal plain study had been the involvement of NGOs in the partnership. Farmers tend to express greater trust for civil society organizations than they do for government agents (Franks and Emery 2012), and NGOs have built trust within communities, and hence farmers rely on them for advice. About 500 UK charities provide local or national environmental services of varying degrees (in Cook et al. 2012), while large conservation charities (NGOs) play an instrumental role in balancing conservation and agricultural production. For example, the Royal Society for the Protection of Birds (RSPB), although originally founded (in 1889) to halt the enormous wholesale destruction of birds for their plumage, today manages 200 reserves, owns farms, and is supported by over a million members (RSPB 2013). It has teams of dedicated advisers who work with farmers to promote farm management in support of declining farm-dependent species of bird, and maintains a research department.

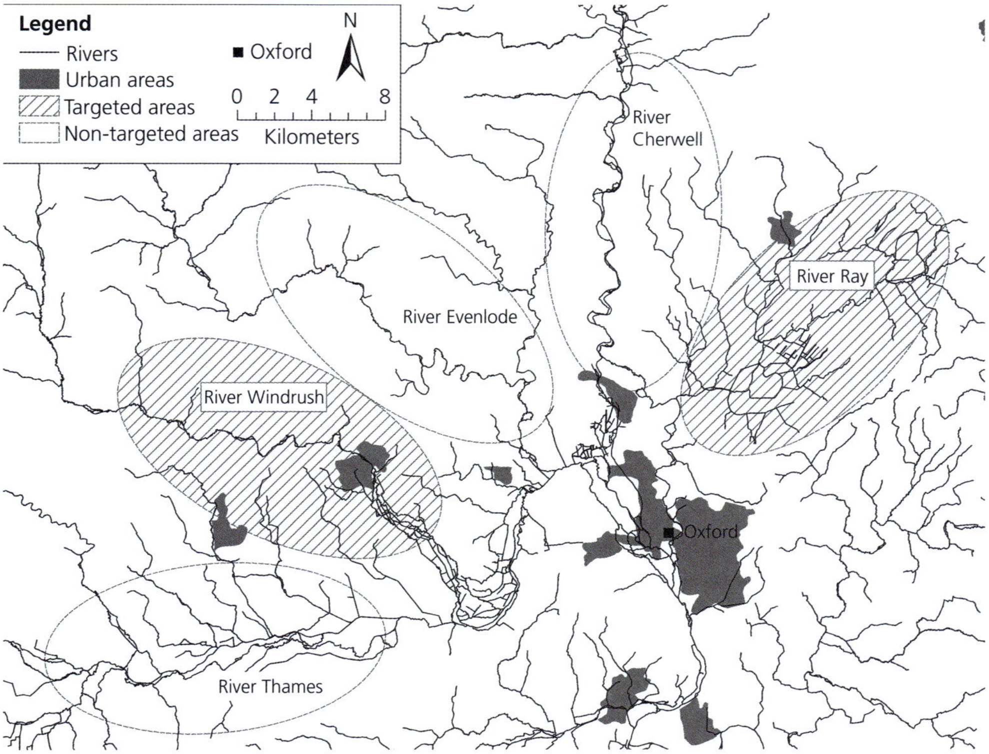

Figure 14.2 Map showing study areas for the Upper Thames Project. In targeted areas farmers were proactively encouraged to join AES. In non-targeted areas there was no attempt to influence uptake of AES.

Numerous other NGOs, such as the Freshwater Habitats Trust and the Wildlife Trusts, also offer advice to farmers. A key organization has been the Farming and Wildlife Advisory Group (FWAG), formed in the 1960s by a group of farmers who realized that conserving the environment was essential to the success of their farming business (FWAG 2013). FWAG enjoys recognition by the government, the National Farmers Union, and landowners, and has changed farmers' attitudes towards conservation by offering impartial advice. Cook et al. (2012) characterized FWAG as amending the damage to biodiversity that originates from omissions in the planning process. FWAG went into administration in 2011, but it is now regionally back by farmers' demand.

In the Upper Thames project, in order to encourage farmers' trust, and hence participation, we collaborated with two local NGOs: the local Bedfordshire, Buckinghamshire, and Oxfordshire Farming and Wildlife Group (BBO FWAG) and the Bedfordshire, Buckinghamshire, and Oxfordshire Wildlife Trust (BBOWT). In addition, we formed a project steering group chaired by the Environment Agency and including Natural England, RSPB, and Oxfordshire Nature Conservation Forum. We also identified and worked with c. 50 collaborators across the catchment who were actively pursuing biodiversity and landscape conservation interventions, as independent organizations, with separate funding streams and operational themes.

Central to the aims of the UTP was for FWAG to persuade farmers whose farms may have been key to achieving better habitat connectivity, and, as a result, more effective biodiversity conservation (see Box 14.4), to use best management practices of habitat features on their land. Whether farmers were not in AES because they had not previously been eligible, or were reluctant to apply to an agri-environment scheme because they lacked knowledge or understanding, our intervention circumvented these problems by targeting individual landowners and guiding them through the application process, including more complex applications (e.g. HLS). Since our success hinged on

Box 14.3 Benefits to moths of landscape connectivity

One of the aims of the UTP was to provide empirical evidence of the biodiversity benefits of targeting farmers to increase landscape connectivity. Merckx and colleagues (Merckx et al. 2009; Macdonald and Feber 2015: Chapter 8) led a landscape-scale study to assess the effects of two key features of arable farms, grassy field margins (6 m-width versus 1 m-width) and hedgerow trees, on species diversity and abundance of 'larger moths' (macro-moths). Larger moths were chosen because they are a sensitive indicator group of terrestrial farmland landscapes, provide ecosystem services, and are nationally declining (see details in Macdonald and Feber 2015: Chapter 8). The authors contrasted these effects both in the UTP targeted areas, where we had encouraged farmers to take up AES for two years (2004–2006), and in two of the control non-targeted areas, where uptake was diffuse. Merckx et al. sampled moths at the end of our intervention, in 2006, by using light traps in sixteen farms (sampled 11 times each; see Merckx et al. 2009 and Macdonald and Feber 2015: Chapter 8 for full details). A total of 27,071 macro-moths from 270 species were recorded.

Locally, Merckx et al. (2009, 2012) found that both wide field margins (6 m) and hedgerow trees increased the species number of macro-moths, by an average of 15%. However, crucially, they showed that the benefits of hedgerow trees were significantly greater only in landscapes where AES uptake had been experimentally increased: in non-targeted areas there was an overall increase of 15% in species richness and 20% in abundance, while in targeted areas the increase was 38% and 60%, respectively. They proposed that the underlying mechanism could be explained by the higher proportion of land covered by AES that was present in the targeted areas, which resulted in the joining up of habitat resources across the landscape. Indeed, while at the start of the project the density of hedgerow trees did not differ between sites[3], after two years of targeting landowners, the density of hedgerow under the enhanced management AES option was c. 219 km in targeted areas versus c. 83 km in control areas.

The results of Merckx and colleagues demonstrated that the effect of hedgerow trees, a prominent feature throughout the landscape that provides shelter, food, and roosting for many species, largely depends on the landscape context, suggesting that beneficial effects of other landscape features and scheme options might only become apparent after a substantial increase in landscape connectivity. The evidence from this project provided strong support that an approach which targets farmers to join up whole landscapes for AES uptake, rather than the classic scattered approach focused on single sites/individual fields, can have positive biodiversity gains by strengthening the diversity within landscapes and whole ecosystems. The macro-moths study gave an empirical measure of the success of the UTP approach, and the benefits for biodiversity on farmland of complementing the entry-for-all approach to AES with a system that targets specific areas and/or landscapes (Merckx et al. 2009).

engaging farmers, and initial financial costs of applying to AES can be a deterrent, our team offered a funded conservation assessment service—WFCP—including advice on grants that were available.

Our experience from the Chichester plain project demonstrated that farmers could best be persuaded to join AES if they were given help to apply and, particularly, to access financial incentives (Dutton et al. 2008). Consequently, in the UTP we (i) first approached farms whose position or features gave them the potential to improve habitat connectivity and, if the farmer was interested, we (ii) arranged to conduct an audit of habitats and biodiversity, along with an assessment of farm management practices. Our team then (iii) produced a WFCP, including a map of sensitive habitats and suggestions for habitat enhancement. The WFCP was based on the funding realities open to each farm, offering advice on grant schemes to which they might be entitled for implementing our recommendations[4].

An important aspect of successfully engaging farmers was the incentive of the free WFCP (the cost per farm for a WFCP carried out by FWAG averaged £1200 in 2004). We had sought funds to enable us to offer this (with some direct funding of capital works

[3] Mean count ±S.E.: targeted, 155 ± 35.3 versus non-targeted, 167 ± 12.9; $t_{8,84} = 0.32$, $P = 0.76$.

[4] As habitat enhancements were carried out, we tested their effectiveness through comparative trials that measured their impact on biodiversity (e.g. Macdonald and Feber 2015: Chapters 8 and 14).

Box 14.4 The common toad, beyond the pond: landscape genetics in the Upper Thames area

The common toad *Bufo bufo* (Box Fig. 14.4.1) is a species of concern in the UK since experiencing dramatic and largely unexplained population declines (Carrier and Beebee 2003). Amphibians are often considered particularly sensitive to the effects of habitat change: loss, degradation and fragmentation. There is a consensus among herpetologists that the effects of habitat change are the most important contributor to the global amphibian extinction crisis (Wake 1991; Houlahan et al. 2000; Stuart et al. 2004). As a result of agricultural conversion and intensification, the landscape which the toad inhabits in the UK has changed dramatically over the past century through, for example, widespread pond drainage (Williams et al. 2010) and the loss and fragmentation of natural and semi-natural habitats through conversion (Blaxter and Robertson 1995).

The toad is reliant on both terrestrial and aquatic environments for different life stages; it requires ponds for breeding and development of tadpoles, and terrestrial habitat for foraging, migration and hibernation (Beebee 1985). The adult toad will use ponds for only a few weeks each year, during the breeding season, and so spends the majority of its life in the terrestrial environment. The quality of the landscape between ponds therefore is important in influencing the toad's ability to forage, hibernate (Beebee 1981) and, crucially, migrate between breeding populations maintaining gene flow across the larger population. Previous habitat suitability modelling of toads in agricultural environments predicted low occurrence in crops and a preference for woodland and meadows (Janin et al. 2009). In the Upper Thames area, where this study was conducted, the landscape consists largely of farmland, the majority of which is under agri-environment scheme agreements (Box Fig. 14.4.2).

To determine the importance of different habitats to the common toad, WildCRU's Rosie Salazar and colleagues used Passive integrated transponder (PIT) telemetry to search for toads in the farmland surrounding a large breeding pond in Garford, Oxfordshire (Latitude, Longitude: 51.6633,-1.3919). Between April and October of 2012 and 2013 we caught 90 common toads and used their location information to develop a Resource Selection Function (RSF) model to predict the relative probability of occurrence across the wider landscape. The model showed increased probability of toad occurrence with decreasing distance from woodland habitats and surface water and with increasing distance from the nearest urban environment (Salazar et al. submitted-b).

We collected genetic material from eight ponds in the Upper Thames area (Box Fig. 14.4.2) to study the effect

Box Figure 14.4.1 Common toad (*Bufo bufo*) breeding pair in amplexus. © Rosie Salazar.

continued

Box 14.4 *Continued*

of landscape connectivity on the genetic structure of the Oxfordshire common toad population. Common toads are expected to be sensitive to isolation of their breeding populations as they show a high level of fidelity to their natal breeding pond (Reading et al. 1991) and have relatively moderate migration capacity (maximum recorded migration distance of 3.6 km, Smith and Green 2005). Colonisation of new ponds is influenced by the distance to the nearest occupied pond (Baker and Halliday 1999). Using microsatellite markers established for the common toad (Brede et al. 2001) we calculated the genetic distance (how genetically dissimilar populations were from each other) between all possible pairings of our eight study ponds to compare with the Euclidean distance between these pairs. Our results showed no relationship between the genetic and Euclidean distance (Salazar et al. submitted-a). Using the RSF model developed in the previous study, we applied it to the landscape covering the area between all ponds and identified those areas with

Box Figure 14.4.2 Ponds sampled for genetic material and major habitats. Water, woodland, roads and named places from OS Mastermap (Crown Copyright). Agri-environment scheme agreements layer from Natural England.

continued

Box 14.4 *Continued*

0% probability of common toad occurrence. We then defined these areas as barriers to common toad migration and calculated the cost distance between all ponds to compare with genetic distance. In contrast to the result comparing genetic and Euclidean distance, genetic distance was significantly and positively correlated with the calculated cost distance between ponds (Salazar et al. submitted-a). Increasing cost distance (number and magnitude of barriers between ponds) was associated with an increase in the genetic dissimilarity between pond pairs. This indicates an effect of isolation by barriers between common toad breeding ponds. The barriers identified by the RSF model were those areas close to urban habitats and further from water sources and woodland. We can therefore recommend that restoring connectivity between ponds in farmed landscapes can be achieved by increasing the density of woodland habitats and water sources (e.g. ponds and wet ditches), thereby reducing the distance between these habitats which improves the probability of common toads using these areas in the terrestrial phase.

in collaboration with the Environment Agency). We also made use of other sources of advice, such as Farm Environment Plans (FEPs, for Higher Level Scheme) or Farm Environment Records (FERs, for Entry Level Scheme), which were partially funded by Defra through Natural England. These documents were equivalent to a WFCP, for example, the FEP required management options to be specified, thereby providing the basis for UTP target delivery.

14.4.2 Uptake and habitat delivery

In total, the UTP had contact with 325 landowners within the Upper Thames Tributaries region, through written reports, advice, and discussion. The partnership's advice encompassed 93.6 km^2 of land, with most engagement with 58 landowners on 88.1 km^2 of the Rivers Ray or Windrush catchments. The AES options that were most widely adopted in the UTP were minimal or zero cost ELS options, such as hedgerow management, with 83% of farms (n = 36) including these in their agreements. After five years, we exceeded our targets for habitat creation when options were available through ELS: for example, riparian fencing, in-channel enhancement, and hedgerow restoration (Table 14.1).

By contrast, HLS options were perceived as difficult, even when the financial rewards would be greater than with other options. By 2008, the highest cumulative

Table 14.1 Target delivery during the Upper Thames project, specified as quantities given as part of advice to farmers and amounts in place or in active development as of the end of 2008 (Actual). Uptake is specified as the percentage of UTP advice that developed into actual interventions.

Targets	Target Delivery	Advice given	Actual active*	Actual (%)	Uptake of advice (%)
New wet grassland	200 ha	198 ha	146 ha	73%	74%
Grassland restoration	1000 ha	629 ha	432 ha	43%	69%
Riparian buffer strips	100 km	86 km	77 km	77%	90%
Riparian fencing	50 km	78 km	62 km	124%	79%
In-channel enhancements	10	19	12	120%	63%
Hedgerow restoration	50 km	380 km	298 km	596%	78%
New ponds created	30	28	18	60%	64%
Existing ponds enhanced	150	98	65	43%	66%
New reedbed	5 ha	4.5 ha	1.4 ha	28%	31%
New linking woodland	20 ha	14 ha	12.2 ha	61%	87%

*further habitat work developed as a consequence of advice given during the later stages of the project.

HLS coverage (Windrush) was still only 4.6% of the total study area (ELS: 5200 ha versus HLS: 650 ha). Why was this figure so low? Policy interacts significantly with landowner decisions and some explanation for low overall uptake rates may be provided by the complexity of the scheme changeovers that took place during the intervening years between surveys. Several key changes occurred during this time in AES and farming in general. Firstly, CSS and the Environmentally Sensitive Areas (ESA) schemes were replaced by ESS in 2005 and farmers were caught in the transition between schemes. In September 2006, Defra announced changes to HLS which were not published until late 2008, causing confusion among farmers in terms of their HLS eligibility (Farmers Weekly 2006), which may have discouraged some farmers from entering the often lengthy and costly process of applying for HLS. This general discontent with HLS applications was confirmed when questioning farmers from 42 farms: 40% considered the HLS scheme to be 'too restrictive or overly burdensome'; 14% considered the payments to be insufficient (Table 14.2); and 79% of potentially eligible landowners (i.e. those not in existing schemes; and not having already been rejected from HLS) would not consider HLS. In addition, the survey revealed that 47% (16/34) of farmers that were part of CSS did not go on to enter either ELS or HLS.

Our results suggested that, in the absence of advice, farmers chose the simplest options. Two outcomes exemplify how, without the collaborative partnership and the facilitation of alternative sources of income, it would not have been possible for us to deliver more complex conservation practices at a landscape scale. The first was the creation of wet grassland, supported through HLS, for breeding wading birds. Even though our surveys had revealed that the percentage of farmers in the Upper Thames region that would consider HLS wetland options had declined significantly, from 50% in 2003 to 30% in 2008, the partnership with BBOWT and RSPB helped to establish, as part of HLS agreements, 146 ha of new wet grassland in our targeted areas. Second, we found that, while new woodland was difficult to create under HLS, by advising farmers and facilitating them access to Defra/Forestry Commission grants, the partnership's intervention led to planting 12.2 ha of woodland, primarily under the Woodlands Grant Scheme (WGS) and the English Woodland Grant Scheme (EWGS).

Our intervention also highlighted that farmers' values of what is important for wildlife may override financial intervention. For example, from 18 ponds that were created after our advice (financial support was available through HLS), over half (10) were created outside of AES agreements. Positive attitudes among farmers towards pond creation increased during the time of our project, with 65% (26/40) in favour in 2003 compared with 80% (40/50) in 2008, possibly due to the various pond creation/restoration campaigns that were undertaken at the time by us and various agencies. Indeed, many landowners in the study area took particular pride in their ponds and maintained them in the absence of AES funding. However, although several landowners were sufficiently motivated to work with us on habitat creation, even in the absence of AES incentives, these overall rarely reached delivery in the absence of payments.

Table 14.2 Reasons given by farmers questioned in the UTP area in 2008 for not entering ELS and HLS agreements.

Reasons for not entering ELS/HLS	Number of farms	
	ELS	HLS
Unaware of scheme	1	3
Too restrictive or burdensome	5	17
Insufficient money from scheme	2	6
Previous application rejected	0	2
Already in existing scheme	1	5
Other	3	7
Considering	2	2

14.4.3 Behavioural models can help scheme adoption

When the overall aim of a landscape-scale conservation initiative is to achieve linked clusters of participating farmers, then it is important to be able to identify those who are most likely to adopt conservation schemes. One way to do this is through a survey, to quantify farmer attitudes to AES. Dutton et al. (2008) showed how to do this by adapting an existing model based on farmer behaviour and applying it to farmers in the UTP to predict the likelihood of them joining a conservation scheme. If successful, the model could provide a tool to increase, first, the efficiency of landscape-scale planning, by not wasting money targeting farmers that had a low probability of adoption, and second, the effectiveness of the UTP planning, by creating a landscape plan focused on the land owned by farmers likely to adopt AES.

Dutton et al. (2008) contacted farmers in the Upper Thames catchment to describe the purpose of the UTP project and 36 respondents were consequently interviewed to (i) gain information on their farm, management of each field, and their financial returns; (ii) ascertain the farmers' previous experience of AES (ESA, CSS) and for those that had not joined, offered five typical reasons[5], and (iii) asked farmers to rate a series of attitudinal statements on a five-point Likert scale (see Dutton et al. 2008 for details). These attitudinal questions were based on the Theory of Reasoned Action[6] (see Dutton et al. 2008), in which we asked questions related to 'openness in farming' (high scores denoted the incorporation of new farming ideas into their business and a tendency to seek advice), 'chemical use' (high scores reflected ideas of the importance to reduce chemical use) and 'achievement' (high scores denoted a farmer competitively aiming to have the best run business) (Willock et al. 1999)[7]. Finally, (iv) we asked them if they would be willing to take part in future AES-type practices and participate on the whole farm planning process of the UTP.

The results showed that farmers who reported openness to new ideas and conservatism with respect to their use of chemicals were more likely to respond 'yes' to the central question: would they join the UTP? Using these answers, the model was able to categorize 79% of farmers correctly (84% of 'yes' and 74% of 'no'). When we tested the model retrospectively on the Chichester plains farmers (for which we knew their decision in adopting the Chichester scheme), the results were encouraging: the statistical model correctly predicted six out of seven of these farmers should be willing to participate.

Being able to predict which farmers would be willing to undertake conservation action on their land means that it could be possible for local environmental planners to offer greater incentives to willing landowners of land areas that are most important, for example, to connect populations or support a metapopulation, as successfully exemplified by the water voles in the Chichester Plains scheme (Box 14.2). Given the relatively low costs of data collection (phone, post, or email) a model such as ours could also aid with predicting the efficacy of AES in targeted and/or landscape-scale areas, and importantly, the success of any future changes to current schemes (Dutton et al. 2008).

These ideas have now been shown to deliver on a larger scale. A multidisciplinary study of the UK research councils Rural Economy and Land Use (RELU) programme explored options for making AES schemes better value for money. Focusing on interested farmers and providing face-to-face input delivered improved conservation outputs (Freckleton et al. 2008). Professor Bill Sutherland, who led the Freckleton et al. (2008) study, concluded that targeting the more compliant farmers was better than indiscriminately paying large numbers of landowners.

14.5 Conclusions

Our evidence suggests that intervention by non-governmental organizations can increase uptake of AES and therefore increase habitat connectivity and AES benefits at a landscape scale. Also, we have shown that landowners can be persuaded and coached through the process of choosing and delivering complex options (e.g. wet grassland; in-channel enhancements) that would otherwise have been ignored. Decentralized schemes, like the UTP project, have the advantage of providing the capability for negotiation with landowners running the schemes. Because of this, this approach could also prove useful to inform farmers when undertaking works outside of AES. It has been estimated that the total cost to farmers of landscape management (e.g. hedge trimming, pond creation) outside AES across England is £412 m/year (in 2006; £2410 per farm) (CPRE and NFU 2006).

Information from the Upper Thames project highlights the importance of considering both human and biodiversity landscapes. Within the human sphere, we can see that policy interacts significantly with landowner attitudes and the wider rural economy. The biodiversity gains came about with great effort from, and careful involvement of, the many partners on our team. The work on moths showed that the more *laissez faire* approach in our control sites was demonstrably poorer at delivering biodiversity gains. Exceptions can arise where partners have particular interests and specifically target their resources, for example the RSPB's Lapwing Landscapes project and BBOWT's Living Landscapes project along the Thames. Nonetheless, significant resources beyond those available to the landowners through AES are required.

[5] 'Too much hassle to set up', 'didn't pay enough money', 'ties up the land for too long', 'don't think the scheme would work very well', 'didn't have this farm at the time' or 'other'.

[6] 'An actor's behaviour is a function of the actor's objectives which are themselves informed by the actor's attitudes' (Dutton et al. 2008).

[7] Willock et al. (1999) had used a factor analysis to reveal that these questions were the attitudes that best related to environmental action (Dutton et al. 2008).

Since farmers support the idea of collaborative AES, where landowners are financially rewarded by applying in groups (Franks and Emery 2012), and hence increasing landscape connectivity, the framework we developed could assist with encouraging neighbouring applications, especially within areas with reluctant attitudes. The UTP also highlights, through the projects that progressed within the Upper Thames (which form a number of chapters of this volume, and its sister volume, Macdonald and Feber 2015), that it is essential that a coordinating role is maintained across landscapes. Such a role needs to be aware of and understand the many elements at play within the rural environment; from developing and evaluating scientific projects, to maintaining relationships with conservation practitioners, policy makers and, critically, landowners. Providing linkages between these vital elements, as we have done, will strengthen understanding of the complexities of restoring biodiversity to the British countryside and will lead to the further development of approaches and policies that will safeguard the rural environment in a changing world.

Acknowledgements

We gratefully thank the Tubney Trust for funding the five-year core UTP programme, and particularly remember the role of the late Terry Collins. Many members of WildCRU participated in this project, but we mention in particular Adam Dutton, Tom Moorhouse, Thomas Merckx, Rosie Salazar and the late Rob Strachan and, among our external collaborators, the late Gareth Edwards-Jones. The Upper Thames Project depended, as described in the text, on the collaboration of many non-government and government organizations, without whom the project could not have succeeded. Paul Johnson acknowledges the support of the Whitley Trust.

References

Baguette, M., Blanchet, S., Legrand, D., Stevens, V.M., and Turlure, C. (2013). Individual dispersal, landscape connectivity and ecological networks. *Biological Reviews*, **88**, 310–326.

Baker, J.M.R. and Halliday, T.R. (1999). Amphibian colonization of new ponds in an agricultural landscape. *Herpetological Journal*, **9**, 55–63.

Beebee, T.J.C. (1981). Habitats of the British amphibians (4)—Agricultural lowlands and a general discussion of requirements. *Biological Conservation*, **21**, 127–139.

Beebee, T.J.C. (1985). *Frogs & toads*. Whittet, Stowmarket, Suffolk.

Blaxter, K. and Robertson, N. (1995). *From dearth to plenty: the modern revolution in food production*. Cambridge University Press, Cambridge, UK.

Brede, E.G., Rowe, G., Trojanowski, J., and Beebee, T.J.C. (2001). Polymerase chain reaction primers for microsatellite loci in the Common Toad Bufo bufo. *Molecular Ecology Notes*, **1**, 308–310.

Butler, S.J., Vickery, J.A., and Norris, K. (2007). Farmland biodiversity and the footprint of agriculture. *Science*, **315**, 381–384.

Carrier, J.A., and Beebee, T.J.C. (2003). Recent, substantial, and unexplained declines of the common toad Bufo bufo in lowland England. *Biological Conservation*, **111**, 395–399.

Cook, H., Wem, F.C., Benson, D., Inman, A., Jordan, A., and Smith, L. (2012). Catchment management groups in England and Wales: extent, roles and influences. *Water and Environment Journal*, **26**, 47–55.

Cornulier, T., Robinson, R.A., Elston, D., Lambin, X., Sutherland, W.J., and Benton, T.G. (2011). Bayesian reconstitution of environmental change from disparate historical records: hedgerow loss and farmland bird declines. *Methods in Ecology and Evolution*, **2**, 86–94.

CPRE and NFU (2006). *Living Landscapes: Hidden costs of managing the countryside*.

Doncaster, C.P. (1992). Testing the role of intraguild predation in regulating hedgehog populations. *Proceedings of the Royal Society B*, **249**, 113–117.

Doncaster, C.P., Rondinini, C., and Johnson, P.C.D. (2001). Field test for environmental correlates of dispersal in hedgehogs Erinaceus europaeus. *Journal of Animal Ecology*, **70**, 33–46.

Dutton, A., Edwards-Jones, G., Strachan, R., and Macdonald, D.W. (2008). Ecological and social challenges to biodiversity conservation on farmland: reconnecting habitats on a landscape scale. *Mammal Review*, **38**, 205–219.

Farmers Weekly (2006). <http://www.fwi.co.uk/>.

Feber, R.E., Asteraki, E.J., and Firbank, L.G. (2007). Can farming and wildlife coexist? In D.W. Macdonald & K. Service, eds., *Key Topics in Conservation Biology*, pp. 239–252. Blackwell Publishing, Oxford, UK.

Fitzgibbon, S.D. (1997). Small mammals in farm woodlands: the effect of habitat, isolation and surrounding land use patterns. *Journal of Applied Ecology*, **34**, 530–539.

Franks, J.R. and Emery, S.B. (2012). Incentivising collaborative conservation: lessons from existing Environmental Stewardship Scheme options. *Land Use Policy 2013*, **30**, 847–862.

Freckleton, R.P., Sutherland, W.J., Watkinson, A.R., et al. (2008). Modelling the effects of management on population dynamics: some lessons from annual weeds. *Journal of Applied Ecology*, **45**, 1050–1058.

FWAG (2013). Farming and Wildlife Advisory Group. <http://www.fwagsw.org.uk/>.

Gelling, M., Macdonald, D.W., and Mathews, F. (2007). Are hedgerows the route to increased farmland small mammal density? Use of hedgerows in British pastoral habitats. *Landscape Ecology*, **22**, 1019–1032.

Green, R., Osborne, P., and Sears, E. (1994). The distribution of passerine birds in hedgerows during the breeding season in relation to characteristics of the hedgerow and adjacent farmland. *Journal of Applied Ecology*, **31**, 677–692.

Hanski, I. (1999). Habitat connectivity, habitat continuity and metapopulations, in dynamic landscapes. *Oikos*, **87**, 209–219.

HM Government (2011). *The Natural Choice: Securing the Value of Nature*. The Stationery Office. <http://www.ukeof.org.uk/>.

Hof, A.R. and Bright, P.W. (2010). The value of agri-environment schemes for macro-invertebrate feeders: hedgehogs on arable farms in Britain. *Animal Conservation*, **13**, 467–473.

Hof, A.R. and Bright, P.W. (2012). Factors affecting hedgehog presence on farmland as assessed by a questionnaire survey. *Acta Theriologica*, **57**, 79–88.

Houlahan, J.E., Findlay, C.S., Schmidt, B.R., Meyer, A.H., and Kuzmin, S.L. (2000). Quantitative evidence for global amphibian population declines. *Nature*, **404**, 752–755.

Janin, A., Lena, J.P., Ray, N., Delacourt, C., Allemand, P., and Joly, P. (2009). Assessing landscape connectivity with calibrated cost-distance modelling: predicting common toad distribution in a context of spreading agriculture. *Journal of Applied Ecology*, **46**, 833–841.

Mabey, R. (1980). *The common ground: a place for nature in Britain's future?* Hutchinson in association with the Nature Conservancy Council, London.

Macdonald, D.W. (1984). A questionnaire survey of farmers' opinions and actions towards wildlife on farmlands. In D. Jenkins, ed. *ITE Symposium No. 13: Agriculture and the environment*, pp. 171–177.

Macdonald, D.W. and Feber, R.E., eds. (2015). *Wildlife Conservation on Farmland. Managing for Nature on Lowland Farms*. Oxford University Press, Oxford.

Macdonald, D.W. and Johnson, P.J. (1995). The relationship between bird distribution and the botanical and structural characteristics of hedges. *Journal of Applied Ecology*, **32**, 492–505.

Macdonald, D.W. and Johnson, P.J. (2000). Farmers and the custody of the countryside: trends in loss and conservation of non-productive habitats. *Biological Conservation*, **94**, 221–234.

Macdonald, D.W. and Johnson, P.J. (2003). In *Conservation and Conflict: Mammals and Farming in Britain* (eds. F.H. Tattersall and W.J. Manley), pp. 2–16. Linnean Society Occasional Publication, Westbury Publishing, Yorkshire.

Macdonald, D.W., Johnson, P.J., Feber, R.E. and Tattersall, F.H. (2000). Habitat heterogeneity on farmland. In M. Hutchings., eds., *Ecological Consequences of Habitat Heterogeneity*, pp. 357–378. Blackwell Science, Oxford.

Macdonald, D.W., Tattersall, F.H., Service, K.M., Firbank, L.G., and Feber, R.E. (2007). Mammals, agrienvironment schemes and set-aside—what are the putative benefits? *Mammal Review*, **37**, 259–277.

Merckx, T. and Berwaerts, K. (2010). What type of hedgerows do Brown hairstreak (*Thecla betulae* L.) butterflies prefer? Implications for European agricultural landscape conservation. *Insect Conservation and Diversity*, **3**, 194–204.

Merckx, T., Feber, R.E., Riordan, P., et al. (2009). Optimizing the biodiversity gain from agri-environment schemes. *Agriculture, Ecosystems and Environment*, **130**, 177–182.

Merckx, T., Marini, L. Feber, R.E., and Macdonald, D.W. (2012). Hedgerow trees and extended-width field margins enhance macro-moth diversity: implications for management. *Journal of Applied Ecology*, **49**, 1396–1404.

Moorhouse, T. P., Palmer, S. C. F., Travis, J.M. J., et al. (2014) Hugging the hedges: Might agri-environment manipulations affect landscape permeability for hedgehogs? *Biological Conservation*, **176**, 109–116.

Morris, C. and Potter, C. (1995). Recruiting the New Conservationists—Farmers Adoption of Agri-Environmental Schemes in the UK. *Journal of Rural Studies*, **11**, 51–63.

NE (2010). *Hedges, the Law, Rules and Regulations*. Natural England.

Rackham, O. (2003). *The Illustrated History of the Countryside*. Weidenfeld & Nicolson, London.

Reading, C.J., Loman, J., and Madsen, T. (1991). Breeding pond fidelity in the common toad, *Bufo bufo*. *Journal of Zoology*, **225**, 201–211.

RSPB (2013). <http://www.rspb.org.uk/>.

Salazar, R.D., Eynard, S.E., Montgomery, R.A., Coles, R., Rogers, A., and Macdonald, D.W. (submitted-a). The effect of Euclidean distance and isolation by barriers on common toad Bufo bufo genetic distance.

Salazar, R.D., Montgomery, R.A., Thresher, S.E., and Macdonald, D.W. (submitted-b). Mapping the relative probability of common toad occurrence in terrestrial lowland farm habitat in the United Kingdom.

Slade, E.M., Merckx, T., Riutta, T., Bebber, D.P., Redhead, D., Riordan, P., and Macdonald, D.W. (2013). Life-history traits and landscape characteristics predict macro-moth responses to forest fragmentation. *Ecology*, **94**, 1519–1530.

Smith, M.A. and Green, D.M. (2005). Dispersal and the metapopulation paradigm in amphibian ecology and conservation: are all amphibian populations metapopulations? *Ecography*, **28**, 110–128.

Strachan, R. and Holmes-Ling, P. (2003). *Restoring water voles and other biodiversity to the wider countryside. A report on the Chichester coastal plain sustainable farming partnership*. Wildlife Conservation Research Unit, Oxford.

Strachan, R. and Jefferies, D.J. (1993). *The water vole Arvicola terrestris in Britain 1989–1990: its distribution and changing status*. The Vincent Wildlife Trust, London.

Stuart, S.N., Chanson, J.S., Cox, N.A., Young, B.E., Rodrigues, A.S.L., Fischman, D.L., and Waller, R.W. (2004). Status and trends of amphibian declines and extinctions worldwide. *Science*, **306**, 1783–1786.

Verboom, B. and Huitema, H. (1997). The importance of linear landscape elements for the pipistrelle *Pipistrellus*

pipistrellus and the serotine bat *Eptesicus serotinus*. *Landscape Ecology*, **12**, 117–125.

Wake, D.B. (1991). Declining amphibian populations. *Science*, **253**, 860–860.

Ward, J.F., Austin, R.M., and Macdonald, D.W. (2000). A simulation model of foraging behaviour and the effect of predation risk. *Journal of Animal Ecology*, **69**, 16–30.

Ward, J.F., Macdonald, D.W., Doncaster, C.P., and Mauget, C. (1996). Physiological response of the European hedgehog to predator and nonpredator odour. *Physiology and Behavior*, **60**, 1469–1472.

Ward, J.F., Macdonald, D.W., and Doncaster, C.P. (1997). Responses of foraging hedgehogs to badger odour. *Animal Behaviour*, **53**, 709–720.

Wembridge, D. (2011). *The state of Britain's hedgehogs*. British Trust for Ornithology (BTO), commissioned by People's Trust for Endangered Species (PTES) and the British Hedgehog Preservation Society (BHPS).

Williams, P., Biggs, J., Crowe, A., Murphy, J., Nicolet, P., Weatherby, A., and Dunbar, M. (2010). *Countryside Survey: Ponds Report from 2007*. Technical Report No. 7/07. Pond Conservation and NERC/Centre for Ecology & Hydrology. 77pp. CEH Project Number: C03259.

Willock, J., Deary, I.J., Edwards-Jones, G., Sutherland, A., Dent, J.B., Morgan, O., and Grieve, R. (1999). The role of attitudes and objectives in farmer decision making: business and environmentally-oriented behaviour in Scotland. *Journal of Agricultural Economics*, **50**, 286–303

CHAPTER 15

Many hands make light work—but do they? A critical evaluation of Citizen Science

Christina D. Buesching, Eleanor M. Slade, Chris Newman, Terhi Riutta, Philip Riordan, and David W. Macdonald

In the end we will conserve only what we love. We love only what we understand. We will understand only what we are taught.

Baba Dioum (1968) from his speech to the International Union for Conservation of Nature in New Delhi.

15.1 Introduction

Pressures of human society—population growth, development, and industrialization—are ever increasing, threatening global biodiversity and ecosystems (Cardinale et al. 2012; Hooper et al. 2012). To safeguard not only wildlife, but also society's well-being, it is crucial to inform and engage citizens as well as decision makers in conservation imperatives (McKinney and Harmon 2007). Harnessing capacity of individuals to contribute to conservation has become a priority (Oberhauser and Prysby 2008). Active participation can be particularly beneficial in terms of achieving more in less time (many hands making light work). More importantly, perhaps, is how public involvement in environmental issues increases understanding (Reed 2008), as epitomized by Senegalese environmentalist Baba Dioum (1968): 'In the end we will conserve only what we love; we will love only what we understand; and we will understand only what we have been taught.'

A wealth of national and international policy has been enacted to redress biodiversity loss, though the effectiveness of different strategies can be assessed only through monitoring their outcomes (e.g. Newman and Macdonald 2013). For example, part of the UK government's plan to comply with the Convention on Biological Diversity (CBD 1993), was to initiate a scoping study to monitor (in this instance) the status of Britain's mammals. Macdonald et al. (1998) conceived the Mammal Monitoring Network (MaMoNet) but identified that, while professional ecological surveyors were essential, systematic national monitoring would prove too labour-intensive and costly to be carried out solely by professionals. The solution? To involve a corps of amateur volunteers, or 'Citizen Scientists' (Irwin 1995)—and, in the process, simultaneously deliver on the CBD's Agenda 21 *'3-Cs'*: Conservation, Commerce and Community (also termed the *'3-Es'*: Environment, Economy and Equity). Engaging the general public in conservation initiatives addresses in particular the 'Community/Equity' aspirations of the Convention—e.g. Articles 3.25 (Engaging young people in environmental responsibility); 3.26 (Involving indigenous people and communities); 3.27 (Working with NGOs); 3.28 (Supporting local authority initiatives); and 4.36 (Promoting education, public awareness and training)—while involving the corporate sector in Citizen Science contributes to the Commercial/Economic objectives of the CBD (see Pearce 2009). Under this scenario, companies act as 'biosphere stakeholders', through demonstrating 'Corporate Social Responsibility' (Henningfeld et al. 2006).

Although large volumes of data can be gathered by Citizen Scientists, there has been some derision by the scientific community of the value of the information collected by amateur volunteers (Darwall and Dulvey 1996). In the USA, for example, the government's

Wildlife Conservation on Farmland. Conflict in the Countryside. Edited by David W. Macdonald and Ruth E. Feber.
© Oxford University Press 2015. Published 2015 by Oxford University Press.

assertion that 'volunteers are incompetent and biased', has led to some public monitoring programmes being reverted to projects carried out solely by professional scientists (Darwall and Dulvey 1996). Certainly, many tasks in conservation and ecology are highly technical and skilled (for example sedating and handling wildlife), for which biologists undergo years of training. Activities suitable for delegation to amateurs are thus limited by ethical, legal, and practical constraints. Nevertheless, potentially, non-specialist work can benefit from the support of a large number of even inexperienced participants.

15.2 WildCRU's Citizen Science study

In April 2000, we began a ten-year pioneering study to evaluate the merits and disadvantages of engaging non-professionals in ecological surveying, based on our own volunteer-based monitoring programme. In this context, unlike many volunteer programmes where volunteers contribute to existing research studies, our programme was unusual in that it was deliberately conceptualized and designed from the outset to use volunteers in data collection, within a framework involving engagement and learning.

Using work undertaken primarily through Citizen Science (CS) and Corporate Social Responsibility (CSR), the over-arching scientific goal of WildCRU's volunteer-based studies was to monitor how a variety of taxa respond to climatic and environmental change. The study was located in Wytham Woods; Oxford University's 'living laboratory', and home to a multitude of long-term ecological studies (see Savill et al. 2010; Macdonald and Feber 2015: Chapters 4 and 12, this volume).

In this chapter, we report on two facets of CS: the veracity of the work undertaken by volunteers and the benefits of public and corporate participation for personal development and attitudes. Notwithstanding cascade advantages to society, we consider how investment in training and team management must be balanced against costs; we discuss the motivation corporate sponsors might have for sending their employees on environmental volunteering programmes, and most importantly, we investigate the veracity of the data produced, so-called parameters of Quality Assurance/Quality Control (QA/QC: Wagner 1995; Michener 2006).

From the scientists' perspective, we describe the advantages and disadvantages of using (and therefore managing) inexperienced volunteers and examine whether volunteer-collected data are reliable. Can reliability be built-in to these data through appropriate training, deployment, and supervision? And is this efficient—or should professionals be used for data gathering, confining the engagement of the public to their education? To these ends, we cross-reference to the more general area of ecotourism, which engages the public (and may have economic benefits) but does not involve the collection of scientific data.

15.2.1 The volunteers

Predominantly, our volunteers were recruited by the Earthwatch Institute, an international charity that seeks to 'engage people across the globe in scientific research, and in doing so, promotes the understanding of environmental sustainability' (Earthwatch Institute 2014). Earthwatch has a broad volunteer base, comprising > 40,000 members from around the world. Their volunteers have diverse social, economic, and educational backgrounds, with some (for example, from schools or disadvantaged sectors of society) being supported by corporate sponsorship. Earthwatch also solicits the active participation of corporate partner employees, either on bespoke teams (with a focus on CSR) or among general recruits. The profile of this recruiting agency thus suited our purposes, providing not only volunteers, but also a suitable infrastructure to examine broader facets of Citizen Science (see Henderson 2012).

Between April 2000 and December 2011, we hosted 92 volunteer teams, attended by almost 800 volunteers, within both public and corporate programmes. The 'Public Programme' (PP) comprised 42 teams (April 2000–September 2006), attended by 364 predominantly self-funded volunteers drawn from the general public. Maximum team size was 12, and total length of stay was five days. The remaining 50 'Corporate Programme' (CP) teams (May 2008–December 2011) comprised staff from the Hong Kong Shanghai Banking Corporation (HSBC), also with a maximum team size of 12 but for a period of seven or eight fieldwork days. A total of nearly 400 participants from 24 different countries, from the lowest to the highest levels within the bank, and all roles, from tele-assistants to bank managers and heads of department, participated in this programme.

Analysing data from the years 2000–2001, we found a distinct female bias among PP volunteers (61% F: 39% M), as seen for other ecological volunteer programmes (Coghlan 2005). Although ages of volunteers ranged from 11 to 86 years, men followed an almost bimodal age distribution, with peaks in

20–30 (29.6%) and 40–50 year olds (37%). The predominant age of women was 30–40 years old (35.5%) (Newman et al. 2003). More than 60% of participants were single (i.e. giving their parents, offspring, or siblings as emergency contact, but not mentioning a partner), and over 90% of all team members participated in the project on their own, unaccompanied by friends, spouse, or family, typical of other residential volunteering projects (Coghlan 2005). In contrast, our corporate employee teams were predominantly comprised of 20–50 year olds, and included mostly the lower levels of management.

CS projects can play an important role in involving new (and otherwise unlikely) constituencies in nature conservation and scientific research, as exemplified by our involvement with a local drug and alcohol rehabilitation centre, the Ley Community (Ley 2014). This collaboration, enabled by a grant from the National Lottery Fund, supported the social resettlement and work experience of reformed addicts after a year in rehabilitation, earning WildCRU and the Earthwatch Institute the National Charity Award for Community Involvement in 2002. Although Lottery Funding ended in 2004, WildCRU's relationship with the Ley Community continues, and each year, at least 12 residents, funded by a local charity, help with our long-term badger population study in Wytham Woods (see Chapter 4, this volume).

Although not especially reflected in our volunteer profiles at Wytham, the youth market is an increasing segment within CS. Teenager events are big business, especially in the USA (Paris 2008). Guiney and Oberhauser (2009) describe youth participation in conservation ventures as 'restorative', and 'reconnecting motives, values, and concepts of nature'. Nevertheless, our many 'teen-team' (aged 15–18) expeditions in Canada (Buesching and Newman, unpubl. data), where teens undertook similar ecological monitoring to our UK programme, showed that, because 'teens' are often volunteered by their parents or schools, they may be less motivated. Some teenagers on our Canadian project have reported conservation-related 'epiphanies' (see Driver and Johnson 1984); this extent of personal growth often outweighs the actual scientific rewards to the project.

15.2.2 Costs and benefits of using volunteers

Understanding how the trio of ecotourism, conservation biology, and volunteer benefits (Brightsmith et al. 2008) contribute to CS is essential (Silvertown et al. 2013), and presents an ambiguous marketing phenomenon (Tomazos 2010). By definition, voluntary work encapsulates the notion that volunteers donate their time to help, without monetary recompense. 'Volunteerism' is thus an example of what Clary et al. (1996) term 'prototypic planned helping', requiring planning, sorting priorities, and matching personal capabilities and interests with the type of required volunteer task (Clary et al. 1998; Brown 1999). Crucially, our PP model, typical of projects of this kind, was not perceived as free labour by the participants, since residential expeditions have to be paid for: *c.* £250 at the time, while other projects can cost in the region of £2500 for a two-week period (<http://www.earthwatch.org>). This situation can create two obstacles (see Campbell and Smith 2006). First, some people might object to paying to help others fulfil their jobs (in this case, supporting scientists: see Moon et al. 2002; Benson and Seibert 2009), although including a component of 'training/instruction' may offset this—termed 'commodification' in the tourist industry (Cousins et al. 2009). Second, the experience must be accessible to those who might benefit most from participation (disadvantaged/students/developing countries, etc.). Here, fellowships and sponsorship encourage cross-sector participation (Nixon 1997).

A key point is that volunteers were expected to 'work' on the project (either voluntarily during vacation time, or on corporate secondment)—a specific derivative of ecotourism identified as 'Participatory Environmental Research Tourism'—PERT (Ellis 2003; for perspectives see Cohn 2008; Brightsmith et al. 2008). This leads to divergent core objectives (Kruger 2005). On the one hand, PERT provides a distinct market niche within the competitive ecotourism segment; on the other, delivery of active research participation may involve tasks that are arduous, or repetitive, where this might diminish either the appeal of the expedition, or enjoyment for the participants (relative to pure ecotours/safaris, etc.; see Dolnicar and Randle 2007). This disjunction is termed the 'false dichotomy of participation' by Lawrence (2006), a dilemma often afflicting science undertaken by volunteers (see Moscardo et al. 1996; Manfredo et al. 1996; Lovell et al. 2009). The obvious solution is to collect data within a programme that is not 'boring' (particularly so for those not incentivized by career or salary, as with professional ecologists)—an aspiration more easily conceived than achieved, whereby interspersing educational and cultural activities (e.g. museum visits, lectures; Fig. 15.1) can lead to distractions from core research goals and impact on research time and logistics (Coghlan 2006).

Figure 15.1 Although a 'cost' to the scientist, it is important to motivate CS through teaching project-related background knowledge to enhance data quality and volunteer enjoyment. Photograph ©C.D. Buesching.

15.2.3 What can volunteers contribute to conservation ecology research?

Notwithstanding the benefits of social inclusiveness and corporate responsibility, acceptance of volunteer-collected data within the scientific community has proven difficult (Mackney and Spring 2000; Brudney and Gazely 2006). As a consequence, CS brings with it a set of challenges and potential hazards that may limit progress (Macdonald et al. 2002; Buesching and Newman 2005; Dickinson et al. 2010). For example, some scientists argue that volunteer training and supervision take time away from professional research (Douglas and Rollins 2007; Brightsmith et al. 2008; Silvertown et al. 2013), and that data can be prone to high intra- and inter-observer variability (Darwall and Dulvey 1996; Newman et al. 2003; Buesching et al. 2014). The argument is also made that some tasks will always require professional involvement to conform with government licensing procedures, Health and Safety requirements, animal welfare considerations, or for accurate species identifications (Lorimer 2009; Silvertown et al. 2013). Furthermore, as volunteer support cannot always be guaranteed, CS projects have to be capable of adapting their work schedule accordingly, often on an *ad hoc* basis.

Few studies have analysed the validity and efficacy of data collected by Citizen Scientists (Mackney and Spring 2000; Butt et al. 2013). We therefore undertook our own study to evaluate volunteer efficacy, paraphrasing Juvenal's question: 'Quis custodiet ipsos custodes?'[1] (see Newman et al. 2003).

15.2.4 How proficient were amateur volunteers as ecological monitors, and what factors influenced the quality and validity of the data they collected?

A limited, but increasing, number of studies (e.g. marine: Darwall and Dulvey 1996; Foster-Smith and Evans 2003; woodland mammals: Newman et al. 2003; Buesching et al. 2014; trees: Butt et al. 2013) indicate that, with appropriate training, most volunteers are capable of performing appropriately allocated monitoring techniques proficiently (Box 15.1). Of concern, however, is that inter-observer variation in data quality among volunteers can be up to four times higher than among professional biologists (Newman et al. 2003; Butt et al. 2013).

A variety of different factors affect volunteer performance, and thus data veracity (QA/QC), essential for use in publications and reports, or to underscore policy (see Brandon et al. 2003; Pattengill-Semmens and Semmens 2003). Establishing an acceptable range of data quality is therefore important. Butt et al. (2013), for example, found that, although tree biomass data collected by volunteers was of a slightly lower quality than that collected by professionals, this had little overall impact on carbon biomass estimates over the large scale, and so still provided a useful way of collecting data on carbon stocks across large spatial and temporal scales.

15.2.5 Motivation and volunteer performance

'Can you be bothered?' Lienhoop and Fischer (2009) asked this question of volunteers and established that their motivation played a significant role in their valuation of species conservation measures. Volunteer expectations and motivations range from 'gathering new experiences' (Pearce and Lee 2005) and

[1] i.e. 'Who will take care of the care takers?' Juvenal's Satires.

Box 15.1 Using volunteers to estimate deer population densities

Deer population densities and ranges are increasing in Europe and North America (Fuller and Gill 2001). Unregulated numbers of deer can jeopardize biodiversity and public safety through adverse effects on forest ecosystems (Gill 2006; Buesching et al. 2011; Gill and Fuller 2007), spreading zoonotic diseases (Spielman 1994), and causing serious road traffic accidents. Deer densities, though, are difficult to monitor using visual counts. Estimating densities from faecal pellet (dropping) surveys (Box Fig. 15.1.1) is much more reliable but time consuming, and thus expensive. Utilizing volunteers though, could potentially make this method more cost effective.

Working with over 600 volunteers, we established an ecological monitoring protocol for deer, suited to volunteer deployment (Buesching et al. 2014). We collected data in the UK (Oxfordshire: Wytham Woods; 356 volunteers) and in Canada (Nova Scotia: Cook's Lake Farm; 255 volunteers). We used teams of 5–12 volunteers to survey the standing faecal crop (i.e. the number of deer pellets on the ground) in 120 quadrats of 10 m × 10 m (Box Fig. 15.1.2) (Bailey and Putman 1981) per year, moving through the survey area as an evenly spaced line-up, thus avoiding double counts of pellet groups or un-surveyed stretches.

Because of concurrent deer culling at Wytham Woods, we were able to validate their estimates, and we found that the inter-annual change in deer densities from the volunteers' pellet counts correlated with the reduction in deer numbers achieved by shooting (roe and fallow combined for three successive years: $n = 6$; Pearson's correlation = 0.890, $P = 0.017$).

Overall, pellet group counting with supervised volunteer teams appeared to be a viable time- and cost-effective technique for monitoring deer populations.

Box Figure 15.1.1 White-tailed deer droppings used in faecal pellet counts to estimate deer densities in Nova Scotia, Canada. Photograph ©C.D. Buesching.

continued

Box 15.1 *Continued*

Box Figure 15.1.2 CS volunteers surveying quadrat during faecal pellet counts to estimate deer and hare densities. Photograph ©C.D. Buesching.

'being adventurous' (Moscardo et al. 1996), to 'learning about the environment' (Manfredo et al. 1996) and 'supporting an organization/scientific project' (Bonjean et al. 1994).

Only around 30% of our CP volunteers declared previous knowledge of, or interest in, the natural environment. All saw the opportunity as a bonus within their banking careers; of the first 24 bank employees seconded to assist us, three subsequently left their jobs to start a career in a conservation-related field. In contrast, 60% of the PP volunteers stated on the first evening that they hoped to gain a better understanding of nature and the environment by working alongside like-minded people. Around 25% of these volunteers were at a turning point either in their private (e.g. divorce) or professional (e.g. career change; job loss) lives. Approximately 25% were keen naturalists with lifelong interests in nature, *c.* 60% had some underlying interest in conservation, whereas approximately 15% were embarking on something completely new (Silvertown et al. 2013). In addition, around 10% of PP participants had underlying career-related motivations; for example, biology or veterinary science students seeking work experience and references.

In terms of specific motivation, we found that, if volunteers arrived with a clear expectation of the goals they hoped to achieve (e.g. 'collecting course credits' or 'seeing a live badger'), they were more likely to be hard-working and appreciative of the overall project experience than volunteers with only vague goals[2], although the latter often described themselves as 'easy-going' and 'happy with anything'. This remained true even if the specific expectations were not/could not be fulfilled, but were instead substituted by other valuable experiences (see subsection 'Wishful thinking'). The mismatch between volunteers' skill-sets/experience and the expedition ahead of them is a phenomenon termed the 'expectation disconfirmation paradigm' (Yüksel and Yüksel 2001). This creates a paradox: some participants on expeditions may be signing up for an experience they are unlikely to enjoy for its own merits (Dempsey 2012), but rather benefit from different ancillary perspectives of the experience.

[2] On their feedback forms 96% of volunteers with clear expectations (n = 47) rated the overall experience in the field as 5/5 compared to 76% of volunteers without such expectations (n = 32).

15.2.5.1 Wishful thinking and the effects of boredom on volunteer performance

A content and satisfied volunteer is usually a good volunteer (Caissie and Halpenny 2003), easier to motivate and manage (Campbell and Smith 2006), and also more successful in collecting good quantities of reliable data (Jacobson et al. 2012). Success fosters success and enhances motivation (Prussia and Kinicki 1996; Silvertown et al. 2013), thus preceding surveys with positive training practice is beneficial. For example, we found that volunteer performance in finding deer droppings (Newman and Buesching 2004a,b; Box 15:1) increased three-fold if volunteers first experienced the reward of surveying a few plots containing at least some droppings, instead of starting with a series of 'empty' quadrats (Buesching et al. 2014).

Nevertheless, in order to direct learning and performance (Ballantyne et al. 2007), it proves important to explain the relevance of 'negative' results to volunteers (Silvertown et al. 2013). This is so because many novice volunteers tend to disregard non-events (e.g. not understanding the importance of the absence of field signs/animal sightings as evidence of absence: Manel et al. 2001). Similarly, they may undervalue common events (e.g. the repeated findings of field signs of the same species: McLaren and Cadman 1999; Newman et al. 2003; Danielsen et al. 2005), and, by seeking out the rare and 'more exciting' observations, they can thus induce considerable data skew (Mumby et al. 1995; Brandon et al. 2003; Foster-Smith and Evans 2003; Genet and Sargent 2003).

For example, in the Mammals of Nova Scotia Project (Canada), more than 75% of participants mentioned bears on their wish-list of species sightings (Buesching and Newman, unpubl. data). Subsequently, 21% of the teams who did not find any evidence of bears during their two-week expedition recorded large coyote droppings (common throughout the area, and thus familiar to all volunteers—unlikely to be genuinely misidentified by this stage) as black bear droppings on their last transect surveys. This apparent 'wishful thinking' did not apply to those who had really seen evidence of bears, only 3% of whom misidentified coyote droppings.

In addition, volunteers display a lower 'boredom threshold' than do professionals (Basinger 1998; Blackburn and Frank 2010). In a large-scale study on volunteer-based monitoring of deer densities in the UK and Canada, Buesching et al. (2014) found that, after 60–90 minutes of surveying quadrats for faecal pellets (around ten quadrats; Macdonald and Feber 2015: Chapter 12), volunteer attention dropped dramatically. This resulted in three times more missed droppings (verified by accompanying scientist) and a four-fold increase in the time needed to arrive at and measure out the next survey quadrat. Simultaneously, time needed to survey each quadrat shortened to almost half, indicating a decrease in volunteer survey attention (see also Newman et al. 2003). Since many monitoring tasks require a large number of repetitive surveys, it becomes essential to vary daily volunteer tasks sufficiently to maintain interest, and thus data quality (see Asah and Blahna 2012). For example, we found that combining half a day of tree measuring with half a day of mammal or moth trapping resulted in more motivated and engaged HSBC teams (see Section 15.4). Similarly, analysing survey results with the team participants proved important (Fig. 15.2), as amassing data without seeing the product of their labour left volunteers feeling disenfranchised (Silvertown et al. 2013). For example, we undertook weekly classroom sessions with teams, explaining the mathematics of turning dropping counts into average deer densities. However, such input adds to the investment of scientist time (Box 15.2).

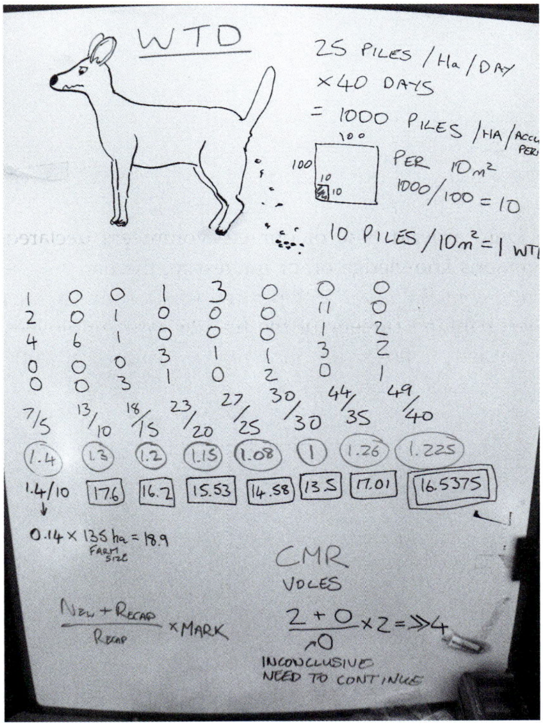

Figure 15.2 Analysis of the data collected by each volunteer team, at a level non-professionals can understand easily, is important for volunteer motivation. Photograph ©C.D. Buesching.

Box 15.2 Banker versus researcher

Based on our experience of running HSBC field teams, we quantified the research time output of two 12-person HSBC corporate volunteer teams, which differed in their durations (one seven-day team, Box Fig. 15.2.1; or two four-day teams running consecutively, Box Fig. 15.2.2), and compared this with the outputs from a scientific team (researcher plus a trained assistant). Research output by volunteers was calculated by multiplying the actual research hours during a typical 7.5 hour day (excluding introduction sessions on day 1 and 'wrap-up' sessions, presenting the results collected by the team and setting these into the context of the wider research programme on day 4 or 7) by the 'efficiency ratio', i.e. how much research a team of volunteers can undertake compared to one researcher. We estimated that a team of 12 volunteers can carry out as much research as one researcher in day one (as most of the time in the field is taken up by training), twice as much in day two, and three times as much in day three (i.e. four volunteers can achieve as much as one professional). In a seven-day team the assumption is that on days 4–7 the volunteer team can do a maximum of 3.5 times as much research as one researcher alone (3.5 volunteers equal the work of one researcher). Lovell et al. (2009) suggested a more generous estimate, that two volunteers can equal one researcher. Obviously, the volunteer efficiency estimated following this formula is very sensitive to the specific assumptions and to the complexity of the research tasks; but the calculations presented here represent a realistic scenario for the teams we were running. Time input by the researcher includes the time spent in the field and time needed for preparation of equipment, planning the logistics, and putting together the research 'wrap-up', based on the data collected by each volunteer team.

Box Fig. 15:2.1 shows that the scientific team of one researcher and a trained assistant provided a steady output of 15 researcher hours per day (i.e. one full working day per researcher; a fairly conservative estimate as field work often requires that researchers spend more than 7.5 hours in the field during field seasons), and resulted in the highest research time output. However, an almost comparably high research output can be achieved with a seven-day volunteer programme that includes one introduction/training day, five days of research, and one rest day in the middle of the programme (Box Fig. 15.2.1). In contrast, short teams result in a loss of research time, as the input in training time by the researcher is a fixed cost that does not depend on the duration of the team. Thus, the two consecutive four-day

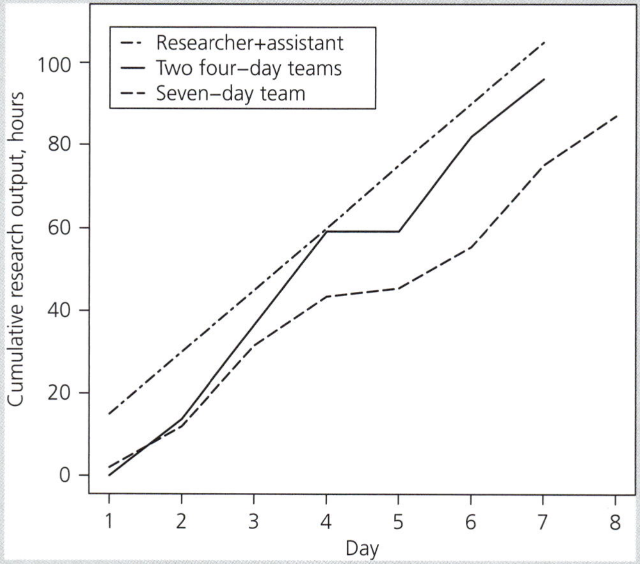

Box Figure 15.2.1 Comparison of the cumulative research time output by two consecutive four-day teams (twelve volunteers), by one seven-day team (twelve volunteers), and by a researcher and one trained research assistant, measured as 'researcher hours', i.e. output equivalent to one researcher.

continued

Box 15.2 *Continued*

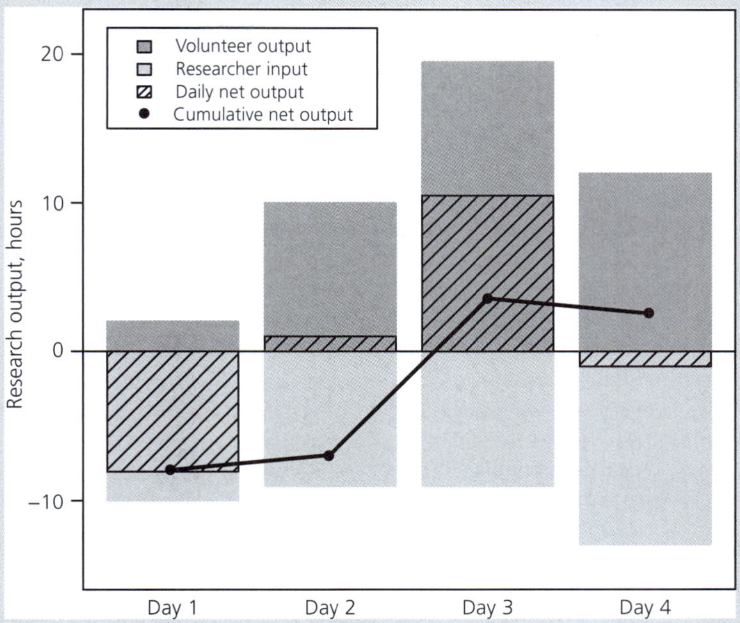

Box Figure 15.2.2 Time input by researcher, research time output by a team of twelve volunteers (measured as 'researcher hours', i.e. output equivalent to one trained researcher), and the daily and cumulative net balance over a four-day volunteer team.

volunteer programmes, which ran for eight days, was the least efficient despite having one more day of research time in its structure (Box Fig. 15.2.1). Eight days of research, by two separate volunteer teams, offer only one productive research day per team (day 3, Box Fig. 15.2.2) due to the large input in time by the researcher in training the team on the first day, and to maximum efficiency in data collection only being reached on day 3, as the volunteers are still familiarizing themselves with the techniques on day 2 (the first full day in the field) (Box Fig. 15.2.2). Thus, having to train and debrief two teams significantly reduces the amount of time spent on research. We conclude that planning volunteer programmes to run for longer periods of time and accounting for the time needed to train teams is essential for volunteers to make a significant contribution to research outputs.

CS volunteers may also be put off when they consider tasks too strenuous to complete, often not truly comprehending the level of fitness or the physical conditions described in the project briefing (e.g. adverse weather, etc.), when professional researchers would expect to continue working. This can result in data loss, often incompatible with sustained monitoring efforts (Macdonald et al. 1998; Silvertown et al. 2013). In this context, the need for responsible risk management to accommodate inexperienced amateurs (public liability) can seriously constrain volunteer deployment during inclement conditions in tough environments (Benson 2010).

15.2.5.2 Individual variation in volunteer performance

Individual volunteers vary according to psychological and motivational factors, as well as physical qualities, such as gender, age, fitness and dexterity (Ericsson and Wallin 1999; Barrett et al. 2002; Genet and Sargent 2003; Buesching et al. 2014). These factors can affect volunteer suitability for specific tasks (Newman et al. 2003; Buesching et al. 2014).

From our evaluation of 155 PP volunteers, made by one male and one female researcher (see Newman et al. 2003), we found that men scored, on average, higher (3.2/5) than did women (2.7/5) on a scale of 0 (= unreliable work/data quality) to 5 (= very efficient/data comparable to professional quality) (Newman et al. 2003; Buesching and Newman 2005). There was, however, task-specific variation in these findings. While men and women scored similarly for deploying small mammal traps, men performed significantly better at finding deer droppings in artificially stocked quadrats (containing a known number of droppings) than did women[3]. Investigating this further, we observed that the gender difference was most pronounced when experience was minimal: after surveying only five quadrats men out-scored women[4]. After 20 quadrats, however, this difference had all but disappeared[5]. This seemed to be because men were initially more confident, while women gained proficiency through practice and experience (see also Buesching et al. 2014). Interestingly, the standard variation among women in task-specific scores was generally much higher than for men[6] (Newman et al. 2003), reflecting the fact that many of our best, but also of our worst, volunteers were women, whereas the majority of men were average. The performance of individuals was influenced by training, gender, and willingness and aptitude to survey (W&A). After five quadrats, men averaged significantly higher effectiveness[7] than women, predominantly due to shorter survey times[8]. After 20 quadrats this gender difference in effectiveness was no longer significant[9], although survey time persisted as the most influential component of gender-related differences[10]. W&A was related significantly to effectiveness[11]. Higher W&A also improved species identification scores significantly by quadrat 20, but not by quadrat 5. Age did not influence ability to perform surveys effectively.

Overall, physical fitness (assessed on a scale of 1 = very unfit/overweight, to 5 = extremely fit and athletic) correlated strongly and positively with overall volunteer performance scores[12], and the perceived usefulness of volunteers by the researchers[13], also taking into account the level of attention and supervision volunteers required to carry out the tasks (Newman et al. 2003).

Age, if corrected for fitness, however, had no significant effect on volunteer ability, although different age classes showed different task-specific strengths. For example, older people often proved more patient and meticulous, and thus better suited to fine-scale monitoring at a leisurely pace (e.g. species lists; bird calls; direct observations), while younger volunteers were often more energetic and thus better suited to more active tasks (e.g. transect walking through difficult terrain: Driver and Johnson 1984; Silvertown et al. 2013). Furthermore, some physical ailments cause certain task-specific limitations, as exemplified by red/green colour blindness. Men thus affected found on average only 16 ± 8% of deer droppings in artificially stocked survey quadrats, compared with 71 ± 22% of droppings found by normal-sighted volunteers (Buesching et al. 2014).

As an interesting footnote, there are also acknowledged health benefits related to participation in fieldwork and countryside projects, both physical and mental (see Miles et al. 1998; Pretty et al. 2005); where Hughes et al. (2013) propose that these could contribute to lowering healthcare budgets. To foster these benefits, the British Trust for Conservation Volunteers (BTCV) runs a 'Green Gym' scheme, specifically aimed at counteracting sedentary lifestyles through active participation in practical conservation tasks (e.g. Birch 2005), where this leisurely form of occupational therapy has been assessed to genuinely benefit participants (Yerrell 2008).

15.2.6 What type of training is best? And how can scientists make the volunteer experience credible and successful?

Leadership is crucial in working with Citizen Scientists (Coghlan 2008), but scientists are not necessarily adept at managing public groups (Cochran-Smith and Lytle 1990). Scientists may be supported by the volunteer recruiting agencies (see Barker et al. 2011) or through training and practice from colleagues (Bickford et al. 2012). In short, scientists get back only what they put in (Newman et al. 2003; Douglas and Rollins 2007; Coghlan 2008).

Certain types of volunteer-based surveys, particularly national ones (e.g. British Trust for Ornithology Breeding Bird surveys; Risely et al. 2008), rely upon written instruction provided by the scientists (e.g. guidebooks, web-based material; Silvertown et al.

3 Mann–Witney U-test: $n_{men} = 17$, $n_{women} = 27$, $p < 0.001$.

4 men: 2.6 ± 0.68; women: 2.2 ± 0.95; $F_{1,88} = 7.39$, $p = 0.008$; GLM.

5 men: 3.47 ± 0.68; women: 3.36 ± 0.95; $F_{1,88} = 2.15$, $p = 0.146$; GLM.

6 ANOVA: $F_{1,58} = 4.59$; $p < 0.05$.

7 (men: 2.6 ± SD 0.68, women 2.2 ± 0.95; $F_{1,88} = 7.39$, $P = 0.008$).

8 (men: 2.77 ± 0.77, women: 1.93 ± 0.73; $F_{1,88} = 23.09$, $P = 0.001$).

9 (men: 3.47 ± 0.68, women: 3.36 ± 0.95; $F_{1,88} = 2.15$, $P = 0.146$),

10 (men: 3.7 ± 0.88, women: 3.36 ± 0.94; $F_{1,88} = 3.82$, $P = 0.050$).

11 (quadrat 5: $F_{1,88} = 5.64$, $P = 0.001$; quadrat 20: $F_{1,88} = 5.39$, $P = 0.001$).

12 $H_{4,56} = 14.20$; $p < 0.01$; Kruskal–Wallis test.

13 $H_{4,56} = 11.82$; $p < 0.02$; Kruskal–Wallis test.

Figure 15.3 Teaching in the field. Seeing field signs in a naturalistic setting is essential for volunteers to learn. Photograph ©C. Newman.

2013). Practical training, though, proves the most important factor affecting performance of volunteers engaged in fieldwork (Fig. 15.3) (Foster-Smith and Evans 2003; Cohn 2008; Blackburn and Frank 2010).

For example, on our own project (Newman et al. 2003), more than 90% of our PP volunteers needed to see most field signs (e.g. droppings, tracks) only once or twice in a naturalistic setting to be able to identify them correctly, with fewer than 10% still being uncertain after five days (< 80% correct identification). Mis-identification rate of deer droppings through confusion of different deer species as well as with rabbits decreased from 19.2% after five quadrats to 6.9% after 20 quadrats (Newman et al. 2003). Perhaps counter-intuitively, however, volunteer efficiency at finding a known number of droppings in an artificially stocked quadrat fell from 75% to 67% with practice, and surveys became more superficial (whereas the research scientists scored a consistent 74%: Newman et al. 2003). We interpret this deterioration as either a symptom of diminished attention, or arising from over-confidence.

A labour-intensive task over a wide area, such as live-trapping of small mammals (Macdonald and Feber 2015: Chapter 4) to survey their numbers, genuinely benefits from having volunteer assistance (see Sibbald et al. 2006). We found that correct operation of the traps (e.g. provisioning and setting traps correctly in appropriate locations) proved easy to teach to novice volunteers, with 50% of volunteers being able to set traps correctly after one demonstration, 81% after two demonstrations, and 98% after three demonstrations, although 2% needed up to seven demonstrations (Newman et al. 2003). Trap-setting improved if volunteers were instructed to 'think like a mouse'. The cognitive associations this evoked led 80% of volunteers to set traps in places well protected from predators, rich in food, and thus generally attractive to small mammals (Fig. 15.4). In contrast, if these preferential habitat types were described in a purely logical matter-of-fact fashion, only 40% of volunteers deployed traps appropriately.

While we discovered that experienced volunteers found background information more interesting than did novices[14] (Newman et al. 2003), in our experience, training should not assume, or rely upon, prior knowledge if there is to be methodological consistency. Perhaps unexpectedly, some of our most successful volunteers came from the Ley Community, Alcohol- and Drug Rehabilitation Centre, who had neither prior interest in, nor involvement with, nature conservation before joining our project, but outshone retired vets and seasoned naturalists with their overall performance[15]. We attribute this to them having no a priori (mis-) conceptions about survey protocols, instead assiduously relying upon their instructions.

Aside from introductory training, we found that considerable on-going volunteer supervision was necessary to minimize volunteer error and to enhance performance (see also studies by Foster-Smith and Evans 2003; Douglas and Rollins 2007; Blackburn and Frank 2010). Providing positive feedback on team

[14] Mann–Witney U-test: $n_1 = 20$, $n_2 = 24$, $p < 0.01$.

[15] Ley Community volunteers: average score over all tasks 76 ±13%, n = 12; compared to highly qualified volunteers: 70 ± 9%, n = 34; $p < 0.05$; Mann–Witney U-test.

Figure 15.4 Success. Having *thought like a mouse*, a volunteer caught a small mammal in her Longworth trap. Photograph ©C.D. Buesching.

and/or individual performance incentivized effective data collection (Genet and Sargent 2003; Macdonald et al. 2002; Waterman 1993), without drawing attention to under-performing volunteers (Macdonald et al. 2002). Where relevant, placing a firm emphasis on animal welfare engendered a greater sense of personal satisfaction (Fennell and Nowaczek 2010).

15.3 Citizen Science and Corporate Sponsors: the HSBC Climate Centre at Wytham

In 2004, Britain's then Chief Scientist, Professor David King, sounded a dramatic warning, identifying climate change as a greater risk to society than terrorism (King 2004). This was echoed in the Stern Report (2006), which emphasized that technological designs driven only by conservation interests often ignore the need to be socially inclusive, or to incorporate adaptive forms of economic development as outlined in the Convention on Biological Diversity (1993).

This political realization precipitated a significant increase in charitable donations by major companies and financial institutions (Philibert 1999; Saha and Darnton 2005). Many of these sponsorship commitments were tied to employees gaining experience in practical conservation work through volunteering. Thus, corporate sponsors aimed to gain environmental credentials, and conservation organizations stood to gain financial support and a potentially large labour force (Rondinelli and Berry 2000; Earthwatch 2014).

In 2002, HSBC launched their $50 million five-year 'Investing in Nature' initiative, and became a corporate partner of Earthwatch, the World Wildlife Fund (WWF), and Botanic Gardens Conservation International. To date, over 2000 HSBC employees have participated in Earthwatch projects around the world. In 2007, the HSBC made a further investment of $100 million over five years in The Climate Group (Earthwatch, the Smithsonian Tropical Research Institute (STRI) and the WWF), launching the 'HSBC Climate Partnership', aiming to 'inspire action by individuals, businesses, and governments worldwide' (Earthwatch 2014).

Our work with CP volunteers at the Climate Centre at Wytham followed on from our experience with PP. In Wytham, the mosaic of small woodland areas typifies woodland structure in the UK, and provided an excellent opportunity for investigating how habitat fragmentation impacts certain woodland species, which may need to shift their distributions to stay within their envelope of tolerance—broad-scale work which could benefit from volunteer support.

For this corporate component of our CS evaluation, we trialled survey methods for several taxa in Wytham Woods and assessed which monitoring techniques were effective alongside logistical implications, and in turn, which surveys proved most popular, influencing data reliability and quantity (Lovell et al. 2009).

Survey protocols for small mammals, moths, and soil macrofauna (such as woodlice and millipedes) proved easy to teach and could be done in most weather conditions. As the timings of surveys for soil invertebrates was flexible, and small mammal and moth surveys were conducted at dawn and dusk, this left time for other activities, including classroom-led education sessions, an integral part of the CP volunteer programme. Furthermore, these labour- and time-intensive surveys typify the kind of task where volunteer involvement extends the professional scientist's capacity to undertake monitoring over large spatial and temporal scales (Lovell et al. 2009; Slade et al. 2013). In the remainder of this chapter we present two case studies to illustrate what can be achieved with volunteers.

15.4 Case studies highlighting how high-calibre science can be achieved through carefully designed volunteer-based research

15.4.1 Case study 1. Using volunteers to investigate the effects of decomposition by woodlice and millipedes

Around 80% of litter in a forest is processed by soil macrofauna (Hassall et al. 1987; David and Handa 2010) and, in their absence, litter disappearance is greatly reduced (e.g. Seastedt 1984; Bradford et al. 2002; Hattenschwiler et al. 2005; Vasconcelos and Laurance 2005). In British forests, woodlice and millipedes (Fig. 15.5) are some of the most abundant macro-decomposers. They are sensitive to temperature and moisture, with species differing in their tolerance to dry conditions (Sutton 1980; Warburg et al. 1984; David and Handa 2010; Hassall et al. 2010). Thus, the abundance, diversity, and species composition of macro-decomposers is predicted to vary with changing climatic conditions, and to differ among edges and interiors of forests, and fragments of different sizes. However, the responses of macro-decomposers and decomposition rates to changes in climate and land use have been studied little in the field (Hattenschwiler and Bretscher 2001; Aerts 2006; David and Handa 2010), in part because invertebrates are time-consuming to sample (Lovell et al. 2009). Thus, our CP programme provided the ideal opportunity to study how soil macrofauna and decomposition processes are affected by fragmentation and changes in climatic conditions over a large spatial scale.

Being larger, slower, and more readily identifiable than many invertebrates, woodlice and millipedes have been used successfully in CS surveys by the Natural History Museum in London, which has produced online identification guides (Natural History Museum 2012). Following their protocols, we focused on five common woodlouse species (common striped woodlouse *Philoscia muscorum*, common pill woodlouse *Armadillidium vulgare*, common shiny woodlouse *Oniscus asellus*, common rough woodlouse *Porcellio scaber*, and the ancient woodland indicator species, *Lygidium hypnorum*); and five millipede species (pill millipede *Glomeris marginata*, white-legged millipede *Tachypodoiulus niger*, common flat-backed millipede *Polydesmus angustus*, blunt-tailed snake millipede

Figure 15.5 Woodlice and millipedes are important decomposers of leaf litter in temperate forests. Photograph ©Earthwatch.

Cylindroiulus punctatus, and *Ophyiulus pilosus*), which are all relatively easy to identify to species level by non-specialists. Moreover, as hand searching through leaf litter for these species is so time consuming, using amateur volunteer teams is an efficient approach.

We searched 25 cm × 25 cm quadrats in nine permanent plots situated within woodland edges, woodland cores, and surrounding woodland fragments, for our chosen species. Each team was instructed on how to hand search the vegetation, deadwood, litter, and the litter–soil interface in the quadrats. Each plot consisted of 10–25 subplots of 20 m × 20 m, and 10 quadrats were surveyed in each subplot. Volunteers used a 'pooter' (designed to 'suck' invertebrates into a tube), to collect all woodlice and millipedes within a quadrat, which were then sorted and identified to species level in the laboratory (Fig. 15.6). Many volunteers were initially unenthusiastic about the proposed tasks, expressing concerns about the sampling equipment and getting dirty. However, once they collected their first woodlice or millipede, even the most apathetic volunteers became engaged and saw the task as a novel experience. Supplying simple equipment, such as kneeling mats and gardening gloves, enhanced the comfort and motivation of volunteers.

Teams of three proved the most effective workgroup size—one to do the 'pootering', one to turn over the litter and instruct the 'pooterer' where to suck, and the third to fill in the data sheet and write the labels on the tubes. Each team had to include at least one willing 'pooterer' (the less keen could undertake the other tasks); although we observed that curiosity usually took over and by the end of the day everyone had at least tried to suck up a woodlouse! One plot took on average one hour to complete, thus a team of 12 could do three of the nine plots in a day, spending the last two hours in the laboratory identifying their samples (typical day: 9 am–4 pm). Thus, a complete survey of all nine plots took three full days.

Correct species identification is crucial, so our volunteers were subjected to half a day of desk-based training on the ecology, biology, and identification of woodlice and millipedes, followed by a half-day practical session on sampling. For quality control, all specimens collected during the project were identified in the laboratory under supervision of the project scientists. Ability to identify invertebrates to species level varied among volunteers, but increased with the number of samples completed, such that by the last day > 75% of samples were identified correctly. However, after two hours of identifying the samples in the laboratory, concentration and enthusiasm deteriorated, and mistakes became more common. In about 10% of cases, repeat identification of samples was required (revealed by spot-checking), and the benefits of volunteer involvement were thus reduced.

Overall, volunteers were most effective during the field sampling phase, enabling us to sample nine plots in three days with a 12 person team, a task which we estimate would have taken a single researcher nine days. As these surveys were repeated every three months, we estimate volunteer participation in the field to have

Figure 15.6 Citizen Scientists sort and identify samples under the supervision of Earthwatch scientists. Photograph ©A. Harrington.

saved us around 24 days of professional researcher time per year. However, the amount of researcher time spent helping with and correcting species identifications (i.e. two hours per day) was equivalent to the time needed by two scientists to identify the specimens on their own. Thus, in the last two years of the surveys, all species identifications were carried out by professional scientists only. Nevertheless, these Citizen Science surveys produced high quality published outputs. We showed that decomposition is moisture-limited but that, in the field, the presence of macrofauna accelerates decomposition rates irrespective of soil moisture conditions, and that macrofauna are particularly important in the decomposition of recalcitrant leaf litter, such as oak (Fig. 15.7) (Riutta et al. 2012; Slade and Riutta 2012). We also found that some macrofauna species exhibit reduced functioning under moisture stress, suggesting that species identity and assemblage composition are more important when moisture is limited, and that complementarity between macrofauna species mitigate these effects (Fig 15.8), such that decomposition rates are maintained even under low moisture conditions (Collison et al. 2013).

15.4.2 Case study 2. A landscape-scale study of woodland moths made possible by volunteer participation

Work undertaken by Merckx et al. in the Upper Thames region showed the importance for farmland moths of habitat features such as hedgerow trees and habitat connectivity (Merckx et al. 2009a, 2012), and how species responses vary with their mobility (Merckx et al. 2009b, 2010; Macdonald and Feber 2015: Chapter 8). We expanded this work by using HSBC volunteers to assist with a mark–release–recapture study on 87 macro-moth species, investigating whether life-history traits and landscape characteristics could predict macro-moth responses to forest fragmentation. This project, one of the largest of its kind, was made possible due to our HSBC team workforce, who were able to deploy and monitor 44 light traps (Fig. 15.9) across the landscape. Benefiting from our volunteer-based labour force we were able to trap and mark nearly 15,000 moths during the study, compared to 1300–1700 trapped over a similar time frame by a professional moth researcher alone (Merckx et al. 2009a,b, 2010).

Volunteers assembled and deployed the 44 six-watt pre-fabricated Heath-style moth traps across Wytham Woods, the surrounding forest fragments, and the matrix landscape in between, covering a 2 km radius from the centre of the Woods. Twenty-two of these battery-operated moth traps ran every night through June and July 2009. Supervised volunteers carried the 7 kg batteries to each trap at dusk and re-visited the traps at dawn to record the night's captures and collect the batteries to be recharged.

Captured moths were identified to species level and marked uniquely and harmlessly on the hindwing before release. Although volunteers were given species identification cards, they were not expected to learn all species. Every team included a supervising expert responsible for species identification, as well as for handling and marking moths. One volunteer helped with these tasks, while the second volunteer recorded species and trap details, and the third volunteer recorded the marks being given. Division of labour, and minimization of boredom, was thus achieved most effectively with teams of three volunteers.

To enhance volunteer performance and understanding, we contextualized the study to the participating volunteers. Links between climate change and habitat modification, declines in moth numbers, and reductions in dispersal capability (Conrad et al. 2006; Fox et al. 2013; Fox 2013) are also sufficiently clear to explain to participants with little scientific knowledge, and can be found in resources aimed specifically at the general public (e.g. mothscount[16], Fox et al. 2013).

The risk of volunteer boredom and the associated risk of reduced data veracity (Lovell et al. 2009; Bueschiing et al. 2014) were mitigated by having a variety of trap locations and the chance of encountering different moth species. With over 900 species of macro-moths in the UK (Fox et al. 2013) a previously unseen species was usually encountered every morning. Moreover, the variety of colours and patterns in moths often surprised and held the group's attention (Fig. 15.10). Many were so captivated and engaged with the project that around one third of the volunteers helped during their time off, which was hardly ever the case with the other projects.

All data from the Wytham study were entered into the National Moth Recording Scheme (2012), and thus volunteers were able to see their own contribution to the research online before they left the project. Overall, despite longer hours in the field and an early morning start, in this case, volunteer feedback suggested higher enjoyment and engagement in those teams involved in the moth project compared with teams involved in mammal trapping. Public negativity towards invertebrates has been noted as one of the key challenges in their conservation (Kellert 1993; Lovell et al. 2009). At

[16] http://www.mothscount.org/.

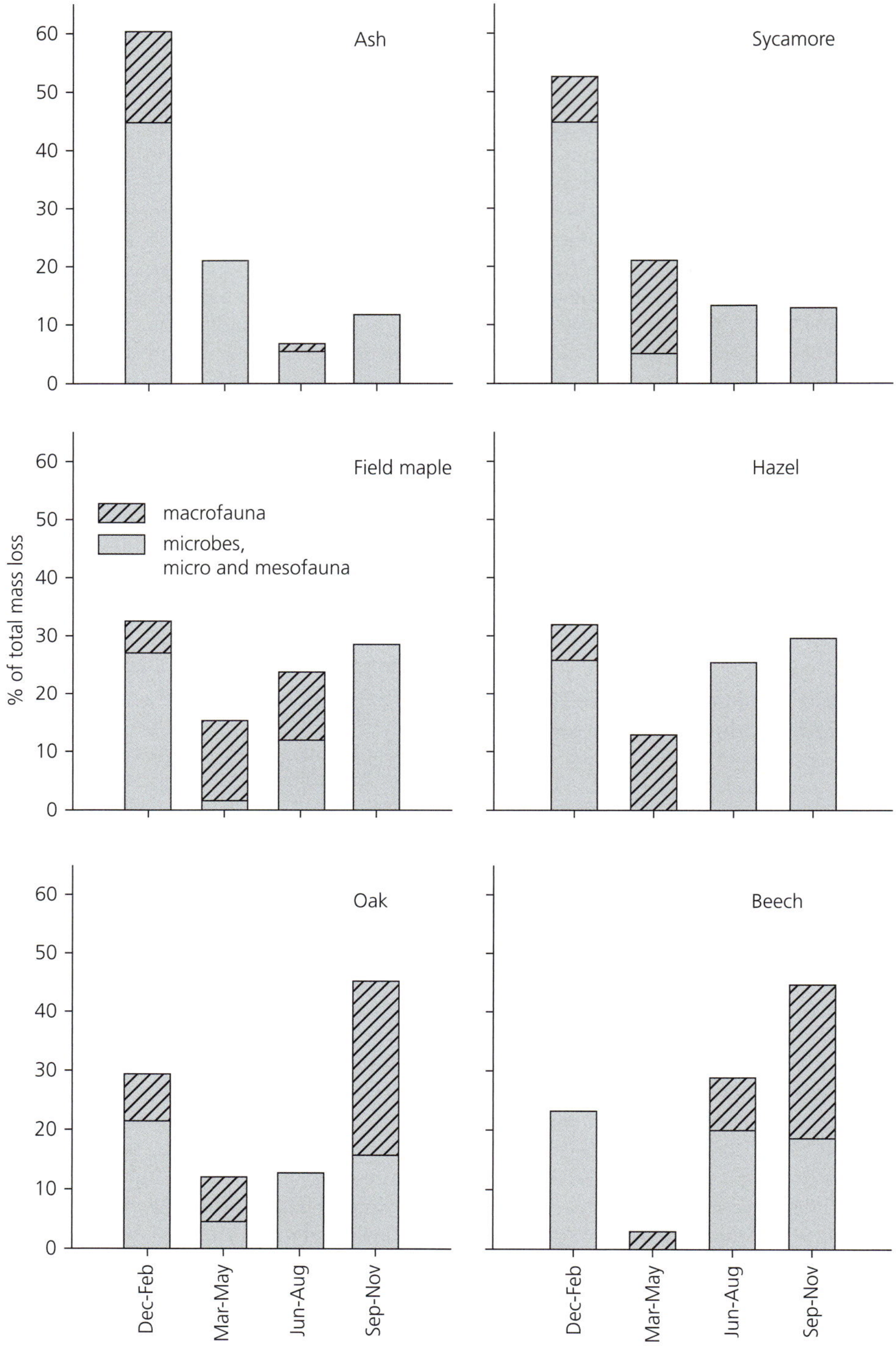

Figure 15.7 Contribution of macrofauna (shaded area) to leaf litter decomposition of six common temperate tree species over a year. Figures have been scaled so that the bars sum up to 100% for each species. From Slade and Riutta (2012).

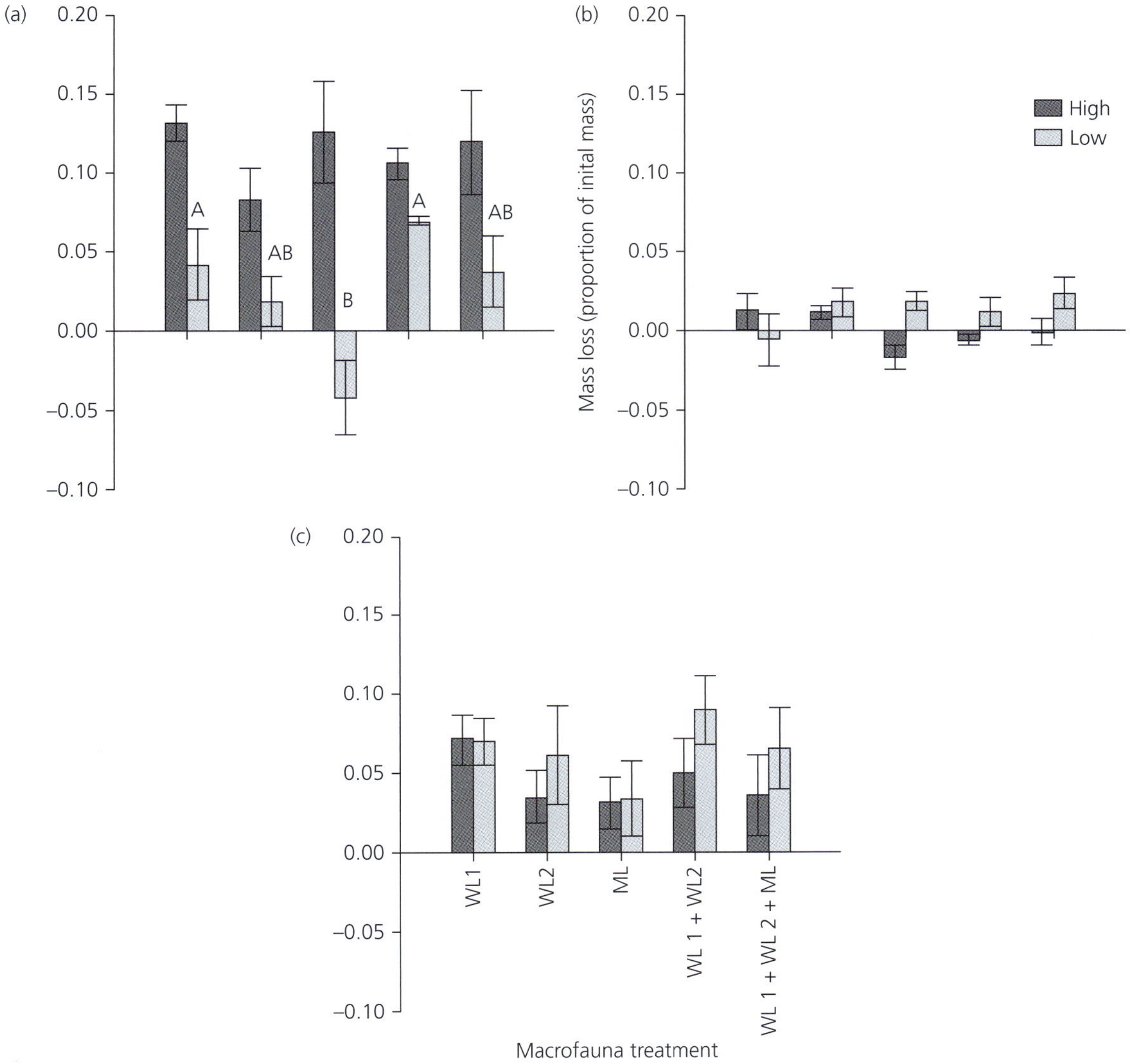

Figure 15.8 Mass loss of (a) ash, (b) oak leaf litter, and (c) moss for different macrofauna treatments (WL = woodlouse, ML = Millipede) at high and low moisture. In (a) macrofauna levels at low moisture not connected by the same letter are significantly different ($P \leq 0.05$) as tested by Tukey–Kramer HSD tests. From Collison et al. (2013).

the beginning of their placement, around half of our volunteers voiced an active dislike of moths, perceiving them as 'small, brown insects flapping around lights and eating clothes'. However, as captured moths can be handled and observed close up, volunteers were able to get 'hands-on' experience which greatly increased their enjoyment and appreciation of moths and enthusiasm for the study (Fig. 15.11). At the end of the fortnight, each volunteer, when asked if they still felt this aversion, thought that their perception of moths had improved and that they now appreciated the role moths play in the ecosystem, as well as appreciating their diversity and beauty. In addition, several of the British HSBC team members reported that they continued to monitor moths after returning home, and one volunteer initiated a programme of moth trapping in a local school.

Again, this study utilizing Citizen Scientists produced high quality published outputs (Slade et al. 2013; Merckx and Slade 2014). Our results showed that

Figure 15.9 Each morning the light traps were checked and the moths identified and recorded with the help of Earthwatch HSBC volunteers. Photograph © Yifie Zhang, Earthwatch.

Figure 15.10 The variety of colours, patterns and sizes, as well as their ease of handling make moths a great taxon for engaging members of the public in Citizen Science projects. Photograph ©Yifie Zhang, Earthwatch.

forest size, as well as connectivity, are important. Solitary trees and small fragments functioned as 'stepping stones', especially when their landscape connectivity was increased, by being positioned within hedgerows or within a favourable matrix. Mobile forest specialists were most affected by forest fragmentation. In spite of their high intrinsic dispersal capability, these species were confined mostly to the largest of the forest patches due to their strong affinity for the forest habitat, and were also heavily dependent on forest

Figure 15.11 Benefiting from our volunteer-based labour force we were able to trap and mark nearly 15,000 moths during the study. Here, a poplar hawkmoth *Laothoe populi* waits to be recorded. Photograph ©Yifie Zhang, Earthwatch.

connectivity in order to cross the agricultural matrix. Our results thus have practical implications, suggesting that forest fragments need to be larger than five hectares and to have interior forest more than 100 m from the edge in order to sustain populations of forest specialists. For full details of the results see Macdonald and Feber 2015: Chapter 12.

Overall, moths proved to be an excellent taxa for Citizen Science programmes, a fact recognized by the organisation 'Butterfly Conservation' with its initiative—'Moths Count' and the National Moths Recording Scheme, which have engaged over 2 million people in recording the distribution of the UK's macro-moths since 2006 (Fox et al. 2011).

15.5 General conclusions

Citizen Science: Panacea or Pandora's box (Scott et al. 2003)? While Citizen Science can deliver a cost-effective way to collect large quantities of good quality data as outlined in this chapter, and can thus be rewarding on many different levels (see Bullbeck 2004), potential problems and disadvantages have to be assessed carefully to maximize the benefits for ecological monitoring programmes (Silvertown et al. 2013). Furthermore, as international demand for nature-based tourism and active participation in conservation initiatives continues to grow, the consequences of providing public access to protected areas need to be assessed (Dolnicar 2006; Miller et al. 2011; Zhou et al. 2013). Once the preserve of charities, Citizen Science now hosts a proliferation of private companies seeking to make money by selling international conservation work to tourists as a commodity. This commodification of CS depends upon balancing the scientific legitimacy of projects against the need to offer desirable volunteering experiences (Cousins et al. 2009).

Consumer choices are increasingly influenced by social and environmental considerations (e.g. Kahn 2007). Corporate images rely heavily on their perceived social inclusiveness and environmental sustainability (Dyllick and Hockerts 2002; van Merrewijk 2003). Whether cynical or heart-felt, 'environmentally-friendly' 'eco-' branding (see Kahn 2007) highlights how businesses have caught on to the sales-pitch of sustainability (Dyllick and Hockerts 2002; Van Marrewijk 2003). Globally, thousands of firms participate in the Environmental Protection Agency's partnership programmes, and many others participate in industry-led environmental programmes such as those of the World Business Council for Sustainable Development, the Chicago Climate Exchange, and the American Chemistry Council's 'Responsible Care' program (Baron 2003).

In spite of these efforts, large portions of the public continue to view business as an enemy of the environment (Lyon and Maxwell 2011). Furthermore, although many companies publicize their environmentally-friendly actions, others are surprisingly hesitant to promote or report their environmental credentials. Part of the reason appears to be that activists react more angrily to firms that lay claim to being virtuous, and then are discovered to have feet of clay, than to firms that never make such claims (Lyon and Maxwell 2011). As for the corporations participating in environmental and CS programmes, an uneasily derogatory term has entered the lexicon associated with CSR: 'Green-washing' (e.g. Laufer 2003). Brian Tokar

(1997) paraphrased: 'Earth for sale: reclaiming ecology in the age of corporate greenwash', leading other authors, such as Ramus and Montiel (2005) to question whether corporate environmental policies are a form of green-washing, too. Lyon and Maxwell (2011) present a complex algebra to induce full environmental disclosure through mandatory audit, which could serve to standardize corporate environmental credentials. Thus, the intrinsic benefits of active corporate engagement in conservation through partnership, rather than simple sponsorship, may serve to obviate sceptical views that stake-holders use CSR for self-promotion (Peloza 2005).

The importance of environmental responsibility in volunteering is widely accepted, where two main paradigms have emerged from the tourism perspective: ecotourism—a demand-driven concept limited to nature-based tourism—and sustainable tourism—a supply-sided view characterized by industry regulations. Unfortunately, ecotourism is limited in size, and supply-sided measures contradict implicitly the short-term aim of profit-maximization of the tourism industry. As highlighted by Dolnicar (2006), alternative ways of integrating nature conservation and tourism are needed. In this context, CS has the potential to contribute to both conservation and sustainable development, and provides an important incentive to generate public support and revenue for land stewardship (Kirkby 2010; Kirkby et al. 2011). To ensure that these benefits, for both nature and society, are maximized, it is important to be vigilant for unintended consequences, and to manage expeditions and access to sensitive areas adaptively to minimize any costs (Klein et al. 1995; Salvador et al. 2011; Zhou et al. 2013). Guttentag (2009) reports numerous possible negative impacts of volunteer-driven tourism that deserve attention from both researchers and project managers: ranging over neglect of locals' desires, hindering of work in progress and completion of unsatisfactory work, a disruption of local economies, a reinforcement of conceptualizations of the 'other' and rationalizations of poverty, and the instigation of cultural changes.

An elephant in the room concerns the international travel often involved in participation in more exotic (and popular) volunteering expeditions, where this recreation perversely carries a substantial 'carbon footprint' (Hunter and Shaw 2005). While it is easy to placate one's conscience by saying the journey was to 'save the planet' (so-called 'ethical consumption'; Starr 2009), the harsh reality is that the measurable benefits may be small, whereas the ecological footprint of trans-continental travel and the consumerism entailed in purchasing gear are often substantial (see Hunter and Shaw 2007). While Earthwatch, and similar organizations, have responsible policies on carbon offset (Gössling et al. 2007; Mair 2011), there remains a personal ethical component to this issue. Clearly, either the more useful or substantial the data collected, or the more life-enhancing the experience of participation, the more defensible is the balance.

Disturbance to ecosystems is a further risk associated with large numbers of volunteers participating in conservation initiatives (Zhou et al. 2013). This is a complex variable (Leighton et al. 2010), subject to non-linear effects (Fernández-Juricic et al. 2003), particularly at the community level (Allen and O'Connor 2000). For example, in a protected region in central China, we found greater trail use to be associated with fewer mammal field signs along transects close to trails, compared with more distant transects (Zhou et al. 2013). Species and guilds exhibited different responses to trail type and use; smaller mammals (< 15 kg) were deterred primarily by the extent to which the trail modified the forest landscape, whereas larger mammals (> 15 kg) were deterred primarily by greater trail use. That larger species are perturbed by simple human presence is of particular concern, as it is these (often charismatic) species that attract public interest, but will prove increasingly hard to observe as tourist numbers increase.

Ultimately, we return to the 3Cs (CBD 1993), where biodiversity is now a business (Steele 1995), and the fact that self-funded volunteers support this industry reflects a genuine desire from the public to participate in nature based activities (for perspective see: Meletis and Campbell 2007; Nowaczek et al. 2007). In overview then, while CS has many advantages and benefits, we are wary of naive advocacy. CS, and related facets of PERT and ecotourism, within the broader remit of purposeful recreation, are a compromised and conflicted tool for conservation, driven by a varied host of personal, commercial, and scientific agendas (Leach et al. 2005). Nevertheless, this tool can be used very constructively, and we hope we have elucidated a little on the craftsmanship needed to do so.

Acknowledgements

We would like to thank foremost our many volunteers without whom this study would have been impossible. The support of the Earthwatch Institute in supplying our volunteer teams over such a long period of time is greatly appreciated. We would also like to thank the volunteers from the Ley Drug and Alcohol

Rehabilitation Volunteers, who continue to help us with the manual aspects of our field work. This collaboration is supported by the DLM Charitable Trust. John Fa made valuable comments on an earlier draft of this chapter. We thank Ruth Feber for helpful input to this chapter.

References

Aerts, R. (2006). The freezer defrosting: global warming and litter decomposition rates in cold biomes. *Journal of Ecology*, **94**, 713–724.

Allen, A.P. and O'Connor, R.J., (2000). Interactive effects of land use and other factors on regional bird distributions. *Journal of Biogeography*, **27**, 889–900.

Asah, S.T. and Blahna, D.J. (2012). Motivational functionalism and urban conservation stewardship: implications for volunteer involvement. *Conservation Letters*, **5**(6), 470–477.

Bailey, R.E., and Putman, R.J. (1981). Estimation of fallow deer (Dama dama) populations from faecal accumulation. *Journal of Applied Ecology*, **18**, 697–702.

Ballantyne, R., Packer, J., Hughes, K., and Dierking, L. (2007). Conservation learning in wildlife tourism settings: lessons from research in zoos and aquariums. *Environmental Education Research*, **13**(3), 367–383.

Barker, E., Phillips, R., Kusek, K., and Thomas, B. (2011). Earthwatch and HSBC: embedding sustainability into the DNA of HSBC's business. *People and Strategy*, **34**(1), 24.

Baron, D.P. (2003). Private politics. *Journal of Economics & Management Strategy*, **12**(1), 31–66.

Barrett, N., Edgar, G., and Morton, A. (2002). *Monitoring of Tasmanian inshore reef ecosystems. An assessment of the potential for volunteer monitoring programs and a summary of changes within the Maria Island Marine Reserve from 1992–2001*. Tasmanian Aquaculture and Fisheries Institute Technical Report Series. p. 53.

Basinger, J. (1998). To scientists who use paying volunteers in fieldwork, the benefits outweigh the bother. *The Chronicle of Higher* Education, **44**, A14–A15.

Benson, A. M. (Ed.). (2010). *Volunteer tourism: theoretical frameworks and practical applications*. Routledge.

Benson, A., and Seibert, N. (2009). Volunteer tourism: motivations of German participants in South Africa. *Annals of Leisure Research*, **12**(3–4), 295–314.

Bickford, D., Posa, M.R.C., Qie, L., Campos-Arceiz, A., and Kudavidanage, E.P. (2012). Science communication for biodiversity conservation. *Biological Conservation*, **151**(1), 74–76.

Birch, M. (2005). Cultivating wildness: three conservation volunteers' experiences of participation in the Green Gym Scheme. *The British Journal of Occupational Therapy*, **68**(6), 244–252.

Blackburn, S. and Frank, L. (2010). Assessment of guide reporting & preliminary results of lion monitoring. Mara Predator Project.

Bonjean, C.M., Markham, W.T., and Macken, P.O. (1994).Measuring self-expression in volunteer organisations: a theory-based questionnaire. *Journal of Applied Behavioural* Science, **30**, 487–514.

Bradford, M.A., Tordoff, G.M., Eggers, T., Jones, T.H., and Newington, J.E. (2002). Microbiota, fauna, and mesh size interactions in litter decomposition. *Oikos*, **99**, 317–323.

Brandon, A., Spyreas, G., Molano-Flores, B., Carroll, C., and Ellis, J. (2003). Can volunteers provide reliable data for forest vegetation surveys? *Natural Areas Journal*, **23**(3), 254–262.

Brightsmith, D.J., Stronza, A., and Holle, K. (2008). Ecotourism, conservation biology, and volunteer tourism: A mutually beneficial triumvirate. *Biological Conservation*, **141**(11), 2832–2842.

Brown, E. (1999). Assessing the value of volunteer activity. *Non-profit & Voluntary Sector Quarterly*, **28**, 3–17.

Brudney, J.L., and Gazley, B. (2006). Moving ahead or falling behind? Volunteer promotion and data collection. *Nonprofit Management and Leadership*, **16**(3), 259–276.

Buesching, C.D. and Newman, C. (2005). Volunteers in ecological research: amateur ecological monitors—the benefits and challenges of using volunteers. *Bulletin of the Ecological* Society, **36**, 20–22.

Buesching, C.D., Newman, C., Jones, J.T., and Macdonald, D.W. (2011). Testing the effects of deer grazing on two woodland rodents, bankvoles and woodmice. *Journal of Basic and Applied Ecology*, **12**(3), 207–214.

Buesching, C.D., Newman, C., and Macdonald, D.W. (2014). How dear are deer volunteers: The efficiency of deer monitoring using volunteer teams to conduct pellet group counts. *Oryx*, 1–9.

Bulbeck, C. (2004). *Facing the wild: ecotourism, conservation and animal encounters*. Earthscan.

Butt, N., Slade, E.M., Thompson, J. Mahli, Y., and Riutta, T. (2013). Quantifying the sampling error in tree census measurements carried out by volunteers and its effect on above-ground carbon stock estimates. *Ecological Applications*, **23**, 936–943.

Caissie, L.T., and Halpenny, E.A. (2003). Volunteering for nature: motivations for participating in a biodiversity conservation volunteer program. *World Leisure Journal*, **45**(2), 38–50.

Campbell, L.M., and Smith, C. (2006). What makes them pay? Values of volunteer tourists working for sea turtle conservation. *Environmental Management*, **38**(1), 84–98.

Cardinale, B.J., Duffy, J.E., Gonzalez, A., Hooper, D.U., Perrings, C., Venail, P., Narwani, A., Mace, G.M., Tilman, D., Wardle, D.A., Kinzig, A.P., Daily, G.C., Loreau, M., Grace, J.B., Larigauderie, A., Srivastava, D.S., and Naeem, S. (2012). Biodiversity loss and its impact on humanity. *Nature*, **486**, 59–67.

Clary, E.G., Snyder, M., Ridge, R.D., Copeland, J., Stukas, A.A., Haugen, J., and Miene, J. (1998). Understanding and assessing the motivations of volunteers: a functional approach. *Journal of Personality & Social Psychology*, **74**, 1516–1530.

Clary, E.G., Snyder, M., and Stukas, A.A. (1996).Volunteers's motivations: findings from a national survey. *Non-profit and Voluntary Sector Quarterly*, **25**, 485–505.

Cochran-Smith, M. and Lytle, S.L. (1990). Research on teaching and teacher research: the issues that divide. *Educational Researcher*, **19**(2), 2–11.

Coghlan, A. (2005). *Towards an understanding of the volunteer tourism experience*. D.Phil.-thesis, School of Business at James Cook University, Australia.

Coghlan, A. (2006). Volunteer tourism as an emerging trend or an expansion of ecotourism? A look at potential clients' perceptions of volunteer tourism organisations. *International Journal of Nonprofit and Voluntary Sector Marketing*, **11**(3), 225–237.

Coghlan, A. (2008). Exploring the role of expedition staff in volunteer tourism. *International Journal of Tourism Research*, **10**(2), 183–191.

Cohn, J.P. (2008). Citizen science: can volunteers do real research? *Bioscience*, **58**, 192–197.

Collison, E.J., Riutta, T., and Slade, E.M. (2013). Macrofauna assemblage composition and soil moisture interact to affect soil ecosystem functions. *Acta Oecologica*, **47**, 30–36.

Conrad, K.F., Warren, M.S., Fox, R., Parsons, M.S., and Woiwod, I.P. (2006). Rapid declines of common, widespread British moths provide evidence of an insect biodiversity crisis. *Biological Conservation*, **132**, 279–291.

Convention on Biological Diversity (CBD) (1993). <http://www.cbd.int/secretariat>.

Cousins, J.A., Evans, J., and Sadler, J. (2009). Selling conservation? Scientific legitimacy and the commodification of conservation tourism. *Ecology and Society*, **14**(1), 32.

Danielsen, F., Burgess, N.D., and Balmford, A. (2005). Monitoring matters: examining the potential of locally-based approaches. *Biodiversity and Conservation*, **14**, 2507–2542.

Darwall, W.R.T. and Dulvey, N.K. (1996). An evaluation of the suitability of non-specialist volunteer researchers for coral reef surveys, Mafia Islands, Tanzania—a case study. *Biological* Conservation, **78**, 223–231.

David, J.F. and Handa, I.T. (2010).The ecology of saprophagousmacroarthropods (millipedes, woodlice) in the context of global change. *Biological Reviews*, **85**, 881–895.

Dempsey, S.E. (2012). Nonprofits as political actors. *Management Communication Quarterly*, *26*(1), 147–151.

Dickinson, J.L., Zuckerberg, B., and Bonter D.N. (2010). Citizen science as an ecological research tool: challenges and benefits. *Annual Review of Ecology, Evolution, and Systematics*, **41**, 149–172.

Dolnicar, S. (2006). Nature-conserving tourists: the need for a broader perspective. *Anatolia*, **17**(2), 235–255.

Dolnicar, S., and Randle, M. (2007). The international volunteering market: market segments and competitive relations. *International Journal of Nonprofit and Voluntary Sector Marketing*, **12**(4), 350–370.

Douglas, C., and Rollins, R. (2007). Motivations, training and supervision of volunteers in conservation. *Environments: A Journal of Interdisciplinary Studies*, **35**(1).

Driver, B.L., and Johnson, L.A. (1984). A pilot study of the perceived long-term benefits of the Youth Conservation Corps. *The Journal of Environmental Education*, **15**(2), 3–11.

Dyllick, T. and Hockerts, K. (2002). Beyond the business case for corporate sustainability. *Business Strategy and the Environment*, **11**(2), 130–141.

Earthwatch (2014). <http://earthwatch.org/> (accessed October 2014).

Ellis, C. (2003). When volunteers pay to take a trip with scientists—participatory environmental research tourism (PERT). *Human Dimensions of Wildlife*, **8**(1), 75–80.

Ericsson, G. and Wallin, K. (1999). Hunter observations as an index of moose *Alcesalces* population parameters. *Wildlife Biology*, **5**, 177–185.

Nowaczek, A.M., Moran-Cahusac, C. and Fennell, D.A. (2007). Against the current: striving for ethical ecotourism. In *Critical Issues in Ecotourism: Understanding a Complex Tourism Phenomenon*, Higham, J. E., ed., pp. 136–156. Routledge .

Fennell, D. and Nowaczek, A. (2010). Moral and empirical dimensions of human–animal interactions in ecotourism: deepening an otherwise shallow pool of debate. *Journal of Ecotourism*, **9**(3), 239–255.

Fernández-Juricic, E., Sallent, A., Sanz, R., and Rodríguez-Prieto, I. (2003). Testing the risk-disturbance hypothesis in a fragmented landscape: nonlinear responses of house sparrows to humans. *The Condor*, *105*(2), 316–326.

Foster-Smith, J. and Evans, S.M. (2003). The value of marine ecological data collected by volunteers. *Biological* Conservation, **113**, 199–213.

Fox, R. (2013). The decline of moths in Great Britain: a review of possible causes. *Insect Conservation and Diversity*, **6**(1), 5–19.

Fox, R., Parsons, M.S., Chapman, J.W., Woiwod, I.P., Warren, M.S., and Brooks, D.R. (2013). *The State of Britain's Larger Moths 2013*. Butterfly Conservation and Rothamsted Research, Wareham, Dorset, UK.

Fox, R., Randle, Z., Hill, L., Anders, S., Wiffen, L., and Parsons, M. (2011). Moths count: recording moths for conservation in the UK. *Journal of Insect Conservation*, **15**, 55–68.

Fuller, R.J., and Gill, R.M.A. (2001). Ecological impacts of increasing numbers of deer in British woodland. *Forestry*, **74**(3), 193–199.

Genet, K.S. and Sargent, L.G. (2003). Evaluation of methods and data quality from a volunteer-based amphibian call survey. *Wildlife Society Bulletin*, **31**, 703–714.

Gill, R. Influence of large herbivores on tree recruitment and forest dynamics. (2006). *Aspen Bibliography*. Paper 3171. http://digitalcommons.usu.edu/aspen_bib/3171.

Gill, R., and Fuller, R.J. (2007). The effects of deer browsing on woodland structure and songbirds in lowland Britain. *Ibis*, **149**(s2), 119–127.

Gössling, S., Broderick, J., Upham, P., Ceron, J.P., Dubois, G., Peeters, P., and Strasdas, W. (2007). Voluntary carbon offsetting schemes for aviation: efficiency, credibility and sustainable tourism. *Journal of Sustainable Tourism*, **15**(3), 223–248.

Guiney, M.S., and Oberhauser, K.S. (2009).Conservation volunteers' connection to nature. *Ecopsychology*, **1**(4), 187–197.

Guttentag, D.A. (2009). The possible negative impacts of volunteer tourism. *International Journal of Tourism Research*, **11**(6), 537–551.

Hassall, M., Edwards, D.P., Carmenta, R., Derhe, M.A., and Moss, A. (2010). Predicting the effect of climate change on aggregation behaviour in four species of terrestrial isopods. *Behaviour*, **147**, 151–164.

Hassall, M., Turner, J.G., and Rands, M.R.W. (1987). Effects of terrestrial isopods on the decomposition of woodland leaf litter. *Oecologia*, **72**, 597–604.

Hättenschwiler, S. and Bretscher, D. (2001). Isopod effects on decomposition of litter produced under elevated CO_2, N deposition and different soil types. *Global Change Biology*, **7**, 565–579.

Hättenschwiler, S., Tiunov, A.V., and Scheu, S. (2005). Biodiversity and litter decomposition in terrestrial ecosystems. *Annual Review of Ecology, Evolution, and Systematics*, **36**, 191–218.

Henderson, S. (2012). Citizen science comes of age. *Frontiers in Ecology and the Environment*, **10**, 283.

Henningfield, J., Pohl, M., and Tolhurst, N. (2006). *ICCA Handbook of Corporate Social Responsibility*. John Wiley and Sons.

Hooper, D.U., Adair, E.C., Cardinale, B.J., Byrnes, J.E.K., Hungate, B.A., Matulich, K.L., Gonzalez, A., Duffy, J.E., Gamfeldt, L., and O'Connor, M.I. (2012). A global synthesis reveals biodiversity loss as a major driver of ecosystem change. *Nature*, **486**, 105–108.

Hughes, J., Pretty, J., and Macdonald, D.W. (2013). Nature as a source of health and well-being: is this an ecosystem service that could pay for conserving biodiversity? In Key Topics in Conservation II. David W. Macdonald and Kathy Willis, eds., Wiley Publishing.

Hunter, C. and Shaw, J. (2005). Applying the ecological footprint to ecotourism scenarios. *Environmental Conservation*, **32**(4), 294–304.

Hunter, C. and Shaw, J. (2007). The ecological footprint as a key indicator of sustainable tourism. *Tourism Management*, **28**(1), 46–57.

Irwin, A. (1995). *Citizen Science: A Study of People, Expertise, and Sustainable Development*. Routledge.

Jacobson, S.K., Carlton, J.S., and Monroe, M.C. (2012). Motivation and satisfaction of volunteers at a Florida natural resource agency. *Journal of Park and Recreation Administration*, **30** (1).

Kahn, M.E. (2007). Do greens drive Hummers or hybrids? Environmental ideology as a determinant of consumer choice. *Journal of Environmental Economics and Management*, **54**(2), 129–145.

Kellert, S.R. (1993).Values and perceptions of invertebrates. *Conservation Biology*, **7**, 845–855.

King, D. (2004). Climate change science: adapt, mitigate, or ignore? *Science*, **303**, 176–177.

Kirkby, C.A. (2010). *The market triumph of ecotourism: a social cost-benefit analysis of a rainforest ecotourism cluster in Amazonian Peru*. Doctoral dissertation, University of East Anglia.

Kirkby, C.A., Giudice, R., Day, B., Turner, K., Soares-Filho, B.S., Oliveira-Rodrigues, H., and Yu, D.W. (2011). Closing the ecotourism-conservation loop in the Peruvian Amazon. *Environmental Conservation*, **38**(01), 6–17.

Klein, M.L., Humphrey, S.R., and Percival, H.F. (1995). Effects of ecotourism on distribution of waterbirds in a wildlife refuge. *Conservation Biology*, **9**(6), 1454–1465.

Krüger, O. (2005). The role of ecotourism in conservation: panacea or Pandora's box? *Biodiversity & Conservation*, **14**(3), 579–600.

Laufer, W.S. (2003). Social accountability and corporate greenwashing. *Journal of Business Ethics*, **43**(3), 253–261.

Lawrence, A. (2006). 'No personal motive?' Volunteers, biodiversity, and the false dichotomies of participation. *Ethics Place and Environment*, **9**(3), 279–298.

Leach, M., Scoones, I., and Wynne, B., eds. (2005). *Science and Citizens: Globalization and the Challenge of Engagement* (Vol. 2). Zed Books.

Leighton, P.A., Horrocks, J.A., and Kramer, D.L. (2010). Conservation and the scarecrow effect: can human activity benefit threatened species by displacing predators? *Biological Conservation*, **143**(9), 2156–2163.

Ley (2014). <http://www.leycommunity.co.uk> (accessed 5 October 2014).

Lienhoop, N., and Fischer, A. (2009). Can you be bothered? The role of participant motivation in the valuation of species conservation measures. *Journal of Environmental Planning and Management*, **52**(4), 519–534.

Lorimer, J. (2009). International conservation volunteering from the UK: what does it contribute? *Oryx*, **43**(3), 352.

Lovell, S., Hamer, M., Slotow, R., and Herbert, D. (2009). An assessment of the use of volunteers for terrestrial invertebrate biodiversity surveys. *Biodiversity and Conservation*, **18**, 3295–3307.

Lyon, T.P. and Maxwell, J.W. (2011). Greenwash: corporate environmental disclosure under threat of audit. *Journal of Economics & Management Strategy*, **20**(1), 3–41.

Macdonald, D.W. and Feber, R.E., eds. (2015). *Wildlife Conservation on Farmland. Managing for Nature on Lowland Farms*. Oxford University Press, Oxford.

Macdonald, D.W., Mace, G., and Rushton, S. (1998). *Proposals for future monitoring of British mammals*. Defra, London.

Macdonald, D.W., Newman, C., Buesching, C.D., and Tattersall, F.T. (2002). Volunteer mammal monitors: measuring the professionalism of amateurs. In *ENSUS 2002: Marine science and technology for environmental sustainability*. University of Newcastle Publishing.

Mackney, P. and Spring, N. (2000). *The Earthwatch perspective on volunteers and data validity*. Earthwatch Institute.

Mair, J. (2011). Exploring air travellers' voluntary carbon-offsetting behaviour. *Journal of Sustainable Tourism*, **19**(2), 215–230.

Manel, S., Williams, H.C., and Ormerod, S.J. (2001). Evaluating presence–absence models in ecology: the need to account for prevalence. *Journal of Applied Ecology*, **38**(5), 921–931.

Manfredo, M.J., Driver, B.J., and Tarrant, M.A. (1996). Measuring leisure motivation: a metaanalysis of the recreation experience preference scales. *Journal of Leisure Research*, **28**, 188–213.

McKinney, M., and Harmon, W. (2007). Governing nature, governing ourselves: engaging citizens in natural resource decisions. *International Journal of Public Participation*, **1**(2).

McLaren, A.A. and Cadman, M.D. (1999). Can novice volunteers provide credible data for bird surveys requiring song identification? *Journal of Field Ornithology*, **70**, 481–490.

Meletis, Z.A. and Campbell, L.M. (2007). Call it consumption! Re-conceptualizing ecotourism as consumption and consumptive. *Geography Compass*, **1**(4), 850–870.

Merckx, T. and Slade, E.M. (2014). Macro-moth families differ in their attraction to light: implications for light-trap monitoring programmes. *Insect conservation and diversity*, *7*(5), 453–461.

Merckx, T., Feber, R.E., Dulieu, R.L., et al. (2009b). Effect of field margins on moths depends on species mobility: field-based evidence for landscape-scale conservation. *Agriculture, Ecosystems & Environment*, **129**, 302–309.

Merckx, T., Feber, R.E., McLaughlan, C., et al. (2010). Shelter benefits less mobile moth species: the field-scale effect of hedgerow trees. *Agriculture, Ecosystems & Environment*, **138**, 147–151.

Merckx, T., Feber, R.E., Riordan, P., Townsend, M.C., Bourn, N.A.D., Parsons, M.S., and Macdonald, D.W. (2009a). Optimizing the biodiversity gain from agri-environment schemes. *Agriculture, Ecosystems and Environment*, **130**, 177–182.

Merckx, T., Marini, L., Feber, R.E., and Macdonald, D.W. (2012). Hedgerow trees and extended-width field margins enhance macro-moth diversity: implications for management. *Journal of Applied Ecology*, **49**, 1396–1404.

Michener, W.K. (2006). Meta-information concepts for ecological data management. *Ecological Informatics*, **1**(1), 3–7.

Miles, I., Sullivan, W.C., and Kuo, F.E. (1998). Ecological restoration volunteers: the benefits of participation. *Urban Ecosystems*, **2**(1), 27–41.

Miller, T.R., Minteer, B.A., and Malan, L.C. (2011). The new conservation debate: the view from practical ethics. *Biological Conservation*, **144**(3), 948–957.

Moon, W., Florkowski, W.J., Brückner, B., and Schonhof, I. (2002). Willingness to pay for environmental practices: implications for eco-labeling. *Land Economics*, **78**(1), 88–102.

Moscardo, G., Morrison, A.M., Pearce, P.L., Lang, C., and O'Leary, J.T. (1996). Understanding vacation destination choice through travel motivation and activities. *Journal of Vacation* Marketing, **2**,109–122.

Mumby, P.A., Harborne, A.R., Raines, R.P., and Ridley, J.M. (1995). A critical appraisal of data derived from Coral Cay conservation volunteers. *Bulletin of Marine Science*, **56**, 737–751.

National Moth Recording Scheme (2012). http://www.mothscount.org/.

Newman, C. and Buesching, C.D. (2004a). Name droppings. *Deer: The Journal of the British Deer Society*, **13**, 12–15.

Newman, C. and Buesching, C.D. (2004b). Getting on the right track. *Deer: The Journal of the British Deer Society*, **12**, 20–22.

Newman, C., Buesching, C.D., and Macdonald, D.W. (2003). Validating mammal monitoring methods and assessing the performance of volunteers in wildlife conservation – 'Sed quis custodiet ipsos custodes?' *Biological Conservation*, **113**, 189–197.

Newman, C. and Macdonald, D.W. (2013). The implications of climate change for terrestrial UK mammals. Terrestrial biodiversity climate change impacts report card technical paper. Living with environmental change partnership. http://www.lwec.org.uk/.

Natural History Museum (NHM) (2012). http://www.nhm.ac.uk/.

Nixon, R.A. (1997). EARTHWATCH expeditions impact science education: exposing teachers to science and technology in the field and in the classroom. *Journal of Science Education and Technology*, **6**(3), 213–229.

Oberhauser, K., and Prysby, M.D. (2008). Citizen science: creating a research army for conservation. *American Entomologist*, **54**(2), 103–104.

Paris, L. (2008). Through the looking glass: Age, stages, and historical analysis. *The Journal of the History of Childhood and Youth*, *1*(1), 106–113.

Pattengill-Semmens, C.V., and Semmens, B.X. (2003). Conservation and management applications of the reef volunteer fish monitoring program. In *Coastal Monitoring through Partnerships*, pp. 43–50. Springer, Netherlands.

Pearce, J.M. (2009). Appropedia as a tool for service learning in sustainable development. *Journal of Education for Sustainable Development*, **3**(1), 45–53.

Pearce, P.L. and Lee, U. (2005). Developing the travel career approach to tourist motivation. *Journal of Travel Research*, **43**, 226–237.

Peloza, J. (2005). *Corporate social responsibility as reputational insurance*. Working Paper Series. Center of Responsible Business, UC Berkeley.

Philibert, C. (1999). The economics of climate change and the theory of discounting. *Energy Policy*, **27**, 913–927.

Pretty, J., Peacock, J., Sellens, M., and Griffin, M. (2005). The mental and physical health outcomes of green exercise. *International Journal of Environmental Health Research*, **15**(5), 319–337.

Prussia, G.E., and Kinicki, A.J. (1996). A motivational investigation of group effectiveness using social-cognitive theory. *Journal of Applied Psychology*, **81**(2), 187.

Ramus, C.A. and Montiel, I. (2005). When are corporate environmental policies a form of greenwashing? *Business & Society*, **44**(4), 377–414.

Reed, M.S. (2008). Stakeholder participation for environmental management: a literature review. *Biological Conservation*, **141**(10), 2417–2431.

Risely, K., Noble, D.G., and Baillie, S.R. (2008). The breeding bird survey 2007. *BTO Research Report*, **508**, 3.

Riutta, T., Slade, E.M., Bebber, D.P., Taylor, M.E., Malhi, Y., Riordan, P., Macdonald, D.W., and Morecroft, M.D. (2012). Experimental evidence for the interacting effects of forest edge, moisture and soil macrofauna on leaf litter decomposition. *Soil Biology and Biochemistry*, **49**, 124–131.

Rondinelli, D.A. and Berry, M.A. (2000). Environmental citizenship in multinational corporations: social responsibility and sustainable development. *European Management Journal*, **18**(1), 70–84.

Saha, M. and Darnton, G. (2005).Green companies or green con-panies: are companies really green, or are they pretending to be? *Business and Society Review*, **110**, 117–157.

Salvador, S., Clavero, M., and Leite Pitman, R. (2011). Large mammal species richness and habitat use in an upper Amazonian forest used for ecotourism. *Mammalian Biology-Zeitschrift für Säugetierkunde*, **76**(2), 115–123.

Savill, P., Perrins, C., Kirby, K., and Fisher, N., eds. (2010). *Wytham Woods: Oxford's Ecological Laboratory*. Oxford University Press, Oxford.

Scott, A., Christie, M., and Tench, H. (2003). Visitor payback: panacea or pandora's box for conservation in the UK? *Journal of Environmental Planning and Management*, **46**(4), 583–604.

Seastedt, T.R. (1984). The role of microarthropods in decomposition and mineralization processes. *Annual Review of Entomology*, **29**, 25–46.

Sibbald, S., Carter, P., and Poulton, S. (2006). *Proposal for a national monitoring scheme for small mammals in the United Kingdom and the Republic of Eire*. Mammal Society.

Silvertown, J., Buesching, C.D., Jacobson, S., and Rebello, T. (2013). Citizen science for nature conservation biology. In Key Topics in Conservation II. David W. Macdonald and Kathy Willis, eds., Wiley Publishing.

Slade, E.M., Merckx, T., Riutta, T., Bebber, D.P., Redhead, D., Riordan, P., and Macdonald, D.W. (2013). Life-history traits and landscape characteristics predict macro-moth responses to forest fragmentation. *Ecology*, **94**, 1519–1530.

Slade, E.M. and Riutta, T. (2012). Interacting effects of leaf litter species and macrofauna on decomposition in different litter environments. *Basic and Applied Ecology*, **13**, 423–431.

Spielman, A. (1994). The emergence of Lyme disease and human babesiosis in a changing environment. *Annals of the New York Academy of Sciences*, **740**(1), 146–156.

Starr, M.A. (2009). The social economics of ethical consumption: theoretical considerations and empirical evidence. *The Journal of Socio-economics*, **38**(6), 916–925.

Steele, P. (1995). Ecotourism: an economic analysis. *Journal of Sustainable Tourism*, **3**(1), 29–44.

Stern, N. (2006). *UK Government Report on Climate Change*.

Sutton, S.L. (1980). *Woodlice*. Pergamon Press, New York.

Tokar, B. (1997). *Earth for Sale: Reclaiming Ecology in the Age of Corporate Greenwash*. South End Press, Boston, USA.

Tomazos, K. (2010). Volunteer tourism: An ambiguous marketing phenomenon. *Innovative Marketing*, **6**(4), 42–47.

Van Marrewijk, M. (2003). Concepts and definitions of CSR and corporate sustainability: between agency and communion. *Journal of Business Ethics*, **44**(2–3), 95–105.

Vasconcelos, H.L. and Laurance, W.F. (2005). Influence of habitat, litter type, and soil invertebrates on leaf-litter decomposition in a fragmented Amazonian landscape. *Oecologia*, **144**(3): 456–462.

Wagner, G. (1995). Basic approaches and methods for quality assurance and quality control in sample collection and storage for environmental monitoring. *Science of the Total Environment*, **176**(1), 63–71.

Warburg, M.R., Linsenmair, K.E., and Bercovitz, K. (1984). The effect of climate on the distribution and abundance of Isopods in England. Adapting conservation to a changing climate. Charles Darwin House, London. *Symposia of the Zoological Society of London*, **53**, 339–367.

Waterman, A.S. (1993). Two conceptions of happiness: contrasts of personal expressiveness (eudaimonia) and hedonic enjoyment. *American Psychological* Association, **64**, 678–691.

Yerrell, P. (2008). *National Evaluation of BTCV's Green Gym*. School of Health and Social Care, Oxford Brookes University.

Yüksel, A. and Yüksel, F. (2001). The expectancy-disconfirmation paradigm: a critique. *Journal of Hospitality & Tourism Research*, **25**(2), 107–131.

Zhou, Y., Buesching, C.D., Newman, C., Kaneko, Y., Xie, Z., and Macdonald, D.W. (2013). Balancing the benefits of ecotourism and development: the effects of visitor trail-use on mammals in a Protected Area in rapidly developing China. *Biological Conservation*, **165**, 18–24.

Index

C